12 BIOLOGY

PNG UPPER SECONDARY

Takis Solulu
Beatrix Marjen-Waiin
Martin Hanson
Maria Sinclair
Terry Bunn
Anna Roberts

OXFORD

Oxford University Press is a department of the University of Oxford. It furthers the University's objective of excellence in research, scholarship, and education by publishing worldwide. Oxford is a registered trademark of Oxford University Press in the UK and in certain other countries.

Published in Australia by
Oxford University Press
Level 8, 737 Bourke Street, Docklands, Victoria 3008, Australia

First published 2015
Reprinted 2016, 2017, 2018, 2020, 2021 (twice), 2024 (twice), 2025

This book was originally published by ESA Publications, Auckland, New Zealand. This edition, specially adapted for the Grade 11 syllabus in Papua New Guinea, is published by arrangement with ESA Publications. Authors of the original work were Martin Hansen and Maria Sinclair and this adaptation has been produced with the assistance of Takis Solulu, Beatrix Marjen-Waiin and Diati Zure.

ISBN 978 0 19 557878 2

Illustrated by Guy Holt
Typeset by diacriTech, Chennai, India
Printed in China by Golden Cup Printing Co Ltd.

Oxford University Press Australia & New Zealand is committed to sourcing paper responsibly.

Contents

Acknowledgments

There are many people to be thanked for their help and assistance in enabling the publication of this series to happen. First, we acknowledge the cooperation and generosity of Mark Sayes at ESA Publications in New Zealand, who responded with interest and support when the proposal to adapt his Study Guide series was put to him.

Authors of the original edition were Martin Hanson and Maria Sinclair. This adaptation has been produced with the assistance of Takis Solulu and Beatrix Marjen-Waiin to provide Grade 12 Biology students in PNG with a book that they can use as a compact summary of content and skills related to the PNG Grade 12 syllabus.

We would also like to acknowledge many other individuals who have been happy to advise and assist in different ways: Joy Sahumlal, Greg Kapanombo, Anne Sangi, Safak Deliismail.

The publishers would like to acknowledge and thank ANU E Press for permission to reproduce edited sections from Bourke, R.M. and Harwood, T. (eds) (2009). *Food and Agriculture in Papua New Guinea*. ANU E Press, The Australian National University, Canberra, in Unit 12.1 Topic 9 and Unit 12.2 Topic 3. The original book is available online at http://epress.anu.edu.au/food_agriculture_citation.html. Copies may also be in school libraries or available through UPNG Bookshop in Port Moresby.

The author and the publisher wish to thank the following copyright holders for reproduction of their material.

© ESA Publications (NZ) Ltd, pp. 19 (all), 20 (all), 81 (both), 82, 97, 119, 136, 144, 145, 146, 153, 156, 157 (both), 158, 199, 265; Getty Images/Stuart Wilson, 134 (centre); Shutterstock/Tony Campbell, 129/kzww, 134 (right)/Sekar B, 134 (left).

Every effort has been made to trace the original source of copyright material contained in this book. The publisher will be pleased to hear from copyright holders to rectify any errors or omissions.

Introduction

This book has been published to provide the information required by students in order to successfully complete the Upper Secondary course in Biology at Grade 12 in Papua New Guinea.

The book is written in a manner that best develops an overall understanding of the biological concepts and processes set out in the PNG Grade 12 Syllabus. The order of Units in this book follows the order of Units presented in the Biology syllabus for Grade 12:

Unit 12.1 Ecology

Unit 12.2 Population

Unit 12.3 Genetics

Unit 12.4 Evolution

Within each Unit there are a number of Topics, which follow the main sub-headings and bullet points set out within each Unit in the syllabus. The aim is to provide structure and content in a concise and compact format for students to use as an effective resource to support the classroom experience. It is acknowledged that Biology is more interesting and meaningful when students are exposed to a variety of resources and materials and we encourage students and teachers not to rely on this book as a sole source of information.

If you have any suggestions about how this book might be improved in future editions, please contact Oxford University Press:

Fax: 00 61 3 9934 9100

Customer Service Email: exportsales.au@oup.com

We wish you every success in your studies.

The authors

Unit 12.1 Ecology

Topic 1: Biomes and habitats

The word 'biomes' refers to the major communities of plants and animals that live in a particular type of environment, such as forests, deserts, grasslands, the aquatic environment and tundra (which is the large areas in the north of the northern hemisphere where no trees grow and the soil below the surface of the ground is always frozen).

This first Unit in the Grade 12 Biology syllabus focuses on ecology – the scientific study of the relationship of plants and living creatures to each other and to their environment. It describes the concepts and processes relating to ecology. Topic 1 deals with:

- Ecological niche – environment, habitat, adaptations, niche.

Ecology is the field of science that studies the relationship between organisms and the environment.

- **Organism** refers to any living thing.
- The **environment** refers to the surroundings of the organism and includes all the factors that act upon it. These factors may be **abiotic** (non-living) or **biotic** (living).

Abiotic factors make up the *physical* characteristics of the environment.

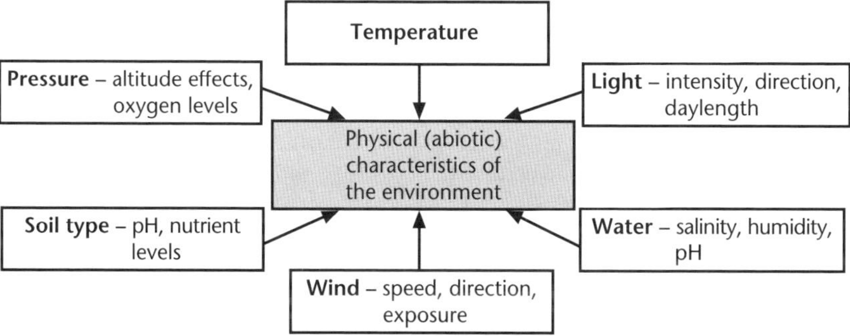

Biotic factors are other organisms (eg **predators**, **parasites**, competitors and herbivores).

Habitat

The **habitat** of an organism is the place where the organism lives.

Habitat is usually defined in terms of the physical characteristics of the environment, eg the habitat of crabs is in the mangroves along the PNG coastline.

Adaptation

Adaptations are inherited characteristics that enable organisms to survive or to reproduce more effectively. Every adaptation has a *purpose*.

Adaptations may be:

- **Structural** – eg wings on a cockatoo are used for flight; thorns on a bougainvillea bush deter animals.
- **Physiological** – eg the blood of tuna fish is quite warm, enabling them to swim very quickly; digestive enzymes in the intestines of animals allow food to be broken down.
- **Behavioural** – eg the sucking behaviour of young calves ensures they get a supply of milk.

Ecological niche

Ecological niche refers to the *role* or *way of life* of an organism in its biological **community**. The niche is a combination of *where* the organism lives (its **habitat**) and *how* it lives there (its **adaptations**). In describing a niche, it is necessary to:

- Name the organism.
- Give its way of life – feeding habit (eg herbivore/**carnivore**) and activity pattern (eg **nocturnal/diurnal**).
- Give its adaptations to its feeding habit and activity pattern.
- Describe other adaptations to its particular habitat and way of life.

The New Zealand kiwi

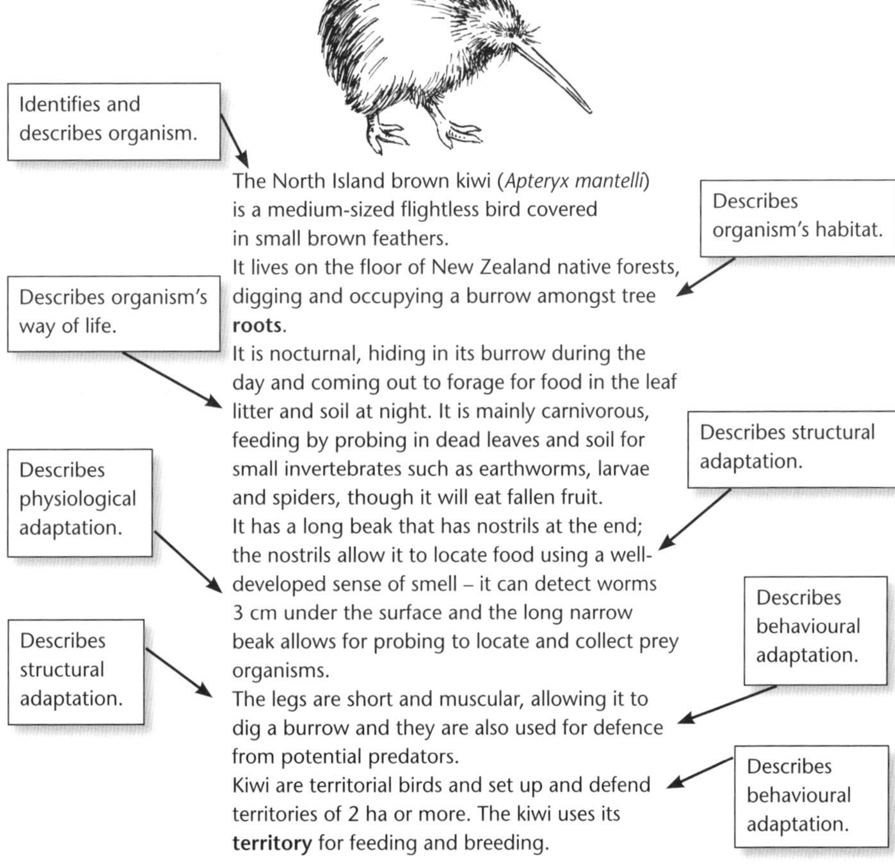

The North Island brown kiwi (*Apteryx mantelli*) is a medium-sized flightless bird covered in small brown feathers.
It lives on the floor of New Zealand native forests, digging and occupying a burrow amongst tree **roots**.
It is nocturnal, hiding in its burrow during the day and coming out to forage for food in the leaf litter and soil at night. It is mainly carnivorous, feeding by probing in dead leaves and soil for small invertebrates such as earthworms, larvae and spiders, though it will eat fallen fruit.
It has a long beak that has nostrils at the end; the nostrils allow it to locate food using a well-developed sense of smell – it can detect worms 3 cm under the surface and the long narrow beak allows for probing to locate and collect prey organisms.
The legs are short and muscular, allowing it to dig a burrow and they are also used for defence from potential predators.
Kiwi are territorial birds and set up and defend territories of 2 ha or more. The kiwi uses its **territory** for feeding and breeding.

More descriptive detail on the kiwi's adaptations to its way of life could follow if needed; a niche description can be brief or extensive, depending on the requirements.

Each species has a unique niche

Where niches of different **species** overlap, **competition** between them (**interspecific competition**) occurs. The greater the overlap, the greater the competition. If the niches become sufficiently similar, then the species cannot co-exist; one species will out-compete and eliminate the other – this is known as **Gause's principle**, or the **competitive exclusion principle**.

Example

Two species of caterpillar are found living on aibika plants in Papua New Guinea. They have the same habitat (the leaves of the aibika) but are able to co-exist because they are adapted to feeding on different parts of the leaf – one caterpillar has biting/chewing mouthparts and eats 'notches' from the edge of the leaf, while the other has rasping mouthparts and scrapes holes or 'windows' from the middle of the leaf.

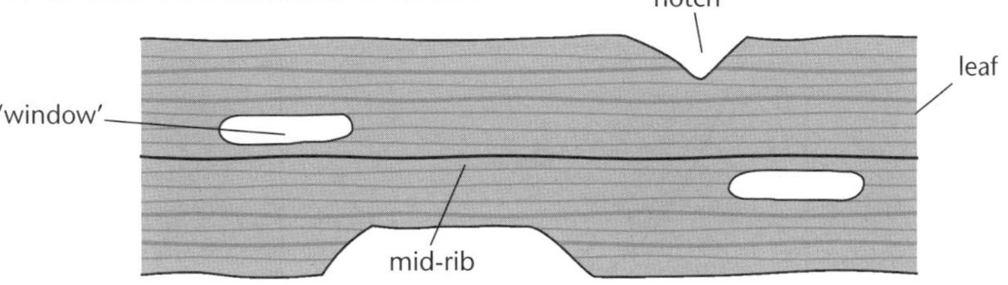

The **niche** that an organism occupies (its *actual* or *realised* niche) may be different from its *potential* niche. The potential niche is that which the organism is adapted to occupy but is prevented from occupying because of interspecific interactions with other organisms (such as competition for **resources** or preferential predation).

The niche may change during an organism's life cycle.

Example

The niche of butterflies (eg Monarch butterfly) changes as they metamorphose from caterpillars into flying adults.

Tolerance

Tolerance is an organism's ability to withstand *variation* in the environmental conditions.

When an environmental factor exceeds the **tolerance limits** of an organism, the organism suffers **physiological stress**. If the level of stress is too high, the organism will die.

Between the tolerance limits lies the **optimum range**; ie the range in which an organism functions at its best.

Example

Temperature tolerance limits of humans

The optimum environmental temperature range for humans is 20–30°C. Outside this range, humans show signs of physiological stress as their bodies try to cope with the adverse temperatures.

- Above 30°C the body begins to perspire rapidly. At even higher temperatures, panting, breathlessness, coma and even death can occur.

- Temperatures lower than 20°C cause shivering and goose bumps (in the *naked* body) as the body tries to produce and conserve heat energy. At even lower temperatures, blood circulation to the skin and then the limbs will be stopped. Eventually, if the body's core temperature drops too much, a coma, then death, will result.

Stress levels and tolerance limits

When stress levels are plotted on a graph, they are seen to increase rapidly when tolerance limits are exceeded.

Relationship between tolerance limits and stress levels

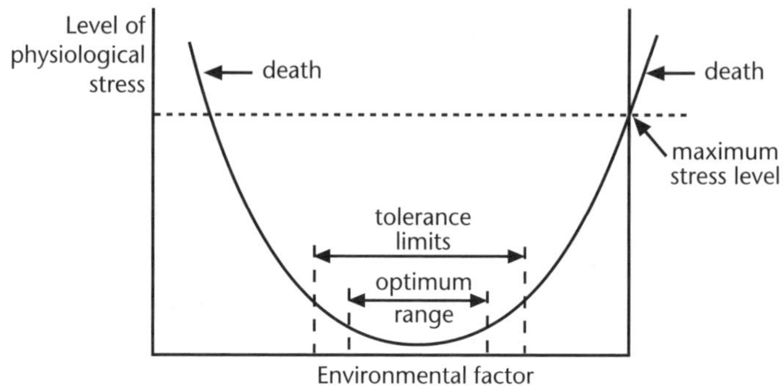

Acclimation

When environmental conditions change *slowly* (eg with changing seasons), organisms may adjust their tolerance limits – this is **acclimation**.

Example

Human skin darkens with regular exposure to the sun (sudden, long exposure to fierce sunlight can result in death from severe sunburn).

Mammals' hair or fur grows thicker and denser with falling temperatures and is shed with rising temperatures (sudden exposure to cold temperatures without thick coat hair can result in death from **hypothermia**; sudden exposure with a thick coat to hot temperatures can result in death from **hyperthermia**).

Liebig's Law of the Minimum

Liebig's Law of the Minimum states that:

'The functioning of an organism is controlled by whatever (essential) environmental factor is present in the least amount.'

Example

An environment may supply sufficient light intensity, water and CO_2 to meet the needs of plants, but if there is insufficient magnesium, Mg, in the soil, then this would limit the growth of plants in that area, because Mg is needed for **chlorophyll** production.

In the temperate and sub-temperate regions of the world, the soils are deficient in the element cobalt, Co. Cobalt is essential for the healthy growth of sheep and cattle. Cobalt is therefore a factor limiting these animals' ability to live there successfully; farmers have to add Co to the soil in fertilisers.

Unit 12.1 Activity 1A: Ecological niche

1. Distinguish between the terms:
   a. Environment and habitat.
   b. Biotic and abiotic.
   c. Tolerance and acclimation.

2. Define the terms:
   a. Ecological niche.
   b. Tolerance.
   c. Adaptation.
   d. Competitive exclusion principle.

3. The graph below shows the changes in biomass (dry weight) of two species of clover grown together in equal numbers in the same plot. Equal numbers of each species germinated.

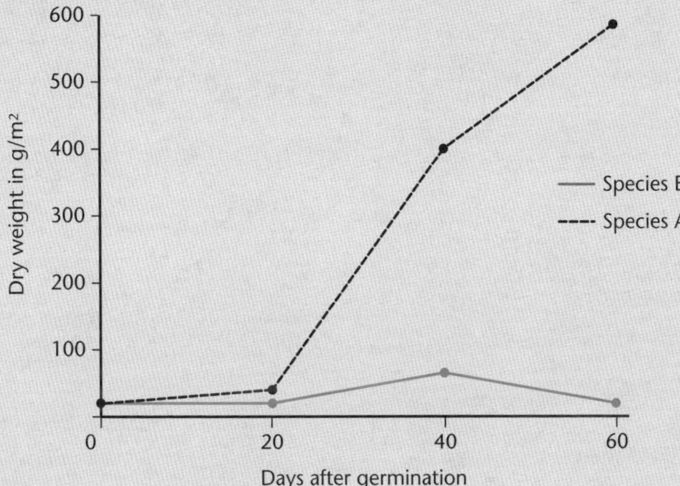

Explain how these results support Gause's principle.

4. Surveys of some of the creeks of the Southern Highlands Province produced the following data:

Creek	pH of creek water	Number of plant species	Number of animal species
1	4.0	8	7
2	8.1	19	13
3	4.4	11	6
4	6.6	23	19
5	5.7	16	9

   a. Explain the relationship between the pH of the creek and the number of animal species.
   b. Explain why *Creek 4* has the greatest biodiversity.
   c. Explain which creek ecosystem would be expected to be the least stable.

Unit 12.1 Ecology

Topic 2: Biological communities

Authors: Martin Hanson with Beatrix Marjen-Waiin

The major communities of plants and animals that live in a particular type of environment are referred to as 'biomes'. In Topic 2, we describe the ecological characteristics found in two biological communities by looking at:

- Defining a food chain.
- Food chains and food webs.
- Trophic levels and food pyramids.
- Investigating 'what feeds on what'.
- Lifestyles in a community.
- Why communities are different from one another.
- Comparing communities – this requires you to compare two communities and give reasons for similarities or differences in ecological characteristics found in the two communities.

A community

A **community** consists of all the organisms living in a particular area, eg a stretch of native bush, a rocky shore, a pasture.

An **ecosystem** is more complicated than a community – it includes all the non-living constituents as well, such as the water in a lake and the mud at the bottom. Thus a bush community is part of a bush ecosystem.

Communities are very diverse. Some are *natural* in the sense that they do not require human interference, eg a wetland, a rocky, sandy or muddy shore. Others are *managed* – these only exist because of human interference, eg a mown lawn, a planted pine forest. Before the arrival of humans, most of Papua New Guinea was covered with native forest, which still today is the most familiar natural community.

Communities are characterised by their **ecological characteristics**. In this Topic, a forest community is used to illustrate the various ecological characteristics found in communities.

A forest community

The dominant organisms in a forest are the large **plants**, the trees. Like other plants, they use solar energy to make carbohydrate from carbon dioxide and water in the process of **photosynthesis**. Since all the other organisms in the community depend on them for their food, the plants are called **producers** or **autotrophs** ('self-feeders').

Directly or indirectly, all the other organisms in the forest depend on the producers for their energy, and are called **heterotrophs**.

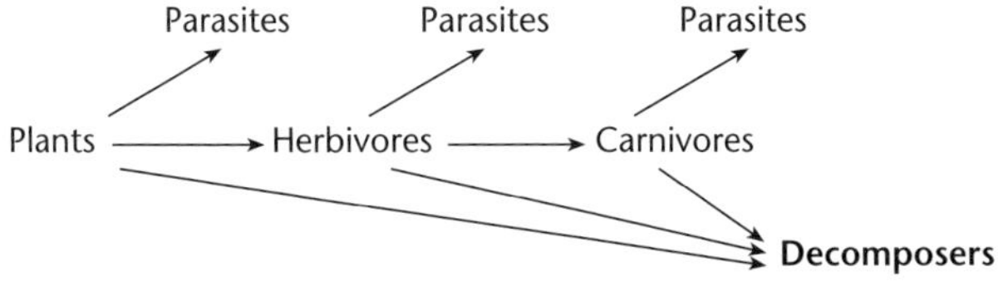

***Consumers** and decomposers depend on producers.*

Animals feeding directly on the live parts of plants, such as the **leaves**, are called **herbivores**. Herbivores are in turn eaten by **carnivores**, or meat-eaters. Herbivores and carnivores feed on living matter, which is killed in the process. These are collectively called **consumers**.

Decomposers feed on dead matter and live inside it. Animal decomposers eat dead matter and digest it in a special digestive cavity, the **gut**. **Bacteria** and **fungi** secrete enzymes into the food surrounding them and absorb it through their body surfaces. **Detritus feeders** are animals that feed on dead vegetable matter.

Parasites feed off another organism, called the **host**, harming it but not usually killing it – eg a tapeworm lives in the small intestine of a mammal, feeding on digested food.

Though parasites are less obviously visible than their hosts, they are actually more abundant, since most animals and plants have more than one species of parasite – eg Polynesian rats are hosts to two species of lice, six species of flea, and 10 other species of parasite; wild goats are hosts to over 25 species of parasite.

To make things even more complicated, many parasites have parasites feeding off them, and these *hyperparasites* may even have parasites feeding off them!

Some feeding relationships benefit both organisms – eg the roots of most trees and many other plants contain a fungus that helps the plant to absorb phosphate from the soil; in return, the fungus obtains **organic** materials such as **sugar** from the plant. This kind of relationship, in which both partners benefit, is called **mutualism**.

Another example of mutualism is the relationship between leguminous plants and the bacteria living in their roots. The bacteria convert nitrogen gas, which plants cannot use, to **ammonia**, which they can use. The plant combines the ammonia with organic compounds made by the plant, to make **amino acids**, which the plant then uses to make **proteins**. The **microbes** living in the guts of many herbivores digest **cellulose** (which the animal cannot do), and at the same time gain the food, warmth and **anaerobic** (oxygen-free) conditions they need.

Feeding relationships between different organisms can be represented by a **food chain**.

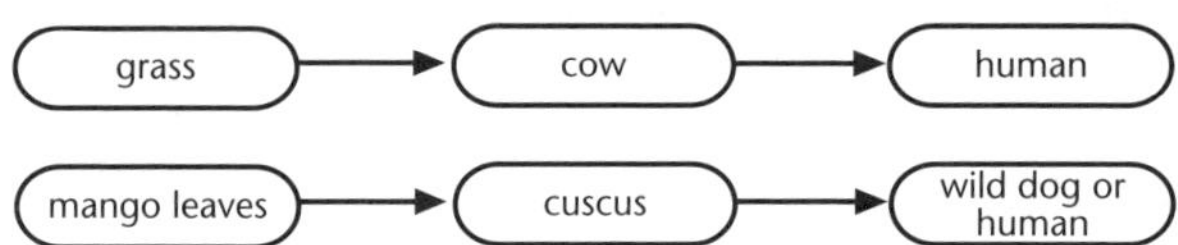

Arrows indicate the direction of energy flow – thus energy in grass enters the cow, and then the human; similarly, energy in mango leaves enters the cuscus and then the wild dog or human.

Food chains.

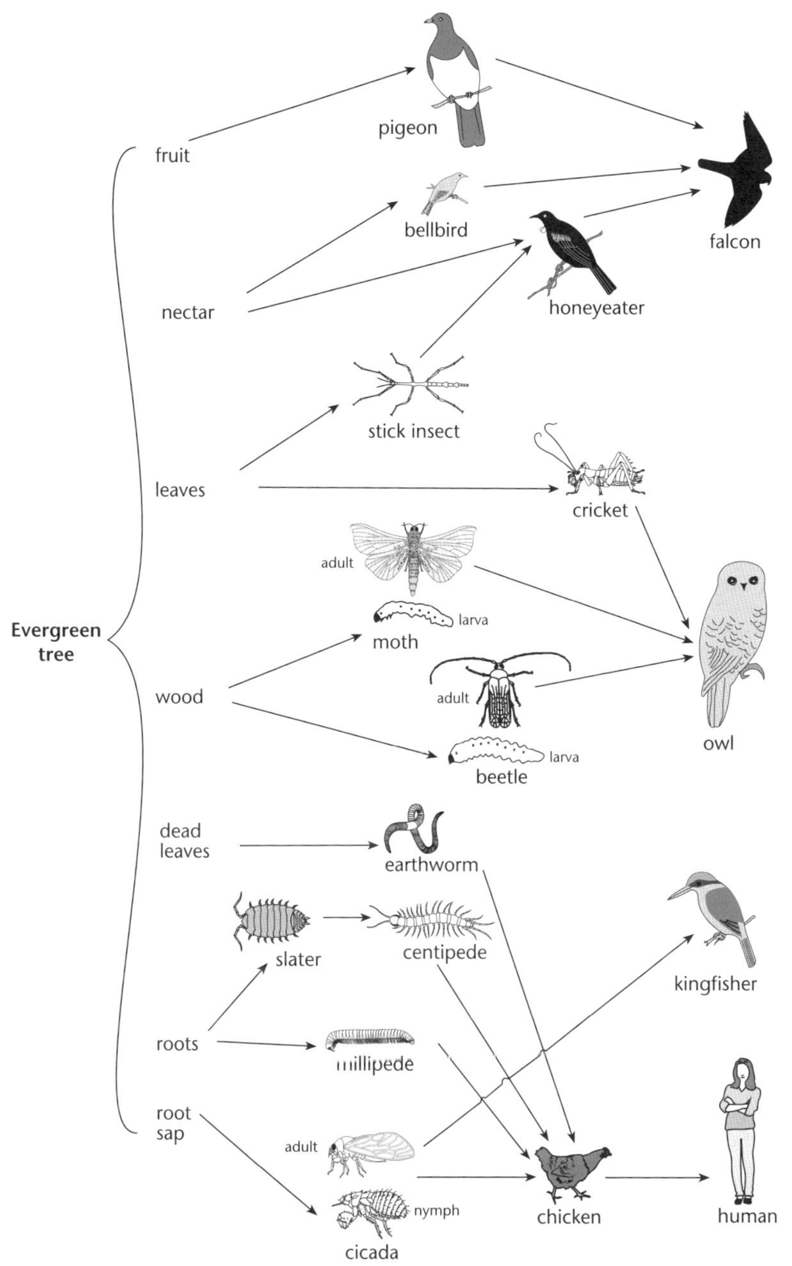

Simplified food web based on an evergreen tree.

Food chains are almost always much more complicated than those shown in the diagram on the previous page.

- Most organisms are eaten by more than one kind of consumer (eg mice are eaten by cats, snakes, owls and falcons).
- Most consumers eat more than one kind of food organism (eg red deer have been recorded as feeding off over 20 species of plants).

Food chains *branch* and interconnect, so the result is a **food web**.

Food webs are usually themselves very complex.

- In the case of a food web based on a forest tree, there are usually far more kinds of organism involved – only a tiny proportion of possible links is usually shown.

For the food web based upon the tree on page 9, pigeon range far and wide beyond the one tree, and feed on many other kinds of fruit besides puriri berries.

- Decomposers such as bacteria and fungi are not shown.
- Many animals change their diets as they get older, sometimes with a complete change of habitat.

Adult mosquitoes feed on **nectar** (and females also feed on blood), but the young stages feed on bacteria and other suspended microscopic life in fresh water.

- In regions where there are distinct seasons, feeding habits may change with the seasons.

Many trees produce berries only at certain times of the year (usually autumn), and so other foods are eaten at other times of the year. Some animals feed mainly on insects in winter, but eat mainly grass seeds in summer.

- Food webs in adjacent communities are often interconnected.

Adult dragonflies feed on land-living insects but the young stages live in fresh water, feeding on small animals.

- Some communities depend entirely or almost entirely on food produced elsewhere, and are thus importers of food.

Organisms living in deep caves depend on organic matter produced by plants above ground and washed down in streams.

Even where there is plenty of light, some members of a community depend on food imported from elsewhere. The barnacles and mussels on a rocky shore feed on tiny plankton (drifting life) produced in the open sea.

Trophic levels

One way of simplifying the feeding relationships in a community is to consider all the plants together as occupying the same **trophic level** (trophic means 'feeding').

The animals feeding directly off the plants collectively occupy the next (herbivore) trophic level.

The animals feeding off the herbivores occupy the carnivore trophic level.

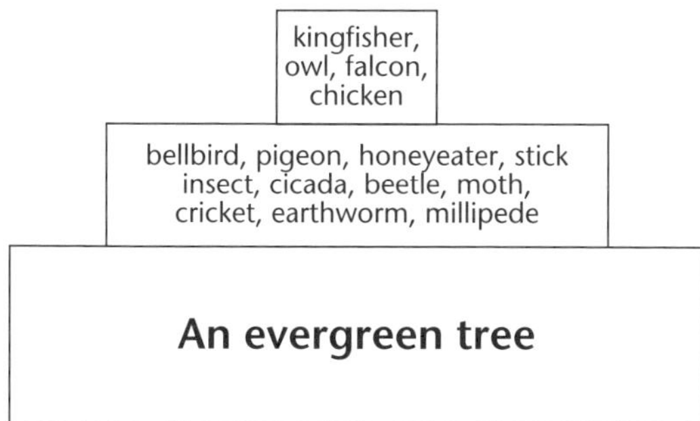

The organisms are grouped together into trophic levels. Each block represents the total mass of organisms (the biomass) in that trophic level but not to scale.

A simplified food pyramid based on the food web of a tree.

The most important thing about a food pyramid is that the total mass (**biomass**) of each trophic level *decreases* moving *up* the pyramid. (The decrease is actually much greater than that shown above.) Each kilogram of carnivore thus needs *much* more than a kilogram of herbivore to support it, and each kilogram of herbivore feeds off a *much* larger mass of plant.

Why the total biomass of each trophic level decreases

When food is consumed, *some* of it is not digested and absorbed, but passes out as **faeces**, and of the food that is absorbed, only a *small* proportion is used in growth, most being used as fuel and converted into carbon dioxide and water in **respiration**.

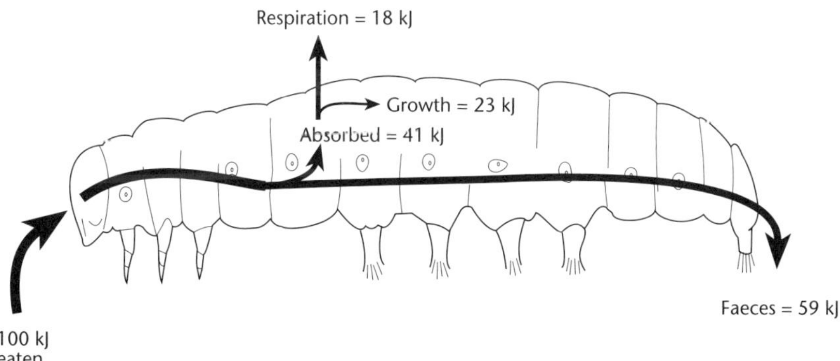

Less than a quarter (23 kJ) of the food energy eaten (100 kJ) is used in growing new flesh, and it is only this that can enter a carnivore, such as a bird, in the next trophic level.

The food energy eaten by a caterpillar.

In homeothermic ('warm-blooded') animals, the proportion of energy converted into new **tissue** is even less than in cold-blooded animals (such as caterpillars), because of the large amount of energy needed by warm-blooded animals to keep warm.

Example

An eagle takes in many times its own mass of animals as food in its lifetime, and each of these prey creatures needs many times its own body mass of plant food.

This loss of energy that occurs moving up food pyramids helps to explain why carnivores are generally rarer than herbivores. A tiger needs about 100 km^2 of forest, because each of the many deer it feeds off during its lifetime needs several hectares of plants to support it.

Food pyramids are rarely as simple as outlined above:

- Omnivores belong to more than one trophic level.
- Some carnivores feed off other carnivores – these *secondary carnivores* occupy an even higher trophic level than the *primary carnivores*, which feed on herbivores.

How to determine 'what feeds on what'

Food webs are most easily discovered through observation.

Herbivores, which spend a high proportion of their time feeding and remain close to or on food plants, are easily observed.

Carnivores are more difficult to observe because they tend to feed intermittently and of short duration.

Building up a picture of what feeds on what is time consuming, but fortunately several things will help to make things easier:

- The mouthparts of insects often give a clue as to their kind of diet – centipedes and carnivorous beetles have large jaws; millipedes and slaters, which are herbivores, have very small jaws.
- Keeping animals in captivity and offering a range of possible food organisms gives an idea of food preferences. (What happens in captivity doesn't necessarily happen in the wild – eg prey animals in captivity may not have the same freedom to escape that they would in nature.)
- Animals can be dissected and their gut contents examined – sometimes food organisms have hard parts that resist **digestion** and may be seen under the microscope (eg skeletons, **exoskeletons** of arthropods, and plant cell walls.)

How heterotrophs and decomposers depend on the sun

Directly or indirectly, animals and decomposers depend on the sun for their energy.

Plants depend directly on the sun, and animals depend either directly (herbivores) or indirectly (carnivores) on the plants. The energy in the dead matter that decomposers feed on has also come ultimately from the sun.

How plants depend on heterotrophs

Although plants do not need animals or decomposers for their energy, they do depend on them for raw materials. To obtain energy from organic matter, consumers and decomposers break the organic matter down into simple substances such as carbon dioxide, water and mineral salts.

These are the raw materials plants use. Without the heterotrophs, **autotrophs** would run out of raw materials. Between them, autotrophs and heterotrophs are responsible for the recycling of matter.

Lifestyles in the community

The most obvious feature of any community is the variety of species (kinds) of organism present. Each species lives in a particular way, which differs from the lifestyles of other species.

An organism's lifestyle, or the way it lives, is its **niche**. This is quite different from an organism's **habitat**, which is the place where it lives. There are many different niches in a forest community, resulting in a large variety (or **diversity**) of species.

Terms such as producer, carnivore, herbivore and decomposer only partly describe an organism's niche. Different herbivore species may specialise on different plant foods.

Example

In the food web shown on page 9, stick insects and honeyeaters both feed off the tree, but they do not compete with one another for food because the stick insects feed off leaves and the honeyeaters feed on nectar (as well as other foods).

Carnivores also specialise on particular foods.

Example

Kingfishers are diurnal (day-active) birds and feed on diurnal animals such as cicadas. Owls are nocturnal (night-active) and feed on night-active animals such as crickets. Much smaller carnivores, such as centipedes and spiders, feed on correspondingly smaller prey such as slaters.

Decomposers also tend to specialise.

- Slaters and millipedes tend to eat softer food such as dead leaves and small roots, whilst harder food (wood) tends to be attacked by fungi and wood-boring insects (eg the grubs of huhu beetles).
- Some fungi feed on recently dead leaves, concentrating on easily digested **starch** as their main food, while other fungi get established later and utilise the cellulose, which is more difficult to digest. The last decomposer fungi involved are those that attack **lignin**, the material that stiffens the cellulose walls of strengthening tissues of plants.

Plants also reduce competition by occupying different niches. A clear example of this is in layering or **stratification** of the vegetation in a forest.

- The tallest trees of the forest, such as rimu, rata, totara and kauri, protrude above the others and are called **emergents**. These plants experience the brightest light, strongest winds, lowest humidities and widest fluctuations in temperature.
- Slightly smaller trees, such as karaka and puriri, form a continuous layer, the **canopy**.
- Smaller trees, such as mahoe (whiteywood) and various species of *Coprosma*, form the **sub-canopy**.
- Ferns and other smaller plants form an even lower, **shrub layer**.
- Mosses and liverworts hug the ground itself (where conditions are cooler and more humid, and the light is dimmer) and form the **herb layer**.

The plants with the largest biomass in a community are the **dominant species**. The dominant species in a community provides food and shelter for many other species and so determines the other species that live there.

Example

Bellbirds and honeyeaters depend on the forest trees that provide them with nectar. Pigeons feed largely on fruits of trees.

Organisms may also depend on other plants for shelter.

Example

Ferns depend on the taller trees for shelter from drying winds and the higher temperatures experienced by the plants forming the canopy. Most of the birds of the forest depend upon the safety of the trees to build their nests.

The dependence is not all one-way – many forest trees depend on insects and birds to carry pollen, and later they may depend on birds to disperse their **seeds**.

Organisms within a community show different types of **density** and **distribution patterns**. Density refers to the numbers of organisms in a certain area.

Example

Ferns often have quite a high density, ie many of them would be found growing very close to each other. Many native birds such as the kokako have a relatively low density with only a very few birds found in a certain area.

Most organisms within a community are **distributed** in a certain way, and the **distribution pattern** is a result of certain environmental factors. It would be unusual for organisms to be scattered randomly throughout an area.

- Distribution can be uniform, where individuals are spread evenly over the area. This pattern is characteristic of organisms that defend an area from others for certain **resources** such as space for breeding, or food. For example, nikau trees that compete against each other for light, soil **nutrients** and water will be evenly spread, resulting in each having the best chance of gaining sufficient resources.
- Distribution can be **clumped**, where groups of individuals live or grow very close to each other, with large spaces between groups. This type of distribution is found in organisms where the benefit of the protection given by living in a clump outweighs the disadvantage of increased competition for resources. For example, slaters living in a decomposing log clump together to avoid drying out.

Why communities are different – environmental factors

Every organism is adapted (suited) to living in a particular range of conditions. The conditions in which an organism lives make up its **environment**.

'Environment' is quite different from 'habitat', which is simply the place an organism lives (its 'address'). While a puriri tree's habitat remains the same, its environment is constantly changing (eg day and night and the seasons).

Environmental factors are of two kinds:

- **Biotic** factors are other organisms, such as food, predators, parasites and competitors.
- **Abiotic** factors are physical things such as temperature, light intensity, humidity, pH, mineral concentration, wind speed and direction and wave action. (Obviously not all of these operate in a given environment – eg wave action applies only to aquatic habitats.)

Role of climate and soils

The kind of plants the community contains is determined by climate, and the material on which or in which the plants grow (ie the soil, or in some cases, bare rock).

In natural soils (ie those not deliberately modified by humans), the kind of soil depends on the:

- Rock from which it had been formed.
- Climate.

These two factors have interacted to produce the variety of soil types in Papua New Guinea.

As far as plant growth is concerned, the most important properties of a soil are the:

- pH.
- Mineral content.
- Water content – this is determined by the average mineral particle size (larger particles tend to favour good drainage), the proportion of humus (partly decayed organic matter) which retains water, and rainfall.

Role of humans and other animals

When a paddock is left ungrazed for more than a few months, the vegetation doesn't just get longer – some plant species eventually disappear.

Though the continual **grazing** and trampling by stock represents a considerable drain on the resources of the plants, not being grazed is an even greater hazard to a pasture plant species in the long run. This is because plants that compete more strongly for light begin to take over and, if left ungrazed, a pasture slowly reverts to scrub, then to forest.

Grazing mammals are thus essential for the survival of many pasture plants.

A lawn is a pasture 'grazed' by mowers and trampled by human feet. Plants growing on lawns can survive grazing and trampling because of their low-growing habit.

- Plants such as dandelions and daisies have stems with such short **internodes** that the leaves lie flat on the ground in a *rosette*.
- Plants such as white clover and creeping buttercup have horizontally growing stems that send out roots at the nodes.
- Plants such as grasses have leaves that, when removed, are replaced by new ones produced at the growing point, which is sufficiently close to the ground to escape damage.

Example

Grasses are particularly well adapted to resist grazing and trampling. Unlike nearly all other plants, a grass leaf does not cease growing when it reaches a certain size, but can continue to grow at its base. Thus, leaf tissue eaten by grazing mammals can be continually replaced – provided grazing is not too heavy. Grasses are also unusual in the way they are able to recover from trampling. Each node retains the capacity to grow. When a grass stem is trampled, the cells on the lower side of the node expand, causing the node to bend, raising the internode above.

Comparing communities

When comparing communities, first decide what communities you want to study and compare. Your chosen communities should be reasonably close by and easily accessible. Be careful to avoid meaningless comparisons, such as attempting to compare a pine forest and a rocky shore.

There should be clear-cut differences within the abiotic (physical) environment for your two communities. For example, an exposed rocky shore and a sheltered rocky shore differ in the intensity of wave action, and a mud flat differs from a sandy shore in that the mineral particles of mud are much finer (there may also be other differences). A lawn and an infrequently mown grass verge would also be suitable communities.

Whatever you choose, you will need to be able to obtain measurements of various physical factors, especially the one you think is likely to be most responsible for the differences between the two communities.

Some abiotic factors vary greatly from day to day, eg wind strength. With factors like these you must either arrange to take measurements at the two communities as nearly simultaneously as possible, or take measurements over extended periods so that you can work out and compare averages.

Describing ecological characteristics in two communities

Ideally, you will have the opportunity to go 'into the field' to gather actual data for one or both communities. However, you may also complete an assessment from other resources such as resource sheets, graphs, tables, photos, videos or websites.

To compare your communities:

- *Describe the ecological differences between two communities.* This means that you must give an account of the ecological characteristics found in both communities and it must be a full description.

Example

In a comparison of the diversity of plant species in native bush and a pine forest, it is not sufficient to say 'The native forest had a high diversity of plant species.' You also need to describe the diversity in the pine forest, *and* give an indication of what the diversity was, eg 'The native bush had a much higher diversity of plant species than the pine forest. I found a total of 23 different species of plants in the native bush, but only ten in the pine forest.'

- *Discuss the reasons for the similarities or differences in the ecological characteristics in the two communities.* This means you must refer to environmental factors in your discussion, and for 'Excellence' you should also make links to the biology or adaptations of the organisms.

Example

In a study of the density of fern plants in a native bush community and a pine forest community, the density in the native bush was found to be much less than that in the pine forest. Data collected relating to the environmental factors found the pine forest floor had a higher light intensity and higher wind speeds than the native bush. These two environmental factors could be used to discuss this difference in relation to the biology of the fern plant, eg 'Ferns require low light, high humidity and low wind as their foliage is delicate and easily damaged and does not have a thick **waxy cuticle**. Therefore, in the pine forest community, the high density of the fern plants would provide protection against the higher wind speeds. In addition, the greater overlapping of the leaves would provide leaves with greater protection from burning by the high light intensity and, by trapping water vapour released by the leaves through **transpiration**, would increase the humidity of the air surrounding the leaves. The native bush community has many larger canopy and sub-canopy trees and shrubs that provide shelter from light, and decrease the wind speeds at the forest floor level. These plants also help trap water vapour, so that the humidity of the forest floor layer (where the ferns are found) is high. Therefore, it is not necessary for the fern plants to grow at such a high density.'

Unit 12.1 Activity 2A: How organisms depend upon one another

1. Match the phrases **1–20** with the terms **A–T**.

Phrase	Term
1. All the factors in the surroundings that influence an organism	A. Abiotic
2. All the living organisms in a particular area	B. Autotroph
3. Alternative name for autotroph	C. Biotic
4. Animal feeding on dead plant matter	D. Carnivore
5. Animal that feeds on other animals	E. Community
6. Animal that feeds on plants	F. Consumer
7. Collective name for carnivores and herbivores	G. Decomposer
8. Community plus all the non-living components of the habitat	H. Detritus feeder
9. Organism that feeds at the expense of another organism, usually without killing it	I. Dominant species
10. Influences that take the form of other organisms, such as predators	J. Ecosystem
11. Organism that lives inside the dead matter on which it feeds	K. Environment
12. Non-living influences, or physical factors, such as light intensity	L. Habitat
13. Organism that depends on other organisms for its supply of energy	M. Herbivore
14. Organism that does not depend on others for its supply of energy	N. Heterotroph
15. Relationship between two species in which both benefit	O. Mutualism
16. Layering of vegetation in a forest	P. Niche
17. Place in which an organism lives	Q. Parasite
18. Species with the greatest total biomass in an area	R. Photosynthesis
19. Use of light energy to make organic matter from inorganic matter	S. Producer
20. How an organism lives; the sum total of its needs	T. Stratification

2. The diagram shows the flow of energy (solid lines) and matter (dotted lines) in a community. W, X, Y and Z represent decomposers, plants, carnivores and herbivores, but not necessarily in that order. Identify W, X, Y and Z.

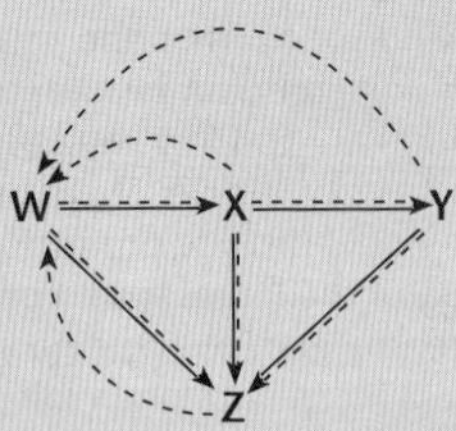

3. Suppose that the sun suddenly ceased to shine so that the world was plunged into permanent, total darkness. Which kinds of organisms would you expect to die last – decomposers, plants, herbivores or carnivores?
4. Describe ways in which animals depend on plants, apart from providing food.
5. Explain how plants depend on:
   a. Heterotrophs.
   b. Insects.

Unit 12.1 Ecology

Topic 3: Population and communities

In Topic 3 we continue the description of the concepts and processes relating to ecology and the focus on communities. This Topic deals with:

- Succession, zonation, stratification.

A biological **community** is *all the organisms* that live within a defined area (eg lake, forest, beach, grassland) and their *interrelationships*. Communities may be large and long lived (eg the community of Lake Kutubu or of the Sepik plains), or they may be small and short lived (eg the community of a cow pat or of a shallow pond). Usually, a community is named after the:

- Dominant species that lives there (this also usually has the largest biomass) – such as a kunai grassland community.
- Most important physical feature of the community – such as a rocky shore community, a freshwater stream community.

Communities may have well-defined boundaries, such as a lake, or they blend into each other across a transition zone (eg a sand dune community may blend into a coastal forest community).

Sand dune community

Wetland community

(Soft-rock) rocky shore community

Succession

The development of a mature community from bare land is ecological **succession**. This is usually a long slow process (taking hundreds to thousands of years for our native forest communities). The development of a mature community from bare land that has not been inhabited before (eg the moraine formed by glacial action; new sand dunes on the seashore; cooled lava from a volcanic eruption) is called **primary** (1°) **succession**. More common is **secondary** (2°) **succession**, which occurs when previously inhabited land is cleared by an (often catastrophic) environmental event such as a landslide, fire, flood or human activity. The process of succession is the same in both, but the formation of the mature community in secondary succession is faster because the ground is already fertile.

The first organisms to colonise the new habitat are **pioneer plant species**. These hardy plant species are tolerant of exposure, high temperatures, **dehydration** (loss of moisture), low soil fertility, etc. Pioneer plant species trap nutrients and utilise inorganic compounds and provide habitats for the first consumers by providing shelter and food. As the pioneer plant species and their associated organisms die and decompose, they increase soil fertility. The changed environmental conditions allow other species to grow up and above the **pioneer species**, killing the pioneer species by shading them out. The succession of establishment, competition and replacement continues, providing a variety of distinct successional communities. The final community is called a **climax community**.

Climax communities are stable. No further succession occurs because climax species provide the conditions necessary for the survival of their own species.

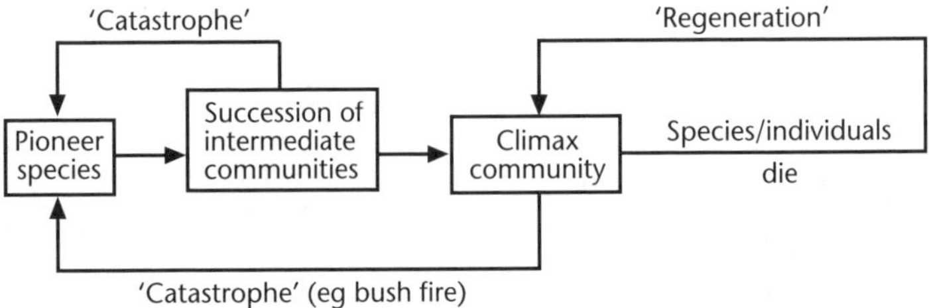

Many pastures in temperate and sub-temperate regions represent an early stage in succession. Weed species (gorse, tea tree, thistles etc) represent members of the next successional stages, so these species are continually invading pastures, and, if unchecked, replace pasture species.

Secondary succession after land slip

Primary succession – lava

A distinct *pattern over time* seen in the species distribution of a community is another way of defining **ecological succession**.

When a new habitat becomes available (eg after the flooding of a valley or after a lava flow from a volcanic eruption), it may take many years, maybe up to thousands, before the community reestablishes itself. Whole sequences of communities replace one another as species within each community change the physical environment, making conditions unfavourable for other species and even for themselves. Interspecific competition is a major contributory factor.

Primary succession

Primary succession occurs in coastal habitats such as sand dunes and mangrove swamps and in the colonisation of bare rock. Though the species involved differ in each case, the environment is initially so unfavourable to plant growth that only a few **pioneer species** can survive.

Examples

On bare rock the problem is lack of water, and only lichens are able to survive. Their fungal threads penetrate the rock by entering minute cracks that retain water. This enables very small plants such as mosses to get established. Mosses compete with the lichens, and as the organic matter builds up, conditions change increasingly in favour of the mosses. The small amounts of humus formed enable slightly larger plants such as small herbs to grow. These deposit more humus and trap more water, allowing larger plants such as shrubs and eventually small trees to become established. These overtop the other plants and thus compete more successfully for light. Eventually, over centuries, the community becomes mature forest. When there is no further change in species composition a **climax community** exists.

Example

Over most of New Zealand the climax community is forest. In sand dunes, only certain grasses such as *Spinifex* and marram grass can survive the constantly shifting, salty sand, which can become very hot and dry. The pioneer grasses stabilise the sand and help to deposit enough humus for other plants, such as the introduced lupin, to become established. These eventually competitively replace the grasses, but are themselves later replaced by small trees such as certain *Coprosma* species.

Secondary succession

Secondary succession occurs when fertile soil exists from the beginning, and is consequently much more rapid than primary succession. An example is the re-growth of forest after fire or felling; the early stages can be seen in a neglected garden.

Whilst primary succession takes centuries or longer, secondary succession is rapid. The order in which plants are established is governed by their relative rates of arrival as seeds or **spores** (or the speed of germination of seeds already present), and their relative rates of growth. The first species to be established thus tend to produce large numbers of seeds, which are often wind dispersed. As in primary succession, competition plays a key role, each species being replaced by slower-growing but eventually taller plants, forest being the final result.

Stratification

Within many plant communities, especially forests, **stratification** occurs. This is the distinct vertical *layering* ('strata') of plants that occurs as a response to *light intensity*.

For any particular community, layers may be identified and named – this can vary from community to community:

Examples

Stratification is typical in most plant communities.

A community of birds living in a single tree can exhibit stratification, especially during the breeding season and if the different species are closely related.

Stratification is seen in lakes where different fish species distribute themselves into vertical layers as a result of oxygen and food requirements.

- **Canopy** – these are the tallest trees, which must be adapted to/tolerant of high light intensities, high temperatures, high wind speeds, low humidity.
- **Sub-canopy** – these are shorter trees which must be adapted to/tolerant of reduced light intensities, reduced temperatures, reduced wind speeds, increased humidity.

Trees in the canopy and sub-canopy layers are adapted by having tall trunks with a crown of branches and leaves (or **fronds**) at the top.

Sometimes, tree ferns (and nikau) are put into a separate layer called the *tree fern layer*.

- **Shrub layer** – bushy branching plants that must be adapted to low light intensities, low temperatures, low wind speeds, higher humidity.
- *Ground layer* – a diverse layer that includes juvenile forms of the upper layers. Species that live here permanently must be adapted to the lowest light intensities, temperatures, air movement, highest humidity. Damp conditions can be common.
- *Leaf litter* – covers the forest floor and is the fallen leaves, fronds, branches and seeds from all plants. Fungi live here and decompose dead material; seeds **germinate** in the soil and seedlings grow in the sheltered conditions. A rich collection of invertebrates makes its home here.

All species of **invertebrates**, including centipedes, millipedes, insects, spiders, snails, slugs and earthworms, may be found in the litter layer. Geckos are usually found on the bark and branches of the trees. Birds may be found associated with particular layers, depending on their adaptations and feeding preferences; some birds are found in shrubs, at the ground layer and on the leaf litter, others in the canopy and sub-canopy.

Vines or **lianas** are common in PNG forests. They have roots in the ground and stems that climb up other plants to get to the upper layers, especially the sub-canopy and canopy (for increased intensity, eg rattan).

Epiphytes (or *perching plants*), especially the perching lilies or *Sungi* and lichen, may be found on the trunks and in the notches of branches of the large canopy trees. They spend all their life here and have no connection with the ground.

This is an example of what stratification may look like:

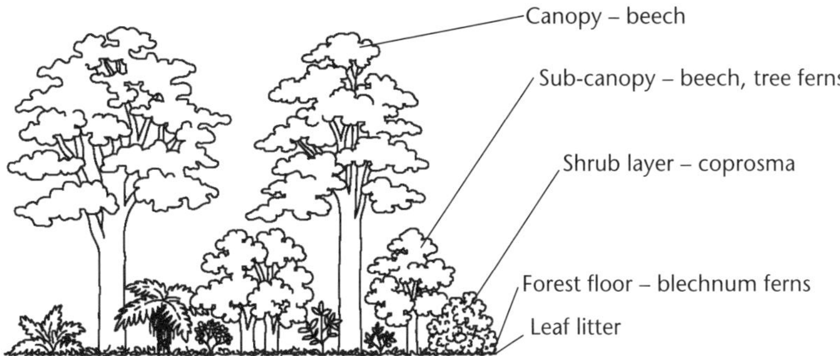

Zonation

Zonation is the distinct horizontal bands of life forms that occur across a particular habitat or area. Zonation results from the *constant change* (or *gradient*) of some environmental factor across the community.

As environmental conditions exceed the tolerance limits of a species, the zone of distribution for a species finishes. Another species, *adapted* to the new environmental conditions, forms another zone until it too can no longer cope with conditions. Competition between species where they meet means zones are usually quite distinct.

Zones are often named for the dominant/characteristic species found there. When environmental conditions in a zone are harsh or extreme (eg high up a mountain, high tide area of the rocky shore), the **biodiversity** of the zone is typically less than that of a zone where conditions are more moderate or not so extreme (eg lower parts of a mountain, low tide area of the shore) – organisms that live in harsh conditions must have high physiological tolerances. However, the greater the biodiversity, the greater the interspecific competition that is likely to occur.

Zonation

A distinct *horizontal pattern* seen in the species distribution across a community is another way to define **zonation**.

The species best adapted to the particular environmental conditions present in a zone tends to be the dominant species. Once environmental conditions exceed the limits of tolerance for that species, another species forms the next zone.

Examples

'Bands' of **populations** are seen on the rocky shore from the high tide mark down to the low tide mark.

Zonation patterns are also present on mountains where bands of plant populations can be seen, each adapted to the specific environmental conditions present as a result of change in altitude. Competition becomes more intense at lower altitudes where there are not the extremes of abiotic factors such as temperature and wind.

Example

Zonation on Mt Wilhelm

The temperature change from the bottom of Mt Wilhelm to the summit causes zonation in the plant species. Moving down the mountain, competition between plants for light, space and nutrients increases markedly.

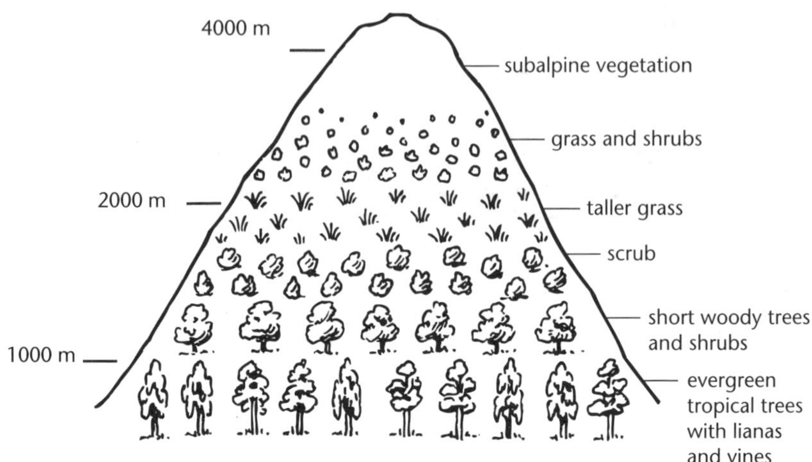

Zonation on the rocky shore

The environmental factor that causes zonation on the rocky shore is tidal movement (ie exposure to sea water and air). Competition for space and food increases down the shore from high to low tide.

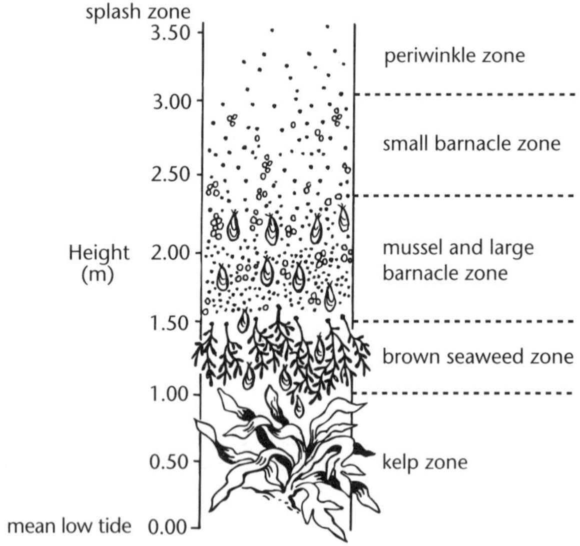

Unit 12.1 Activity 3A: Communities

1. Distinguish between the following pairs of terms:

   a. Epiphytes and lianas.

   b. Zonation and stratification.

   c. Primary and secondary succession.

2. Compare the typical growth form of a plant in the canopy layer with that in the shrub layer.

3. The diagram below shows the distribution of vegetation on a mountain such as Mt Gilluwe.

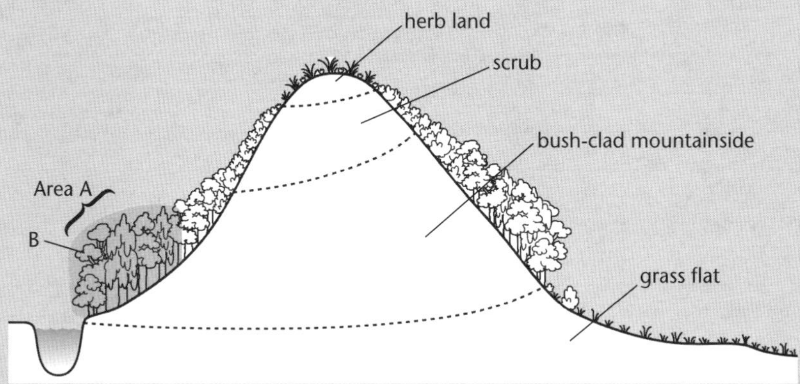

   a. Name the community pattern shown in *Area A* and the main environmental factor responsible for it.

   b. Name the layer of trees labelled *B* and identify three abiotic factors that they must be tolerant to.

   c. Name the community pattern shown by the change in vegetation with altitude up the mountain. Identify two abiotic factors that may be responsible for this pattern.

   d. Suggest reasons for differences shown in this pattern between one side of the mountain and the other.

4. The following flow chart relates to an area of native bush following a landslide:

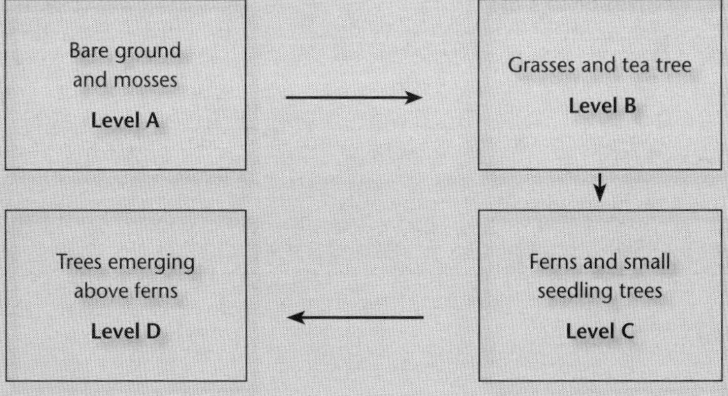

a. Identify the type of succession occurring here.
b. Give the name for the first plants that occur and explain any adaptations they must have for living here.
c. Name the mature community that results.
d. Explain how three physical factors would change from level A to level D.
e. Explain why animals do not arrive until the area is well colonised (eg level B).

5. The table below shows the percentage abundance of some of the plants growing on an area of sand dunes.

Plant species	Percentage abundance									
	Distance from high tide mark (m)									
	0	50	100	150	200	250	300	350	400	450
Pingao	0	12	16	14	8	0	0	0	0	0
Spinifex	0	0	9	18	26	14	8	0	0	0
Marram grass	0	3	8	15	16	16	7	2	0	0
Hare's tail	0	0	0	0	6	6	3	0	0	0
Cassinia	0	0	0	0	8	13	15	4	1	0
Coprosma	0	0	0	0	0	0	0	0	6	10

a. Identify the community pattern shown in this data.
b. Describe environmental factors that may cause this pattern.
c. Explain ways in which the plants may change the environment of the sand dunes.

6. A student set aside a piece of moist pumpkin under a trough in the laboratory, and, over a period of three weeks, observed the growth of a number of different moulds. The student recorded her observations as in the diagrams that follow:

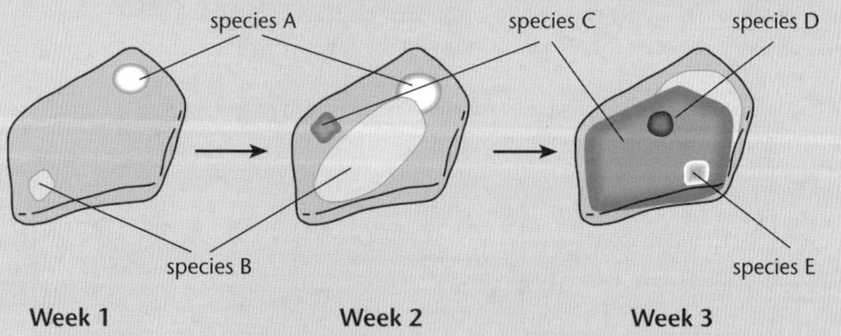

a. Give a reasoned opinion as to what the community pattern shown in the diagram is.
b. Provide an explanation for the disappearance of *Species A*.

Unit 12.1 Activity 3B: Competition between species

1. **a.** What is succession?
   b. Distinguish between primary and secondary succession and explain why secondary succession is so much faster than primary succession.
   c. Describe the role that competition plays in primary sucession.
   d. An investigation was carried out to find out the effects of grazing by deer on an area of grassland they frequented. From within a 22-year study period, a three-year period was chosen when the deer population was small (as indicated by low catch numbers). During this three-year period, the percentage frequency of a number of plant species was determined, as well as the mean height of the vegetation. The results are shown below.

Species	Percentage frequency		Mean height of plant species (cm)
	Start of 3-year period	End of 3-year period	
Fescue grasses (*Festuca*)	51	32	5.1
Sedges (*Carex*)	11	8	3.7
Creeping bent grass (*Agrostis stolonifera*)	10	5	1.4
Wild thyme (*Thymus drucei*)	15	4	0.7
Gutierrezia sarothrae	13	45	18.6
Cirsium flodmanii	0	6	25.6
All vegetation: Start of 3-year period End of 3-year period			 2.3 17.5

   i. Explain the results shown in the percentages of the grass species over the three-year period.
   ii. Discuss how the vegetation in the area might change further if deer numbers remained low over the next 10 to 20 years.

2. The following diagram shows the vertical distribution of two species of European filter-feeding barnacles, *Semibalanus balanoides* and *Chthamalus stellatus* on rocky shores. *S. balanoides* inhabits most of the intertidal zone, while *C. stellatus* is restricted to the extreme upper end of the shore. Very young *Chthamalus* occur down to mid-tide level, but they grow more slowly than *Semibalanus*. It was suggested that the reason for the absence of older *Chthamalus* individuals was competition for space by the more rapidly growing *Semibalanus*.

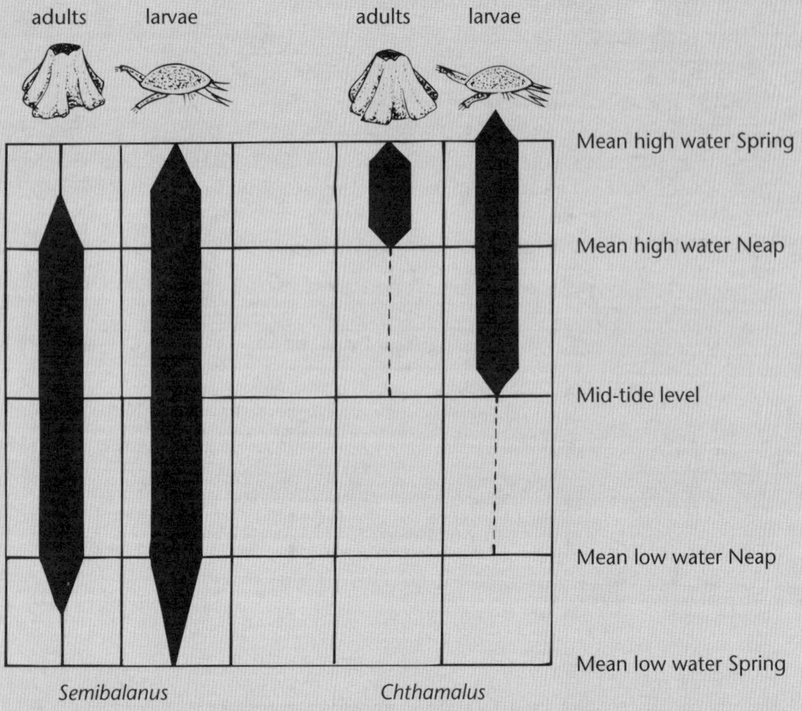

adults
larvae
adults
larvae
Mean high water Spring
Mean high water Neap
Mid-tide level
Mean low water Neap
Mean low water Spring
Semibalanus
Chthamalus

Suggest two possible reasons for the inability of *Semibalanus* to grow on the upper shore.

3. What does the *competitive exclusion principle* state?
4. Nuthatches are small insectivorous birds of the genus *Sitta*, native to Europe and Asia. The diagram below shows the heads of two species, *S. neumeyer* and *S. tephronota*. They have very similar beak lengths where they are **allopatric** (occupy different areas), but are clearly distinguishable where they are **sympatric** (ranges overlap).

Allopatry
Greece
Central Asia
Sympatry
Iran
S. neumayer
Iran
S. tephronota

   a. What name is given to this phenomenon?
   b. Suggest a mechanism by which the similarities of **i.** beak; **ii.** eye stripe have probably arisen.

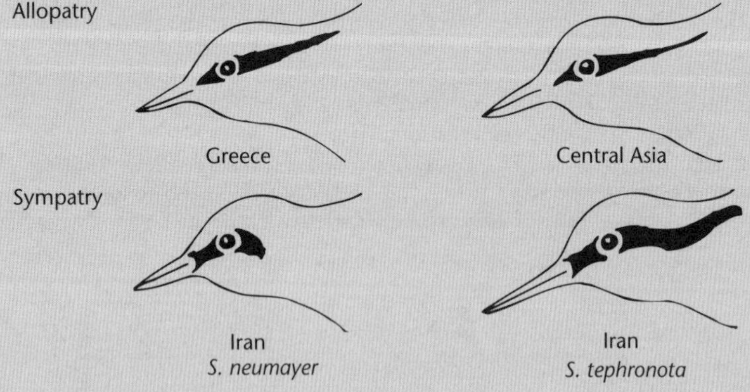

5. Kawakawa, hangehange, nikau palm and pongas are small subcanopy trees.
Suggest how they may continue to co-exist in the same patch of forest.

6. *Asterionella formosa* and *Synedra ulna* are two species of freshwater diatom.
These are small single-celled plants with cell walls consisting largely of silica, the raw material for which is obtained from the surrounding water as soluble silicates.
In an investigation into factors affecting the abundance of diatoms, these two species were maintained in small culture flasks. Each species was grown in isolation (Graphs A and B) and also in mixed culture with the other species (Graph C). Each flask was inoculated with a small number of diatoms and the growth in numbers was monitored by regular sampling during a period of several weeks. The concentration of silicate was also monitored.

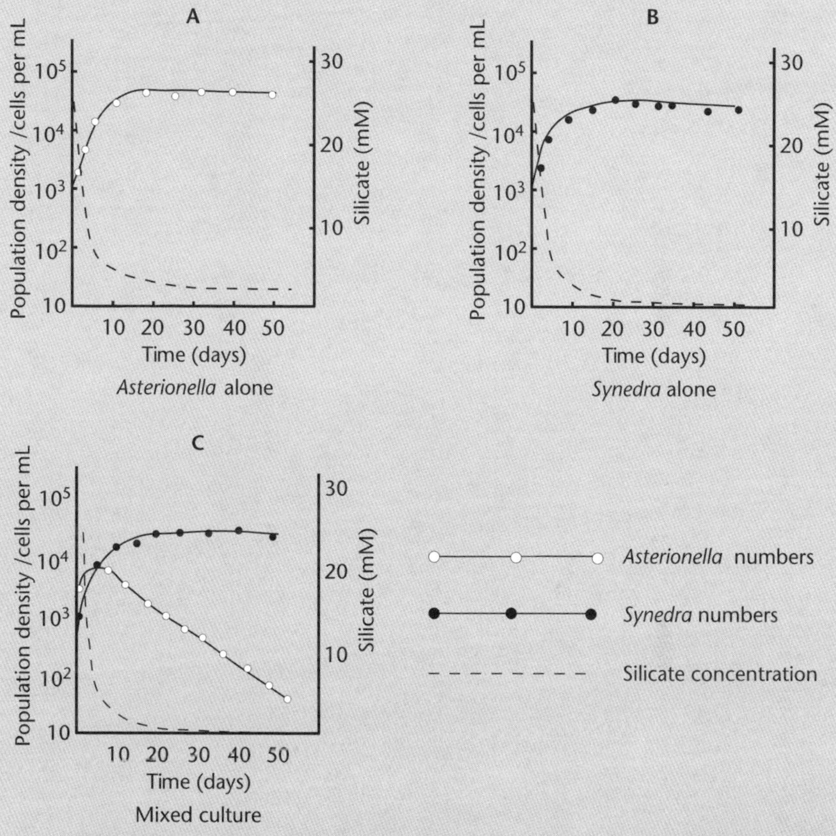

a. i. What can you conclude regarding the interaction between the two populations in Graph C? Justify your answer with reference to population changes shown in Graph A and Graph B.

ii. Suggest an explanation based on the changes in silicate concentration in Graphs A and B.

b. What general ecological principle is illustrated by these data?

c. In most natural environments several closely related species of organism frequently co-exist. Suggest one reason why.

Unit 12.1 Ecology

Topic 4: Interrelationships – interspecific competition

In this Topic we extend the focus on communities and the interrelationships that exist between species and individuals. In particular, this Topic deals with:

- Competition, exploitation and mutualism.

A key aspect of biological communities is the *interrelationships* that exist between species and individuals. These may be beneficial, harmful or neutral.

Competition

Competition occurs when resources are in short supply.

- **Intraspecifc competition** occurs between members of the *same* species and is the more fierce of the two types of competition, as the individuals have the same resource requirements.
- **Interspecific competition** occurs between members of *different* species. It can be important in determining an organism's *actual niche* as well as its *actual zone* in the community (an organism's tolerance and adaptations will give it a potential niche and set the absolute boundaries for its life zone, but competition with other species may reduce these boundaries).

Competition is *harmful* to *both* individuals/species involved, and can induce physiological stress, resulting in increased **mortality** and reduced **natality**.

- Plants are most likely to compete for light, space, water, nutrients.
- Animals are most likely to compete for space, food, water, shelter, nest sites.

Interspecific competition can be detected by removing one species. It is highly likely that two species were in competition for a resource if the:

- Population of the remaining species increases (over a short time frame).
- Actual niche of the remaining species expands (over a long time frame).
- Zone boundaries of the remaining species increase (over a long time frame).

Example

Effect of interspecific competition on species population and growth

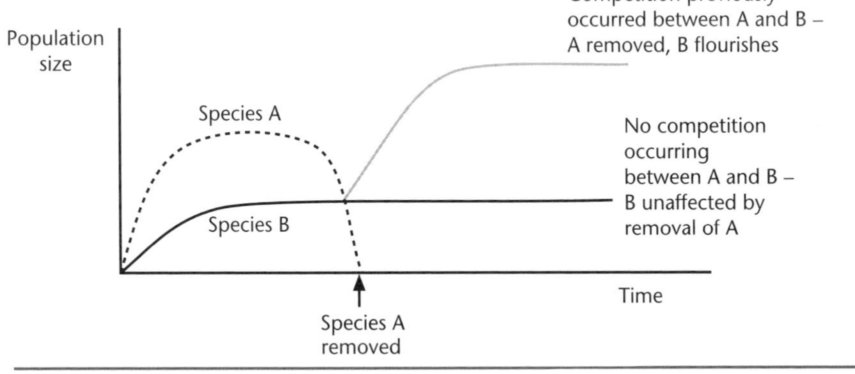

Evidence that interspecific competition does occur is of two kinds – laboratory experiments and field observation.

Laboratory experiments on competition

Because interspecific relationships are so complex, laboratory experiments – in which environmental factors can be controlled – are set up to study them. Early experiments on competition carried out by the Russian scientist Georgyi Gause involved setting up mixed and separate cultures of two closely related freshwater protozoa, *Paramecium caudatum* and *P. aurelia*.

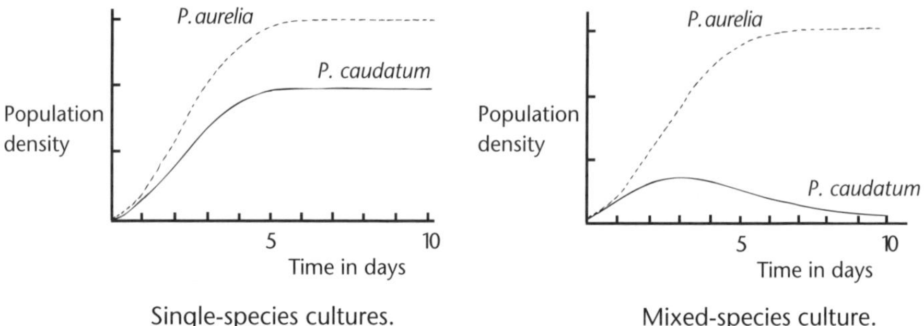

Competition between Paramecium species.

Although *P. caudatum* flourished on its own, it was eliminated in the presence of the smaller *P. aurelia*. Other pairs of related species gave similar results; under a given set of conditions, one species was always eliminated. Gause's experiments also showed that the species that 'won' depended on the conditions (eg when the water was regularly changed, *P. caudatum* 'won'). Experiments by other scientists on grain beetles showed that even quite small changes in temperature or humidity could alter the result.

Gause concluded that *two species could not share the same habitat if they had similar requirements.* This became known as **Gause's principle**, or as it later became known, the **competitive exclusion principle**.

Non-competing organisms can also affect the outcome of competition.

Example

The grain beetle *Tribolium castaneum* is more sensitive to a certain protozoan parasite than a close relative, *T. confusum*. In physical conditions in which *T. castaneum* would normally 'win', the presence of the parasite caused *T. confusum* to 'win'.

The sum total of an organism's requirements constitute its **niche**, which can be likened to its 'profession', in contrast to its *habitat*, which is its 'address'. The competitive exclusion principle in effect states that *no two species can occupy the same niche indefinitely in the same habitat.*

Example

A sea anemone and a mussel may inhabit the same rock pool but they clearly occupy different niches; the anemone feeds on small fish and crustaceans, the mussel feeds on plankton.

Field evidence for competitive exclusion

Many animals appear to live very similar lives, but upon closer study they are seen to be utilising different resources.

Example

Among the grazing mammals of the Serengeti plains of East Africa, there are subtle differences in diet. The dominant grazing mammals are zebras, Thomson's gazelles and wildebeeste, all of which migrate across the plains in the dry season. Zebras feed on grass stems, which are high in lignin but low in protein. Since they are non-ruminants, they cannot process food as efficiently as the other two species; they compensate for this by consuming about twice as much food. Wildebeeste and Thomson's gazelles feed mainly on grass leaves, but do not compete since the Thomson's gazelles migrate about a month later than the wildebeeste, by which time the vegetation has recovered from the effects of the wildebeeste.

Some of the strongest evidence for interspecific competition is **character displacement**.

Character displacement occurs when the differences in a character common to different species is exaggerated where the species habitats overlap.

Example

Two species of ground finches (*Geospiza*) living together (ie in **sympatry**) on one island of the Galapagos Islands differ markedly in body size and beak proportions, enabling each species to exploit slightly different feeding niches and reduce interspecific competition. However, where the two species occur separately (ie in **allopatry**), the beak size and body proportions are very similar – each species is intermediate in the two characteristics compared to when they coexist. Interspecific competition appears to be a strong selection pressure in this case.

Extension: Some problems with competitive exclusion

If the competitive exclusion principle is valid, then in any given habitat there must be as many niches as there are species. How do they all manage to co-exist? Animals have very diverse feeding habits, but plants seem to present problems. For example, there may be a dozen species of plant living in a lawn, yet all require only light, CO_2, water and a variety of mineral salts. There may not seem to be enough different ways of obtaining these resources to account for the observed variety of species. However:

- Unlike most laboratory experiments, real environments are seldom, if ever, constant. It may be that at any given time of the year one species is in the process of gaining the ascendancy over another, but conditions do not remain favourable for long enough for it to 'win'. In parts of the world where there are four distinct seasons, for instance, clover continues to grow throughout the autumn, winter and spring, but makes little growth in the hot, dry conditions of summer. On the other hand *Paspalum*, an introduced tropical grass, dies back in the winter and makes rapid growth in the summer.
- Besides showing variation in time, many natural environments also vary greatly in *space*. Even though in a given area species A may be in the process of eliminating species B, there may be constant immigration by species B from areas in which species B is eliminating species A, and vice versa.

- In real environments, predation may prevent populations reaching levels at which competition becomes significant. Most predators show some prey specialisation, so this would favour greater **diversity** of species, each with a lower density.

Niches change

An animal's diet may change considerably with its stage of development. This is especially true of animals that undergo a metamorphosis from a larva to the adult, as in many insects and amphibians. Even within the adult stage, an organism's niche is not static, but may change with the seasons.

Example

In places where there are four seasons in a year, insects may account for over half of some birds' diets in summer. But during the rest of the year these birds may feed mainly on plant material such as leaves and berries. This largely vegetarian diet for much of the year brings those birds into competition with other animals, and may be one reason for a decline in bird numbers.

The competitive exclusion principle helps to explain why there are so many species (estimates range up to 10 million). For sympatric species (ie those sharing the same habitat), the more similar their niches, the more intense the competition and the greater the evolutionary pressure for them to diverge to minimise competition.

Examples

Plants living in native forest have differing light requirements. The large trees forming the canopy are 'sun plants', and are able to utilise high light intensities. The **shade plants** of the understorey and forest floor, such as tree ferns, can utilise less intense light.

Many flowers are so constructed that their nectar can only be reached by certain kinds of insect. Clover is pollinated almost exclusively by bees. The nectar is concealed within a tube that has to be opened by depressing the 'keel' petals. Only bees are heavy enough to depress the keel and have a **proboscis** long enough to reach the nectar; they tend to specialise on those flowers over which they have a semi-monopoly.

Strong and weak competitors

First to become established on recently exposed soil are plants such as groundsel, shepherd's purse and other 'weeds'. Although poor competitors for light and soon replaced, they are nevertheless successful because they grow very quickly and produce large numbers of widely dispersed seeds in the short time available. Since these plants utilise habitats that are short-lived, selection is for a very short life cycle and high seed output. Plants that form climax vegetation have a different strategy. They grow more slowly, taking much longer to begin seed production, and put a greater proportion of their resources into competing with other plants.

These differences in life cycle strategy have parallels in the animal kingdom. At one end of the spectrum are species that can take advantage of abundant but briefly available resources, such as houseflies which lay their eggs in a corpse. For such so-called '*r*' species, the emphasis is on reproductive output rather than competitive ability. At the other extreme are species known as '*K*' species that live in more stable environments, in which competition is more severe.

These two life cycles strategies are not sharply distinct, but are the opposite ends of a continuous spectrum.

Contest or interference competition

Contest or *interference competition* occurs when some individuals of one species actively prevent individuals of another species from obtaining a resource.

Examples

Tuis chase bellbirds away from clusters of flowers containing nectar; a group of hyenas drives a lion away from its kill.

Contest competition also occurs *within* many animal species (resulting in social hierarchy and territoriality).

Contest competition differs from *scramble competition*, in which there is no active interference, some individuals getting less of a resource only because others get more.

Allelopathy and antibiosis

A number of plants reduce competition by **allelopathy** – producing chemicals that inhibit growth of other species.

Example

The leaves of the American sagebrush contain high concentrations of volatile chemicals called *terpenes*. When the leaves fall they inhibit the growth of other plants and the germination of seeds. Many other plants are rich in terpenes and probably reduce competition in this way. This is probably the reason for the lack of vegetation under many gymnosperms such as *Macrocarpa*.

A similar phenomenon occurs in many fungi, which produce *antibiotics* – substances that inhibit the growth of bacteria. **Antibiosis** also occurs in many marine animals, such as sponges.

Exploitation

Exploitation occurs when an individual/species exploits another in some way and so *harms* it. The main examples of this are *predation*, *herbivory*, *parasitism*.

- **Predators** hunt, kill and eat *other animals* (their *prey*).
- **Herbivores** eat *plants*.
- **Parasites** feed from other *living* organisms (their **host**).

In exploitation, one species obtains food from another.

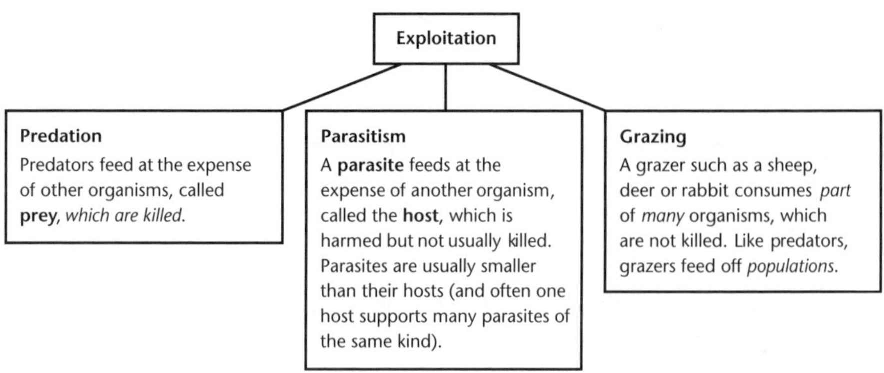

Parasitism

A parasite depends for its food on one organism, the *host*, which is not usually killed (although the parasite may so weaken the host as to render it more vulnerable to other hazards). There are probably more parasitic than free-living species, so parasitism can hardly be considered to be an 'abnormal' way of life. Animal parasites can be **ectoparasites**, which live on the outside of the host, and **endoparasites**, which live inside the host's body (eg tapeworms and flukes).

A well-adapted parasite does not kill the host, for in doing so it would kill itself.

Animal parasites may be **ectoparasites** ('ecto' means *external*) or **endoparasites** ('endo' means *internal*).

Ectoparasites

These are less intimately associated with the host than endoparasites, and in some the relationship is so brief that some would not consider them to be parasites (eg mosquitoes). Many bloodsucking ectoparasites are important as **vectors** in the transmission of disease (eg certain species of *Anopheles* mosquitoes transmit malaria, and body lice transmit typhus). Many ectoparasitic insects are wingless (eg lice and fleas). Both fleas and lice have flattened bodies, which give some protection against the scratching of a mammalian host, and lice have strong claws which enable them to hold on tightly to their host's body.

Ectoparasites live on the skin of the host and feed on its blood. They have adaptations such as hooks or claws for holding onto skin and piercing/sucking mouthparts for obtaining blood (eg fleas, mosquitoes, leeches, ticks).

Ectoparasites may be able to transmit diseases, eg fleas (plague) and mosquitoes (malaria).

Animal endoparasites

Endoparasites live inside the host – either in the blood vessels feeding from the blood (eg liver flukes), or in the intestines feeding from the digested food (eg hydatid tapeworm). When present in large numbers (an over-infestation), endoparasites can severely harm the host. Endoparasites typically have a complex life cycle that involves two different hosts (often a carnivore and a herbivore). Adaptations to the lifestyle usually involve the reduction of body systems (eg reduction in locomotion, digestive, sensory systems) and the development of hooks and suckers to hold themselves in place. Large numbers of eggs are produced to improve the chances of finding another host.

Endoparasites typically inhabit internal spaces such as the gut (eg tapeworms and some roundworms), **bile** ducts (eg liver fluke) and blood vessels (some flukes and roundworms). The most extreme adaptations to parasitism are found amongst endoparasites such as tapeworms, which show the following adaptations:

- *Reduction of structures compared with free-living animals*. Endoparasites do not have to find food or escape from enemies, and most have reduced sensory, muscular and **nervous systems**.
- *Presence of structures for holding on*, such as hooks and suckers in tapeworms, and clawed legs of lice.
- *A highly developed reproductive capacity*. Since all hosts ultimately die, the parasite must somehow get its offspring into new hosts. Reproduction is therefore linked to the problem of transmission to new hosts. Adult parasites are incapable of surviving in the world outside, so it is the eggs or young that have to get to a new host. The chances of successful transmission are extremely low, and the parasite has to channel most of its resources into egg production.
- *A complex life cycle*. The problems of transmission are not solved merely by producing many offspring. Many endoparasites have a life cycle in which there are several quite different young stages, each living in a different host.

The successful reproduction of an endoparasite depends on the survival of its host, so it is in the interest of the parasite to harm the host as little as possible; the most successful endoparasites live in a reasonably harmonious relatisonship with the host.

Example

The evolution of mutual adaptation between host and parasite can be seen in myxomatosis. The introduction of the virus into the United Kingdom and Australia was followed by decreased **virulence** of the virus and increased resistance by the rabbits.

In evolving very close relationships with hosts, most endoparasites are adapted to living in a very limited range of hosts. Like all specialisation, there is a cost attached to this *host-specificity* – the evolutionary future of an endoparasite is dependent on that of its host.

Example

The liver fluke (*Fasciola hepatica*) endoparasite

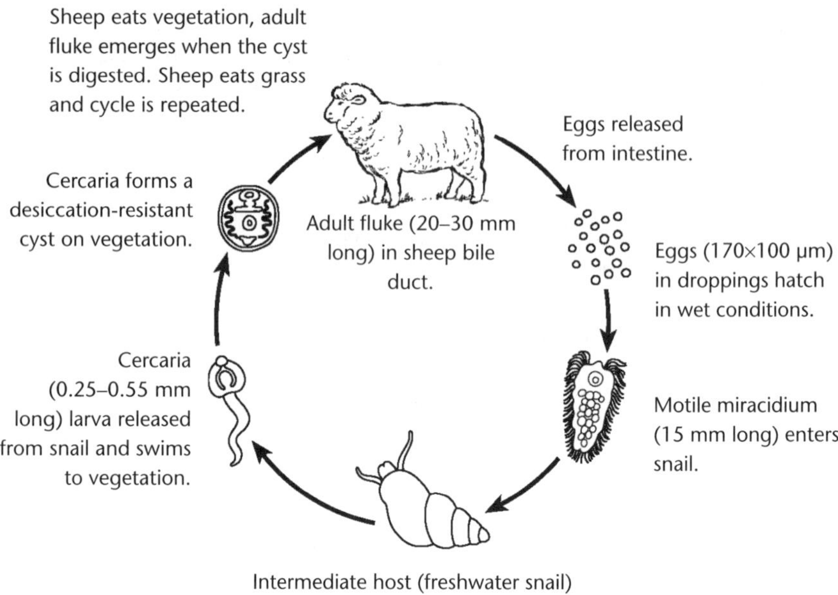

Fasciola hepatica is common in sheep in countries that have a temperate climate. It causes loss in stock condition, **liver** damage, lowered fertility and even death.

Plant parasitism

Plants may also be parasites (though the habit is rare – unlike in animals, where parasites are found in most animal groups).

Plant parasites tap into the host's phloem, using the host's food made from photosynthesis.

Compared with animals, a much smaller proportion of flowering plants (and even fewer non-flowering plants) are parasitic. All connect with the vascular tissues of the host by structures called **haustoria**.

Plant parasites are quite distinct from **epiphytes**. Though an epiphyte is attached to a larger plant and has no connection with the ground, it has no vascular connection with the larger plant, obtaining only support.

Total parasites depend on the host for all their raw materials and energy, and lack both roots and chlorophyll.

Example

There are several New Zealand species of total parasite, but all are inconspicuous, locally distributed or uncommon. The most likely total parasite to be encountered in New Zealand is broomrape, an introduced plant common on verges and embankments and parasitic on roots of clover and other pasture plants. Only the scaly flower shoots appear above ground, in spring.

Another root parasite is the native *Dactylanthus taylorii.* It parasitises the roots of forest trees throughout the North Island. Where the host tissue meets the parasite, it forms peculiar radiating furrows, forming a so-called 'wooden rose'. The plant is entirely below ground except for the brown flowering shoots.

A 'wooden rose' left by haustoria of Dactylanthus.

Broomrape.

Partial parasites have chlorophyll and depend on the host only for water and minerals.

Example

The native mistletoe *Ileostylus micranthus* is a partial parasite, and grows on a wide variety of trees, including some conifers. It has small green flowers and yellow berries.

Native mistletoe Ileostylus.

Parasitism in fungi

There are a great many parasitic fungi, many of which do considerable damage to crops.

Examples

Ceratocystis ulmi is the cause of 'Dutch elm' disease, which has wiped out most of the elm trees in Europe and in recent years has become established in New Zealand. The fungus is spread by a bark beetle and infests the **xylem** and phloem, rapidly causing the death of the host by preventing transport of water and organic **solutes**.

Armillaria is a 'toadstool' that parasitises kiwifruit. Like a number of other fungi it can live a non-parasitic life as a saprobe, feeding on dead matter in the soil. Since it does not have to be parasitic, it is said to be a *facultative* parasite, in contrast to *obligate* parasites (which are exclusively parasitic).

Saprolegnia is another facultative parasite, and is the cause of a fungus disease of fish in aquaria. After killing the host, the fungus lives as a saprobe on the corpse.

Predation

Predation is a relationship, so predators and prey must be considered together. Predator–prey relationships can be considered over two very different time scales:

- Over the short term, predators and prey may influence each other's *numbers*.
- Over a very much longer time scale, predators and prey shape each other's *evolution*. Evidence for this lies in the adaptations of predators for obtaining food and of prey for avoiding being eaten.

Predator strategies

There are many predator strategies. Some rely on speed; for instance, the cheetah. Other cats stalk their prey and then pounce (eg leopards), or make a short rush (eg lions). Many predators rely on their ability to be undetected against their background until the prey is close enough to be caught (eg mantids and chameleons). Some predators rely on the ability to paralyse the prey with poison, as in sea anemones and many snakes.

Some plants produce defences against animal predators by growing thorns, for example, or being poisonous.

Belonging to a group

Despite the fact that food has to be shared, membership of a group can increase food intake in two quite different ways:

- Locating food – when food items are difficult to find *but locally abundant* (as with some seeds and small insects), group foraging pays, because the 'cost' of sharing is small.

Example

Many small birds such as sparrows often forage in flocks.

- Catching food – group hunting can be advantageous when food is sufficiently active to resist and escape.

Examples

- Larger, more heavily armed prey can be killed; a pack of wolves can kill a moose, and two or more lions can tackle a wildebeest with less risk of injury. Some predators are exclusively group-hunters (eg African hunting dogs).

- A few predators cooperate in a co-ordinated fashion. Chimpanzees cooperate to hunt monkeys. One strategy involves one chimpanzee individual distracting the attention of a monkey while another stalks it. Alternatively, some chimpanzees act as beaters, driving a monkey into a position in which another chimpanzee can catch it.

Snares

Some predators combine a snare with a lure.

Example

The New Zealand 'glow worm' is the larva of a fly and lives in caves and holes in steep banks.

The insect first makes a horizontal silk tube in which it spends most of its time, and then adds vertical trap-threads covered with sticky droplets. The larva has a luminous **organ** and light is directed onto the droplets by a bowl-shaped reflector in its abdomen. The droplets reflect the light and small insects are attracted to them, become trapped and are eaten when the larva hauls up the thread.

The New Zealand glow worm.

Mimicry

Some predators obtain a meal by deception.

Examples

- Cleaner fish obtain food by removing parasites from the gills of other fish, which open their mouths for the cleaner. Cleaners are recognised by their 'clients' because of their distinctive colour pattern and mode of swimming. Some predatory fish mimic the cleaner's appearance and movement, and use this deception to get close enough to bite chunks out of the fins and gills of client fish.

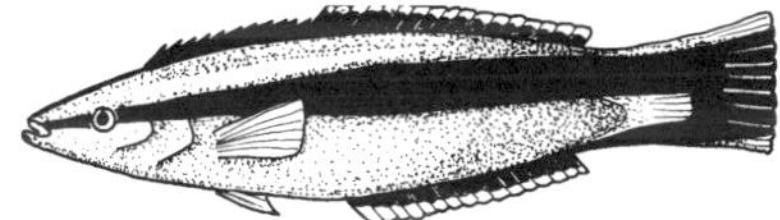

mimic – the sabretooth blenny

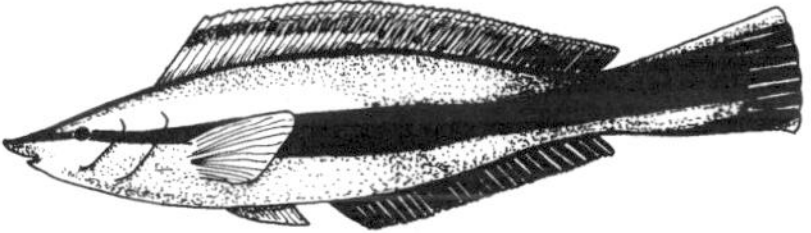

model – cleaner wrasse

Mimicry by a predator.

There are limits to how successful predatory mimicry can be, as the deception will only work if the mimic is rarer than the cleaner.

- Female fireflies attract males by emitting a species-specific pattern of flashes. Females of some predatory fireflies mimic the flash patterns of other species, luring the prey males to their deaths.

Prey defences

Living in groups

Many animals obtain safety in numbers, through greater vigilance, increased 'armaments', dilution and confusion effects, and concealment of young by adults.

Greater vigilance

For animals that rely on speed for escape, such as antelopes and ostriches, early warning is crucial.

Living in groups increases the likelihood of early detection of a predator.

Example

The larger the size of a pigeon flock, the lower the rate of predation.

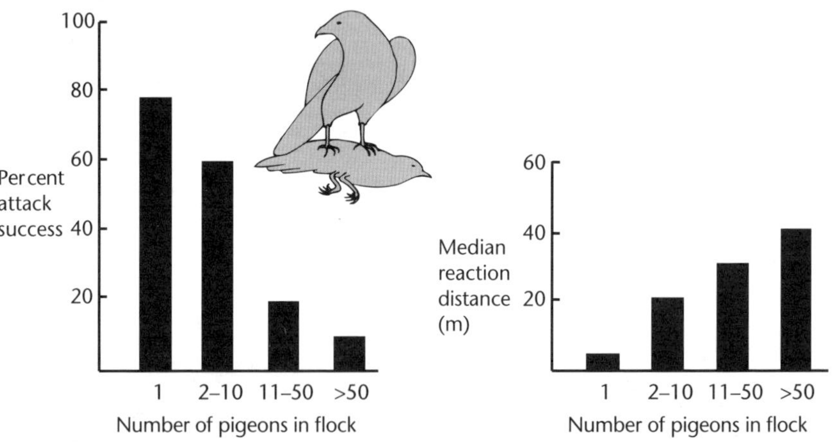

Effect of flock size on predation in pigeons.

Some animals increase their vigilance by combining the sensory capacities of different species.

Examples

Impala (a kind of antelope) have a keen sense of smell and spend much time near baboons, which have excellent colour vision.

A mixed group of ostriches and zebras combine the keen sense of smell of the zebras with the acute colour vision atop the high head of the ostriches.

In addition to increased vigilance, group-living allows longer feeding time. An animal feeding on food close to the ground has its head too low to spot an approaching predator. From time to time therefore it must stop feeding and raise its head. A member of a group needs to raise its head less often because the cost of vigilance is shared.

Increased armament

Some large herbivores combine to deter predators.

Example

A leopard will not attack baboons if there are several males present, but will do so if there is only one male. Many sea birds such as terns and gannets nest in large colonies and make mass attacks on a predator.

Dilution effect

In any given predator attack, the larger the group, the less likely any particular prey will fall victim. The presence of other prey 'dilutes' the effect of the predator. Although larger prey groups are more conspicuous and are more often attacked, this does not fully counteract the dilution advantage – a herd of 100 antelope is not attacked 100 times as often as a single animal. Even if every predator attack were successful, the chances of any given prey being eaten in an attack are smaller if it is a member of a large group.

Confusion effect

Most predators prefer to attack isolated prey because the predator finds it difficult to focus its attention on an individual prey when there are many others, particularly if they are rushing in all directions.

Example

Most fish schools show a 'fountain effect', the shoal splitting into two every time the predator rushes into it.

Most commercial fish are caught in large numbers precisely because they swim in schools. Thus behaviour that is successful against most predators can render them vulnerable to human hunters. It may be that as fishing pressures increase, selection will operate against schooling within these species.

Concealment of young by adults

The more adults present in a group, the greater the protection they offer in terms of concealment.

Example

Adult wildebeeste are not attacked by hyenas, so their visibility is no disadvantage. The young, however, are heavily preyed upon. When hyenas are nearby, female wildebeeste take their calves to the far side of the herd where they are unlikely to be seen.

Deception

Deception can involve camouflage, counter-shading, Batesian and Mullerian mimicry, diversion of attack and **autotomy**.

Camouflage (crypsis)

To be effective, crypsis must be linked with appropriate behaviour. The animal must select the appropriate background, and move as little as possible. This may be incompatible with feeding, and some cryptically coloured animals are **nocturnal**, remaining motionless during the day, eg the forest gecko.

Examples

The eggs and chicks of most ground-nesting birds rely on crypsis to avoid detection, as do many insects and bottom-living fish such as flounders.

Katydids (long-horned grasshoppers) and mantids are difficult to see against green leaves, and against lichen-covered bark a forest gecko may be almost invisible.

Counter-shading

Pelagic (open water) fish such as kahawai have silvery undersides and are dark on top. This renders the animals less visible against the light background above when seen from below and the darkness below when seen from above.

Batesian mimicry

A palatable animal (the mimic) resembles an unpalatable one (the model), thus gaining protection.

The success of **Batesian mimicry** as a defence strategy is self-limiting – if a mimic becomes too common, not enough predators will learn to avoid the model, and any effectiveness is decreased.

Example

In North America the distasteful monarch butterfly (*Danaus plexippus*) is the model for the more palatable viceroy (*Limenitis archippus*).

Monarch (model)

Viceroy (mimic)

Mimicry in North American butterflies.

Mullerian mimicry

There is no distinction between model and mimic, several unpalatable species resembling one another. It is easier for predators to learn to avoid a group of unpalatable species, since a lesson learned after a 'bad' taste of one species will also apply to other species of similar appearance.

Diversion of attack

Many butterflies (eg yellow (*Vanessa itea*) and red (*Vanessa gonerilla*) admirals) and moths (eg gum emperor *Opodipthera eucalypti*) have large eye spots on their wings. These may have the effect of diverting the attack of a predator away from the body to the wings, which can withstand more damage.

Autotomy

This is the shedding of part of the body when a prey is attacked, with the result that the predator may be diverted by the lost part long enough for the prey to escape.

Example

All New Zealand lizards and also the tuatara (*Sphenodon punctatus*) can autotomise their tails, which continue to wriggle and are later regenerated. Many crabs can also autotomise their chelae (pincers).

Synchronised breeding

In gannets and many other communally nesting birds, individuals of both sexes are stimulated sexually by neighbours as well as their own partners. The result is that egg-laying is more synchronised in large colonies than in smaller ones, so the breeding season for the colony as a whole is shorter. By concentrating egg-laying over a shorter period, the birds lose fewer eggs to predators, since each predator can only take a limited number of eggs or young in a given period. A number of other animals 'saturate' the feeding capacity of predators by synchronising their breeding, for instance the palolo worms of the Pacific.

Chemical/physical defence

The most familiar example of chemical defence is the venom injected by bees and wasps, but chemical defence is far more widespread than this. Plants in particular produce a wide range of toxic and distasteful chemicals, for instance:

- *Cyanogenic glycosides*, which release hydrogen cyanide when the plant's tissues are damaged. Examples are clover leaves and karaka seeds.
- *Cardiac glycosides* (heart poisons). They are produced by *Coriaria arborea*, a very toxic plant, by the swan plant (food of the monarch butterfly larva), and also by oleander (a common garden plant, and *extremely* poisonous), and by foxglove.
- *Insect hormones*. Ferns and conifers are generally resistant to insect attack because they contain high concentrations of insect hormones. These disrupt the growth of any insects that feed on them. Many podocarps and ferns contain moulting hormone, and some conifers contain juvenile hormone (which inhibits metamorphosis in early larvae).
- *Tannins*, which not only make plants bitter, but also make leaf proteins less digestible and so reduce their food value to animals. Gorillas reject species whose leaves are rich in tannins.
- *Alkaloids* are responsible for the distastefulness and toxicity of many plants, eg ragwort, a weed toxic to cattle, and hemlock. Deadly nightshade berries are extremely toxic to mammals but not to birds, which eat them and disperse the seeds. One of the most deadly alkaloids is strychnine, present in the seeds of a tropical Asian tree.

Herbivores can overcome this battery of defences because:

- In many leaves the concentration of defensive chemicals increases as the leaf ages, so leaf-eating animals limit their dose by concentrating on young leaves.
- Different species of plant usually make different kinds of defensive chemical. By eating small amounts of leaves from many species of plants, animals such as the woolly monkey limit their daily dose to sub-toxic levels.
- Not all plants produce **toxins**. In grasses, for example, the leaves continue to grow at the bases, and the plant relies on rapid regenerative growth to overcome grazing.
- Plants also have physical defences such as thorns and spines on leaves to prevent predation by animals.

Extension: How has chemical defence evolved?

If an animal dies as a result of eating a poisonous plant, how could it learn to avoid the plant? One hypothesis follows. Plants make large numbers of chemicals that fulfill diverse functions. Besides their primary functions, some of these chemicals may originally have *just happened* to be toxic. At first, such chemicals may have been present in such small amounts that an animal would have to eat a considerable amount for the chemical to be lethal, so the defence would not be very effective. Distastefulness is not a property of the substance itself – rather it is a sensation within the brain of the animal that eats it. Suppose that grazing animals varied in how distasteful they found toxic substances. Animals that found the substances most distasteful would be least likely to eat enough to be harmed, and so would leave more offspring. There would thus be selection for finding toxic substances distasteful. Selection for toxicity has probably also occurred in the plants – those producing the greatest quantities of poison would be most distasteful and would be most avoided by herbivores.

Extension: Warning colouration

Most distasteful animals advertise the fact by warning or *aposematic* colouration. The larva of the monarch butterfly has black, yellow and white stripes across the body, and the cinnabar moth larva has yellow and black stripes.

For warning colouration to evolve, the genes responsible must be handed on to the next generation. But a predator cannot learn that the prey is distasteful without killing it. How then could genes for warning colouration be perpetuated? One possibility is that distasteful animals tend to remain close to where they were born, so that many of the individuals in an area are related and thus share many of their genes. An animal that is eaten thus helps to 'teach' a predator that its relatives are distasteful, so the copies of the genes are more likely to be handed on. It has also been observed that instead of investing everything in reproduction and then dying, many distasteful animals live for some time after they have reproduced. During this post-reproductive life there is a strong chance they may be eaten by a predator, which would learn to avoid the descendants of the distasteful animal.

Defenceless native plants

New Zealand plants (and those of New Caledonia) evolved in the absence of browsing mammals. This is probably the reason why the only native forest plants with spines are those in which the spines are used for climbing, such as the bush lawyer (*Rubus cissoides*) and its relatives. Native trees have few chemical defences against introduced mammals, which has resulted in severe damage by possums.

Mechanical protection

This takes two common forms:

- Spines, eg sea urchins, hedgehogs, gorse, roses, cacti, bougainvillea.
- Shells, eg barnacles, many molluscs, hermit crabs, and the protective layer around some seeds, as in almond 'nuts'.

Animal countermeasures

A number of animals have evolved defences against plant toxins. Sheep can detoxify the glycosides of clover. Animals such as the larva of the monarch butterfly have gone a step further, not only tolerating a poison but accumulating and retaining it into the adult stage. The result is a double advantage – the larva has a near-monopoly of the food plant, and both larva and the adult gain protection against predation by birds. The cinnabar moth is another example, introduced into New Zealand because of its ability to feed on ragwort.

Extension: Efficient predators

Before a predator can obtain energy from another organism, it has to spend energy to capture and digest it. To make a net profit, the energy obtained from the prey must exceed the energy used to get and consume it. The time and energy invested can be divided into two stages:

- Searching, which involves all the time and energy spent in locating the prey. In web-weaving spiders and other predators that trap their prey, the equivalent of search time is the time spent waiting for prey to be snared in the trap.
- Handling, which involves pursuit, capture and eating. Where prey is large enough to inhibit further search until it has been digested, handling also includes digestion. In animals that construct traps that have to be repaired after each capture (eg many spiders), repairing the web is effectively part of the handling time.

Predators that make the biggest energy 'profit' will be the most successful and will therefore make the greatest contribution to the gene pool of the next generation, so there will be strong selection for efficiency.

Example

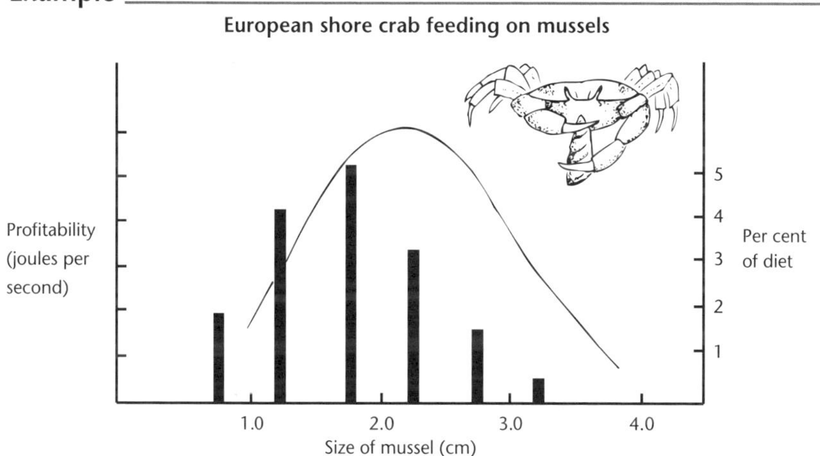

Solid line = energy gained per second of feeding time. Bars = per cent of diet.

Predators select the most profitable prey.

The crabs easily open very small mussels with their pincers, but the energy in the flesh is very small. Large mussels, although containing more energy, take much more time and effort to open. Somewhere between the two extremes is an optimum prey size.

Optimum prey size was determined in the following way. First, the energy content of mussels of different sizes was measured. Next, the average time taken by a crab to open mussels of different sizes (and thus of known energy content) was determined. From these measurements, the profitability of each mussel size was calculated and plotted. A crab was then presented with a variety of prey sizes, and the size of each prey eaten was recorded. The bar graph shows the observed prey preference. The peaks of the two graphs are similar, showing that the crabs were selecting the most profitable prey.

These results do not mean that predators consciously choose prey of optimum size. Selection has simply favoured those individuals that select prey of the 'best' size.

Specialists and generalists

Predators vary according to the breadth of their diet. At one end of the spectrum are **monophages**, which eat only one kind of prey. At the other extreme, and far more common, are **polyphages**, which eat a wide variety of prey.

Example

Brown and black rats are among the most omnivorous of mammals, eating practically anything. They entered New Zealand with the European settlers and, as expert climbers, have taken a heavy toll of the eggs and nestlings of native birds. Mustelids (stoats, weasels and ferrets) were introduced to control rabbits, and without them it is likely that the rabbit problem would be even worse. However, they are opportunist predators and do not confine themselves to rabbits, but also eat many native birds.

Search time and diet

Why does it pay for some predators to eat a wide variety of prey and for others to concentrate on a few? The answer may lie in the relative time and energy spent in searching for prey and handling it.

Example

Some birds eat a wide variety of small insects. They spend a long time searching for prey, which individually contain small amounts of energy. However, once prey has been spotted, the time and energy costs of capture (ie handling) are negligible. It thus pays to eat most kinds of small insect encountered.

It pays a specialised predator (such as a lion) to ignore some prey. In the time a predator would have spent dealing with an unprofitable prey item, it would normally encounter and be able to capture and eat a more profitable one.

Generally speaking, therefore, predators that make a large investment in searching for prey tend to have wider diets than those that have short searching times.

Prey numbers influence predator numbers

A predator can only reproduce if it eats sufficient prey, so it is to be expected that predator numbers are influenced by prey numbers. There are many cases in which increases in prey numbers are followed by increases in predators.

Examples

Kiore

A detailed study of kiore (*Rattus exulans*, the Polynesian rat) on Tiritiri Island in the Hauraki Gulf in New Zealand showed that rat numbers fluctuated in a cyclic manner, with peaks every autumn. Kiore eat small invertebrates, leaves and grass seeds, but in the grassland part of the island the numbers of kiore peaked regularly about two months after the peaks in grass seed production. Examination of stomach contents at intervals during the year showed that only grass seed consumption varied in a way that correlated with the reproduction of the rats.

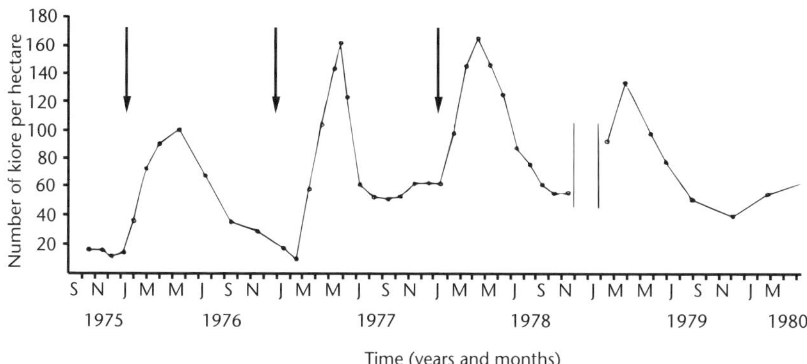

Arrows represent peaks in grass seed output.

Changes in numbers of kiore on Tiritiri Island.

Canadian lynx

The main prey of this medium-sized cat is the snowshoe hare. Records of pelts over two centuries show that the populations of lynx and hare undergo regular oscillations every 9–10 years. After a short delay, each hare peak is followed by a lynx peak. It seems that lynx numbers are fluctuating in response to changes in the supply of prey.

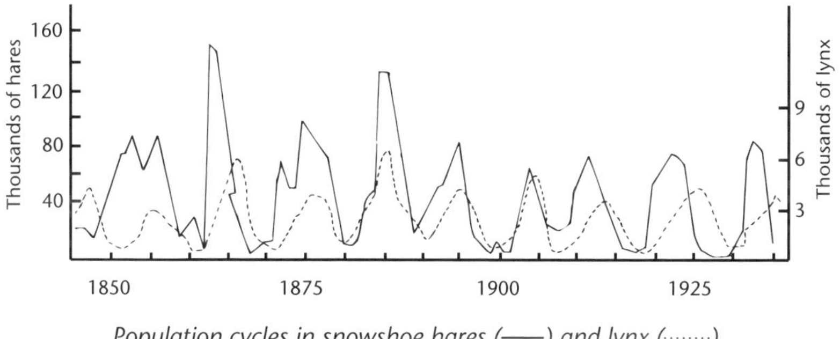

Population cycles in snowshoe hares (——) and lynx (········).

Many predators do control their prey

The results of human interference suggest that in many cases predators control the numbers of their prey. Animals and plants have often become pests after introduction into a country in which there are few if any predators. Rabbits, goats, possums and wasps, broom and gorse, have all been introduced into New Zealand, with disastrous results. In some cases introduced pests have been brought under control by the deliberate introduction of a predator.

Examples

- **The cabbage white butterfly**

  This insect is a pest of cabbages and was accidentally introduced into New Zealand in 1930. It spread rapidly and within a few years had devastated cabbage crops over the entire country. From over 30 species of insects that attack the cabbage white in its native Europe, scientists chose two, *Apanteles glomeratus*, which lays its eggs in the larva, and *Pteromalus puparum*, which lays its eggs in pupae. Both were introduced, and by the mid-1930s had brought the butterfly under control.

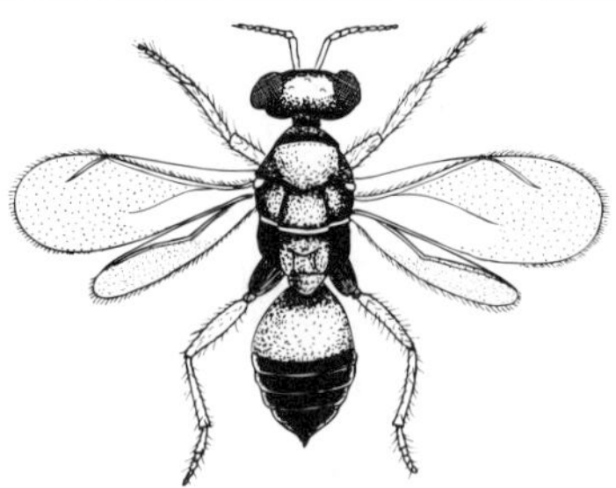

Pteromalus puparum*, a parasitoid wasp introduced to control the cabbage white butterfly in New Zealand.*

- **The cactus moth**
  In the early 1920s, huge parts of Australia were put out of agricultural production by the introduced prickly pear cactus. The problem was solved by the introduction of *Cactoblastis*, a small moth that feeds on the cactus in its native Argentina. The result was dramatic – within two years vast areas of what had previously been dense growths of the cactus were eaten away by moth caterpillars. Since then both cactus and the moth populations have stabilised at much lower levels.

Some predators probably don't control their prey

In other cases, other factors, especially competition, appear to be more important population regulators than predation.

Examples

- The main factor limiting barnacle population size is competition for space. Although vast numbers of their planktonic larvae are eaten, many cannot find space to settle on the rocks and are thus doomed anyway.
- Competition for territory sets an upper limit to numbers of many animals. For the Canadian muskrat, holding a territory is essential for obtaining sufficient food. Animals without a territory, although more likely to be eaten by mink, would usually starve in any case.

Predator–prey cycles

The cycles of the Canadian lynx population strongly suggest that the supply of hares controls the number of lynxes. It is *not* true, however, that the lynx population is *also controlling the hare population*.

- In some parts of Canada (eg Anticosti Island in the Gulf of St Lawrence) where there are no lynxes or other large predators, hare numbers show the same oscillations.
- The snowshoe hare has a much greater reproductive potential than the lynx, so provided that food is not limiting, hare population growth will always exceed that of the lynx.

Regular oscillations of predator and prey populations are not common in nature, being restricted to simple ecosystems with few species, such as the Arctic. Oscillations can also sometimes occur in agricultural and horticultural systems, in which the number of species is kept low by human activity.

In most temperate and tropical ecosystems, populations remain stable and fluctuations are irregular because food webs are complex. Since there are many more predator species, an increase in the numbers of one kind of herbivore can be absorbed by predators switching their choice of prey.

Extension: Investigations into predation

Most communities contain many species and are thus more complex than the Canadian Arctic or an orchard. It is therefore difficult to investigate how the numbers of different species influence each other in most communities. Setting up artificial predator–prey systems in the laboratory permits individual factors to be controlled and studied separately.

The ability of predators to control prey depends on the way the predator populations respond to changes in the numbers of the prey. Any factor which regulates a population at a steady level operates on a **density-dependent** basis. This means that as numbers of prey rise, the *proportion* of the prey population eaten increases – ie the average life expectancy of each individual must decrease.

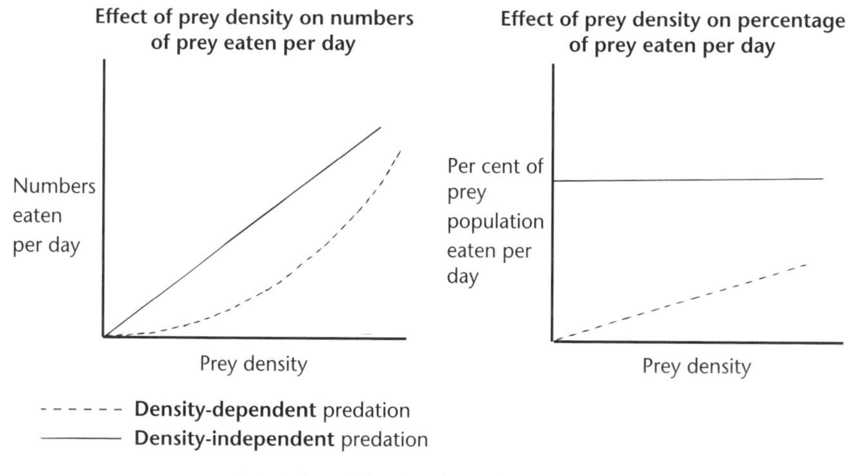

Principles of density-dependent predation.

Functional responses

A series of classic experiments by Holling showed that changes in prey numbers could influence predators in different ways.

An increase in prey density results in each *individual* predator increasing its consumption of prey.

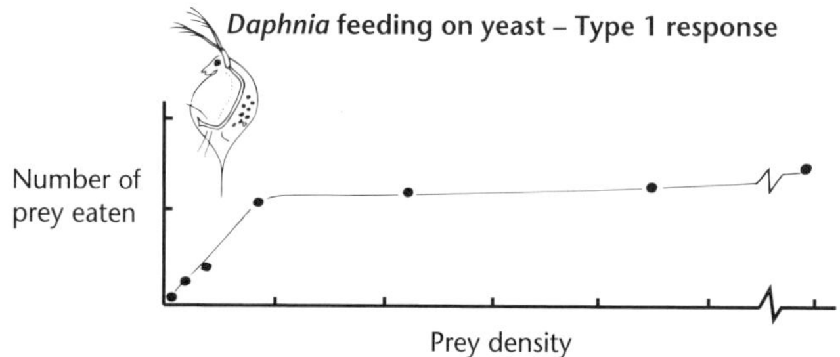

Dragonfly nymph feeding on *Daphnia* – Type 2 response

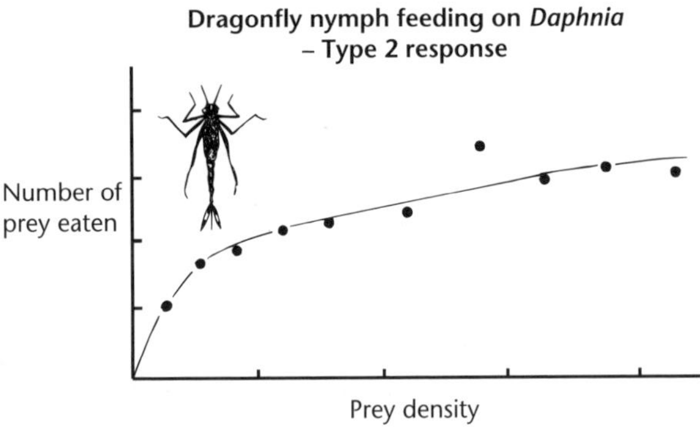

Shrew feeding on sawfly cocoons – Type 3 response

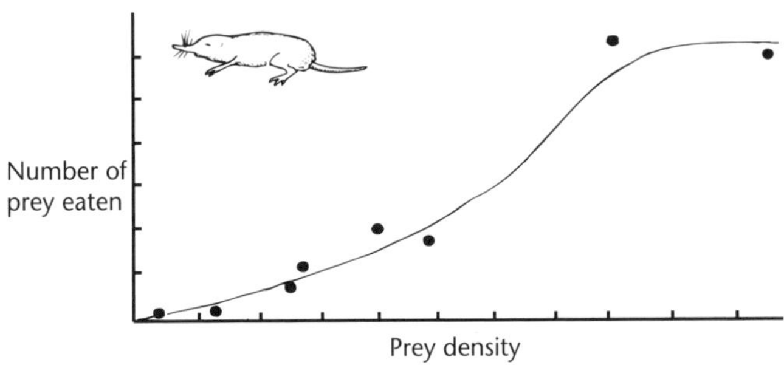

Only a predator showing a Type 3 functional response can *by itself* control a prey population.

Ways in which increasing prey density can influence the rate at which an individual predator consumes prey.

- *Type 1* responses are shown by sessile (anchored), filter-feeding animals, which have no search time because the prey is brought to the predator.

Example

Up to a point at which the filter of a barnacle or mussel becomes saturated, the rate of predation is directly proportional to the prey density, ie the *proportion* of prey eaten does not change. Beyond this point the rate of feeding is limited by the rate at which food can be removed from the filter, and the proportion of prey eaten actually *decreases*.

A predator showing a Type 1 response cannot therefore on its own regulate a prey population.

- *Type 2* responses are characteristic of animals that have to search or lie in wait for their prey. Either they feed on one type of prey or their prey preference does not depend on how common it is.

Example

For a mantid, an increase in prey density has two effects. First, search (actually waiting) time decreases and is eventually insignificant. Handling time (time needed to kill, eat and digest the prey) remains constant, so above a certain saturation prey density the mantid cannot feed any faster. Second, the predator has to ignore an increasing proportion of prey because it is already occupied in handling others. This latter effect causes the graph to flatten out, the proportion of prey eaten *decreasing* as prey density increases.

A predator showing a Type 2 response cannot therefore by itself regulate a prey population.

- *Type* 3 responses occur when a predator changes its preference as the prey becomes more common, so that an increasing proportion of prey is taken as its density rises. The result is a **sigmoid** ('S'-shaped) curve. Clearly, predators with Type 3 responses could regulate a prey population. Type 3 responses are probably most common when the prey depend on cryptic colouration. This is because the ability of many predators to pick out concealed prey improves with practice and therefore with the abundance of the prey. Type 3 responses are obviously more important in opportunist predators with wider diets such as rats, cats and stoats.

Numerical responses

In numerical responses, the density of predators increases by immigration and/or increased reproduction.

Example

On Tiritiri Island the rise in kiore (*Rattus exulans*) numbers each autumn was followed by an increase in the numbers of harriers as a result of immigration.

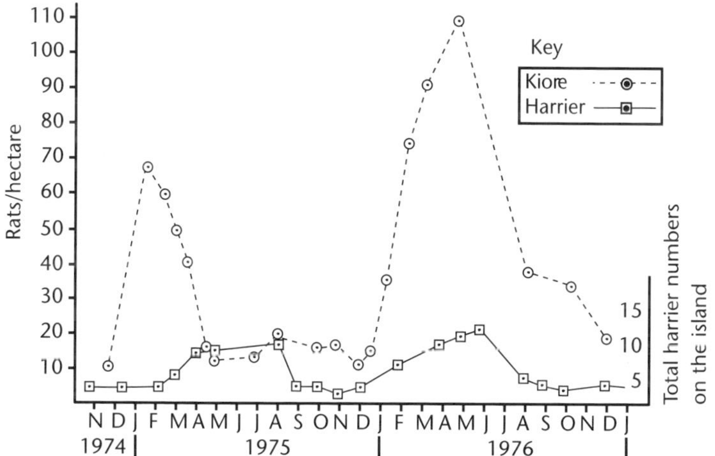

Numerical response of harrier to changes in kiore density on Tiritiri Island, New Zealand.

Numerical and functional responses normally occur together. Provided that predators do not interfere with each other, the combined effect is obtained by multiplication. A five-fold increase in the number of predators (numerical response), and a doubling of their rate of predation (functional response), will result in a 10-fold increase in the rate of predation. Had prey density risen eight times, the 10-fold increase in predation would tend to regulate the prey; if prey density had risen 12-fold, the increase in predation would not. The most effective predator in controlling prey is thus one which shows a strong tendency to shift its prey preference, and at the same time can rapidly build up its numbers, by either immigration or reproduction or both.

Commensalism

Commensalism is a relationship that occurs where one species benefits from another species but does not harm it in any way. Usually, one member of the relationship obtains food from the other.

Example

- Pea crab in mussel – the pea crab obtains food and protection from living inside the mussel shell; the mussel is unaffected.
- Remora (suckerfish) and shark – the remora fish snatches scraps of food from the shark while the shark is feeding; the shark is unaffected.

Commensal relationships may lead to:

- Host/parasite relationships, where one species benefits at the expense of the other.
- **Mutualistic** relationships, where both species benefit from the relationship.

Epiphytes and **vines** are both examples of commensalism.

Mutualism

Mutualism is a relationship that occurs between two species with *both benefiting* from it. In some examples of mutualism, the association is not essential to the two species.

Example

- A crab with sea anemones on its shell – the crab obtains camouflage; the sea anemones gain food scraps.
- Lichen (a fungus and an algae in close association) – the fungus provides support for both as well as obtaining water from **substrate**, and algae photosynthesis produces food.
- Myna birds and cattle – the cattle get ticks and other ectoparasites removed from their skin, which are a source of food for the mynas.

Other organisms in mutualistic relationships *need* each other to successfully survive.

Example

Rhizobium bacteria living inside the root **nodules** of clover provide nitrogen for the clover; the *Rhizobium* obtain a place to live.

Many species of bacteria, fungi and protozoans live inside part of the gut (the **rumen**) of animals such as cows, where they release digestive enzymes that break down **cellulose** into glucose. The cows obtain sustenance (food) from the glucose; the bacteria obtain a suitable habitat inside the cows' rumens.

Symbiosis (meaning 'living together') is used for any close association between two individuals, so includes parasitism, commensalism and mutualism.

Antibiosis or antagonism

Antibiosis is a relationship in which one species *benefits* by releasing a substance that inhibits the growth of another species, so *harming* it.

Example

The mould *Penicillium* produces the antibiotic **penicillin** that inhibits the growth of bacteria. Some plant species release toxic substances into the soil which prevent other species growing there (this is also known as **allelopathy**). The hawkweed species mouse-ear (*Pilosella officinarum*) and king devil (*P. praealta*), introduced to New Zealand in the 1920s, have rapidly increased in abundance in farmed short-tussock grasslands of dry regions of the South Island through ousting agriculturally useful species using allelopathy.

Relationship	Species A	Species B	Explanation of association
Interspecific competition	–	–	Both species compete for same resource.
Antibiosis	+	–	Species A benefits, B harmed.
Parasitism	+	–	Species A is a parasite on the host species B. A benefits, B harmed.
Exploitation	+	–	Species A benefits, B harmed or killed.
Commensalism	+	0	Species A benefits, species B unaffected.
Mutualism	+	+	Both species benefit.

Key: + = benefits, 0 = unaffected, – = harmed

Unit 12.1 Activity 4A: Interrelationships

1. Distinguish between the following pairs of terms:

a. Interspecific and intraspecific competition.

b. Commensalism and mutualism.

c. Predation and parasitism.

d. Endoparasite and ectoparasite.

2. The table below shows one way of classifying relationships between organisms:

	Species A	Species B
(1)	+	+
(2)	+	0
(3)	+	–
(4)	0	0
(5)	0	–
(6)	–	–

+ indicates species benefits,
– indicates species harmed,
0 indicates species unaffected by relationship

Identify the numbers which best show the following relationships (where the first species mentioned is 'Species A') and *name* each relationship.

a. Liana vine on a rain tree.
b. Blood-sucking lice on human hair.
c. Tick-eating birds on a cow.
d. Sea birds diving for fish.
e. A fungus and an alga that together form a lichen.
f. Barnacles attached to the backs of whales.
g. Walnut trees produce a toxic substance that prevents seeds of other plants from germinating.
h. Two species of clover growing together in the same field.
i. Nitrogen-fixing bacteria living in the root nodules of clover plants.

3. The root systems of radiata pines are associated with fine colourless strands of mycorrhiza fungus. These strands extend from within the root tissues of the pine trees out into the soil and may reach the surface over time. The fungus lives on carbohydrates from the pine, but their presence in the roots allows the tree to take up water and nutrients more effectively.

Explain the relationship between the fungus and the pine.

Unit 12.1 Activity 4B: Exploitation of one species by another

1. Describe what is meant by:

a. Batesian mimicry. **b.** Mullerian mimicry. **c.** Parasite.

2. The European red mite is a pest of apples, and preyed upon by another mite, *Typhlodromus pyri*. The graph shows some hypothetical (made up) data relating to changes in numbers of the red mite and its predator in a given area.

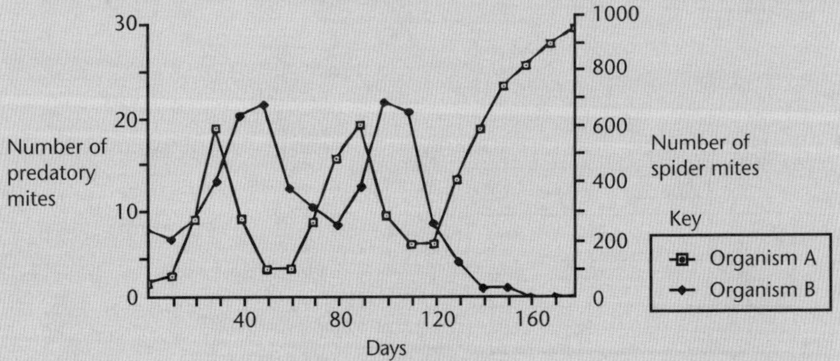

a. Explain the relationship between Organism A and Organism B. Use the graph to justify your answer.

b. At what point during the period in question could the orchard have been contaminated by a drift of chemical spray from a neighbouring orchard? Justify your answer.

c. Explain the different effects of the spray on the two species.

3. The graph below is supposed to show how the variation in the density of prey (moth pupae) affects the rate of predation by mice. The number of mice is constant throughout. The solid line shows the effect on the number of pupae eaten per day, whilst the dashed line shows the effect on the percentage of pupae that are eaten.

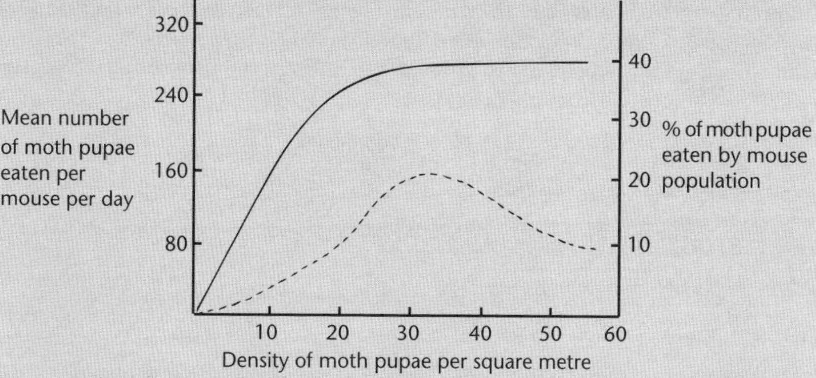

a. Assuming the *solid* line to be correct, over what range of prey density could the mouse population exert some control over the moth population?

b. Assuming the *dashed* line to be correct, over what range of prey density would the mice be able to control the moth population?

c. In fact, one of the two curves has to be incorrect because they are mutually contradictory. Redraw the dashed line so that it is compatible with the solid line.

d. What is meant by saying a predator can only regulate a prey population if it acts in a *density-dependent* manner?

4. Describe *four* quite different ways by which an animal may avoid being eaten.

5. To be successful, crypsis must be accompanied by appropriate behaviour. Explain, giving an example.

6. Describe *two* quite different mechanisms by which plants defend themselves against herbivores, giving an example in each case.

7. Amongst mammals, carnivorous species generally have eyes facing forward, while in most herbivores the eyes are on the sides of the head. Suggest an advantage in each case.

8. The graph on page 49 shows changes in the numbers of lynx and snowshoe hare pelts sold by the Hudson Bay Company in Northern Canada over a period of about a century. Suppose that the graph accurately represented relative numbers of hare and lynx, but that they were not labelled. Give two ways in which you could distinguish which line represented which species.

9. Most insects that are distasteful to predators are diurnal (day-active). Suggest why.

10. Dogs gobble their food, while cats eat more slowly. How might this be related to the different hunting strategies of wolves and most cats?

11. Garden leaf vegetables such as lettuce and cabbage suffer more severely from slugs than do most wild plants. Suggest an explanation.

12. Why do many endoparasites have highly developed reproductive systems?

13. Graphs A–C show some of the ways in which the food preference of a predator may be influenced by relative abundance of *one* of its prey species. In each experiment the total numbers of prey remain the same – it is the abundance of one prey species *relative to other prey* that is varied experimentally.

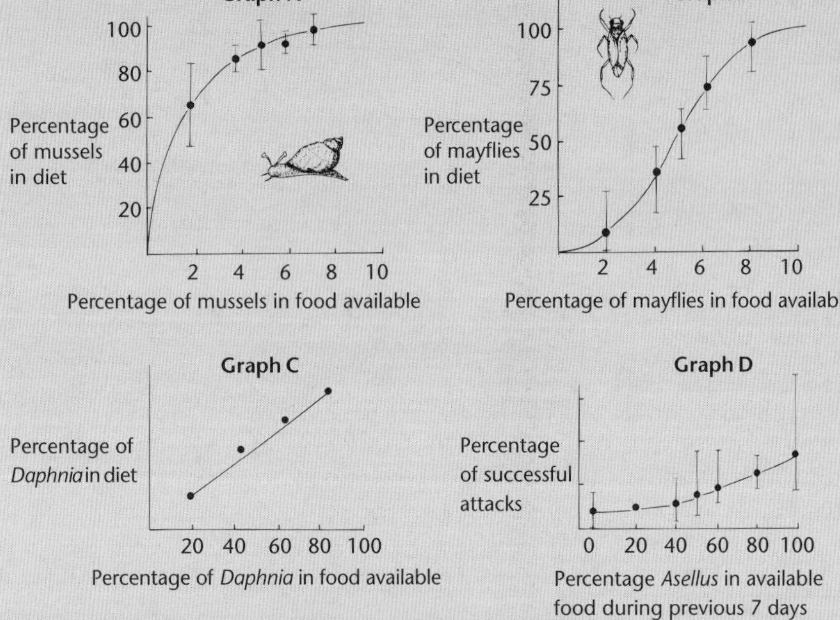

Graph A shows the effect of relative abundance of the mussel *Mytilus edulis* in the diet of a predatory snail. Prey availability is expressed as a percentage of the total food available.

Graph B shows the effect of relative abundance of mayflies on the feeding preference of the predator *Notonecta* (a carnivorous freshwater bug).

Graph C shows the effect of varying the proportion of the small freshwater crustacean *Daphnia* in the environment relative to other food animals, on the proportion of *Daphnia* caught by *Hydra*, a small freshwater coelenterate.

a. Giving a reason in each case, in which of these graphs does the predator show:

i. No particular food preference?

ii. A consistent preference for the prey species regardless of its availability?

iii. A preference for prey that is more common, relative to other prey species?

b. Graph D shows the results of an investigation into the effect of previous experience on the feeding success of *Notonecta* on the small crustacean *Asellus*. What do you conclude from this graph?

14. Guppies are small freshwater fish that feed on small invertebrates such as worms and crustaceans. In a series of experiments on prey preference, guppies were provided with maggots and small freshwater worms in varying proportions. In each case the proportion of worms in the food actually eaten was recorded. The results are shown in the graph.

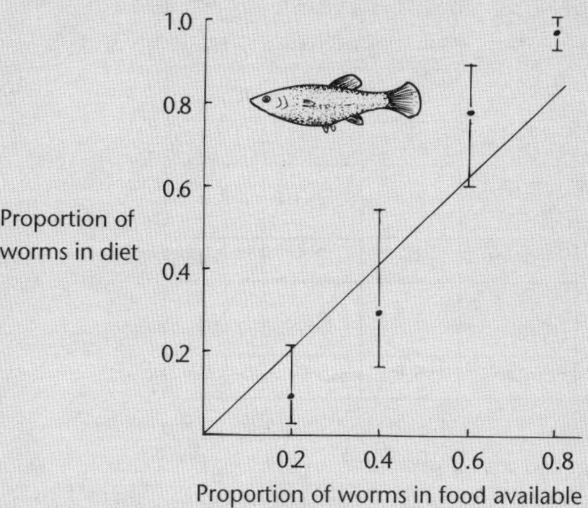

a. What happens to the preference for worms as they become more common?

b. Suggest why the guppies change their preference.

c. What significance does prey switching have for the control of populations by predators?

15. Ostriches have acute vision and when standing with their heads about 2 m above the ground, they have 360° vision and cannot be stalked or rushed by lions, their chief predators. Ostriches feed by pecking their food from the ground and, while their head is down, cannot see a lion until it gets close. It is during such head-down periods that a lion may attempt to stalk an ostrich and, if it can get close enough, rush it and kill it. An ostrich can run faster than a lion and, provided that it sees the lion early enough, it can always escape. Ostriches therefore have to ensure that they balance the need to feed against the need for vigilance.

The following graph shows the results of studies on the behaviour of ostriches.

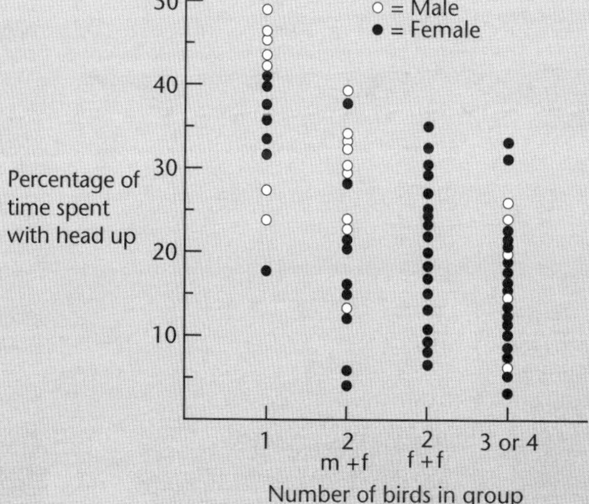

a. What is the relationship between median percentage time spent with the head up and the:

i. Number of birds in the group?

ii. Sex of the birds?

b. Compare the vigilance of female ostriches when in single-sex and mixed groups.

c. The table shows the results of further studies on vigilance in single-sex groups.

		No. of birds in group		
		1	2	3 or 4
Mean no. of head raises per minute	Males	3.7	3.0	2.0
	Females	3.7	2.5	2.0
Mean duration of head raises (seconds)	Males	6.8	5.9	5.5
	Females	5.9	5.0	5.0

How is vigilance affected by the size of the group and the sex of the birds?

d. Male ostriches are territorial and search their surroundings for other males, which they drive out of their territory, but they display to females and mate with them. How do the previous data relate to this sexual behaviour?

Unit 12.1 Ecology

Topic 5: Intraspecific competition

The previous Topic examined competition, exploitation and mutualism between species and individuals. In Topic 5, we look at intraspecific competition in more depth. The Topic describes animal behaviour and plant responses in relation to environmental factors by:

- Describing intraspecific relationships (territoriality, cooperative interactions, reproductive behaviours, hierarchical behaviour, competition for resources).

Because individuals of the same species resemble each other more closely than members of different species, **intraspecific competition** would be expected to be more intense than interspecific competition.

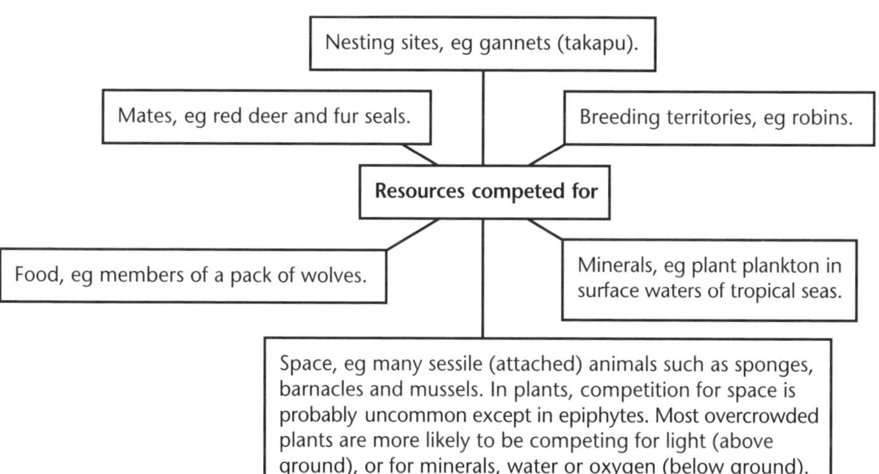

Intraspecific competition has the following important features:

- Directly or indirectly, it always results in a decrease in the reproductive rate of the population, though some individuals (the 'winners') may suffer very little. When competition is for mates or breeding sites, the link with reproduction is clear. If food is being competed for, then the effect upon reproduction is indirect.
- Only resources that are in limited supply relative to their requirements are competed for. Oxygen is never competed for by land animals living above ground since it is so abundant. Most often it is resources contained within a space that are the source of competition, rather than the space itself.
- It operates in a **density-dependent** manner – as numbers increase, either **mortality** (**death rate**) increases or **natality** (**birth rate**) decreases, or, more often, both.

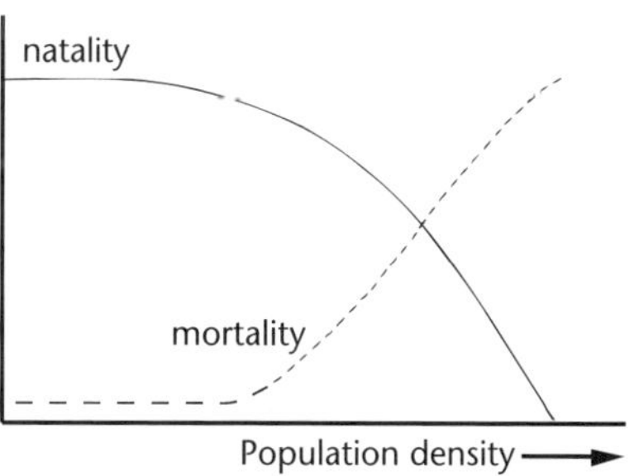

Principle of density-dependent mortality and natality.

Competition for mates

The ultimate 'aim' of all organisms is to pass on their **genes** to the next generation. Except in **hermaphrodite** organisms, male and female gametes are produced by different individuals. Though both sexes invest a good deal of energy in reproduction, they do so in very different ways. A male produces vast numbers of minute **sperm**, while the female typically produces a much smaller number of larger eggs. Since an egg costs more energy to produce than a sperm, it represents a bigger investment.

Because of this, male and female reproductive strategies typically differ in two ways:

- It is in the interests of the female to ensure that the sperm which fertilises an egg is of the highest quality. Females therefore tend to be more selective about the male with whom they mate. As a result of this, in many species there is *competition* by males for females.
- In many species the female continues to invest energy in caring for the eggs after they are fertilised, and in some groups she feeds her young (eg mammals).

Sperm are individually very cheap to produce, so it is in the interest of the male to ensure his sperm fertilise as many eggs as possible. Males therefore typically mate with many females.

Competition by males for females takes two forms – fighting and display. In both cases, it results in strong selection pressure in the males for the ability to compete for females, with the result that males often differ markedly from females. This is called **sexual dimorphism**. In most birds, there is little or no sexual dimorphism, probably because most species compete for mates through reproductive territory rather than by directly fighting for mates.

Fighting

In many animals, competition for females is aggressive. In such cases, winning contests requires physical vigour and is the means by which the quality of the male's genes is 'judged'. When the winner takes 'possession' of many females, a male's contribution to the next generation is very large. There is correspondingly strong selection for fighting ability and thus large size in the males.

Examples

In red deer, the stags (males) are about half as large again as the hinds (females) and have antlers. Fighting during the 'rut' is usually preceded by roaring; it seems that in many cases a stag senses the superiority of its opponent and withdraws from a contest without actual fighting. When bluster fails, the stags lock antlers and fight by pushing against each other. More often than not the contest is settled without serious injury. The winning animal now has possession of a harem of up to 20 hinds.

In fur seals, which breed on the coast of the South Island of New Zealand and especially the subantarctic islands, contests are often bloody; few bulls maintain a harem for more than two or three seasons.

Elephant seals.

In seals and red deer all the females come into season at about the same time. Lions, on the other hand, breed continually, and a female only comes into season when she stops suckling. As a result, a male that has just taken possession of a pride by evicting the previous male cannot reproduce until the lionesses have stopped suckling the cubs fathered by the evicted male. The quickest way for the male to begin reproduction is to kill the existing cubs. Within a day or two of the male killing the clubs, the lionesses come into season and begin mating with the new male.

Display

In some animals the female 'judges' the quality of a male's genes by his appearance.

Example

While it might be thought that the gaudiness of a peacock's feathers would not be a good guide to his ability to survive the rigours of life, it seems that indirectly it does. When a bird is in poor health, its plumage is the first to suffer. If a cock bird looks attractive to a hen because his plumage is bright it can indicate that he is relatively free of internal parasites, he will thus be more successful in handing on his genes.

There is an obvious disadvantage of bright colours – they render a male more conspicuous to predators. However, provided that the advantage given in attracting a mate outweighs this, bright colours will still be selected for. The risk of predation to a brightly coloured bird may explain why in such species the female is usually dull coloured. For her, there is no advantage in being brightly coloured, as this is more likely to reveal the presence of her nest to a predator.

Territory

A **territory** is an area occupied by an animal and defended against others. It is distinct from a **home range**, which is not defended. Territorial behaviour is very widespread in the animal kingdom, especially amongst **vertebrates**. A territory can serve a variety of purposes and exist for varying duration.

- A territory may represent a source of food and be defended all the year round.

Examples

In the North Island of New Zealand, the territory of saddleback birds (*Philesturnus carunculatus*) is defended by both sexes as long as they remain together, which is usually for life. Most fish that live in tide pools, such as many of the cockabullies (eg the variable triplefin), are territorial.

- It can be a site in which to build a nest and rear young, eg gannets.
- It may simply be a place where males mate with females, eg red deer.

Territorial behaviour is a special case of contest competition – the area containing the resource is defended, rather than the resource itself.

Low

A 'keep out' signal. In most cases, this low-level warning is sufficient, and holders of neighbouring territories avoid each other.

Examples

Song in saddlebacks and in many other birds, or scent marking in dogs and many other mammals, is a low-level warning. Bush babies urinate on their hands so that the scent of the **urine** is distributed through the branches.

Various forms of threat display may be employed.

Example

In saddlebacks, territorial threats take the form of head-bowing, tail-fanning and warbling displays; at the same time the wattles become dilated.

As a last resort, fights may occur. In most territorial animals the intensity of the defence increases with nearness to the centre of the defender's territory, so the 'rightful' owner rarely loses.

Example

In saddlebacks, each bird attempts to grapple with the wattles of the other.

High

Increasing level of territory defence

Group territories

In some animals, a territory may be defended by a *group*.

Example

Some birds live in or along the edges of swamps and marshes in groups of up to a dozen individuals. The main requirement of a good-quality territory is that it provides a concealed area surrounded by water in which incubation can be undisturbed. Territory is defended mainly by the breeding (high status) males. Females and non-breeding males take little part in defence. In pukeko, the economics of territorial defence are different from most other territorial species. Whereas in most species a food-rich territory tends to be smaller, for pukeko, a good-quality territory is larger and is defended by more birds.

Costs and benefits of territory defence

Defending a territory costs energy, and also time that could otherwise be spent on feeding. It may be for this reason that so many birds declare their territories by song early in the morning; small animal prey is least active and thus hardest to find at this time.

Defending a feeding territory is not 'worth it' unless the benefits exceed the costs. The benefits depend on how the resource is distributed. When it is patchy, animals have to spend too much time and energy finding it for defence of the area to be economic.

Example

In Tanzania, spotted hyenas are territorial in the Ngorogoro crater where the food is abundant all the year round. On the Serengeti plains, food is seasonal – the hyenas cover wide areas and are not territorial.

The optimum size of a feeding territory is a compromise between the energy gained from the food in the territory, and the cost of its defence.

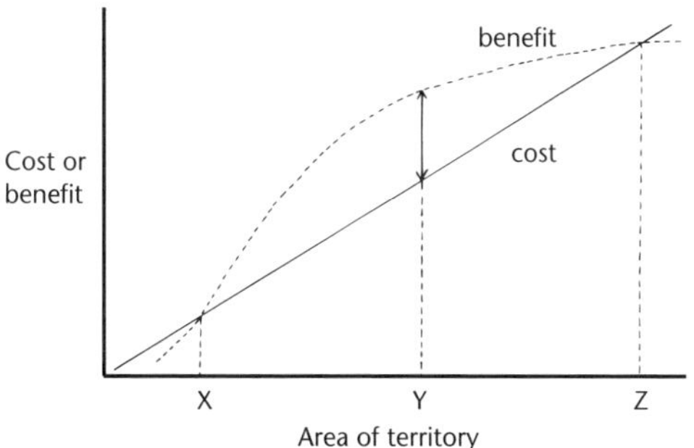

A territory is worth defending if its area lies between X and Z, the optimum being Y.

Optimum territory size.

Observations and experiments show that within a given species, territory size tends to be smaller where the food supply is richer.

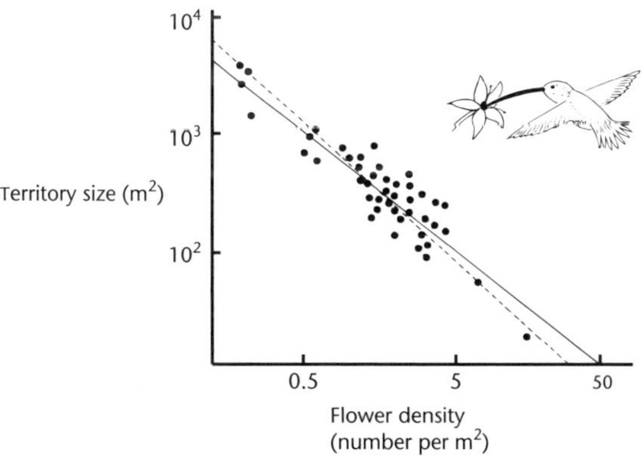

———— line of best fit

------------ calculated assumption that each territory contains a constant number of flowers

Size of a feeding territory depends on the richness of the food supply.

The non-territory holder

When a territory is purely for breeding, non-holders cannot reproduce. With feeding territories, non-territory holders are displaced to marginal areas where food is less abundant, and are consequently less likely to survive. The territory holder is thus at a considerable reproductive advantage over the non-holder. If not holding a territory means considerably reduced reproductive success, why do the losers in territorial disputes usually give way? The answer is probably that even for the non-territory holder there is *some* possibility of gaining a territory in the future, for instance when a holder dies. It may be that the injuries that would be sustained in fighting would reduce the chances of eventual reproduction even more.

Leks

In a number of birds and mammals, the males gather in a small area called a **lek** and compete with one another to establish a mating territory within the lek. A dominance hierarchy results, with the most dominant males obtaining the best territories within the lek. Each territory contains no resources of use to the females – it is purely a mating site. Females are attracted to the lek when then they become fertile, and choose the males with the best territories, which are usually those nearest the centre of the lek. The males chosen by the females can mate with several females, and so pass their 'good' genes on to more offspring, helping to maintain a strong gene pool.

After mating, the females leave the lek and rear their young on their own.

Example

In New Zealand there is a type of parrot (*Strigops habroptilus*) that is the heaviest parrot in the world. It doesn't fly. During the breeding season it forms leks with other males. The lek is typically on a ridge top, on which each male excavates a shallow bowl in the ground. In the bowl he emits a characteristic booming call to attract females. The kakapo is not just the only flightless, nocturnal and world's largest parrot – it is also the only parrot to use a lek.

Territory and population regulation

The limit to the number of territories available limits the size of the population. Since territory tends to prevent overcrowding, it results in population size being regulated at a level that the food supply can usually support. However, *any benefit to the population is incidental* – territorial behaviour can be explained in terms of advantages to the individual.

Social dominance

Territory is only one way by which resources are distributed within a population. In many species which live in groups, resources are shared, but not on a 'fair' basis. Certain individuals are of higher *rank* because they consistently have the best access to food and mates. The commonest form of **social dominance** is the *linear hierarchy*, in which an 'alpha' individual dominates all other members of the group, while a 'beta' animal dominates all except the alpha, and so on. Dominance is not always confined to males (in the pukeko, males and females each have their own hierarchy).

Determining rank

In most animals, the position in the hierarchy is determined early in life during brief fights, after which status is seldom contested except by newcomers. In a newly established group, there may be considerable fighting, but once the hierarchy is established there is usually so little conflict that it is often hard to detect. Only the fact that certain individuals consistently give way to others shows that a hierarchy exists.

Indicating rank

The signals by which animals indicate their status are very diverse.

Examples

In wolves, a dominant animal holds its tail high and its ears erect, while a submissive individual cringes with its tail and ears down.

In the purple swamphens found in PNG, dominance is indicated by a bird holding its head high and pointing its beak towards its opponent. Submission is indicated by turning away, keeping the head low and displaying the white tail feathers.

Dominance and submission postures in wolves and the purple swamphen.

Significance of social dominance

In an established hierarchy, there is minimal conflict, and it is often implied that the function of social dominance is for 'the good of the species'. However, it is *individuals* that are selected in evolution, not groups.

How then does social dominance evolve? The advantages of high rank are obvious – with more food and better access to mates, a dominant male leaves more offspring than a subordinate one. If the prospects for the omega animal at the bottom are so much poorer, why do they remain in the group? The answer is that even for the omega individuals, there is a slim prospect of eventual reproduction if one or more of the dominant animals die. For an outcast, the prospects of reproduction may be nil.

Occurrence of social dominance

Because of the lack of obvious conflict in most social animals, dominance is far more widespread than previously realised. It is particularly common in birds and mammals, but also occurs in many reptiles, fish and some invertebrates.

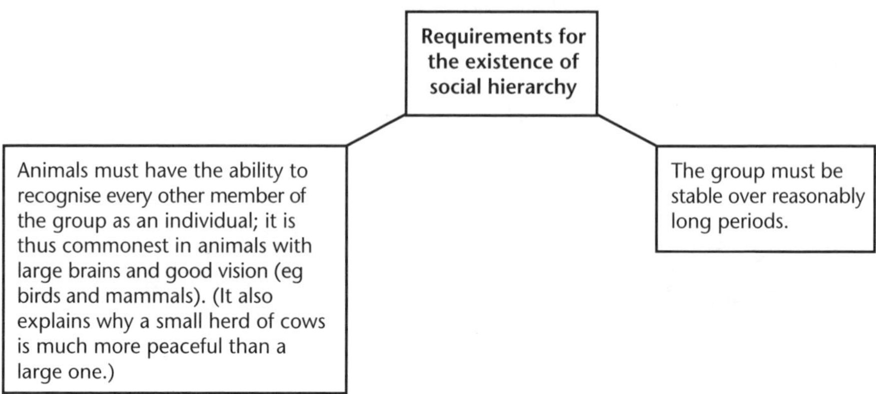

Hierarchical and territorial behaviour are not exclusive of each other.

Example

Silvereyes are a type of bird native to Australia and New Zealand. They are territorial in the breeding season, but in winter they form flocks in which there are well-marked male and female hierarchies.

Avoidance of intraspecific competition

Organisms have two general strategies for avoiding intraspecific competition – dispersal and niche differences.

Dispersal

In plants and in animals that move little (eg barnacles, pipi and marine worms), competition is reduced by dispersal of young stages away from the parent.

In land plants the dispersal stages are inactive (spores or seeds), and can only develop if they land in a suitable habitat. Since the majority land in unsuitable places, by far the greatest mortality occurs in these stages. In flowering plants the seeds or the entire **fruit** are adapted for a specific mode of dispersal, such as wind (eg willow, dandelion), or by some explosive mechanism (eg lupin), or by animals. (The dispersal of fruits and seeds by animals may form the basis for mutual dependence – the plants and animals thus have important influences in each other's evolution.)

The larvae of many marine animals are adapted for dispersal, and often have some control over the habitat in which they settle. Nevertheless, the mortality rate within larval stages of marine animals is enormous.

Differences in niche

Members of the same species with different feeding habits reduce competition.

Examples

- A yellow admiral caterpillar (*Vanessa itea*) feeds on nettle leaves and thus occupies a completely different niche from that of the adult butterfly, which feeds entirely on nectar.
- In some birds, the two sexes have different feeding habits. In the extinct huia (*Heteralocha acutirstris*), the male used his stout beak to find insects by hammering into bark, while the female used her long thin beak to probe into cracks. In bellbirds (*Anthornis melanura*), males feed largely on nectar, while the females feed more on insects. In the saddleback (*Philesturnus carunculatus*), the male is predominantly a ground feeder, the female feeding at a range of heights.

Intraspecific competition in plants

Intraspecific competition affects plant growth in two quite distinct ways:

- It influences average growth per plant.
- It results in considerable variation in the growth of individual plants.

Effects on average growth

Effects of intraspecific competition on average growth of plants are usually determined by using dry mass as a measure of growth because live (ie wet) mass can vary independently of growth – when a plant wilts, for instance, it loses water and therefore live mass.

Example

Effects on average growth was investigated by Ivan Palmblad, a scientist in the US, who sowed seeds of a number of different species of weeds in pots at different densities, each density being replicated in three pots. At the end of the summer the total dry mass of the plants in each pot was determined, as were the number of seeds produced in each pot and the number of plants that had died. The results for one species, *Conyza canadensis* (Canadian fleabane, a type of weed common in New Zealand), are shown below (data are the average for three individual pot values).

Sowing density	1	5	50	100	200
% germination	100	87	56	54	52
% mortality	0	0	1	4	8
% reproducing	100	87	51	42	36
Total dry mass (g)	12.7	17.24	17.75	16.66	18.32
Mean dry mass per plant (g)	12.7	3.5	0.355	0.166	0.092
Total number of seeds	55 596	59 625	40 845	35 264	38 376
Mean number of seeds/ reproducing individual	55 596	13 710	1 602	836	534
Number of seeds produced/ gram of plant dry mass	4 378	3 917	2 301	2 117	2 094

The results are typical of 'weeds'.

- Percentage germination is significantly reduced at higher densities (though not in some other species in the study).
- A 200-fold increase in sowing density resulted in only a 1.4-fold increase in final total mass of plant material $\left(\frac{18.32}{12.7}\right)$. Most of this is accounted for by a 138-fold decrease in mean mass per plant (from 12.7 to 0.092). Only at higher sowing densities was there significant mortality.
- Whereas the 200-fold increase in sowing density resulted in an increase in total dry mass produced, there was a decrease in total seed output.
- Assuming that seed size remained constant as crowding increased, there was a steady decrease in the *proportion* of organic matter devoted to seed production (from 4378 to 2094 seeds per gram of plant dry matter).

Effects on variation in growth of individuals

Competition also results in marked differences in growth of individual plants.

Example

In an investigation into competition in European flax (*Linum usitatissimum*), seeds were sown at different densities and plants harvested after different time intervals and the dry masses of *individual* plants measured.

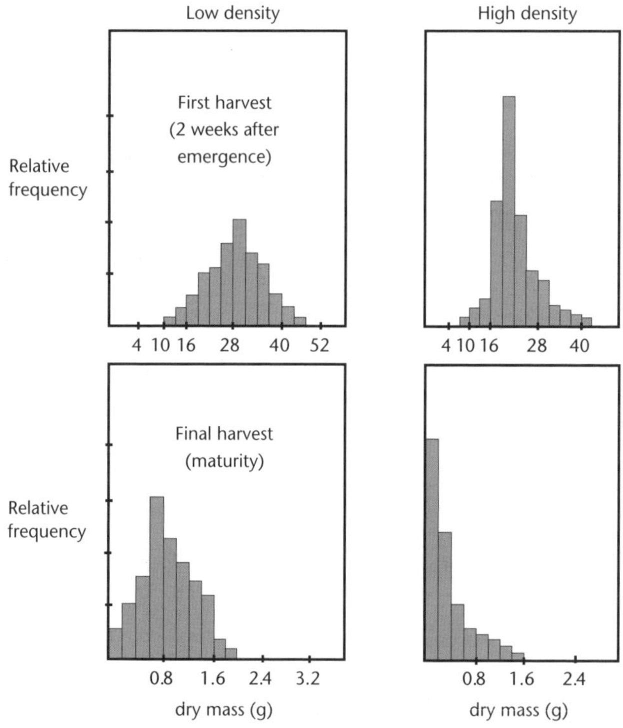

Effect of competition on variation in European flax plants.

Results:

- Although at low densities there was considerable variation in plant mass, distribution was roughly symmetrical, the mean lying approximately in the middle. This situation did not markedly change with time.
- At high densities, by the time of final harvest, the size distribution was highly *skewed*, with proportionately far more undersized plants than at low densities.

Explanation:

- Germination is normally staggered, some shoots emerging earlier than others.
- Those that emerge above the soil first will have no competition for light and will therefore grow rapidly.
- Those emerging later will be over-topped by the early germinators and will therefore receive less light. As a result they grow more slowly. The more slowly they grow, the more they are over-shadowed by the others. As a result, the small differences in germination time and initial growth rate become greatly magnified.

Early competition for light is a problem not confined only to seedlings.

Example

For a plant that spends the winter underground as a bulb, corm or rhizome, it is essential that spring growth be as rapid as possible to avoid over-topping by competitors. This explains the significance of storing energy during the winter, since it permits plants to grow rapidly even when partly over-shadowed by competitors.

Cooperative interactions within species

There are many cases in which organisms benefit from the presence of others of the same species. Courtship and **parental care** are obvious examples.

Organisms of the same species may become clustered together for several different reasons:

- Animals may occur together because they are attracted to each other, forming a *group*.
- Animals may respond independently to the same **stimuli**, forming an *aggregation*.

Example

Certain insects may occur in large numbers under a log, each individual having independently orientated to the damp, dark conditions. There may be advantages in being aggregated – some insects may lose less water when they are close together, but such benefits are incidental.

- Plants may be clustered together because seeds or spores germinated in the same place, or because they were produced by **vegetative** reproduction by the same parent. Though they may benefit from being close to each other by providing mutual shelter from wind, this is incidental.

Groups

Animals in a group actively stay together as a result of *responding to one another's presence*, forming shoals, herds or flocks. In some species, groups are *open* in that membership is temporary, with individuals entering and leaving, for instance most fish schools. In other species, membership is *closed* in that membership is stable enough for each individual to recognise other members. It is in such groups that *social hierarchy* is common.

Group living has been selected for because of its advantages.

- It may increase the chances of finding food.
- It may reduce the risks of being eaten.
- Some animals cooperate in the rearing of young.
- Heat loss may be reduced by huddling together, eg Emperor penguins and over-wintering bees.
- Hydrodynamic and aerodynamic effects.

Examples

Fish swim faster in schools than alone because each swims in the area of reduced pressure generated by the fish in front. Geese and ducks take advantage of a similar effect by flying in a 'V' formation.

- Many animals that are normally solitary migrate in large groups. This may help to cancel out navigational errors made by individuals, besides saving energy.
- **Learning**. In the more highly intelligent animals such as monkeys and the social apes, there are greater possibilities for profiting from the experience of other members of the group. The ability to learn from others is the basis for **cultural evolution**.
- Access to mates is easier when males and females do not have to search for each other.

Groups vary greatly in the permanence of their membership.

Examples

In wolves and most primates, all members of the group know each other as individuals. Other groups are purely temporary assemblages for a specific purpose, such as migratory herds of wildebeest, or large roosting flocks of starlings.

Care by sisters – insect societies

Many ants, bees and wasps live as a *society*, in which all individuals are the offspring of a single female, the *queen*. Males are haploid (developing by **parthenogenesis** without fertilisation) and make no contribution to the colony except in reproduction. Fertilised eggs develop into **diploid** females. Depending on their environment, these develop into either fertile queens or sterile workers. In a number of respects such an ant, bee or wasp society resembles a 'super-organism':

- In a multicellular organism, the **cells** of the body eventually die, many even before the body reproduces (eg skin cells, blood cells). Body cells are dispensable – their role is to get *copies* of their genes into the next generation via the gametes. Sterile workers are like body cells in that they cannot reproduce themselves but cooperate to get copies of their genes (via the queen) into the next generation.
- Like the cells of the body, workers show division of labour, different individuals doing different jobs. The females are of two general kinds or *castes* – *queens*, whose function is to produce eggs, and *workers*, which carry out all foraging, rearing and maintenance duties. In bees, all workers are structurally the same, individuals performing different duties at different stages of their lives. In ants, the sterile females may be structurally differentiated into castes such as soldiers and workers.

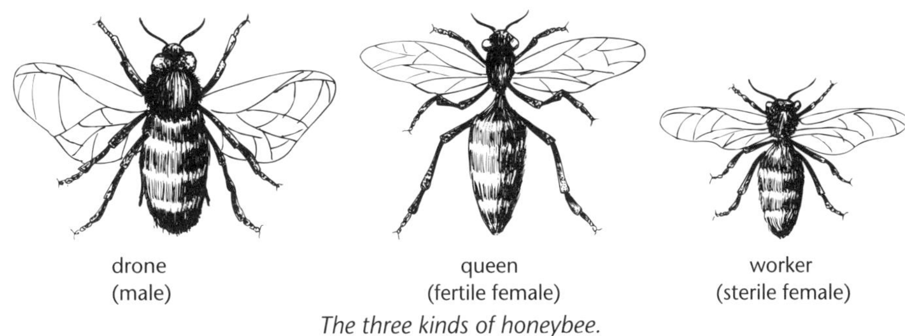

The three kinds of honeybee.

- In a multicellular organism and an insect society, cooperation and division of labour requires some means of *communication*. Members of an insect society communicate by **pheromones**, chemicals that are analogous to the **hormones** used in communication within the body of a multicellular organism. Over 20 pheromones have been identified in honeybees, and some of them function in complex ways. One is the alarm pheromone which results in a frenzied attack by the other workers on intruders.

Honeybee society

A mature colony of honeybees may contain up to 30 000 bees, most of which are workers. There is a high degree of division of labour amongst the workers and the pattern of activity of a worker changes as she ages. A newly emerged worker spends the first few days cleaning cells, but later takes on other duties such as fanning, secreting wax, making orientation flights and finally foraging for nectar and pollen.

It is during orientation flights that a worker learns to use her internal clock to visit flowers that secrete nectar at particular times of the day, and also to use the Sun and the Earth's magnetic field to find her way back to the hive. Mortality is much higher during orientation and foraging, so by delaying the more risky activities till later, a worker increases her average total work output.

Communication

The high degree of integration within a bee colony would not be possible without sophisticated communication.

Example

The best known method used by bees in communication is the 'recruitment dance', by which a worker communicates to other workers the whereabouts of a food source. After a successful foraging trip, a worker performs one of two kinds of 'dance' on the comb in the hive.

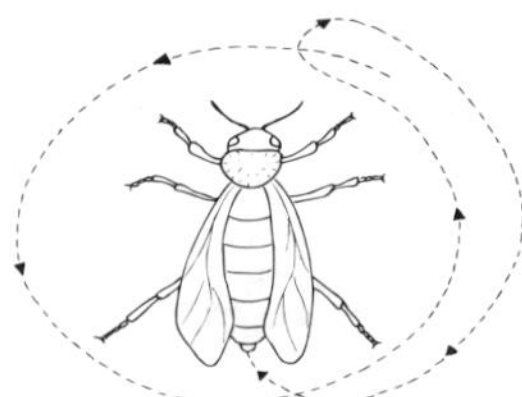

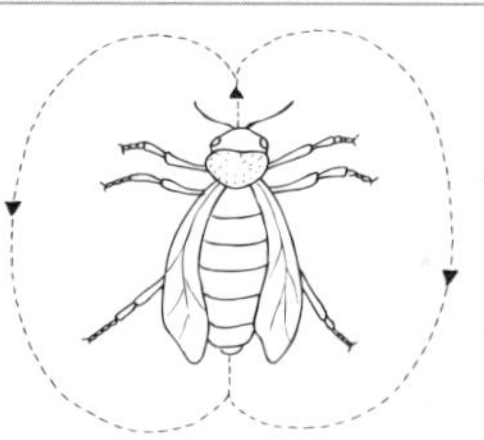

Round dance
If the food source is close to the hive, a bee performs a 'round dance'. The more intensely she waggles her abdomen, the closer the food source.

Waggle dance
With more distant food sources, a bee performs a 'waggle dance'.

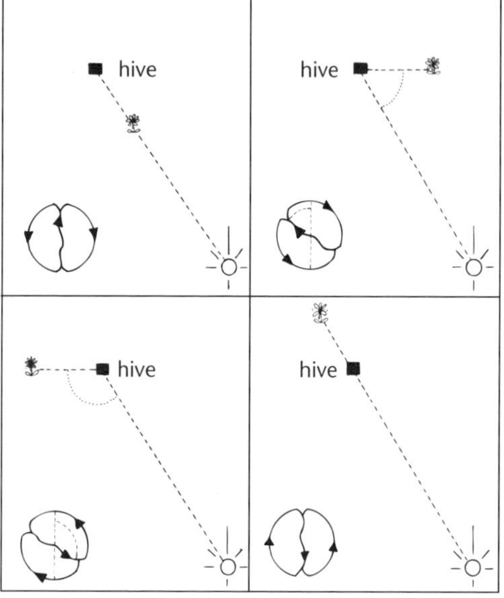

The angle the straight-line part of the dance makes with the vertical is the same as the angle between the Sun and the flowers. To use the Sun in this way, she must make constant adjustment for the apparent movement of the Sun through the sky.

Bee 'dances'

Cooperative interactions between species

Cases in which two individuals of different species behave in a way that benefits *both* are usually referred to as **mutualism**.

Examples

- *Cleaner fish*. In coral reefs, there are fish that specialise in removing the parasites from the gills of other fish, which open their mouths and gill covers to allow the cleaners access. The cleaner fish obtain a supply of food, and the fish being cleaned has its parasites removed.

- *Root nodules of leguminous plants.* This is a partnership between a plant of the pea and bean **family** and bacteria. The bacteria convert nitrogen gas into ammonia (which the legume needs but cannot produce), and the legume converts carbon dioxide into carbohydrate (which the bacteria needs but cannot produce).
- *Lichens.* A lichen consists of a fungus and an alga. Though the alga can and does live independently, the fungus cannot survive without the carbohydrate produced by the alga.
- *Mycorrhiza.* In an association between a seed plant and a fungus, the fungus obtains organic matter from the plant, while the plant obtains mineral salts that the fungus is better able to absorb from nutrient-poor soils. Many forest trees are unable to get established unless infected with the appropriate kind of fungus.
- *Ruminants and many other herbivorous mammals.* In most herbivores, **cellulose** is digested by bacteria and protozoa in a part of the gut specialised for that function. The **micro-organisms** also produce vitamins and amino acids not present in the food; they receive a plentiful supply of food in the warm, **anaerobic** (oxygen-lacking) environment they need.
- **Mitochondria** *and* **chloroplasts**. Evidence has accumulated that these **organelles** evolved from bacteria living in mutualistic relationship with the rest of the cell.

Root nodules of white clover.

Mutualistic relationships show great variety

The degree of interdependence. A relationship essential for survival is said to be *obligate*, eg the micro-organisms that live in the guts of most herbivorous animals. A non-essential relationship is *facultative*, eg the bacteria that live in root nodules.

The degree of specificity of the relationship. At one extreme, each partner is associated with one particular species, eg the root nodules of **legumes**. At the other extreme, one or both of the partners may be represented by a variety of species, eg many flowers and their pollinators.

The degree of intimacy. At one extreme, the relationship may be a physiological one; in some cases, one partner living inside the cells of the other as in the bacteria that live in the cells of leguminous plants. At the other extreme, it may be a more 'long-distance', ecological relationship, as in cleaner fish.

The length of the association, which may vary from lifelong (eg gut **mutualists**) to a short period of the life cycle.

In addition to these cases which are generally agreed to be mutualistic, there are many other, rather looser relationships. Ostriches and zebra for example often live together in temporary associations and probably derive earlier warning of predators.

Extension: Co-evolution in pollination and dispersal

Just as predators and prey have co-evolved (ie have influenced each other's evolution), there has also been co-evolution between many flowering plants and animals that transport pollen and seeds. Many flowers are structurally adapted to be pollinated by a limited range of insects and birds, which in turn have evolved features that enable them to specialise on particular kinds of flower.

Nectar is produced at the base of a long petal tube, only accessible by insects with long proboscides.

Example

A bee has a proboscis that is long enough to reach the nectar deeply concealed within flowers such as clover. Butterflies and moths have even longer proboscides, enabling them to reach nectar at the bottom of long tubular flowers such as honeysuckle.

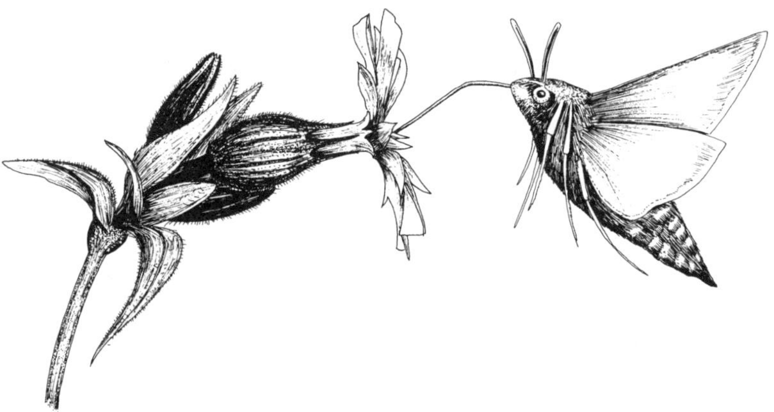

Co-evolution in moths and flowers.

Co-evolution also occurs in the dispersal of many fruits and seeds by birds. Succulent fruits undergo a dramatic change as they develop. While the seeds are developing, the fruit may be laden with bitter-tasting defensive chemicals, but as they mature the **fruit** loses its bitterness and becomes succulent, sweet and conspicuously coloured. Fruit forms an important part of the diet of a number of native birds.

Other kinds of relationship

Relationships in which one animal species gains and the other is neither harmed nor obtains benefit is called **commensalism**.

Example

The small spider crab *Elamena producta*, found living within an abalone shell, does no harm to the mollusc, but obtains shelter and probably small food fragments that drift in with the current of water created by the abalone.

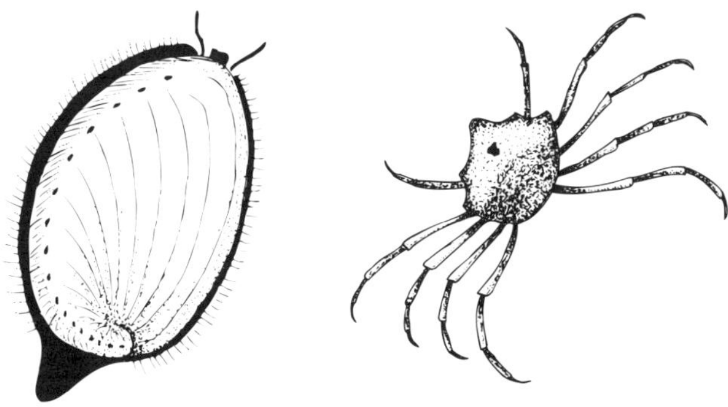

Abalone and a commensal crab, Elamena.

Amongst plants, there are numerous cases of one plant gaining benefit from living in close association with another that is neither harmed nor obtains benefit.

Shade plants

Plants of the understorey and forest floor, such as many ferns, can only survive in the cool, humid conditions created by the forest trees that form the canopy.

Climbers or vines

These plants are rooted in the ground but use other plants for support. A climber thus does not have to use large amounts of energy producing specialised supporting tissue.

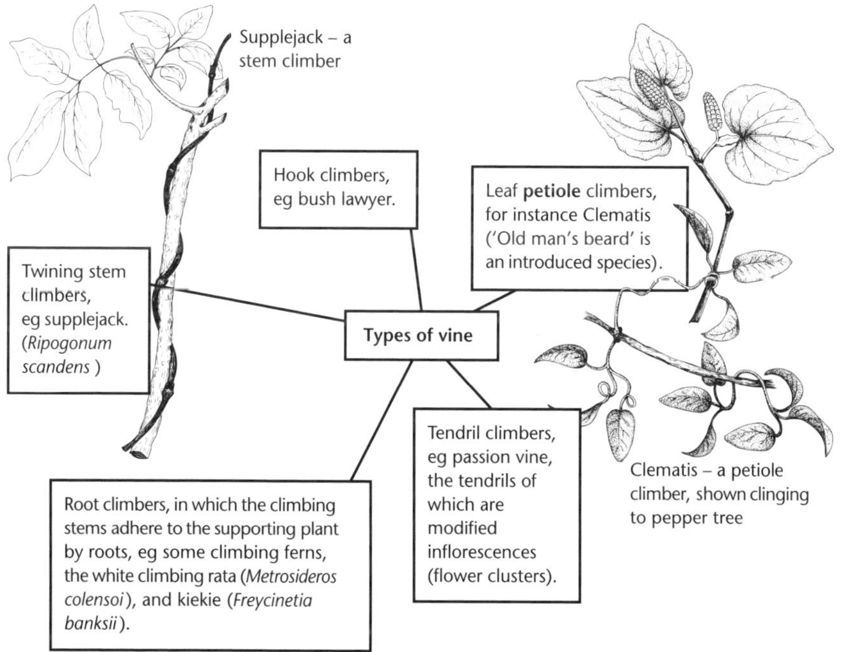

Epiphytes

Epiphytes germinate on the bark of trees and have no attachment to the ground. They obtain no nutrients from the supporting plant. Many lichens, filmy ferns, orchids and other flowering plants are epiphytic. The main problem for these plants is the lack of a reliable supply of water and minerals, and some have water storage tissue or succulent leaves (eg *Pyrrosia serpens*, a very common epiphytic fern).

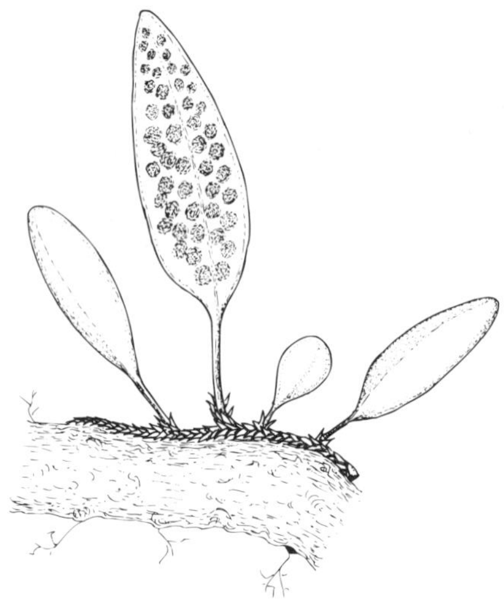

Another common epiphyte is the 'perching lily', *Astelia solandri*.

Unit 12.1 Activity 5A: Intraspecific competition

1. Describe what is meant by:

a. Territory. **b.** Home range.

c. Linear hierarchy. **d.** Dominant.

2. The brown creeper or pipipi is a small bird found in forests of the South Island and Stewart Island in New Zealand. It does not migrate and the birds remain as pairs throughout the year, breeding in spring and early summer. Breeding pairs are territorial. Describe the possible costs and benefits to the brown creeper in not migrating.

3. The takahe is an endangered flightless bird found only in a relatively small area of Fiordland in New Zealand. The birds feed on tussock grass. The male birds are territorial and mate for life. The diagrams below represent summer territories observed for the same area in an isolated valley in Fiordland during 1967 and 1980. Territories held by unmated males are indicated by a black dot.

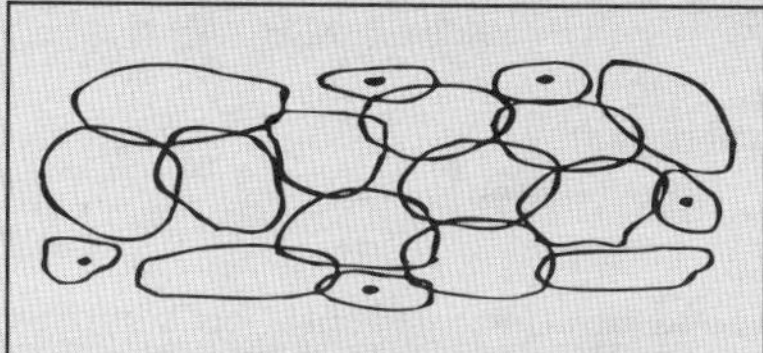

Location of territories established in 1967. Average area two hectares.

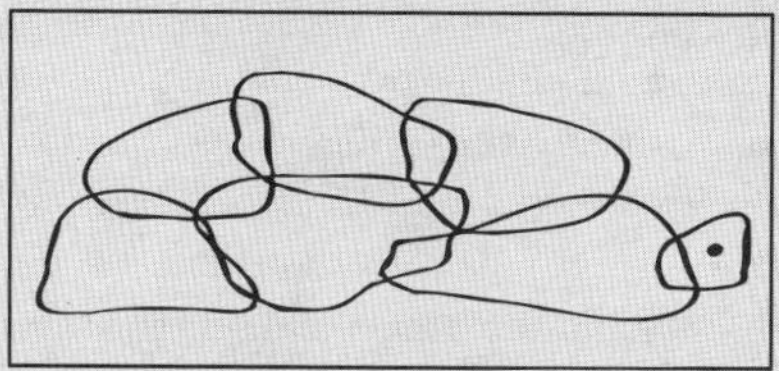

Location of territories established in 1980. Average area nine hectares.

a. How many territories in the study area were held by mated males during:

i. The summer of 1967?

ii. The summer of 1980?

b. Describe the territories held by unmated males and suggest why these territories were held by unmated males.

c. Describe two differences between the territories formed in 1967 and 1980 which are shown by the diagram (include figures to support your answer).

d. State one factor that could account for the differences in territories between 1967 and 1980, and explain how it could cause the differences.

4. Why is intraspecific competition likely to be more intense than interspecific competition?

5. Describe *three* effects that intense intraspecific competition can have on plants.

6. **a.** Distinguish between territory and home range.

b. List the different functions that territory can play in animal life.

7. Why are dominance hierarchies most common in more complex animals?

8. In order to establish if a dominance hierarchy was present in a group of thrushes that fed on the school grounds, a group of students observed interactions that involved challenges between 5 birds at a feeding station. The table shows the raw data for a total of 50 interactions between the birds. For example, 3/5 means that birds 3 and 5 interacted in a way that involved a challenge and 3 was the *winner* while 5 was the *loser*.

Raw data table of thrush challenges				
1/3	4/2	1/3	4/1	1/3
3/2	4/1	1/5	3/2	2/5
4/1	5/2	4/2	3/1	5/2
1/3	1/3	1/2	1/5	1/3
5/2	3/5	3/5	3/5	5/2
4/3	4/5	4/1	1/3	4/1
1/2	5/2	4/3	1/5	3/5
1/5	2/5	3/5	4/1	4/3
5/2	1/3	4/5	4/5	3/2
4/1	1/3	5/2	3/2	1/3

a. Describe the advantages to the group of thrushes of having a dominance hierarchy.

b. Describe possible behaviour of the thrushes the students would have observed that would indicate dominant status in the hierarchy.

c. Using the data given, discuss the dominance hierarchy that exists in this group of birds.

9. a. What are the costs to an animal of defending a territory?

b. What is the benefit of defending a territory, and why does the benefit not continue to increase with increasing territory size?

Unit 12.1 Activity 5B: Cooperative interactions

1. Describe what is meant by each of the following, giving an example for each:

- **a.** Insect society.
- **b.** Commensalism.
- **c.** Mutualism.
- **d.** Epiphyte.

2. Groups are formed when members of a species gather and remain together. Describe *four* different biological advantages of forming groups.

3. What is the essential difference between membership of a pack of wolves and membership of a flock of geese?

4. What is the essential difference between an epiphyte and a plant parasite?

5. Describe the advantages in plants of the climbing habit.

6. Why is cooperative behaviour more common in family groups?

7. Describe the kind of biological relationship that exists between humans and domestic cattle.

8. How does a bee communicate the:

- **a.** Distance of a food source?
- **b.** Direction of a food source?

Unit 12.1 Ecology

Topic 6: Ecosystems

This Topic continues a description of the concepts and processes relating to ecology. A fundamental concept in the study of biology is ecosystems. Topic 6 deals with:

- Food chains/webs, energy flow, nutrient cycling.

An **ecosystem** consists of a community together with its *physical environment*.

The structure of an ecosystem depends upon the:

- Species present.
- Quality and distribution of abiotic resources, such as water and nutrients.
- Range of physical conditions present, such as temperature and salinity (salt content).

Understanding the structure of ecosystems makes it possible to use resources more efficiently and to conserve resources, and makes it possible to manage artificial ecosystems, such as farms and exotic forests, more efficiently.

Ecosystems vary in size – from the very small (such as the fish, snails, weeds, rocks and water of an aquarium) to the very large (such as the tundra and tropical rainforest). Together, all the ecosystems of the world make up the **biosphere**.

The processes of energy flow and mineral cycling determine how an ecosystem operates.

Trophic levels

Trophic level refers to the *feeding* level of an organism in its community ('troph' is Greek for 'food'). Many different terms are used to describe the feeding habits of organisms:

- **Autotrophs** ('self feeders') – those organisms that synthesise organic compounds ('food') from other chemicals using an energy source. The most common are the **photosynthesisers** (all green plants), which use solar energy to bind together H_2O and CO_2 into carbohydrates. Less common are the **chemosynthesisers** – certain species of bacteria that use energy from chemical reactions (eg **nitrogen fixation**, nitrification, denitrification) to achieve the same outcome (ie producing organic compounds from other chemicals).

Because both photosynthesisers and chemosynthesisers can make their own food, they are also called **producers**, and start off all food chains. Chemosynthetic bacteria are the producers in deep-sea food chains.

- **Heterotrophs** are organisms that depend on organic compounds for food. They are the **consumers** and **decomposers**.

Consumers may be categorised into:

	Examples
• **Herbivores** – eat plants.	Cows, horses, most caterpillars, turtles.
• **Carnivores** – eat animals.	Spiders, cats, snakes, sperm whales, crocodiles, sharks, owls.
• **Omnivores** – eat both plants and animals.	Humans, chimpanzees, pigs.
• **Parasites** – eat living organisms (without killing them).	Fleas, mosquitoes, mistletoe, leeches.
• **Scavengers** – eat food scraps, waste products, dead bodies.	Seagulls, crabs, cockroaches, rats.
• **Carrion feeders** – feed off dead bodies.	Harrier hawk, vultures, buzzards, flies, rats.
• **Filter feeders** – filter microscopic organisms/plankton and organic material from the water.	Baleen whales, barnacles, oysters, mussels, prawns, crabs, clams, shrimp.
• **Detritus feeders** – feed off organic debris in the soil/substrate.	Earthworms, mud snails.

Decomposers are species of **fungi** (eg moulds, mushrooms, 'toadstools') and **bacteria** that feed on waste products and dead bodies (using **extracellular** *digestion*), breaking them down (*decomposition*). This process releases chemicals from wastes/bodies into the soil so that the chemicals can be used In new life forms.

All communities rely upon interactions between producers, consumers, decomposers:

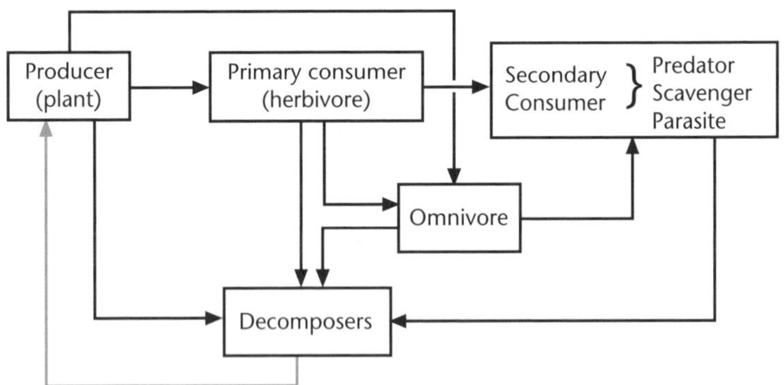

Direction of arrows shows the way food and other nutrients cycle.

Food chains

Within communities, **food chains** may be identified. A food chain is a chain of organisms through which food and energy flow. A food chain begins with a producer.

Example

A simple food chain – pine forest community

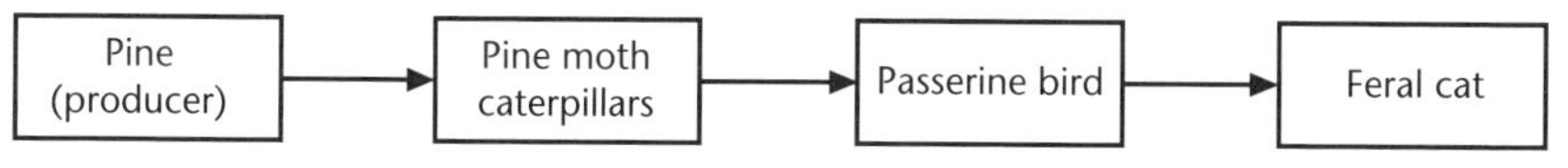

Each link in a food chain represents a trophic level; the connecting arrows show the *flow of energy*. In food chains, the trophic levels given are:

- Producer → herbivore → 1st carnivore → 2nd carnivore → 3rd carnivore.
- Producer → 1st (order) consumer → 2nd (order) consumer → 3rd (order) consumer.

Primary, secondary or tertiary may be substituted for 1st, 2nd or 3rd.

Omnivores will be either 1st- or 2nd-order consumers, depending on the food chain.

Filter feeders and detritus feeders are typically 1st-order consumers.

Parasites, carrion feeders and scavengers are likely to be 3rd- or 4th-order consumers, depending on the food chain.

Many food chains usually exist within a community. Species often belong to more than one food chain and may occupy different trophic levels in different chains.

Example

Series of pine forest food chains – both rats and cats occupy more than one trophic levels

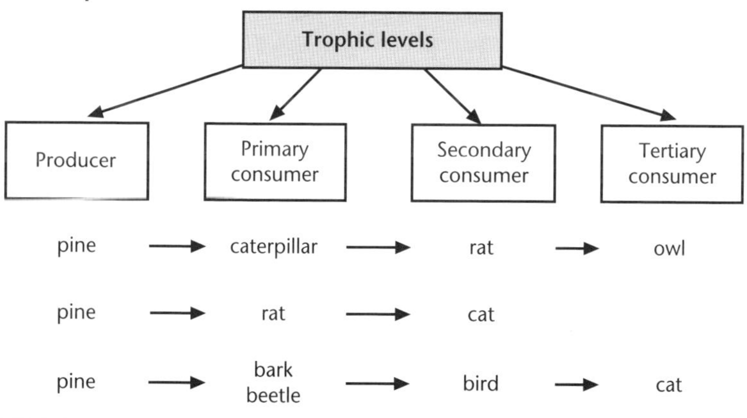

A **food web** can be established when the information from many food chains is put together.

Example

Simple food web from a pine forest community

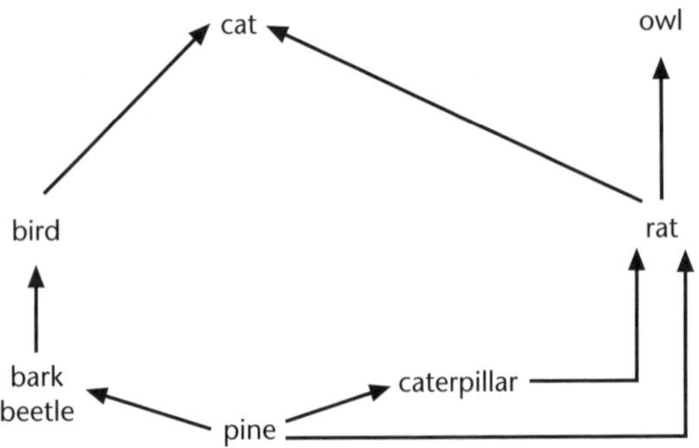

Food webs give a lot more detail than food chains about relationships between organisms in a community.

Energy flow

The *numbers* of organisms at each trophic level in a food chain/web, when represented by a diagram, are often referred to as a *pyramid* because of the typical shape which results.

A **pyramid of numbers** for a dairy farm community might look like the following:

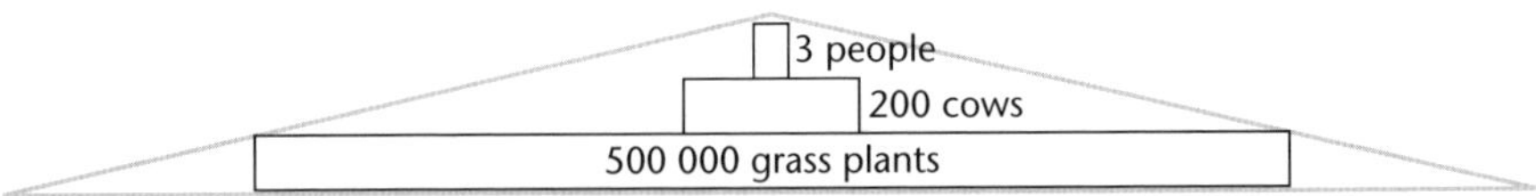

Numbers at each trophic level do not always decrease from producers to carnivores, as the *size* (or more accurately **biomass**) of the organisms is significant.

Example

Pyramid of numbers for a pine forest

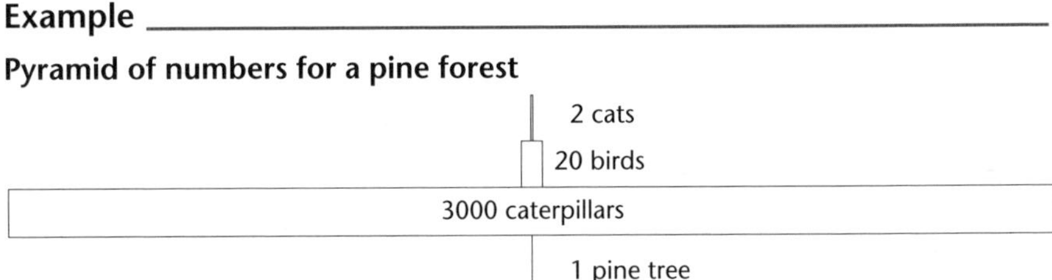

One very large pine tree is able to support many herbivores, etc. The pine tree has a large biomass.

Pyramids of biomass

Biomass refers to the dry mass of an organism; dry mass is used because the water content of organisms varies considerably. The biomass is an indicator of the energy *content* in an organism (or trophic level).

Biomass pyramid for a pine forest community

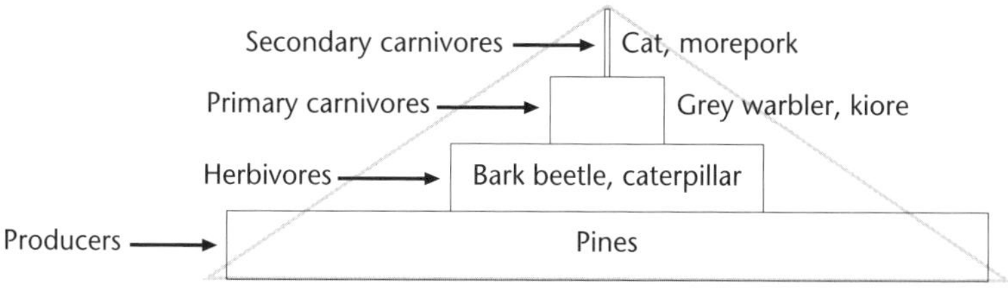

The area of each 'box' represents the biomass of that trophic level.

The units of biomass are mass per unit area (eg $g\,m^{-2}$, $kg\,ha^{-1}$).

For nearly all communities, the biomass of each trophic level is less than that of the level below. This loss of biomass at each level occurs because as food is eaten and digested by animals, much of the energy it contains is lost as heat, used in growth and movement, or used to keep body functions going. This means that most of the energy from the food is not available to be passed on, so the next level has *less* food available to it, and therefore it must have less biomass.

An exception to the typical biomass pyramid shape occurs in the oceans, where the producers (**phytoplankton**) can reproduce so quickly that their biomass is less than that of the primary consumers (zooplankton) that feed on them. In all other communities, however, the biomass of each trophic level is *less* than that of the preceding level.

Energy pyramids

Energy pyramids are very similar in shape to pyramids of biomass. An energy pyramid is really a pyramid of productivity (with units $kJ\,m^{-2}\,year^{-1}$); an energy pyramid indicates energy flow. Most (about 90%) of the energy at each trophic level is lost, either through heat loss from respiration, from **excretion**, body wastes and keeping body functions going, or from death. Because all this energy is not available to the next ('higher') trophic level, each level of the pyramid is progressively smaller.

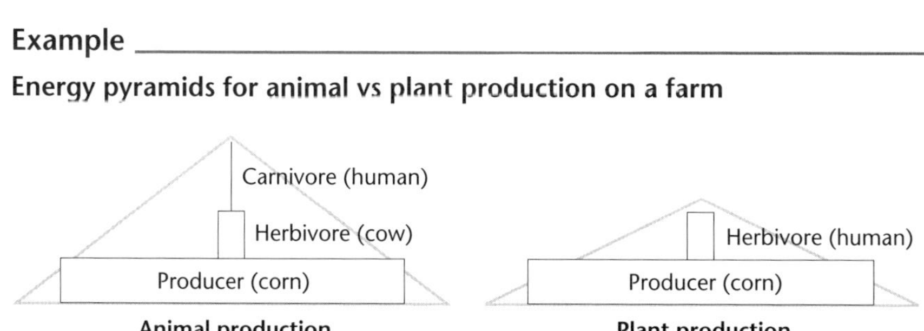

The size of each box represents the amount of energy of each trophic level. The units for energy are kilojoules (symbol kJ).

Comparing the two pyramids in the preceding example shows that when plant material is consumed directly by humans rather than first being turned into animal protein, 10 times more food energy is available. So, to use plant production more efficiently, humans must eliminate as many trophic levels as possible between producers and themselves.

Whatever management practices are used, energy is always lost at each trophic level in ecosystems.

Example

Energy flow through a pine forest ecosystem

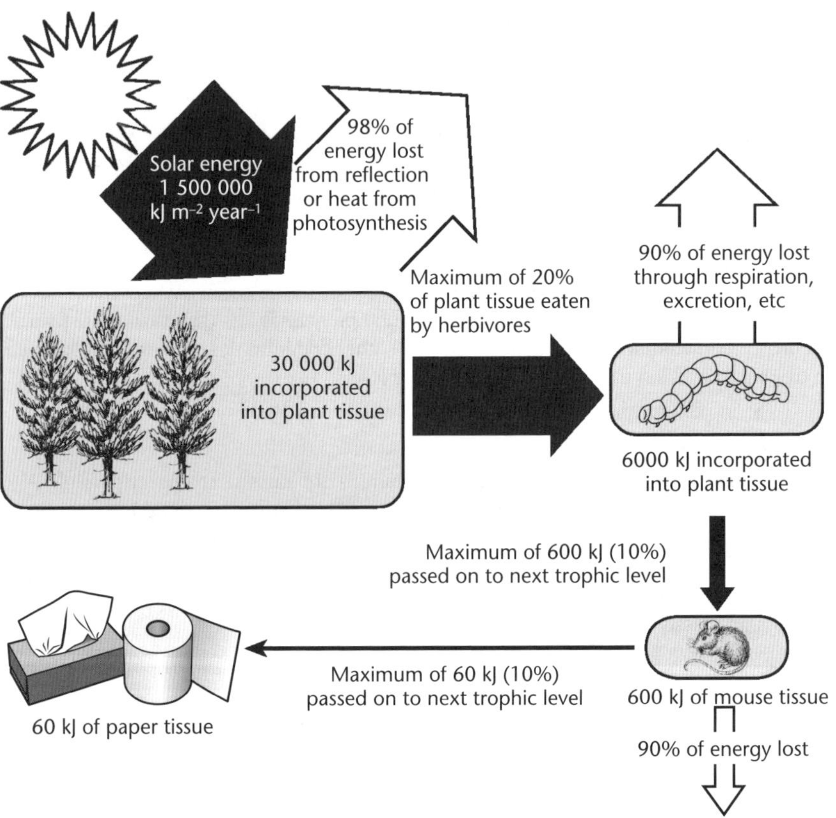

In the pine forest ecosystem shown above, the 30 000 kJ of energy incorporated into the pine trees in photosynthesis gets reduced to 60 kJ in the tissue paper (= 0.2% of the 30 000 kJ in the pines).

Energy enters the food chain in the process of **photosynthesis**, in which producers convert solar energy into chemical energy. This chemical energy is passed down the food chain. Energy is lost at each step as heat in the process of respiration. Up to 90% of the energy at each trophic level is lost through respiration (in homeotherms – warm-blooded animals – the energy loss is even greater than 90%).

Ultimately, *all the energy that enters ecosystems in photosynthesis is lost as heat radiated out into space*. Energy must continually enter the food chain through the producers. Therefore, energy flows through ecosystems; it is *not* **recycled**.

Biomagnification

Biomagnification is also known as 'biological magnification' or 'bioamplification'. It describes the process where the concentration of a substance (usually a poisonous chemical such as mercury or **DDT**) in the living tissues of an organism increases as it works its way up through the higher levels in a food chain. The result of biomagnification is that organisms at the top of the food chain generally suffer more harm from the poisonous chemical than organisms at the lower levels of the food chain.

Certain chemical elements have the tendency to accumulate within food chains. It is this *increase* in the concentration of a substance within a living organism that is called biomagnification. A good example is DDT in food chains.

DDT (**d**ichloro**d**iphenyl**t**richloroethane) was first made in 1874. It was recognised for its properties as an **insecticide** in 1939 and used for agricultural purposes. It was used in human health programs after World War II to control mosquito-borne malaria.

However, DDT is not easily degraded by microbes, or by light or heat, and it can persist in soil for up to 10 years. It has low solubility in water and high solubility in fats and lipids, allowing it to settle in living tissues of biological systems for longer than in the soil. Therefore, DDT tends to concentrate in organisms starting from the bottom of the food chain.

Because biomagnification occurs at each step of the food chain, organisms at the top of the food chain or higher trophic levels accumulate large concentrations of DDT in their lipids. The long-term consequences are usually detrimental. Therefore, in a typical aquatic food chain, large concentrations of DDT can be found in the diving birds that feed on the fish that feed on small insects containing persistent DDT. In diving birds DDT affected eggshell formation. This resulted in thin eggshells that often broke before maturation, leading to a decline in that bird species population.

In addition, edible fish tested positive for high levels of DDT and affected market sales as the fish became unsafe for human consumption, leading to a decrease in the availability of fish protein in the diets of humans.

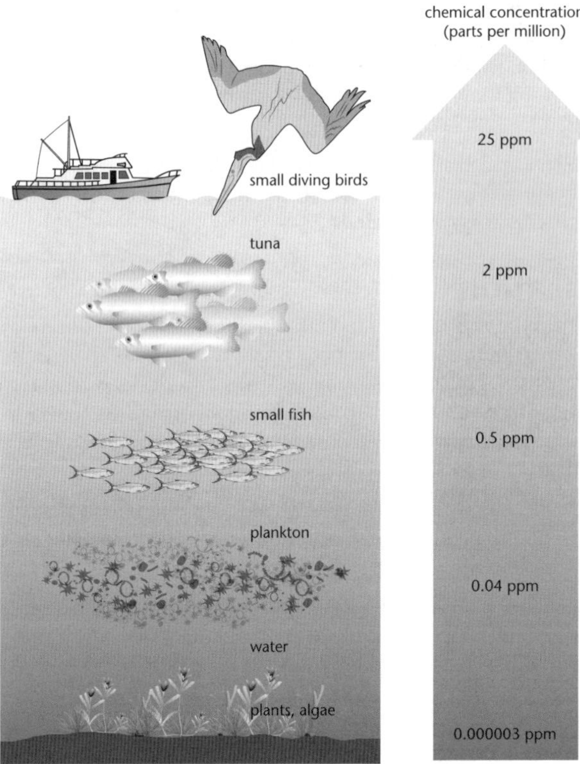

The diagram shows how a substance such as DDT or mercury accumulates as it moves up the food chain. Poisonous substances find their way into oceans and inland waterways when contaminated water from agriculture or industry finds its way into our rivers, lakes and seas. These substances are absorbed in small quantities by algae and plankton. But the small amounts of poison are concentrated in larger amounts in the small fish that eat the plankton and algae, and in turn in the larger fish that feed on small fish. The concentration of toxins is increased in the birds that eat the larger fish. Humans can also be affected.

Nutrient cycles

Nutrients are *recycled* through ecosystems. Although nutrients are lost at each trophic level through excretion, **egestion** and death, they re-enter the food chain through the action of **decomposers**, which break down organic materials so the nutrients enter the soil and are available to the producers.

The important nutrient cycles involve the elements carbon, C, hydrogen, H, oxygen, O, nitrogen, N, and phosphorus, P, as these elements combine to form the giant molecules (proteins, carbohydrates, fats, **nucleic acids** – DNA, RNA) that all organisms are made of and that are essential for growth and **metabolism** of all life forms.

The water cycle

Hydrogen and oxygen are recycled as water. The **water cycle** mainly involves the physical environment.

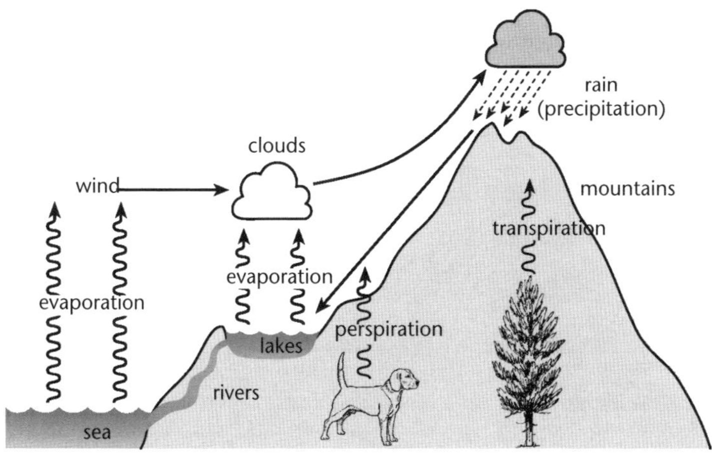

The water cycle.

Most water returns to the atmosphere by **evaporation**, but some water is returned through **transpiration** from the leaves of plants and a small amount of water is returned by animals through **perspiration** or from **gas exchange** surfaces.

The carbon cycle

The **carbon cycle** mainly involves the exchange of carbon dioxide, CO_2, and oxygen, O_2, between plants and animals during photosynthesis and respiration.

Photosynthesis

$$\text{carbon dioxide} + \text{water} \xrightarrow[\text{chlorophyll}]{\text{solar energy}} \text{glucose} + \text{oxygen}$$

$$CO_2 + H_2O \xrightarrow[\text{chlorophyll}]{\text{solar energy}} C_6H_{12}O_6 + O_2$$

Respiration

$$\text{glucose} + \text{oxygen} \longrightarrow \text{carbon dioxide} + \text{water} + \text{energy } [\textbf{ATP} + \text{heat}]$$

$$C_6H_{12}O_6 + O_2 \longrightarrow CO_2 + H_2O$$

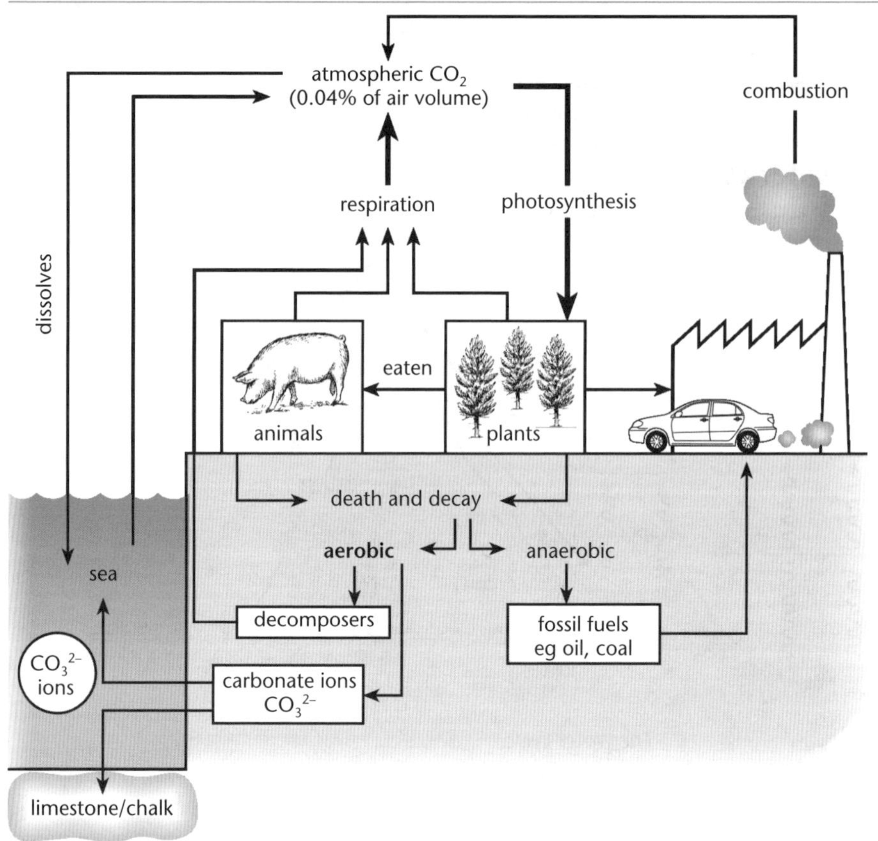

The carbon cycle.

Carbon exists in animal and plant tissues as proteins, fats and carbohydrates. It is transferred between organisms as one organism consumes another. Carbon is lost from each trophic level through respiration and decay.

If the decay of plants and animals occurs without oxygen (ie in **anaerobic** conditions), the remains may turn into fossil fuels, such as coal and oil. The production of fossil fuels takes thousands of years. Humans are consuming these fuels at a rate far faster than they are being made. Unless efforts are made to conserve these fossil fuels, they *will* run out.

Combustion of fossil fuels produces billions of tonnes of CO_2 each year. Although plants use some of the CO_2 produced and a lot of CO_2 is dissolved in the sea to produce carbonate ions, CO_3^{2-}, a significant amount of CO_2 is building up in the atmosphere, contributing to the **Greenhouse Effect**.

Carbonate ions in the sea are used to produce the shells of molluscs, the hard skin of crustaceans and the framework of corals. When these organisms die, their shells may accumulate and turn into **limestone** rock.

The nitrogen cycle

Nitrogen is used to produce **proteins**, which form tissue structures (eg **collagen**) and control body processes (eg enzymes).

Plants produce protein from combining carbohydrate (from photosynthesis) and **nitrate ions** (from the soil). Animals obtain protein by eating plants or other animals.

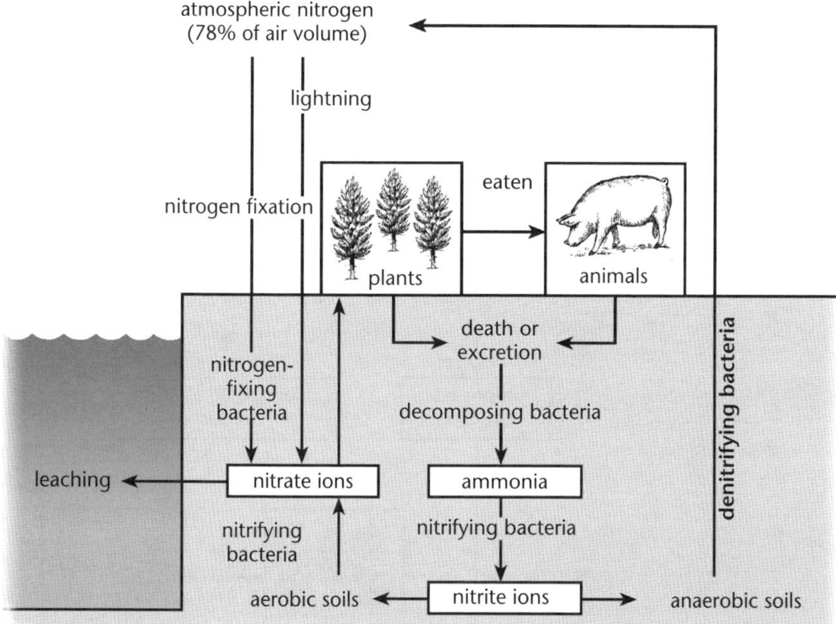

The nitrogen cycle.

The decomposition of excretory products and plant and animal remains produces **ammonia**, NH_3, which is also excreted directly by many animals.

Ammonia is converted by **nitrifying bacteria** (*Nitrosomonas*) into nitrite ions, NO_2^-:

- In very wet soils, anaerobic conditions occur and nitrite is converted into nitrogen gas, N_2, by denitrifying bacteria. Plants growing in these soils will be taking in little or no nitrogen and their leaves often appear yellow.
- In **aerobic** soils, nitrifying bacteria (*Nitrobacter*) convert nitrite into nitrate ions, NO_3^-, which are used by plants.

Some nitrogen is converted into nitrate by lightning, but most nitrogen is **fixed** (converted into nitrogen-containing compounds) from the atmosphere by free-living bacteria (*Azotobacter*) present in the soil, or by bacteria (*Rhizobium*) in the roots of **legumes** such as peas, gorse and lupins.

Nitrate is very soluble, so it is readily **leached** (lost from soils). Leached nitrates carried by streams to the sea may accumulate near the sea floor. Upwelling currents (usually close to the coast) return nitrates to the surface, providing nutrient-rich seas and productive fishing grounds.

The phosphorus cycle

Phosphorus is used by organisms in respiration, to produce protein and in bone formation in animals. It is also an essential element for DNA, RNA and **ATP**.

Phosphorus is the nutrient most likely to be in short supply in an ecosystem, because it is sometimes not readily available for recycling. Phosphorus may become 'locked up' in volcanic ash or in bones, because they are not easily decomposed or broken down by weathering. Phosphate **run-off** from the land can form **sedimentary rocks**.

Growing crops draws nutrients from the soil. In PNG local farmers can apply phosphate fertilisers that are sold in various stores, but more generally, a variety of techniques are used to slow the loss of soil and nutrients. Techniques include fallowing (not planting anything to let the soil recover between crops), rotating crops, planting trees, creating terraces to slow erosion, constructing mounds and beds, and composting.

However, not all phosphorus is lost from an ecosystem. It is present in soils as phosphate ions, PO_4^{3-}, which are readily taken up by plants and returned to the soil by decomposers.

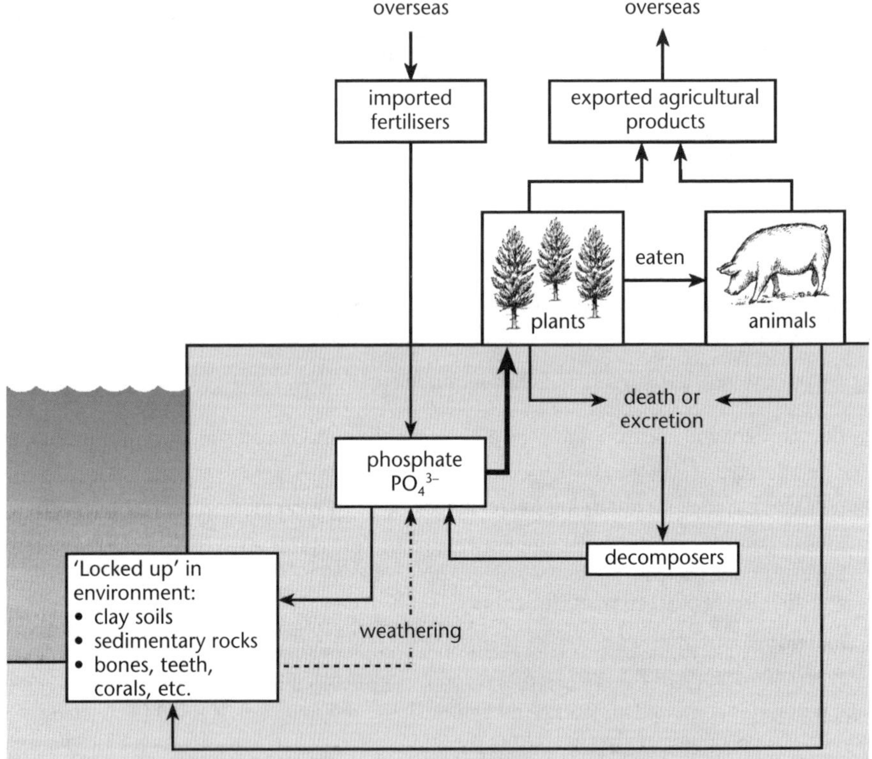

The phosphorus cycle.

The phosphorus, nitrogen and carbon cycles are *linked* together because the same decomposers are common to all cycles.

Unit 12.1 Activity 6A: Ecosystems

1. Distinguish between the following pairs of terms:

a. Autotroph and heterotroph.

b. Photosynthesiser and chemosynthesiser.

c. Producer and consumer.

d. Herbivore and carnivore.

e. Filter feeders and detritus feeders.

2. Read the following extract and then answer the questions that follow:

> Approximately 30 years ago a small flatworm native to New Zealand was accidentally introduced into Ireland. The flatworm lives in the soil where it catches live earthworms. It secretes digestive enzymes onto them and then sucks up the soluble digested material. The flatworm takes about half an hour to digest a single earthworm.

(Adapted from: *Ecology*, Peter Chenn, 1999, London.)

a. Explain the feeding relationship between the flatworm and the earthworm.

b. The feeding method of the flatworm is an example of extracellular digestion. Explain why.

3. Broad bean plants (*Vicia faba*) often have populations of aphids living on them. Aphids are small winged insects with piercing sucking mouthparts they use to suck glucose from the stems of plants. Aphids are eaten by small red and black beetles called ladybirds.

a. Explain the feeding relationship between **i**. the aphid and broad bean **ii**. the aphid and the ladybird.

b. The data below was collected in a glasshouse where broad beans were being grown. It shows the mean population densities of the organisms concerned.

Organism	Number of organisms per m^2
Broad bean	13
Aphid	6500
Ladybird	32

i. Draw a labelled pyramid of numbers for this data.

ii. Sketch a pyramid of biomass for this food chain.

iii. Explain any differences in the two pyramids.

4. The diagram below shows a pyramid of biomass from a mountain stream.

0.2 tonnes	*Galaxias vulgaris*
6.5 tonnes	Aquatic insects and larvae
100 tonnes	Algae and diatoms

Discuss why the biomass of *Galaxias vulgaris* is 0.2 tonnes, while the biomass of the population of algae and diatoms is 100 tonnes.

5. The diagram below shows part of a food web in Lake Taupo, New Zealand:

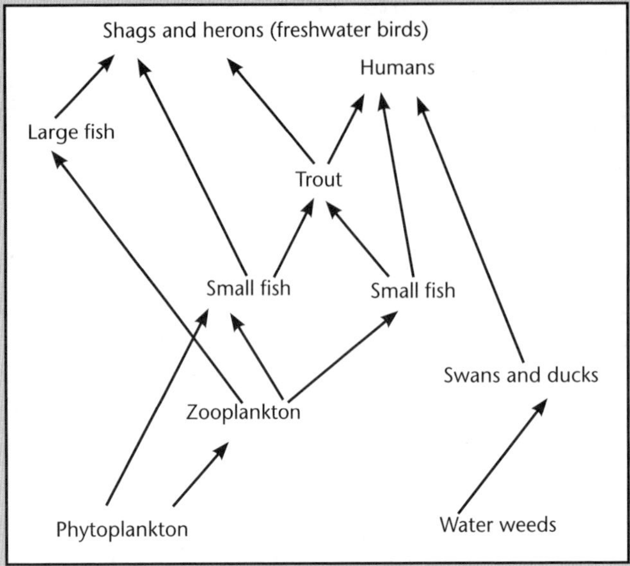

a. Identify organisms from the food web that are:

- **i.** Producers.
- **ii.** 1st-order consumers.
- **iii.** 2nd-order consumers.
- **iv.** 3rd-order consumers.
- **v.** Herbivores.
- **vi.** 1st carnivores.
- **vii.** 2nd carnivores.
- **viii.** 3rd carnivores.
- **ix.** An omnivore.

b. Trout were introduced into the lake during the 1890s. Explain the effect this would this have had on shags.

c. Small fish were introduced into the lake during 1934 and 1940. Explain the effect that this would have had on trout.

d. Sketch a pyramid of energy for this food web.

e. Phytoplankton use only 1–5% of the incoming radiation from the sun for photosynthesis. Describes what happens to the other 95–99%.

f. The surface waters of the lake were sampled over a period of one year and information collected on the numbers of phytoplankton. The data recorded is in the following table.

Month	Phytoplankton numbers/mL of water	Month	Phytoplankton numbers/mL of water
June	40	December	350
July	30	January	590
August	25	February	530
September	20	March	255
October	30	April	95
November	220	May	60

i. Describe three environmental factors that may have caused the rapid increase in phytoplankton from October to January.

ii. Explain what trend you would expect to see for zooplankton numbers during the time period October to January.

6. The following diagram represents part of the carbon cycle:

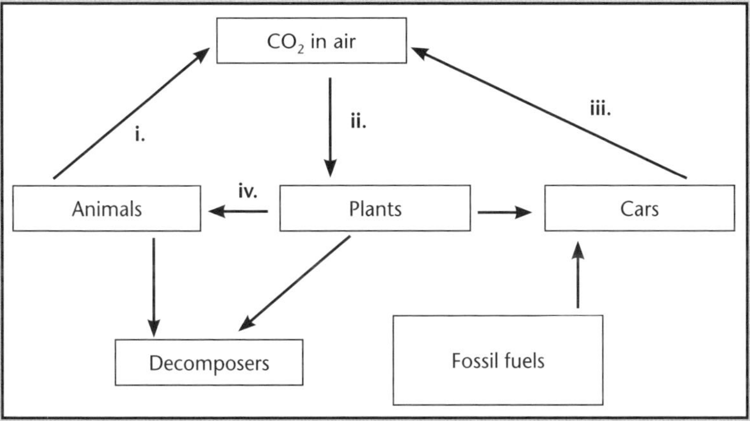

a. Identify the processes labelled **i**, **ii**, **iii**, **iv**.

b. Explain why carbon is such an important element to all organisms.

c. Describe two differences between the carbon cycle and the phosphorus cycle.

d. Describe the main use of phosphorus in animals.

7. The following diagram is a simplified version of the nitrogen cycle. The boxes labelled **V** to **Z** represent different types of bacteria.

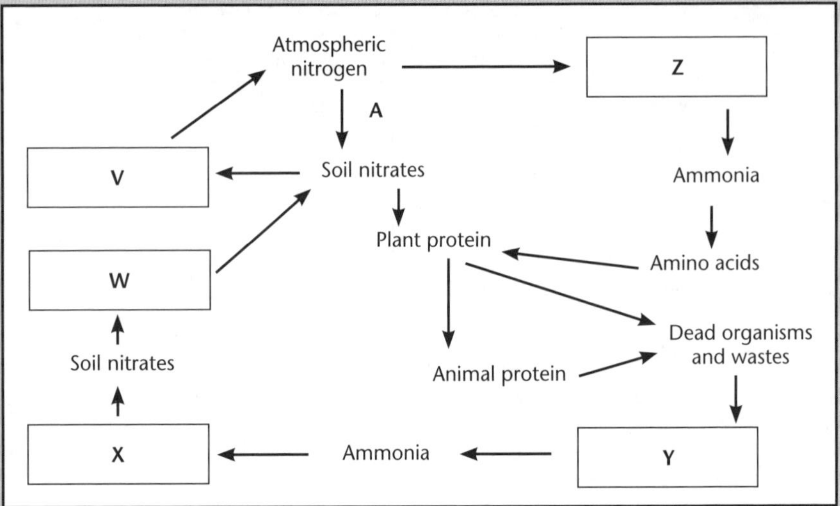

a. i. Identify the types of bacteria that would be found in the boxes labelled **V** to **Z**.
   ii. Many of these bacteria are chemosynthesisers; explain why.

b. Explain what is happening at the arrow labelled **A**.

c. Identify two giant molecules found in all organisms that need nitrogen.

d. Farmers who have a paddock not being grazed by livestock may plant the field in clover and then plough the clover into the soil at the end of the growing season. Explain why.

8. a. The bacterium *Rhizobium* lives in the root nodules of clover.
   i. Identify this relationship.
   ii. Describe how each organism is affected by the relationship.

b. The photograph below shows the results of an experiment on clover growth.

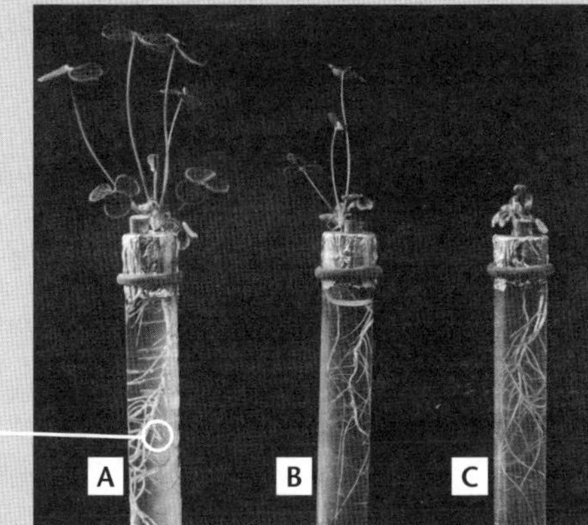

(A) A clover plant inoculated with an effective *Rhizobium* strain. Note the nodules on the root.
(B) A clover plant inoculated with a less-effective *Rhizobium* strain.
(C) A clover plant that was not inoculated.

Discuss these results.

9. Use the diagram below to discuss how energy flows through an ecosystem:

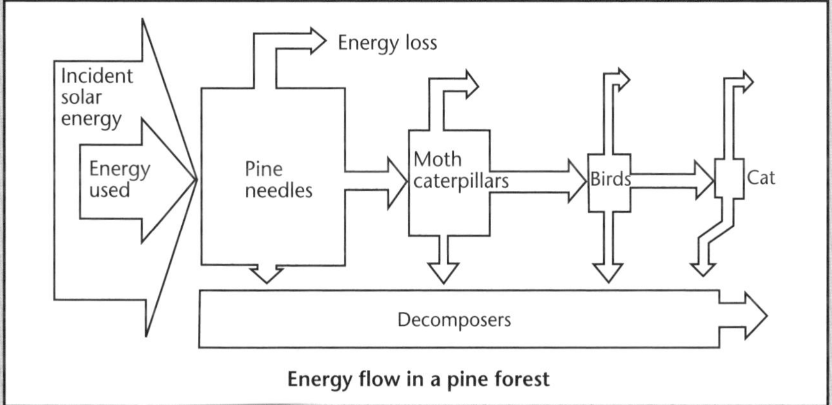

Energy flow in a pine forest

10. Discuss the importance of nutrient cycles in an ecosystem.

11. Discuss the following statement:

'Green plants play a vital role in the flow of energy and materials through an ecosystem'.

Unit 12.1 Ecology

Topic 7: Biomes and habitats

Other Topics in this Unit have looked at various aspects of biomes and habitats, for example, definitions of particular terms such as 'species', 'population', 'community', 'habitat' and 'niche'; also describing the climatic regions and factors influencing plant and animal life on our planet.

In Topic 7 we consider the major different biomes: in other words, the major communities of plants and animals that live in a particular type of environment, such as forests, deserts, grasslands, the aquatic environment and tundra.

Introduction to biomes

Natural selection determines whether or not an organism is able to survive and reproduce in any environment. Both living (**biotic**) and non-living (**abiotic**) factors act to eliminate the weaker individuals in a given population. Over quite some time, the succeeding generations of organisms that live in a particular ecosystem become better adapted to those respective environmental conditions (biomes).

The *Oxford Advanced Learner's Dictionary* defines *biome* as 'the characteristic plants and animals that exist in a particular type of environment, for example in a forest or desert'. The word describes a large region that has a similar climate, soil, plants and animals regardless of where it is located on Earth, and is comparatively different from other regions.

Temperature and precipitation are the most important factors that determine the boundaries of biomes. Towards the poles, temperature becomes the determining factor, while in temperate and tropical regions precipitation is more significant. Light is usually an abundant abiotic factor except on forest floors. Rapid temperature changes, fires, floods, drought and strong winds are the other abiotic factors that can affect biomes.

Many different systems can be used to classify biomes but ecologists usually agree that there are at least ten different biomes. Major biomes include:

- Tropical rainforest.
- Tropical dry forest.
- Tropical savanna.
- Desert.
- Temperate grasslands.
- Temperate woodlands and shrublands.
- Temperate forests.
- Northwestern coniferous forests.
- Boreal forest.
- Tundra.

Every biome has its own set of abiotic factors – particularly climate – and has its own characteristics. In this Topic we will look at the following biomes:

- Tundra.
- Taiga.
- Temperate rainforest.
- Temperate deciduous forest.

Biomes of the world. Different systems are used by ecologists to classify biomes of the world. Some classifications are more detailed than others and there can be differences in terminology. This map shows 18 different terrestrial biomes; it does not show the aquatic biomes. The map is intended to give you a sense of the variety of different environments in the different countries on the Earth. Remember, the actual boundaries between biomes are not as clearly defined as shown on the map.

- Temperate grasslands.
- Chaparral (Mediterranean).
- Hot deserts and arid regions.
- Savanna.
- Tropical rainforests.
- Aquatic biomes (intertidal zone, coral reef, open ocean, freshwater lakes and rivers, and wetlands).

Small maps show the approximate, natural, geographic location of each biome. But the boundaries of each biome are not as clearly defined as those shown on each map. On the ground, there is usually a gradual transition from one biome to another. Plants and animals change gradually as you move from one biome to another. The text that follows is intended only to give you a general understanding of the concepts of biome and habitat, and the maps are approximate.

Tundra

The tundra is the large flat Arctic regions of northern Europe, Asia and North America. It is also called the Arctic tundra because it occurs in extreme northern latitudes where snow melts seasonally.

The tundra has a layer of permanently frozen ground called permafrost, which limits the depth to which roots can penetrate. Tundra is characterised by low species diversity and low primary productivity. Plants grow no taller than 30 cm and the growing season can be as short as 50 days.

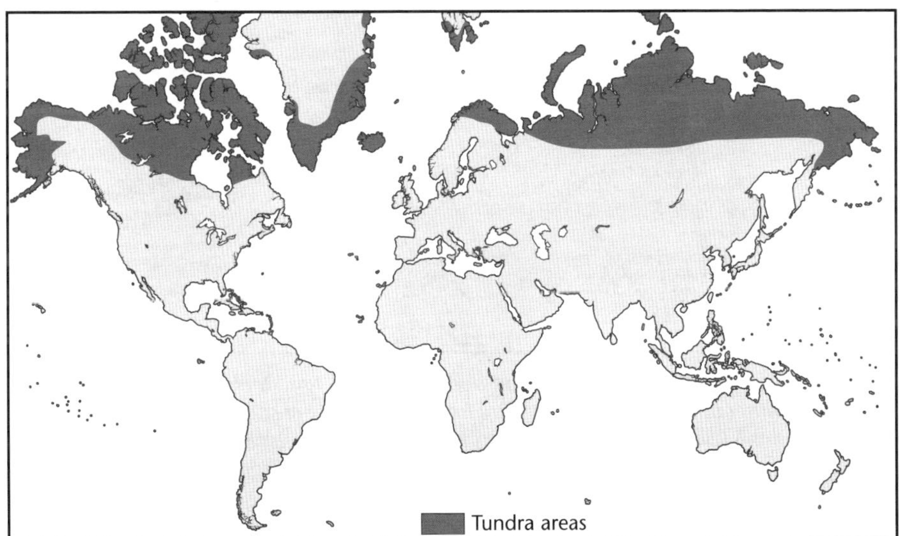

Tundra is found mainly in the large land masses of the high northern latitudes of the northern hemisphere. Parts of Greenland extend far enough north that tundra is replaced by snow and ice. The areas of the southern hemisphere at high latitudes are small by comparison and their climate is affected by oceans that surround them.

The southern hemisphere has very little tundra because the Antarctic has comparatively little ice-free land and the climatic conditions at extreme southern latitudes are different. The Antarctic is much colder than the Arctic and the ground is always covered with snow and ice. There is little permafrost land, which means the conditions are not right for tundra landscape to form.

Taiga

The taiga (pronounced *ty*-guh) is the forest that grows in wet ground in far northern regions of the earth. Taiga is also called boreal forest. It has vast numbers of pine trees, spruce trees and larch trees growing there. The taiga is the largest terrestrial biome on the planet. It stretches (south of the tundra) across North America, Europe and Asia, covering an estimated 11% of the Earth's land surface.

Winters in the taiga are cold and severe with lots of snow. The soil is usually acidic and mineral poor. Permafrost is patchy with warm, humid summers and an annual precipitation around 50 cm a year. Taiga has numerous ponds and lakes and woody trees, and relatively high species diversity and primary productivity.

Like tundra, this type of biome is not found in the southern hemisphere because there are no vast tracts of land in the corresponding southern latitudes.

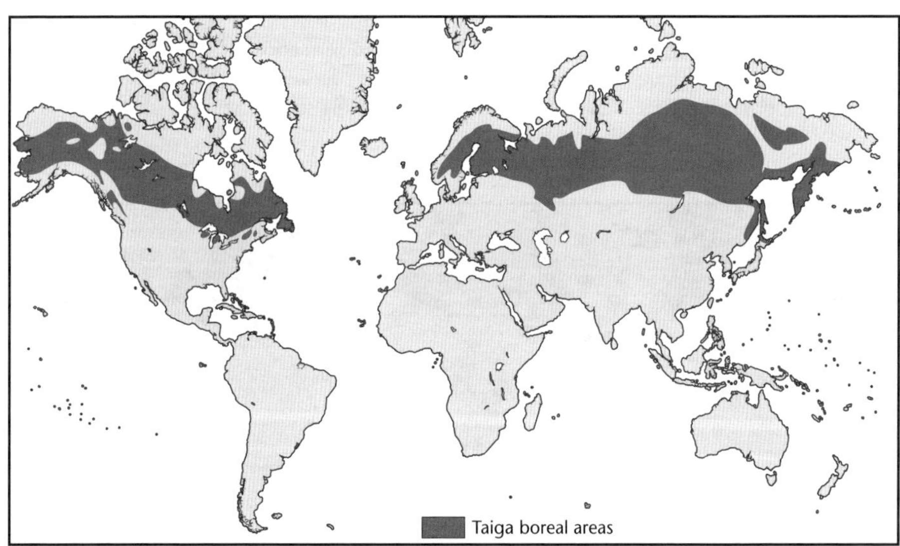

The taiga is located near the top of the world, below the tundra biome. It has two main seasons: winter and summer. Winters are freezing with lots of snow. Summers are warm, rainy and humid, which creates conditions for millions of insects. Birds migrate there every year to nest and feed. The name taiga comes from the Russian word for forest.

Temperate rainforest

Rainforest is unique because of the high levels of rain it gets every year. Temperate rainforest is often referred to as coniferous temperate rainforest. It occurs on the north-west coast of North America. Vegetation similar to temperate rainforest occurs in southeastern Australia, New Zealand and southern parts of South America.

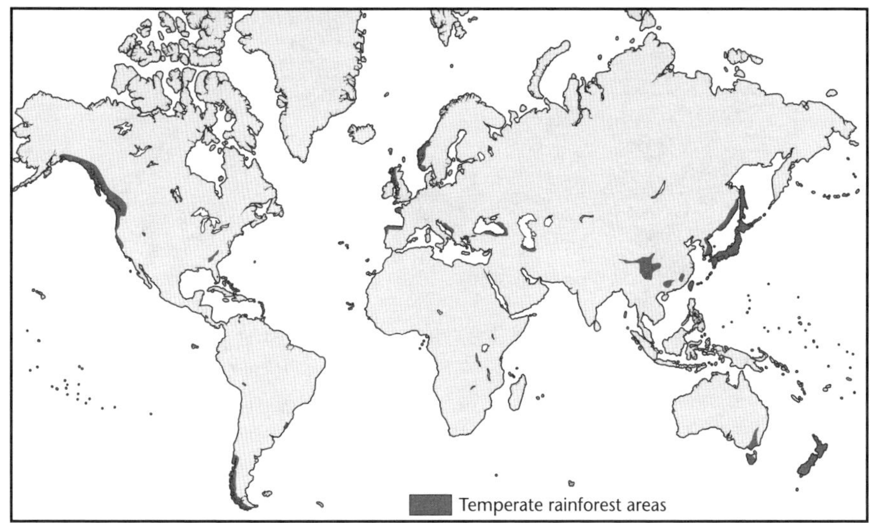

Temperate forests cover a large part of the globe but temperate rainforest is found in only a few areas around the world: mainly western North America, the south-west of South America (parts of Chile and Argentina), small areas of Europe (southern Norway, northern Spain and Portugal), southeastern Australia and in the west of the South Island of New Zealand.

Annual precipitation is usually high (200–380 cm a year) and temperatures are cooler. The low temperatures inhibit microbial decomposition rates and although the organic content of soils is quite high, the soils are relatively nutrient poor. The dominant vegetation is usually large evergreen trees (fir, spruce, cedar and hemlock) with epiphytes (mosses, lichen, ferns) and shrubs. Deer, mules, birds, invertebrates, reptiles and amphibian species are found in this region.

Temperate deciduous forest

Seasonality is a characteristic of the temperate deciduous forest biome. There are four seasons: spring, summer, autumn and winter. The leaves change colour in autumn and trees lose their leaves in winter. The forests are dominated by broad-leaf hardwood trees (oak, hickory, maple, magnolia, beech and gum trees).

Temperate deciduous forest gets less rain than temperate rainforest. The precipitation in these temperate regions range from 75 to 126 cm annually and the soils are usually nutrient rich.

Large mammals, including wolves, bears and deer, can be found in the main regions, as well as many small mammals, birds and invertebrates.

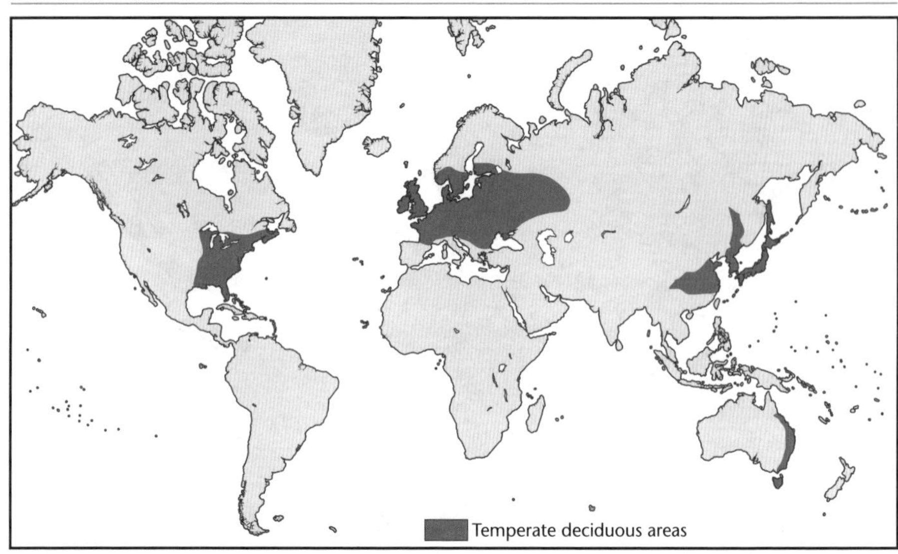

Temperate deciduous forests are found in the east of North America and middle Europe. Some areas are found in Russia, Japan, east China, south Chile and parts of Paraguay. There are also small areas of temperate deciduous forest in New Zealand and south-east Australia. Much deciduous forest has been cleared for farmland and settlement, and the wood harvested for building, heating and cooking. Animals have lost their homes as people move in to build their own homes.

Temperate grasslands

Grasslands (as the name suggests) are tracts of land where the dominant vegetation is grass. They are found on all continents except Antarctica. Grasslands are divided into different categories, among them temperate grasslands, savanna and others.

Temperate grasslands are characterised by hot summers, cold winters and unseasonal rainfall. Annual precipitation ranges between 25 and 75 cm. The soils tend to contain large amounts of organic matter due to grass dying regularly with seasonal changes involving warmer temperatures.

This type of biome is found in the mid-western USA, Ukraine and parts of southern Europe. It has the ideal conditions for many grass species, especially crops such as corn and wheat.

Temperate grasslands were the natural home to many large grass-eating animals, including bison, gazelles, zebras, rhinoceroses and wild horses, and they have become important for farming and raising livestock

Chaparral

Chaparral biomes are associated with a Mediterranean climate and are found in the area around the Mediterranean Sea, California, Western Australia, parts of Chile, and South Africa. The soils are thin and fertile, characterised by frequent fires in late summer or autumn. Chaparral biomes have mild winters with abundant rainfall and extremely dry summers.

Vegetation appears to be similar in the different chaparral sites around the world but the species types are different. Vegetation comprises dense growth of evergreen shrubs with drought-resistant pine and shrub oak trees.

Varieties of deer, wood rats, chipmunks, lizards and many bird species occupy this biome.

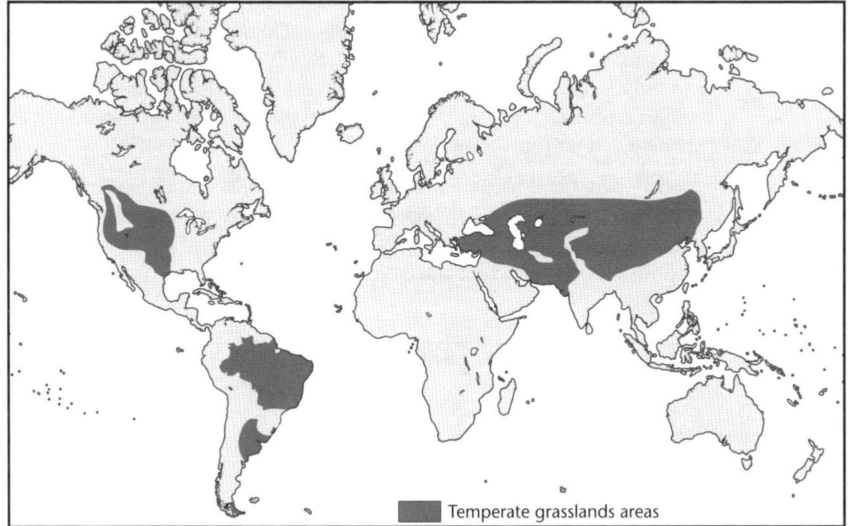

Temperate grasslands are found in the mid-latitudes of North America (the prairies and the Pacific grasslands), Argentina, Brazil and Uruguay (the pampas), and Europe (the steppes).

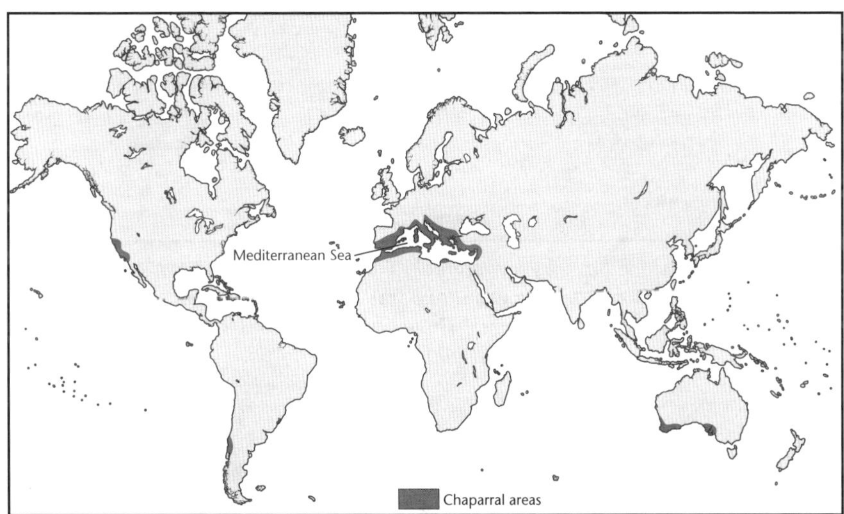

The chaparral biomes all have a climate similar to that of the regions surrounding the Mediterranean Sea. They are found in the mid-latitudes: the Mediterranean Basin, California, Central Chile, south-west Australia, the Cape Province of South Africa and parts of the west coast of North America. This biome is small in area compared to others, but it is home to almost 20% of the world's plant species, with high numbers of rare plants.

Hot deserts and arid regions

This very dry biome is found in temperate desert areas and tropical and sub-tropical regions. The daily temperature extremes of heat and cold are associated with the low water vapour content of the desert atmosphere. Precipitation is less than 25 cm annually.

A few deserts are so dry that plants cannot thrive there but small burrowing animals and insects are adapted to those conditions. Desert vegetation generally includes **perennials** and flowering **annuals** such as cacti, yuccas, Joshua trees and sage bush. Desert plants are characterised by having few or no leaves to conserve water, or else they shed their leaves for most of the year, growing only in the short moist season of the year. The cactus usually photosynthesises from the stem as the leaves are modified into spines for defence.

Vertebrates and invertebrates in the desert biomes have specialised adaptations to deal with the extreme temperature fluctuations and very low moisture content. Some of these animals include species of lizards, tortoise, snakes, frogs and toads, rodents, rabbits, kangaroo, foxes, camels, birds of prey, owls and various tunnelling mammals and insects.

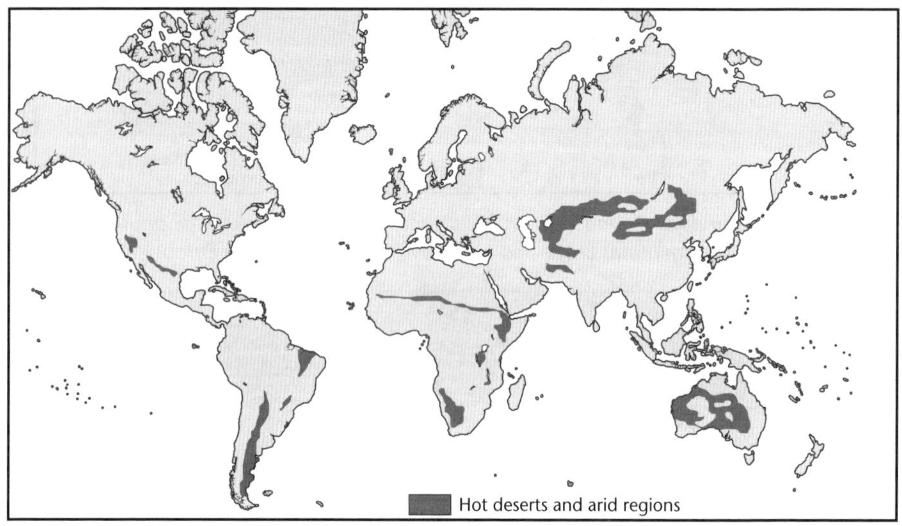

A desert is an area which has very low annual rainfall. Most deserts occur at low latitudes. The Sahara of North Africa is the world's largest desert but other hot deserts occur in south-western USA, Mexico and Australia. There are also cold deserts in parts of central USA (Utah and Nevada) and western Asia.

Savanna

The savanna biome is a tropical grassland area with widely scattered clumps of trees. It has relatively low annual rainfall and prolonged dry **periods**. The temperature does not vary much but precipitation does. Annual precipitation ranges from 85 to 100 cm.

Savanna soil is usually nutrient poor but some studies have shown savanna soil to be rich in aluminium. Savanna vegetation includes acacia, eucalypt and woody shrubs and grasses.

Savanna is found in parts of Africa, southern America, western India, northern Australia and southern parts of Papua New Guinea.

The African savanna is the most studied. It is home to a variety of hoofed and herbivorous mammals such as wildebeest, antelopes, giraffe, zebra, rhinoceros, hippopotamus and elephant. Also found are predators such as lions and hyenas. In many parts of the African savanna, people have started using it to graze their cattle and goats. **Grazing** can mean that the savanna may turn into a desert. Huge areas of savanna are lost to desert each year.

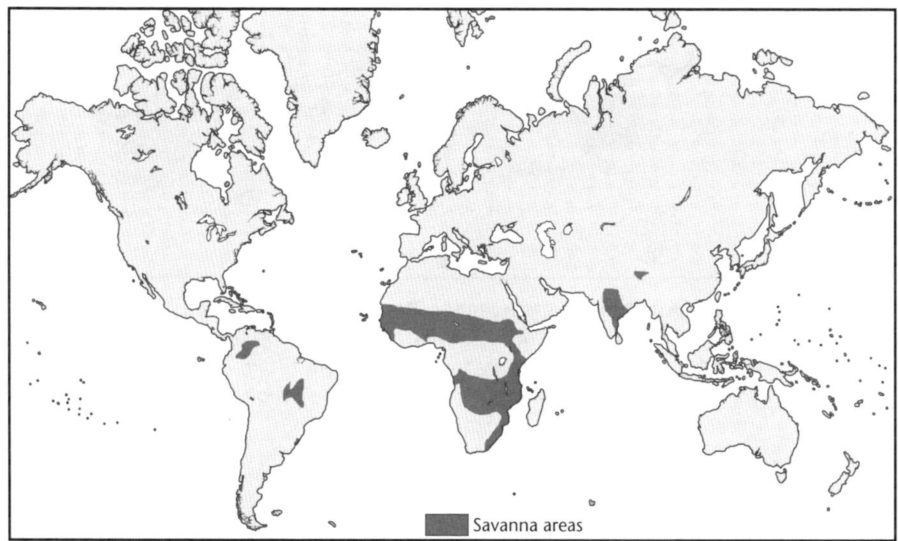

The savanna biome is grassland that is found between a tropical rainforest and hot desert biomes. It has a long dry season (winter) and a wet season (summer). There are several different types of savannas: the East African savanna and the Serengeti Plains of Tanzania are among the best known. South America also has savannas (in Brazil, Colombia and Venezuela) and there is also a savanna in northern Australia.

Tropical rainforest

Tropical rainforests are found in tropical regions where temperatures are high throughout the year and precipitation occurs daily with annual rates of 200–450 cm per year. The precipitation is said to come from locally recycled water entering the atmosphere via plant transpiration processes.

Decomposition rates are quite high with the ideal warm and wet conditions, but mineral absorption by plant roots is also quite high, leaving the soil relatively mineral poor. Despite this, the tropical rainforest is a very productive biome, and compared with all other biomes it has the highest diversity index for animals, plants and microorganisms.

Rainforests are full of evergreen flowering and non-flowering plants that include trees, shrubs, vines and lianas, mosses and ferns. The epiphytic communities are abundant and made up of smaller plants such as orchids and bromeliads.

Abundant animal species include various invertebrates, reptiles, amphibians and birds. Larger animals include mammals, marsupials and rodents.

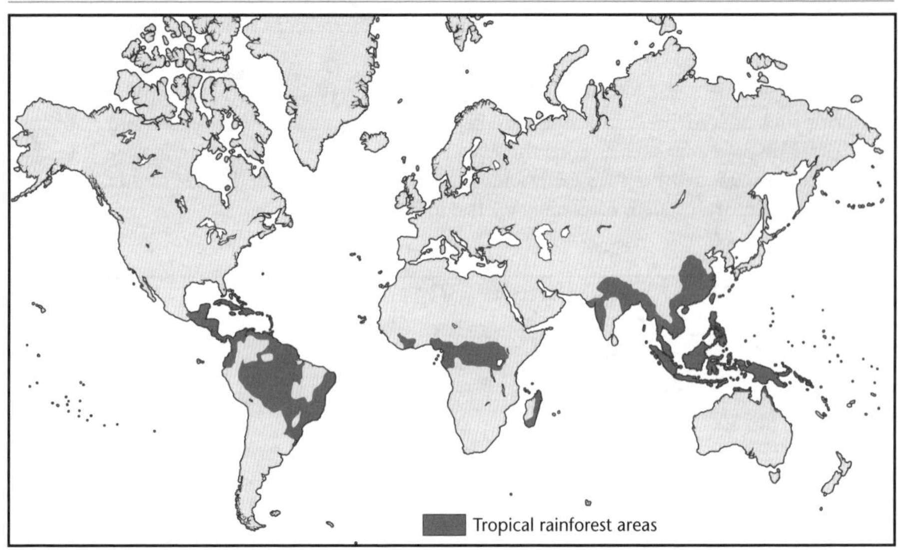

Tropical rainforest occurs mainly between the Tropic of Cancer and Tropic of Capricorn or approximately between the latitudes 28° north and 28° south of the equator. This is called the equatorial zone. Rainforests are found in Asia, Australia, Africa, South America, Central America, Mexico, and on many islands of the Pacific Ocean, Caribbean Sea and the Indian Ocean.

Tropical rainforests are found in the equatorial regions of south-east Asia, South America and New Guinea.

Aquatic biomes

In the aquatic environments, several different biome types exist. These include marine aquatic biomes (intertidal zone/mangroves, coral reefs and the open ocean) and freshwater lakes, rivers, streams and wetlands.

These biomes are distinguished by differences in abiotic parameters such as salinity, oxygen content, depth, current strength and availability of light.

Intertidal zone

The intertidal zone is the area where land meets sea and has a daily cycle of being alternately submerged and exposed by daily tides, but the waters are nutrient rich. Resident organisms are subjected to daily variations in temperature, light intensity and availability of seawater.

Around the world's coastlines, the intertidal biome consists of sandy shores, mudflats or rocky shore. Mudflats in tropical areas are colonised by mangroves, while mudflats in temperate regions are occupied by salt marshes. Plant life on rocky shores consists of seaweeds and green algae, limpets, sea stars, sea urchins, sponges, tube worms, isopods, chitons and stray small fish and sea cucumbers.

Coral reefs

Coral reefs are found in warm (20–30°C) tropical water with lots of sunshine and a lot of solid substrate for attachment by marine invertebrates. The water is usually quite clear and nutrient rich, and is home to microscopic organisms, plankton, bacteria and algae, and a wide range of invertebrates and fish. Studies have shown that 30–40% of all fish species are found in coral reefs.

Australia's Great Barrier Reef is the most famous but others exist on south-east Asian coastlines, Papua New Guinea coastlines, the Pacific Islands and the Caribbean Islands.

Open ocean

The oceans cover about 70% of the Earth's surface. The open ocean is also known as the pelagic zone, ie those parts of the oceans that are not too close to the ocean floor or too near to a coastline. The open ocean has an average depth of 4000 metres. Usually the open ocean is nutrient poor or has relatively low primary productivity but it is home to all the larger marine species (such as tuna, marlin, dolphins, sharks, stingrays and whales).

The upper layers of the open ocean contain the zooplankton and phytoplankton, while the deepest parts containing hydrothermal vents (water temperatures about 350°C) are rich in hydrogen sulfide and are home to chemoautotrophic bacteria and large polychaeter worms.

Freshwater lakes and rivers

These biome types are found on land and can be divided into lentic water habitats (such as lakes and ponds, where the water is standing still) or lotic water bodies (such as rivers and streams, where the water is running).

Lentic water bodies may be nutrient poor (oligotrophic) or nutrient rich (eutrophic) and are dominated by phytoplankton, algae and floating aquatic plants or deeper underwater aquatic plants.

Frogs, turtles, crayfish, insect larvae and various insect species, alligators, crocodiles and fish species of many types can be found in these standing waters.

Lotic water bodies occur on all continents except Antarctica. The flowing action prevents nutrient accumulation that may lead to phytoplankton blooms as seen in lentic water bodies. The flow rates in rivers and streams may be slow or extremely fast, especially in steep countryside with waterfalls.

Various types of well-rooted plants and algae occupy these biomes. Falling leaves from overhanging trees or storm waters provides much of the food source for the aquatic animals in this biome. Although various adapted crustaceans and insects are found in these running waters, fish species such as carp, trout, piranhas and catfish tend to dominate.

Wetlands

Wetland biomes border between lentic and lotic habitats. Wetlands range from marshes and swamps to bogs. They are regularly saturated by surface or ground waters. Some wetlands are flooded when rivers overflow their banks. A few wetland areas develop from estuaries and get flooded by ocean tides. In tropical regions, mangroves develop, while in temperate regions salt marshes develop.

Wetlands are high in nutrient levels and as a consequence low in oxygen concentrations. Temperatures vary with location. Wetlands are the most productive and species-rich freshwater biomes, dominated by lilies, sedges, cattails, cypress, gum trees, diving birds, insects (mosquitoes, dragonflies etc), amphibians, reptiles, otters, alligators and crocodiles.

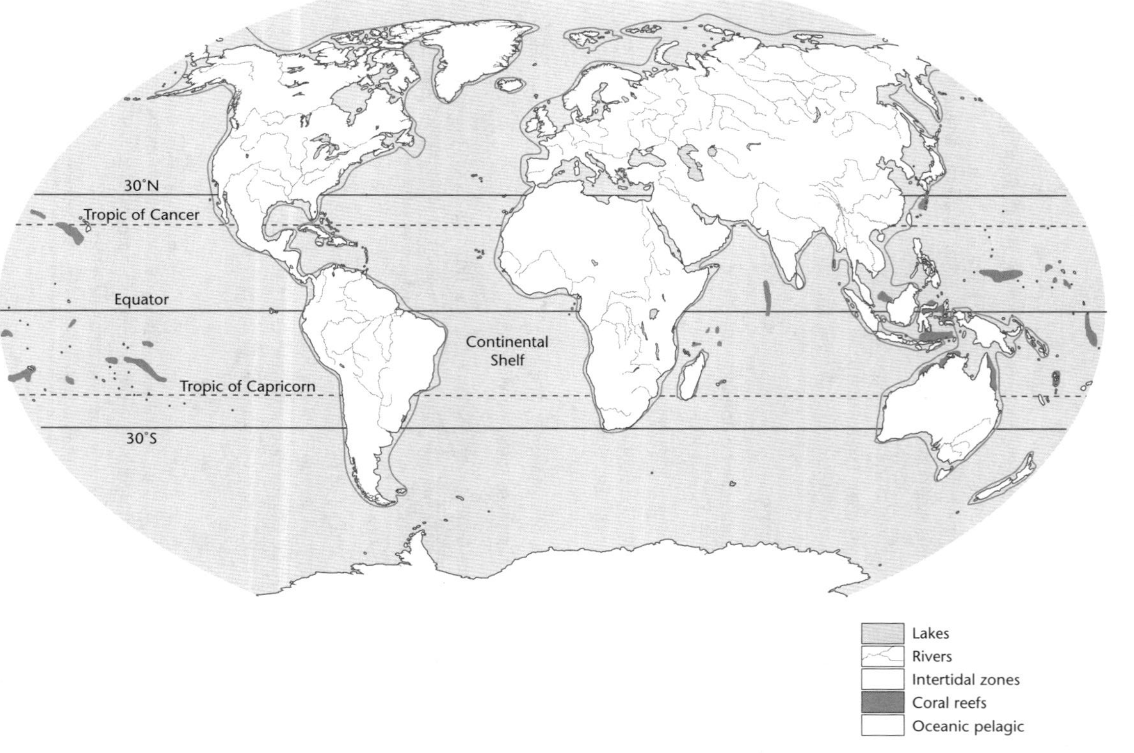

Water covers about three-quarters of our planet. From the largest oceans to small streams and creeks, aquatic biomes are home to an enormous variety of different forms of life. Plants and animals range from large sea creatures to microscopic organisms, from complex plant life to simple algae. But these different organisms, large or small, simple or complex, depend on other biotic and abiotic things in the biome in some way. No organism exists on its own.

Topic 8: Human impacts on the environment

This Topic examines some of the environmental issues that have impacts on the environment and the ways in which more sustainable management practices may be implemented.

In particular, Topic 8 describes:

- Deforestation.
- Overfishing.
- Ozone depletion.
- Global warming.
- Pollution.
- Invasive pests.
- Conservation of endangered species.

Biodiversity refers to the *number of different species* in an ecosystem. It may be used for small or large ecosystems, or for countries (eg the biodiversity of Lake Kutubu; the biodiversity of Papua New Guinea – our ecosystems are among the most diverse in the world).

'At present, the greatest threat to Earth's biodiversity is from the felling of forests and fishing of oceans. Both destroy environments and strip them of key species. Other threats include pollution and invasive introduced species.' (*National Geographic*, 2000.)

Every year, the world's population increases by about 100 million people – 13 times that of the total population of PNG. A report, released by the British Government at the beginning of February 2007, predicts that the world population may reach an 'unsustainable' level of 10.5 billion by 2050. As the population increases, so does:

- Pollution.
- The demand for living space.
- The demand for resources, especially food and (clean) drinking water.

Pollution and the demand for living space and resources have caused environmental issues that affect all humans, especially:

- Deforestation.
- Overfishing.
- Ozone layer depletion.
- Global warming.
- Pollution.

Deforestation

Clear-felling of temperate and tropical forest throughout the world has occurred rapidly during the past century to provide farms, towns and roads. Nearly half of the Earth's original forest cover has been cut, cleared or burnt. Nearly half of what is left has been altered significantly by human activities. Especially hard hit are the tropical rainforests of the Amazon, Central Africa and Asia (Malaysia and Indonesia). It has been estimated that every year about 50 million acres of tropical rainforest are cleared; if this continues, all these forests will be gone in the next 25 years.

Destruction of forests:

- Reduces biodiversity – species become extinct (tropical forests are inhabited by more than half of all living species, so are very rich in biodiversity).
- Reduces genetic biodiversity – species become **extinct** and their genes are lost with them.
- Collapses ecosystems.
- Promotes erosion.
- Promotes the greenhouse effect, as trees absorb CO_2 in photosynthesis.
- Causes large-scale climatic effects – eg rainfall decreases, temperatures rise.
- Increases runoff into rivers as watersheds lose their protection, promoting pollution and/or eutrophication.
- Increases loss of nutrients from an ecosystem.

However, a report to the US National Academy of Sciences in 2006 indicated that many of the world's forests are making a comeback and are more thickly forested than they were 200 years ago. Biomass measurements showed the number of timber-sized trees increased between 1990 and 2005 in 22 of the 50 countries with the most forest. This trend related to the prosperity of the country; those with a GDP per capita over US$6900 have moved from deforestation to reforestation. The USA had the greatest gain in forest over the last 15 years, while Brazil and Indonesia had lost the most forest area – unfortunately these are the two countries with vitally important tropical rainforests.

Overfishing

The depletion of fish stocks in the oceans has been a matter of global concern for a while. A report in 2006 based on global fishing catch data over more than 50 years predicted that by the middle of this century there will be almost no fish left to catch.

Over the past three decades, the global fish export trade has grown four times to 30 million tonnes (US$71 billion). Over 90% of the oceans' big predators have already been fished out of existence. The study's data suggests that 29% of fished species have collapsed and the trend is accelerating. Overfished species listed include Atlantic cod, bluefin tuna, red snapper (not Pacific), swordfish, grouper, sturgeon, Pacific salmon and sharks.

Many current fishing practices are very destructive, with up to half the marine life caught being discarded (often dead) as by-catch; coral reefs can be stripped bare by dragnets.

Stripping ecosystems of their biodiversity makes them prone to collapse – and such collapses are faster, with slower recovery, in areas of lower biodiversity.

If humans do not switch to sustainably harvested seafood (wild and farmed), the ocean's ecosystems will collapse and there will be far less food for the world's ever-increasing population.

It is important that countries introduce a *quota management system* for all commercially and recreationally important marine species, which, together with research and monitoring, aims to sustain fishing for the future. Quotas are not perfect and are subject to human greed. Overfishing exists and our stocks are not yet sustainable (eg in New Zealand a commercial fisher was prosecuted at the beginning of February 2007 for illegally taking over 190 tonnes of orange roughy). Fishing boats from other countries, especially boats from south-east Asia, fish our waters. Destructive practices, such as the use of dragnets, still occur. Surface longlines still snare seabirds such as albatross.

Ongoing monitoring, revised quotas, research into life cycles and catch size, international agreements, legislation and enforcement will all be necessary if the oceans are not to be fished out.

Depletion of the ozone layer

Ozone in the **stratosphere** (the upper atmosphere 10–50 km above the Earth's surface) forms a protective shield against harmful UV (ultraviolet) rays from the Sun. UV radiation causes skin cancer and eye cataracts and can damage crops and **phytoplankton** (**unicellular** algae that form the basis of ocean food chains).

Ozone is produced in the stratosphere when UV light splits oxygen molecules into very reactive oxygen atoms. These oxygen atoms combine with other oxygen molecules forming ozone, O_3.

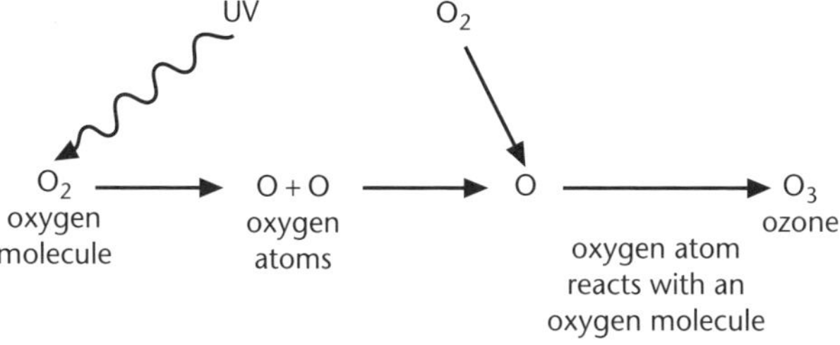

Chlorine gas, Cl_2, in the upper atmosphere breaks down ozone by reversing the reaction above; oxygen, O_2, re-forms from the ozone, O_3.

The chlorine itself comes from the breakdown of **chlorofluorocarbons** (**CFCs**) by UV. CFCs were used in refrigerators, as propellants in aerosol cans, as solvents and as blowing agents to make foam products. CFCs were used primarily on Earth because they are very unreactive – it is only when they reach the stratosphere that UV levels are high enough to break them down into chlorine.

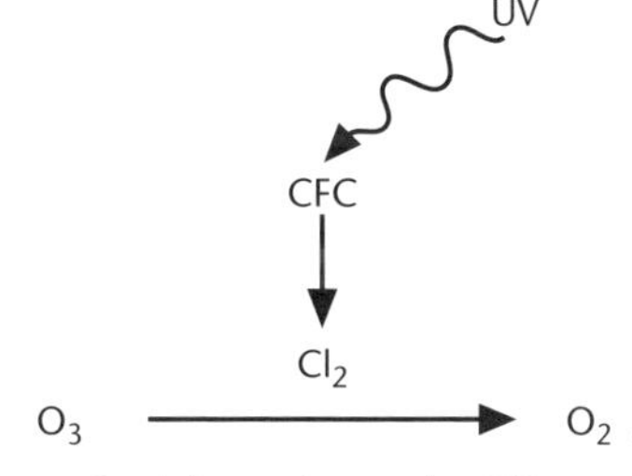

Breakdown of ozone by CFC.

Chlorine itself is not destroyed when ozone is broken down, so that one chlorine molecule can cause thousands of ozone molecules to be destroyed. It is estimated that if all CFC production was stopped today, ozone will continue to be destroyed for the next 100 years from CFCs already present in the stratosphere.

Global warming

'I'm not worried about climate change, I'm absolutely terrified' is what is being echoed around the world today.

The increase of greenhouse gases in the atmosphere is responsible for **global warming**.

The greenhouse effect

Radiation from the Sun that reaches the Earth's surface is re-radiated out from the Earth, mainly as **infrared** (**IR**) or heat radiation.

Effect of greenhouse gases on reflected radiation.

The atmosphere above the Earth is made up mainly of oxygen and nitrogen, but these gases do not affect IR radiation, so it passes through them and is lost into space.

There are, however, other gases that reflect IR radiation back to the Earth, increasing the temperature of the Earth's surface.

These so-called 'greenhouse' gases are all increasing in concentration in the atmosphere through the effect of human activity.

The important greenhouse gases are carbon dioxide, CO_2, methane, CH_4, and CFCs. (CFCs are much lower in concentration than CO_2 and CH_4 in the atmosphere, but they have a much greater effect on trapping heat.) Nitrogen oxides, also at low levels in the atmosphere, are also able to very effectively trap heat.

CO_2 levels have increased from 260 ppm (parts per million) in the 1850s to 350 ppm today. (CO_2 levels in the 1850s can be found by analysing bubbles of air trapped in polar ice of a known age.) CO_2 is increasing because of the large-scale burn-off of forests and the burning of fossil fuels for heating, power generation and transport needs. CO_2 levels would be much higher were it not for plants removing CO_2 through photosynthesis and from CO_2 dissolving in the oceans.

Methane is 20 times more damaging than CO_2 in terms of climate change. Since the 1850s, methane levels in the atmosphere have risen from 760 to 1700 ppb (parts per billion), because:

- During the production of rice, wet soils release methane.
- Drilling for oil and gas releases trapped methane.
- Organic waste in refuse tips decomposing anaerobically produces methane.
- Livestock such as cattle, sheep and deer all belch out methane – New Zealand has 40 million sheep, 10 million cattle and nearly 2 million deer contributing to methane emissions. Livestock also excrete nitrous oxide emissions through faecal matter and urine (nitrous oxide is 330 times more damaging in terms of climate change than CO_2).

The rising levels of the greenhouse gases are predicted to increase the Earth's average surface temperature by between 3°C and 5.5°C over the next 100 years.

The likely effects of global warming are detailed in the US Academy of Sciences report from 2006:

- Melting of the ice caps in Greenland and Antarctica will cause sea levels to rise several metres, perhaps within decades. This will result in tens of millions of displaced people becoming refugees.

- Increased drought, starvation, flood, fire, water shortages.
- New pests, diseases, bugs and weeds threatening biodiversity and changing landscapes.
- 40% of species could die out.

In terms of global warming and Papua New Guinea, there is much evidence from direct recordings and villagers' observations that what happens elsewhere in the world will impact on us. Changes have already been recorded:

- The faster rate of increase in temperature since the mid-1970s in PNG is similar to the global trend.
- In the Highlands, trees such as coconut, betel nut, mango and breadfruit are now bearing fruit where previously the trees grew but did not bear fruit.
- Villagers living at 1900 m altitude near Teptep on the Huon Peninsula believe that the climate has warmed in the decade prior to 1988. By the late 1980s, it had become possible to grow *marita* pandanus and a cultivate palm. Previously, these crops could only be grown at lower altitudes.
- Bird species previously found only at lower altitudes have moved to higher altitude locations.
- An analysis of monthly rainfall patterns at Goroka from 1946 to 2002 found that there had been a shift to longer, but less pronounced, rainy seasons.
- There are indications that rises in sea level are having an impact on shorelines in low-lying areas (eg Gulf of Papua) and the growth of crops on islands.

Other changes that are happening elsewhere may also be experienced in Papua New Guinea, for example:

- Exotic invasive species will increase and thrive – already seen in neighbouring countries, like the spread of ants (such as the Argentine fire ant), also the Australian salt-marsh mosquito. The painted apple moth would likely have not have survived 50 years ago with the colder winters.
- Plants are flowering out of season, birds are nesting earlier.
- Climate change stresses plants; stressed plants are more vulnerable to fungal diseases (eg kauri trees in New Zealand are under attack by a mystery fungus).
- Climate change could change the predator–prey interaction in invertebrates and put stress on ecosystems.
- Increased illness and death in humans from malaria, dengue fever, injuries and heat stress; more pollen and mould spores disrupting respiratory systems.

The increase in the concentration of greenhouse gases in the atmosphere is just one example of pollution.

Pollution

'Natural' pollution results from all biological activities, eg wastes and CO_2 are excreted and heat is lost through respiration.

Nutrient cycles cope with pollution unless the pollutant is:

- 'Unnatural'.
- Too concentrated.

Technology and industry produce pollutants that ecosystems cannot cope with. There is no natural way that nuclear wastes can be disposed of. Oil and other petrochemical products cannot readily be broken down by decomposers.

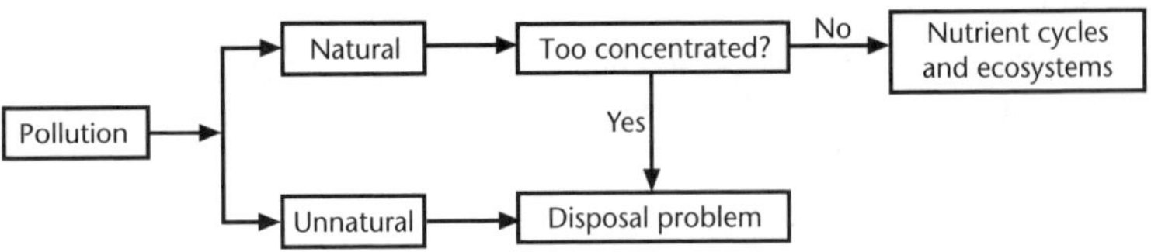

Urbanisation (living in cities) has concentrated pollution problems. There is not enough space or time for 'natural' disposal of wastes to occur. Sewage levels in waterways may become so high that producers and decomposers feeding on the sewage grow rapidly and consume all the oxygen in the water and die. The water then becomes devoid of life. The addition of large amounts of organic matter into waterways is called *eutrophication*.

World's resources

The world's mineral and oil resources are limited and therefore non-renewable (ie they will run out eventually):

- Iron rusts so it is no longer of value.
- Oil is burned to produce electricity and used as fuel for transportation.

To prevent the world's resources literally running out, **renewable** resources must be substituted for non-renewable resources or resources must be **recycled**:

- Hydroelectricity or wind generation can be substituted for power generation from oil or fossil fuels where practicable. Hydroelectricity relies on a renewable resource – water; wind generation similarly relies on a renewable resource.
- Aluminium cans are easily crushed, melted and made back into new cans. Iron cans are difficult to recycle: they rust and contain tin, which is difficult to separate from the iron.

The biotic resources of the world are also endangered. People have over-exploited whales. Whales have been so depleted that some species, such as the blue whale, face extinction.

Unit 12.1 Ecology

Topic 9: Two case studies – New Zealand and Papua New Guinea

In the study of Ecology in Unit 12.1, we have described the different biomes and habitats of the world and the relationships that exist in a natural environment through food chains, food webs and natural cycles. In previous Topics we have looked at the different biomes of the world and factors influencing the availability of plant and animal life and how they are adapted to survive in different conditions. In Topic 9 we now draw together some of the concepts and ideas of previous Topics by looking at two case studies: New Zealand and Papua New Guinea.

The first case study is New Zealand, a developed country. This case study focuses on biodiversity and issues such as invasive species, biosecurity, pest control and why biodiversity matters.

The second case study examines the importance of agriculture in Papua New Guinea, a developing country. The importance of biodiversity as discussed in the first case study is implicit in the second case study, which focuses more broadly on the importance of agriculture in PNG's economy. The second case study also makes the very important point that environmental issues cannot be addressed in isolation but are invariably intertwined with issues of a political and economic nature.

Case study 1: Biodiversity and New Zealand

New Zealand has been isolated from other land masses for about 80 million years. During this time, unique **flora** (plant species) and **fauna** (animal species) have developed through the process of **natural selection**. Any groups of organisms that were not present in New Zealand before separation from **Gondwanaland** are not represented there (eg it has no snakes, no native mammals except bats) unless they have arrived since. Species have been introduced both by accident (eg pukeko, waxeye/tauhou, white-faced heron, spur-winged plover) or by human activities (eg rabbits, possums, cats – this list of organisms introduced by humans is *very* long).

The result of the long period of isolation and the effect of natural selection has produced:

- Large numbers of **endemic** (found nowhere else on Earth) species – 25% of bird species, all 60 reptile species, both species of bats, all four species of native frogs, more than 90% of insect species and marine mollusc species and about 80% of the vascular plants.
- Large numbers of flightless birds (eg weka, kakapo, kiwi).
- Very few native vertebrates apart from birds.
- Only two mammal species – the native bats.
- Species which date back to distant ancestors (eg tuatara are related to reptiles that were common 60 million years ago, there are four species of ancient tree frogs, the podocarp trees of the forest – rimu and totara – are ancient forms with close relatives that lived 190 million years ago).
- Species of trees with distinct juvenile forms (eg lancewood, rimu).
- Many divaricating shrubs.
- Large numbers of lianas.
- A high proportion of dioecious plants (those with male and female flowers on separate plants) – over 12% of native plants are dioecious.
- Few plants have brightly coloured flowers (suggesting a lack of bird and insect pollinators).

The arrival of humans and their activities have caused many native species to become **endangered** or **extinct**. Maori arrived in New Zealand about 1000 years ago and began the destruction of habitat as land was cleared. Maori brought with them the kiore/Polynesian rat and kuri/dog. Native species became extinct – the moa as a result of hunting, the Haast eagle as its main food source (moa) was removed.

Europeans arrived about 250 years ago. Habitat destruction continued as forests, wetlands, mudflats and tussock grasslands were cleared for farms, houses, roads and towns. Many mammal, bird and plant species were introduced. Exploitation of the environment (eg logging, mining, farming, fishing, hydroelectric dams and erosion) occurred.

Species were introduced for farming (eg cows and sheep), sport (eg rabbits, thar and deer), commerce (eg possums), control agents (all mustelids – ferrets, stoats, weasels), pets (eg cats and rosella parakeets), and 'to make New Zealand like England' (eg sparrows, blackbirds, hedgehogs, garden plants and hedgerows). Many introduced species became pests as food was abundant, climate was temperate and there were no natural predators. Amongst the worst pests are:

- Fauna – possums, rabbits, mustelids, feral cats and uncontrolled dogs, rats and mice.
- Flora – gorse, ragwort, thistle, kahili ginger, climbing asparagus, woolly nightshade.

As a result of **predation** (eg mustelids, rats, cats and dogs) and **competition** (eg magpies, mynas, possums and deer) and habitat destruction (eg humans, possums, introduced grasses and ginger), many native species have become:

- **Extinct** (eg moa, huia, laughing owl, New Zealand quail and bush wren).
- **Endangered**. New Zealand has over 500 species listed on the International Red List of Endangered Species, including both species of bat, all four frogs, Hector's and maui dolphins, Hooker's sealion, 25 species of reptiles, 55 species of birds, 20 species of fish, at least 200 invertebrate species and about 300 plant species.

Extensive programs are in place to reduce habitat destruction, control pest species, stop the introduction of new species and to conserve endangered species. These programs may be run by:

- National government (eg Department of Conservation).
- Local government (eg Auckland City Parks Volunteer Program).
- National interest groups (eg Forest and Bird Society, Queen Elizabeth II National Trust).
- Local interest groups (eg Friends of Tiritiri Matangi, Ark in the Park, New Zealand Kiwi Foundation).

These programs include:

- Chemical control of pests (eg dropping of the poison 1080 against possums).
- Biological control of pests (eg cinnabar moth to control ragwort).
- Offshore island sanctuaries (eg Tiritiri Matangi, Kapiti, Codfish, Little Barrier).
- Mainland island sanctuaries (eg Karori, Maungatautari ecological island).
- Sanctuary and zoo-based breeding programs (eg Mt Bruce (takahe), Auckland Zoo (frogs)).
- Corporate sponsorship (eg kiwi – BNZ, hoiho – Mainland, takahe – Mitre 10).
- Habitat preservation and restoration (eg various wetlands and various islands, eg Motutapu and Motuhei in the Hauraki Gulf).

- Marine reserves (eg Long Bay, Goat Island, Poor Knights).
- National and Forest Parks – New Zealand has 14 National Parks (eg Tongariro) and 19 Forest Parks (eg Kaimanawa, Coromandel).
- Volunteer tree planting (eg Hamelin Hill in Auckland, Motuhei Island) and traplines (eg Waikowhai Park in Auckland – possums and rats; Waitakere Ranges – possums and stoats).
- Education.

Many of these programs have been successful (eg rodents have been eradicated on over 100 islands and counts of birds on these islands show increased breeding success, with populations growing). However, a recent report reviewing the first five years of the Government's biodiversity strategy 2000–2005 shows New Zealand is yet to turn the tide on biodiversity loss, and species such as the kiwi, blue duck and yellowhead could become extinct on mainland New Zealand in the next 20 years.

Invasive species and biosecurity

All over the world, animals and plants that evolved somewhere else are turning up where they are not wanted, having been transported by humans deliberately or by accident.

The rapid increase in international travel and trade (especially using containers) has led to the increased risk of unwanted plants and animals spreading from country to country. New Zealand, with its fragile biodiversity and dependence on agriculture, is especially at risk from invading species. Increased biosecurity measures at its borders are in place to reduce risks. Of the 100 least wanted invaders, at least 20 are already in New Zealand (eg Argentine ant, Kahili ginger, gorse, ship rat, rabbit and possum). Of particular concern are those species that carry diseases for humans, crops and livestock (eg Australian salt-marsh mosquito spreads the Ross River virus). The potential damage to New Zealand's ecosystems, economy and health is substantial – hence the stringent checks on baggage of passengers entering New Zealand at its international airports; traps are set in at-risk areas (eg port cities like Auckland) to monitor for the presence of pests such as painted apple moth and fruit fly.

Pest control

Pest control is usually achieved through biological, chemical or mechanical control, or combinations thereof. *Mechanical* control involves shooting or trapping (eg Timms trap for possums).

Examples

1100 predator traps set by the Department of Conservation in the Tasman River Valley over the last two years have trapped 2574 predators, including:

- 230 feral cats.
- 211 ferrets.
- 576 stoats.
- 1293 hedgehogs.
- 207 possums.
- 51 weasels.
- 6 Norway rats.

Gorse is commonly controlled mechanically (by ploughing), or by ploughing in conjunction with biological (gorse mite) and chemical (herbicide) controls.

One less Rattus rattus!

Biological control of pests

Biological control uses *another organism* to control the pest population. Biological control agents rarely eliminate a pest but keep its numbers at manageable levels.

The agents may be:	**Examples**
• Herbivores.	Ragwort in New Zealand is controlled by sheep and goats as well as several imported insect species, such as the cinnabar moth and flea beetle.
• Predators.	Mustelids were introduced to New Zealand to control rabbits (not successful).
• Parasites or parasitoids.	The wasp *Apanteles* (a parasitoid of white butterfly larvae) was introduced to New Zealand to control white butterflies.

Successful biological control is cheap, effective, and does not need repeating. However, large amounts of research and testing need to occur before a control agent is released. Scientists need to be sure that the agent does not 'prey switch' (ie prefer eating other organisms in the food chain rather than the pest species).

Example

In New Zealand, mustelids were introduced to control rabbits; however, these predators switched to preying upon native birds, becoming major pest species.

Chemical control of pests

Chemicals that kill pests are called **pesticides** (eg fly spray, snail bait and 1080 poison). Chemicals used to kill weeds/unwanted plants are called **herbicides** (eg 'Roundup' and 'Answer').

Pesticides and herbicides are very effective and rapidly kill large numbers of the pest species. However:

- Pesticide spraying is costly and time consuming, needing regular applications.
- Pesticides do not usually discriminate between pest species and other related species, eg **insecticides** (pesticides that act against insects) often kill all insects that the spray reaches, not just the pest species, so useful/wanted insects (eg bees, parasitic wasps and ladybirds) are killed (which in turn can severely damage food chains/webs).
- A major side effect of pesticide sprays is the build-up of *resistance* in the pest species, ie populations that are not as affected as much or are no longer killed by the spray. This happens when individuals in the population have a (natural) genetic resistance to the spray. These individuals survive and their offspring inherit the resistant genes. Over time, whole populations develop that are resistant to the spray. Insects that reproduce rapidly and in large numbers have the ability to develop resistant populations quickly (a comparable situation to this is the development of antibiotic resistance in bacterial populations).
- Pesticides may not break down quickly, so remain in the environment for a long time (modern pesticides such as 1080 are designed to break down rapidly).
- Pesticides may *accumulate* along food chains/webs and become concentrated in top carnivores with harmful effects.

Example

The first major pesticide developed was **DDT** (important in killing malaria-carrying mosquitoes and crop pests); however, DDT becomes concentrated along food chains and 2° and 3° **consumers** often showed symptoms of DDT poisoning (eg thin egg shells in birds causing decreased natality) a few years after spraying (see also page 87).

The concentration of DDT is measured in parts per million – 1 ppm of DDT means 1 DDT molecule to every 1 million other molecules in the organism's body:

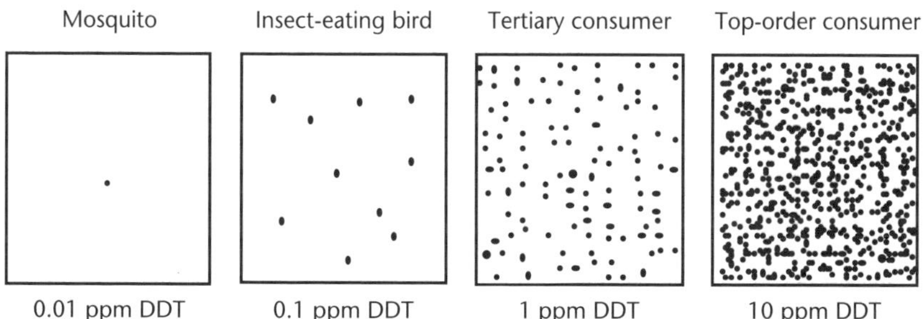

The effects of DDT in ecosystems will be felt for a long time even though its use has been banned in many countries (it is still used because it is often the cheapest way to prevent disease). DDT is not very *biodegradable* – it takes about 50 years to break down into harmless compounds.

Why biodiversity matters

Biodiversity matters because we humans depend on many of the organisms that inhabit the world with us for our necessities of life:

- Food.
- Oxygen.
- Clothing.
- Shelter.
- Soil fertility.
- Nutrient cycling.
- Medicines.

We are as much a part of the ecosystems as all other life forms; we are as interrelated to them as they are to each other. Collapse of ecosystems on Earth will eventually lead to *our* extinction.

Unit 12.1 Activity 9A: Biodiversity

1. Describe the main effects of ozone depletion.
2. Describe the main causes of loss of biodiversity in a country.
3. Explain why ecosystems with low biodiversity are more prone to collapse than those with high biodiversity.
4. Discuss why invasive pest species are so damaging to a country's biodiversity.
5. Compare and contrast biological and chemical control of pests.
6. Describe the likely main effects of global warming.
7. Explain how global warming is occurring.
8. Explain why overfishing is so damaging.
9. Discuss why deforestation is so damaging to biodiversity.

Case study 2: Agriculture in Papua New Guinea

Note: this section is taken from Bourke, R.M. and Harwood T. (eds) (2009). *Food and Agriculture in Papua New Guinea*. ANU E Press, The Australian National University, Canberra. The text is largely based on the Introduction, written by R. Michael Bourke and Bryant Allen, and Section 1.8 Climate Change, written by R. Michael Bourke. We are grateful to the authors and publisher for permission to reproduce their text in this book. The original book is available online at: http://epress.anu.edu.au/food_agriculture_citation.html

Agriculture is the most important activity carried out by the vast majority of Papua New Guineans. For most of us, agriculture fills our lives, physically, culturally, economically, socially and nutritionally. Yet agriculture is the most undervalued and misunderstood part of PNG life.

The reasons for this are partly because mineral and oil exports make PNG comparatively wealthy for a developing country; partly because agriculture is practised in the countryside, away from towns, and is therefore largely 'invisible' to urban people and international visitors; and partly because agriculture is viewed as not being 'modern'.

However, agriculture feeds most Papua New Guineans, houses them, provides a significant amount of food to townspeople and earns a significant amount of foreign exchange. By doing these things, agriculture provides PNG with a social, economic and political stability which has meant that, despite sometimes poor economic management, the country has been able to weather a number of economic and political crises. Furthermore, agriculture will continue in this role, long after all the minerals and oil have been dug up and exported.

Papua New Guinea's ecosystems are some of the most diverse in the world. Our ecosystems range from coral reefs to high montane grassland. Our forests are ranked third in the world. Healthy ecosystems must be maintained to protect agriculture in PNG. It is vitally important that we maintain our healthy ecosystems for the security of our agriculture, and the security of our people, in the long term.

Agriculture in Papua New Guinea

Subsistence food production is the most important part of PNG agriculture. It provides most of the food consumed in the country – an estimated 83% of food energy and 76% of protein. Its importance only becomes apparent when a rare partial failure of food production occurs, as happened in the very severe drought and frosts of 1997.

People's sustenance also comes from hundreds of food crops, farming of domestic animals, fish and locally produced sugar. This is supplemented by imported rice, wheat-based foods, some meat and some fish. Cash income is provided by sales of Arabica coffee, fresh food, cocoa, betel nut, copra, palm oil, firewood, tobacco, fish and many minor products, including vanilla, rubber, balsa and tea.

There is a long history of agriculture in the country that is now called Papua New Guinea. The first settlers arrived about 50 000 years ago. Agriculture began in PNG about 10 000 years ago, at about the same time as it appeared in the Middle East and in China. Some food plants were domesticated in New Guinea and others were introduced from Asia and other Pacific islands. In the 1700s, new crops began to reach PNG from the Americas and one, sweet potato, has become PNG's most important crop. Many new food and potential cash crops were introduced during the colonial period, from the early 1870s onwards.

A feature of the long history of agriculture in PNG is the careful adoption of some new crops and the rejection of others. It is also characterised by innovations in the management of land including draining, fencing, tree planting, composting, soil erosion control and mounding. Many of these changes have been associated with the intensification of land use, in which more food is produced from the same area of land.

Influences of global trends in 2007–2008

Evidence is building that fundamental worldwide changes are occurring, including global climate change, which will pose multiple challenges for agriculture everywhere. Land once used to grow food is growing biofuels; the demand for many products has increased as the economies of China and other rapidly developing nations grow; agricultural land is being lost to urbanisation and industrialisation; fertilisers and herbicides are becoming more expensive; surface water is being reduced in many locations; soil nutrients are being lost; pest and disease problems are increasing; a neglect of agricultural research is reducing technological solutions; marine harvests are declining; and prolonged droughts exist in a number of countries, possibly associated with global climate change. As a result, rice, wheat, corn (maize), vegetable oil, dairy products, meat and fish are all subject to price increases.

These global changes pose challenges for PNG agriculture producers and consumers, including much higher prices for imported foods. Sometimes there are benefits because prices of agricultural commodities that are linked to crude oil can rise sharply (eg palm oil, copra oil and natural rubber). The use of fertilisers and herbicides in PNG agriculture continues to be negligible, with most villagers using little or no inputs into production other than their land and their labour. The increase in exports of marine products, mainly tuna, round logs and palm oil from PNG is the outcome of higher global demand. The changing global economy is reflected in the greater proportion of logs and balsa from PNG exported to China and other emerging economies in Asia.

It is difficult to predict how the far-reaching consequences for the global economy may affect PNG. Rural producers will be affected by these changes, but will be largely sheltered by subsistence food production and their limited dependence on imports.

Strengths of PNG agriculture

PNG agriculture has many strengths. Firstly, by providing most of the food consumed within PNG, it imparts great stability to the nation. Agriculture is also the main source of cash income for a great majority of people. People use money to better their lives, including supplementing protein-poor diets, building better houses, educating their children and accessing medical services. Most rural people are employed as producers, processors or intermediate traders in agriculture. As well, there are vigorous informal businesses where people grow, transport, trade and retail fresh food, betel nut, firewood, fish and other commodities.

Agriculture also provides 17% of PNG's exports by value. Agriculture is a major contributor to development within PNG. A vibrant cash crop component exists in both the formal and informal sectors, in almost all of the more-developed locations. In contrast, the least-developed parts of PNG, where most people are poor, are characterised by subsistence food production only. A number of environmental factors severely constrain agriculture in PNG, including excessively high rainfall, steep slopes, inundated land and extensive cloud cover. Villagers have a detailed understanding of the relationship between crops and the environment.

More important than a good physical environment, however, is the energy, drive and adaptability of the people in more than one million rural households. They respond to new opportunities, including higher prices and better returns to their labour. They form the core of what is a largely unrecognised significant private sector group. Other parts of the private sector consist of people involved in trading and processing commodities for the domestic and export markets. Thousands of people are involved in trading and transporting betel nut alone, as it is moved from lowland locations to the highlands, where it does not grow, and to urban centres and mines. A few small companies participate in the formal economy, trading and processing produce. A number of large firms grow, buy, process and export oil palm, sugar cane, coffee, cocoa, copra, rubber and tea.

Villagers are often portrayed as unresponsive to price or market messages. The devaluation of the PNG currency in the late 1990s is a clear demonstration that this is false. Large increases in the prices for many goods as measured by the consumer price index occurred in the late 1990s including imported rice which, in turn, led to an increase in the price of marketed sweet potato. This was also a time of low returns for coffee in the highlands. Many highlanders responded to the increased price for sweet potato and the reduced price of coffee by growing more sweet potato for sale in urban markets. The increase in production resulted in lower prices for consumers, so that sweet potato became more competitive compared with imported rice, leading to increased consumption of sweet potato. A similar response occurred in 2008 following rapid increases in the price of rice and other imported foods.

Another strength of PNG agriculture is the customary land tenure system. Individuals and companies who wish to access large areas of land for agricultural development can be frustrated by customary tenure, but the system is sufficiently flexible to accommodate increasing population and internal **migration**.

It has been argued that economic development will not occur unless all land is privatised and registered to individuals, but individual titles to land on settlement schemes has often resulted in poor economic outcomes. Nor has the holding of a title helped the plantation sector, where production of all export cash crops except oil palm has declined over the past 30 years. PNG does not have the administrative capacity to survey customary land, identify customary owners, settle the myriad disputes that would result, and issue titles, let alone record the thousands of changes of ownership that will occur. Only where uncertainties in tenure are causing real problems, such as in the areas surrounding larger towns, should attempts be made to interfere with customary tenure.

Marketing systems for oil palm, coffee, cocoa and copra are efficient and return to growers a reasonable share of the international market price. Marketing systems are less efficient for vanilla and most domestically marketed produce, particularly fresh food.

Challenges and constraints

Maintaining biodiversity and monitoring the impact of climate change to ensure the long-term health of PNG's ecosystems is just one of the challenges and constraints facing our country. The environmental challenge is interlinked with challenges of a political and economic nature which are dependent on good governance. Below is a list of those challenges and constraints which presently limit the potential of agriculture in PNG. If the sector is to achieve its full potential, these will need to be addressed by the PNG Government and private sector, supported by international donors.

The major constraints are economic and political as well as environmental:

- Pressure on land associated with rapid population growth, particularly in parts of the highlands and on small islands. This is affecting soil fertility and the ability to grow food.
- Poorly maintained transport infrastructure, particularly roads and bridges.
- Limited new technology being generated from research. Some commodities, particularly palm oil and sugar cane, are supported by excellent research, but most commodities are not adequately supported, including subsistence and marketed fresh food. Research is frequently focused on a crop, rather than the agricultural system of which it is a part.
- Research capacity directed at commodities that have little or no chance of expanding or being adopted, including rice, wheat and other grain crops.
- Very limited effective outreach and agriculture extension capacity. Despite the availability of a moderately large body of information which would advantage producers, little of this is effectively communicated to them.
- Climate change as a significant long-term challenge, particularly rising sea levels, greater rainfall, increasing temperatures and possibly greater frequency of extreme climatic events.
- Mismanagement of the national economy, poor performance of institutions involved in the governance and administration of agriculture, and poor policy making. This includes policies that result in distortions of the terms of trade and large movements in the currency exchange rate.
- Insufficient involvement of women in some aspects of agriculture, particularly trading.
- An HIV/AIDS epidemic, which will severely affect agricultural labour supply in some places.
- Inadequate security for people and property.
- Insufficient access to credit for intermediate traders. Inadequate attention to quality control for some commodities, particularly coffee, fresh food and vanilla.
- Poor communication among growers, middlemen, processors and retailers in the production and marketing chain.
- Insufficient attention given to marketing and promotion of PNG produce.
- Severe poverty for about a sixth of the population, which limits their ability to participate in the formal economy.

Implications of climate change for agriculture in PNG

Climatic change is already influencing agriculture in PNG but, in most cases, it remains difficult to predict what outcomes will be. This is because we do not know enough about probable changes in patterns of rainfall and rainfall extremes. Furthermore, agricultural responses to climate change will be complex, and will depend on how people respond, as well as the responses of plants. For example, an increase in temperature might make coffee rust more severe in the highlands, but this impact could be reduced if the rainfall and humidity was reduced. Conversely, the impact could be greater if the rainfall increased as well as the temperature. This impact would be reduced if villagers adopted improved rust control techniques or rust resistant coffee varieties.

Temperature and agriculture

Increases in minimum and maximum temperature are already having a small influence on agricultural production and will have a greater influence in the future. Some tree crops are bearing at higher altitude in the highlands. Given the direct relationship between altitude and temperature in PNG, it is likely that many crops will be able to be grown at marginally higher altitudes in the future and so the areas where they can be grown will expand. However, the lower altitudinal limit of some crops, such as Irish potato, Arabica coffee and *karuka* (pandanus), will increase because of increasing temperatures.

Sweet potato is the most important food crop in PNG and provides almost two-thirds by weight of the staple food crops. It is an important food in many lowland locations as well as throughout the highlands. Tuber formation in sweet potato is significantly reduced at temperatures above 34°C. Maximum temperatures in the lowlands are now around 32°C, so an increase of 2.0–4.5°C within 100 years could reduce sweet potato production in lowland locations, perhaps within one or two generations.

Daily temperature ranges also have an important influence on productivity. In PNG, the daily temperature range is greater in the highlands than in the lowlands, and is one reason why crop yields are higher in the highlands. For most of the planet, daily range decreased over the period 1950–2004. If this is also occurring in PNG, it will tend to reduce crop yields to an unknown extent.

It is likely that overall temperature increases will marginally reduce productivity in the lowlands and in the main highland valleys, but will marginally increase productivity at locations above 2000 m. Increases in temperature may also change the incidence of some diseases, especially those influenced by rainfall and humidity. Taro blight (caused by a fungus) is less severe in PNG at a few hundred metres above sea level than at sea level, and is rarely found above an altitude of 1300 m. The fungus is sensitive to temperature and a small rise in temperature could mean that the fungus would reduce taro yield at higher altitudes than occurs now.

Similarly, coffee rust is present in the main highland valleys at 1600–1800 m, but does little damage there. It is a problem for coffee production at lower altitudes, below about 1200 m. So again, a rise in temperature is likely to increase the altitude at which coffee rust has a severe impact on coffee production. Changes associated with El Niño Southern Oscillation (ENSO) events are not known. If ENSO events occur more often as models predict, more frosts will result at high-altitude locations (above 2200 m). This will have an adverse impact on agricultural production.

Rainfall and agriculture

Predictions are for higher rainfall in the South Pacific, including PNG. Rainfall patterns in PNG are complex and it is likely that any changes in rainfall patterns will also be complex and so difficult to predict. With a few exceptions, most places in PNG receive high annual rainfall. In the lower-rainfall areas, an increase in rainfall and reduced seasonality of distribution would be beneficial for agriculture.

However, for most of PNG, an increase in total rainfall and a less seasonal distribution would have a negative impact on agriculture. The most vulnerable locations are those where rainfall is already over 3500 mm per year. If ENSO events become more common, more droughts could result as well as more episodes of very high rainfall. However, the likely changes in the pattern of ENSO events are not well understood.

Another outcome of increased rainfall would be greater cloudiness and less bright sunshine. If that was to occur, it would probably reduce crop productivity, especially where cloud cover is already very high.

Sea level rise and agriculture

There are indications that rising sea levels are already having a minor negative impact on very small islands and other coastal locations because of coastal erosion. Rising sea water is also possibly contaminating the freshwater on atolls where swamp taro is grown. Rising sea levels do not necessarily cause very small islands to be covered by sea water: coral reefs respond to rising sea level by growing upwards, and the extra coral sand may cause the islands to rise. Nevertheless, it is quite possible that a rapid rise in sea levels will cause major problems for villagers living on very small islands and in other low-lying coastal locations, such as in the Gulf of Papua.

In the Pacific, attention has focused on small island states such as Tuvalu. However, there are about 100 000 people in PNG living on what have been defined as 'Small Islands in Peril'. These are about 140 islands smaller than 100 km^2 in size and with population densities greater than 100 persons/km^2. It is these people who are likely to suffer the most severe consequences of rising sea levels.

Other implications

Other implications from climate change in PNG include human health, especially malaria incidence in the highlands, water supply, fish availability, and health of coral reefs and other marine ecosystems.

Climate change is going to alter the global economy in coming decades and some of these changes are likely to impact, positively and negatively, on PNG. Such changes could include a reduction in availability of fossil fuels; a greater demand for biofuels from oil crops such as oil palm and coconut; increased demand for hydro-electricity (which is potentially abundant in PNG) for industrial uses; and carbon trading whereby companies pay for trees to be planted on a large scale to absorb carbon dioxide. Some of these changes present economic opportunities that could be beneficial for PNG.

It is known that a higher level of carbon dioxide in the atmosphere results in increased photosynthesis and decreased evaporation and transpiration. This is a positive for crop production. Given that atmospheric carbon dioxide is at record levels and is forecast to keep increasing, plant productivity is likely to increase. However, possible benefits are likely to be offset by decreases in productivity caused by higher temperatures and altered rainfall patterns.

Unit 12.2 Population

Topic 1: Understanding population growth

The study of population is an important aspect of the study of biology. By having an understanding of the total population of organisms, biologists can help us better manage our environment.

In Unit 12.2 students investigate different sampling methods and decide on the best one for any organism of study. The effects of the birth, death, immigration and emigration rates on population and factors that limit population growth are looked at. Human population is examined, as well as the relationship between world population growth and the decrease in the Earth's resources. Students become aware of problems associated with population increase and become better able to make informed decisions later in life.

Topic 1 deals with:
- Growth, survivorship, limiting factors.

Population

In the study of biology, a **population** is a group of individuals of the same species living in the same area at the same time. Populations may be large or small. Population ecology is the study of how populations grow and the factors that promote and limit growth.

Example

This swarm of bees is a population of individual bees of the same species, living in the same area at the same time. Factors that promote and limit the growth of the swarm are of interest to biologists.

A **species** is a group of organisms that can interbreed to produce fertile offspring. Members belong to the same gene pool and are reproductively isolated from other species.

A population may have up to three different age groups in it:

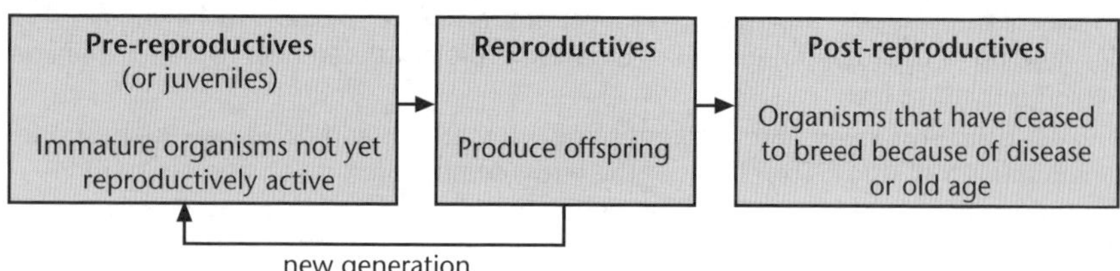

Demography

Demography is the study of birth rates, death rates, age distribution and sizes of population.

Population density is the number of organisms in a given unit area. Population growth affects population density. Knowledge of both can help to make decisions about management of species.

Natality and mortality

Natality means **birth rate**. Human natality is expressed as 'live births per 1000 persons per year'. **Mortality** means **death rate**.

For a population to grow, natality must be greater than mortality.

Age pyramids

Age pyramids display information about the number of organisms alive in particular age groups of a population. Age pyramids can be used to decide whether a population is increasing or decreasing. The width of a pyramid indicates the number of organisms in that age group.

Example

Human age pyramids

For the human population, there are three distinct age pyramids:

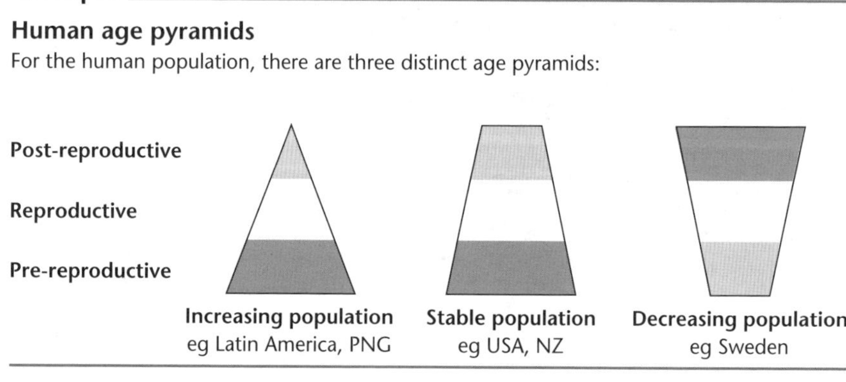

Survivorship

Survivorship is 'the chance of remaining alive'.

If the number of survivors in a population at various age intervals is plotted, a **survivorship curve** results.

Survivorship curves

The three major types of survivorship curve are shown below on the same set of axes.

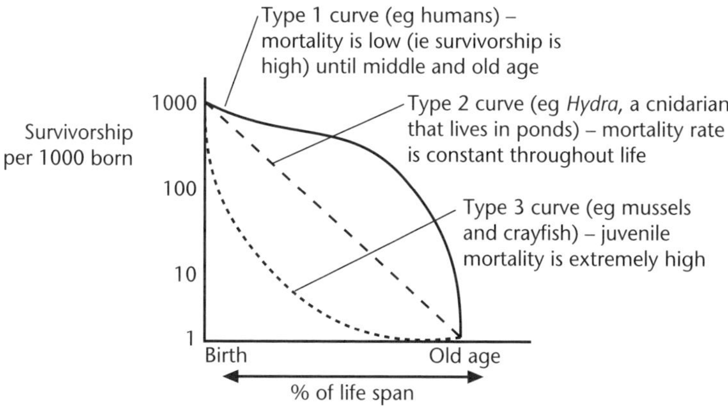

Population growth

When a few individuals are introduced into a new environment, the growth of the population typically gives a **sigmoid** (or **S-shaped**) curve when plotted on a graph.

The curve can be divided up into five distinct regions.

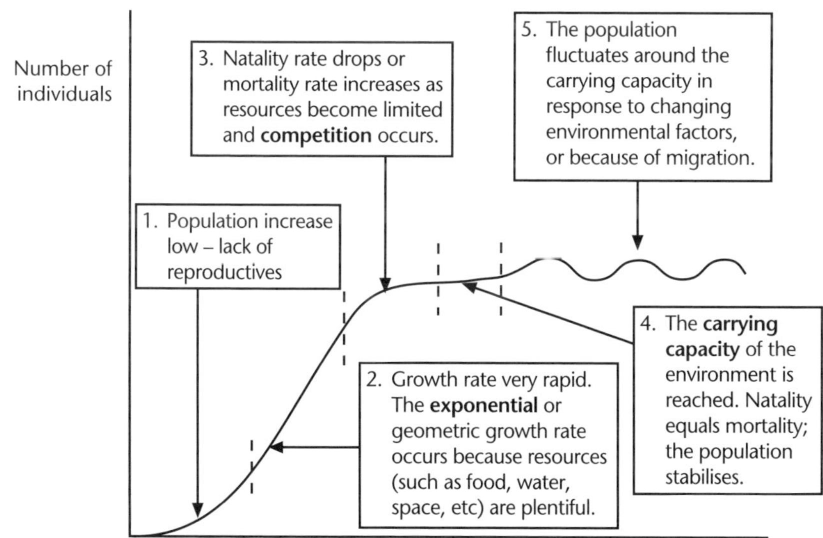

Carrying capacity is the number of individuals in a population that can be supported by the resources available in that environment.

Migration is the movement of individuals into (**immigration**) or out of (**emigration**) a population.

The human population is presently in the *exponential phase* of its growth. The population increase must stop as the carrying capacity of the environment (our world!) is reached.

Limiting factors

Limiting factors are environmental factors (both biotic and abiotic) that act to regulate the growth and limit the size of populations.

- **Density-independent** factors do not depend on the density or size of the population and are typically catastrophic environmental events (eg floods, fires, landslips, severe drought and severe cold). These often one-off or infrequent events may severely reduce the numbers of a population. Provided needed resources remain available, the (now) much smaller population will grow to reach the carrying capacity again.
- **Density-dependent** factors come into effect when the density of the population increases (eg availability of food, water, nutrients, oxygen, mates, nest sites, space and light (plants)). As the density of the population increases, any or all of these factors could become in short supply, causing **intraspecific competition**. Mortality increases and/or natality decreases and/or emigration may occur, so the growth of the population slows and then stabilises at the carrying capacity of the environment.

Disease may be an important density-dependent factor in reducing population size, as spreading of disease increases as populations become denser with individuals living closer together.

Toxic wastes produced by micro-organisms (such as bacteria, yeast and algae) may build up as population density increases, acting to limit the growth/size of the population.

Predation is an important density-dependent factor in controlling the size of populations of prey species (typically herbivores).

Predators usually *benefit* a prey population (but not their unfortunate victims) by:

- Removing the young, old or unhealthy individuals (which can have a beneficial effect on the prey population's gene pool).
- Keeping its size low and so stopping it over-exploiting its environment (overgrazing can lead to large-scale starvation and a **population crash**). An over-exploited environment will usually take a long time to recover and probably be unable support a large population again, as its carrying capacity has been reduced. Removal of a predator by human activities often produces such population crashes.

The populations of predators and their prey typically show *cyclic fluctuations* in numbers, with the predator population growth following *after* the prey population growth.

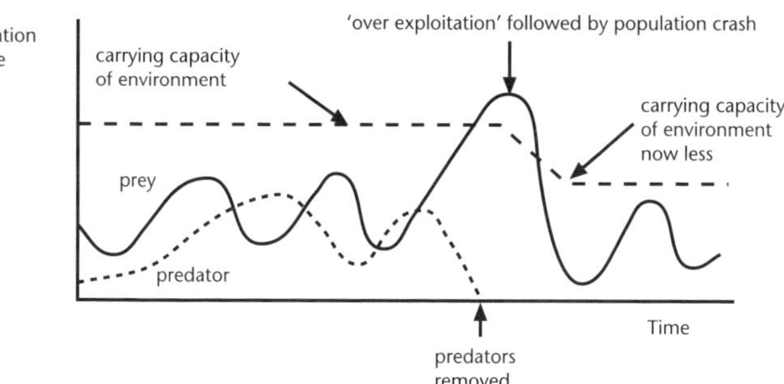

Predator–prey interactions

Physiological stress

Physiological stress occurs as individuals compete for limited resources and space becomes limited. As a result of physiological stress, natality falls and mortality rises; the population size decreases, so stress levels fall and the population stabilises at a lower level.

Most populations will be regulated by a *combination of biotic/abiotic and density-dependent/ independent factors* rather than any one factor.

Territory and home range

Some animals (eg colonial breeding birds such as gannets and black-backed gulls) and groups of animals (eg pukeko and chimpanzees) use territories as a means of limiting access to resources and so regulating populations. A **territory** is a *defended area* with *defined* boundaries set up by an individual, pair or group. The size of the territory depends on the area that the animal(s) can defend successfully. The territory is a place that animals can build a nest site and breed relatively safely. It may provide resources such as food and water, or these may come from a **home range** (the area surrounding the territory where the animals may forage for food, water, nest material, etc). A home range is *shared* by others of the population and so is *not* a defended area. Home ranges may *overlap*.

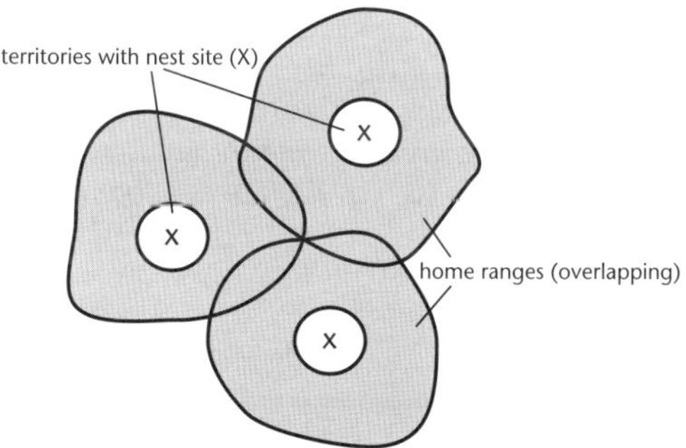

Territories with nest site, surrounded by home range

Animals that are not strong or dominant enough to establish or maintain a territory (usually juvenile, weak or old individuals) do not get to breed. Territories therefore act to reduce competition for resources, spread the population throughout its habitat and limit breeding – all of which combine to regulate the size of the population (and enhance the gene pool, as only the strongest, healthiest and **fittest** get to breed).

Reproductive strategies

Semelparity

Some organisms (such as insects, invertebrates, salmon and bamboo grass) produce all their offspring in a single reproductive time. This pattern is called **semelparity**. Semelparous organisms reproduce once only and then die. However, they may live for many years before reproducing, or they may be annual plants that develop new plants from old seeds.

Iteroparity

Other organisms reproduce in repeated breeding seasons in their lifetime. This pattern is called **iteroparity** and is common in most vertebrates and trees (**perennial** plants). Iteroparity can be seasonal (temperate birds and trees) with distinct breeding seasons, or breeding may happen repeatedly at many times of year (continuous), as with tropical organisms, parasites and mammals.

The uncertainties of an environment are a contributing factor to whether a species reproduces in a semelparous or an interoparous mode.

Comparison of reproductive strategies during lifetimes of representative organisms. The Pacific salmon is a semelparous organism. It lives for many years in the ocean before it swims the freshwater stream where it was born, where it spawns and dies. Squid and octopus are other semelparous organisms that do this. Many insects, including some species of butterflies, cicadas and mayflies, are also semelparous. Iteroparous organisms include pigs, cuscus and humans. Trees such as birch trees or oak trees are iteroparous.

Age classes

The reproductive strategy of a species has a strong effect on age **classes** of a population. Semelparous organisms produce groups of young at the same age (cohorts), which grow at similar rates. Iteroparous organisms have many young of different ages because parents reproduce frequently.

The age classes can be used as specific categories in describing population growth. Age classes may be in stages (eggs, larvae, pupae in insects), years (in mammals) or size (in plants).

It is expected that a population that is increasing in size should have a greater number of young, while a population that is decreasing would have fewer young.

Imbalance in age class is likely to have a serious impact on a population's future. For example, in an over-exploited fish population, there will be no adult fish to produce young and the population may collapse.

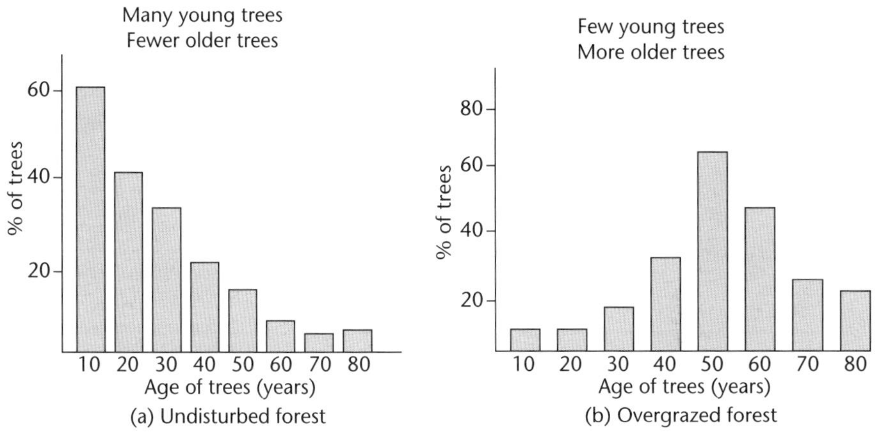

Theoretical distribution of two forest populations.

Unit 12.2 Activity 1A: Population

1. Distinguish between the following pairs of terms:
   a. Territory and home range.
   b. Density-dependent factors and density-independent factors.
   c. Natality and mortality.
   d. Emigration and immigration.

2. The population statistics for a species are as follows:
   - Natality – 67% per annum.
   - Mortality – 31% per annum.
   - Immigration – 20% per annum.
   - Emigration – 27% per annum.

   Calculate the annual increase/decrease per annum for a population of 1000.

3. The following photograph shows part of a population of the common mud snail *Amphibola crenata* on an estuarine mud flat.

a. Identify three characteristics that could be used in describing this population.
b. The snail population is large. Describe any factors that may act to limit the population.
c. Name an important abiotic factor that the snails must be tolerant to.
d. Explain any two adaptations that the snails may have to living in this environment.

4. Saddlebacks are an endangered New Zealand bird, which have been released on offshore islands such as Little Barrier. The data in the table below shows the numbers of saddleback on Little Barrier for the first 12 years after their release on the island in 1980.

Year	1980	1982	1984	1986	1988	1990	1992
Number of saddleback	12	20	38	68	83	91	88

a. Describe why the population increased so rapidly between 1984 and 1986.
b. Explain why the population growth slowed after 1988.
c. Identify the approximate carrying capacity of this environment.
d. Saddlebacks are territorial birds. Discuss how territories can act to regulate population sizes.

5. *Sitophilus oryzae* is a small weevil often found living in bags of rice. Female weevils lay a single egg inside a grain of rice. The larvae feed, grow and pupate entirely within a rice grain. The adult weevils gnaw their way out of the grain. Some students carried out an investigation into the growth of a population of *Sitophilus oryzae*. They placed 6 g of rice into a container and added five male weevils and five female weevils. The adult weevils were counted over the next 200 days.

Time (days)	0	20	40	60	80	100	120	140	160	180	200
Number of weevils	10	15	79	123	119	113	123	139	145	105	97

a. Draw a graph to show the growth of the weevil population.
b. Identify the time when population growth was most rapid.
c. Explain the most likely reason why the population did not keep on growing.
d. Explain what would be expected to happen to the population if records had been kept past 200 days.
e. Sketch a line on the graph then explain what would be expected to happen if 12 g of rice had been added to the container at the start instead of 6 g.

6. The kiore, or Polynesian rat, feeds on grass seeds. The graph below shows the estimated numbers of kiore and grass plants on Tiritiri Matangi Island over a 3-year period in the 1970s.

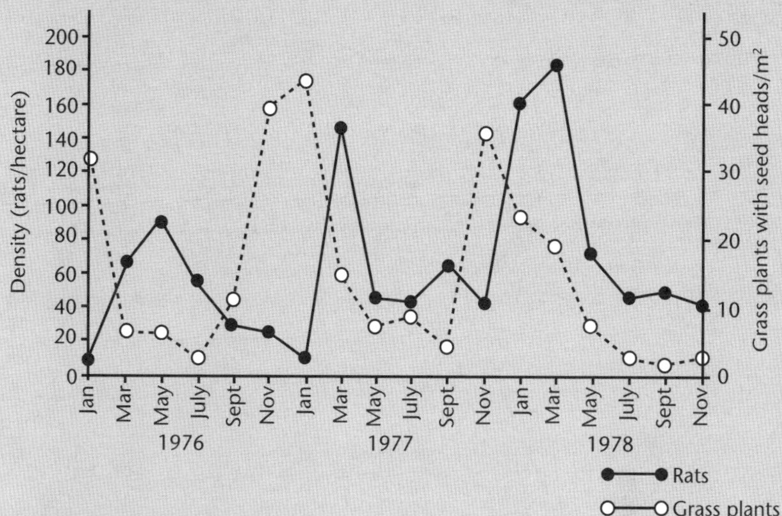

Discuss the population fluctuations shown by kiore and grass plants over the 3-year period.

7. A count of deer on an island forest reserve (of about 500 km²) showed a population of 2000 animals in 1990. The island had vegetation suitable for deer, but food supply was limited and it was feared that overgrazing might result in mass starvation and a population crash. It was decided to bring in wolves, which were natural predators of the deer, in an attempt to limit the deer population. In 1990, 10 wolves were released on the island and the two populations were monitored annually for the next 10 years. The data recorded is given in the table below:

Year	Wolf population	Deer population	Deer offspring	Predation	Starvation
1990	10	2000	800	400	100
1991	12	2300	920	480	240
1992	16	2500	1000	640	500
1993	22	2360	944	880	180
1994	28	2244	996	1120	26
1995	24	2094	836	960	2
1996	21	1968	788	840	0
1997	18	1916	766	720	0
1998	19	1952	780	760	0
1999	19	1972	790	760	0

Use this data to discuss the population regulation of both deer and wolves.

Unit 12.2 Population

Topic 2: Population sampling methods

In Topic 2, students examine different sampling methods and consider the best method for studying particular organisms, population characteristics, density and distribution.

This Topic deals with:

- Sampling populations.
- Processing data.
- Describing patterns.
- Reporting.

Sampling populations

A **population** is a group of individuals of the *same* species living together in a particular area at a particular time.

Example

Populations include guinea pigs in a breeding colony; mud crabs in an estuary; sea urchins on a reef; starfish on the seashore.

It is usually difficult to count all the members of a population (this is called a census) in an area, so a **sample** is taken. This needs to be representative (typical) of the whole population.

To measure or estimate the size of a population of mobile organisms, **mark–recapture** techniques are used. Organisms are captured, labelled (eg ear tags in rodents, leg bands on turtles or birds) and then released. Another sample is taken at a later date. The size of the population is estimated using the following calculation:

$$\text{Size of population} = \frac{\text{Total number in first sample} \times \text{total number in second sample}}{\text{Number of marked animals recaptured in second sample}}$$

Example

Estimating the size of a population

Forty rats are caught, tagged and released. Two months later 28 rats are captured, 14 of which are found to be tagged.

$$\text{Size of rat population} = \frac{40 \times 28}{14}$$
$$= 80 \text{ rats}$$

The population size of rare or shy species, such as the long-beaked echidna or sugar glider, may need to be estimated indirectly from signs of their presence (eg droppings, calls, footprints, tracks, etc).

Transects

A **transect** is a line placed across a habitat. The transect is divided up into intervals (*stations*) and at each interval the population is sampled. Transect lines are most often used where the distribution of a species is affected by an environmental factor – such as tidal movement affecting the distribution of animals on the rocky shore, or trampling by people affecting the distribution of plants in a lawn.

Quadrats

A **quadrat** is a square frame used to isolate an area so that the number of organisms in that area can be counted. Quadrats are often used to sample areas that contain organisms that cannot move (eg sea anemones and plants). Quadrats are often used to take samples along transect lines.

Example

Use of a transect line and quadrats

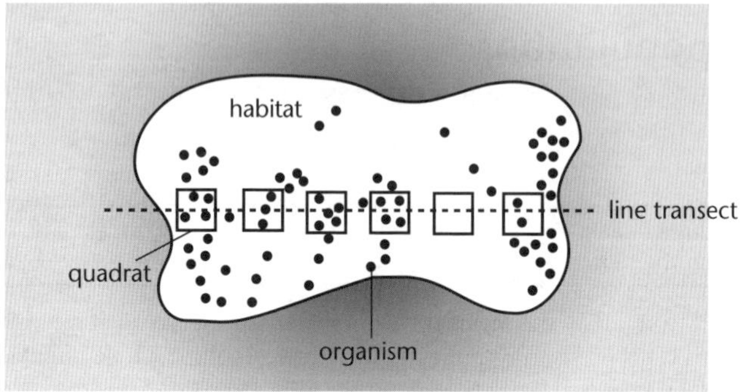

Number of organisms in each quadrant = 4, 3, 5, 4, 0, 2.

Where organisms touch the edge of a quadrat, count only half of those touching the edge as being 'in' the quadrat.

For the quadrat shown: 5 organisms touch the edge of the quadrat; 3 organisms are inside the quadrat.

For the 5 touching the quadrat edge:

Half of 5 ($\frac{5}{2}$) = 2.5, so the quadrat would be regarded as having

3 + 2.5 = 5.5 organisms.

The size of quadrat used is found by taking larger and larger quadrat sizes and counting the number of species found within each quadrat.

When the species number does not increase dramatically, the optimum quadrat size has been found.

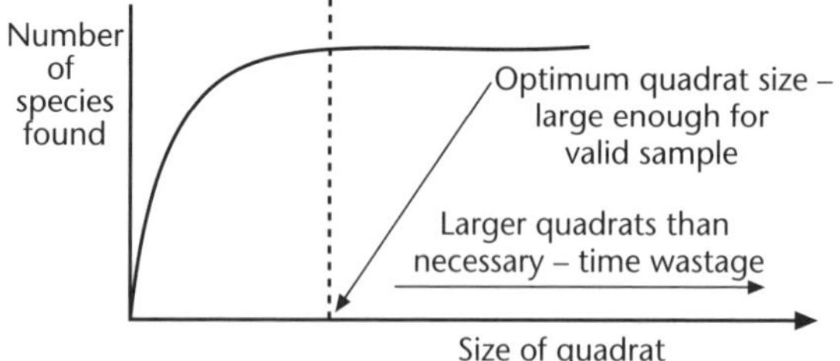

Calculating the optimum quadrat sze.

Example

Using quadrats to estimate populations

frame

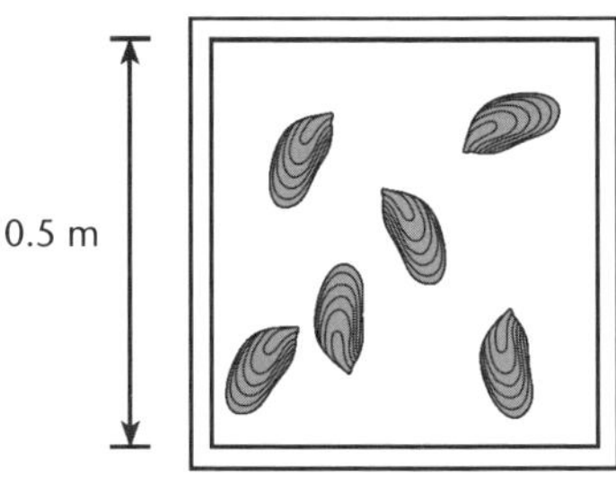

6 mussels

Large or easily distinguishable organisms are counted within quadrats as the '*number* of organisms present'.

small barnacles

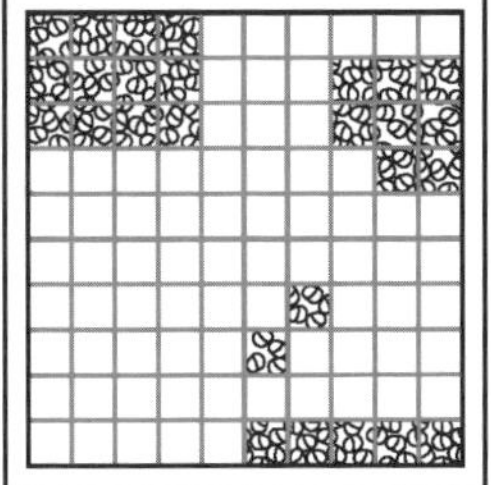

12 + 8 + 2 + 5 = 27 squares with barnacles present

If there are many small organisms or the organisms are clustered together, the quadrat is divided up into a grid and the *percentage cover* is determined.

Percentage cover is calculated by first dividing a quadrat up into a grid:

$$\text{Percentage cover} = \frac{\text{Number of squares organisms present}}{\text{Total number of squares}}$$

Example

Calculating percentage cover

In the small barnacles in the preceding *Example*:

$$\begin{aligned}\text{Percentage cover} &= \frac{27}{100} \times \frac{100}{1} \\ &= 27\%\end{aligned}$$

To prevent **bias**, sample areas must be chosen randomly. This is usually achieved by dividing the habitat up into a grid, and obtaining grid co-ordinates using random numbers. Random numbers can be generated using the random number key on a calculator.

Example

Random sampling a thorn population in a 100 m × 100 m (1 hectare) field

Pushing the random number key 20 times gave 20 random numbers:

1, 2, 4, 3, 8, 7, 1, 6, 2, 6, 9, 3, 2, 4, 7, 7, 8, 0, 3, 2.

These numbers are then paired to give the following grid points:

(1, 2), (4, 3), (8, 7), (1, 6), (2, 6), (9, 3), (2, 4), (7, 7), (8, 0) and (3, 2).

A quadrat is then placed at each grid point and the factor being investigated (number of thorns, size of thorns, number of flowers, etc) is recorded.

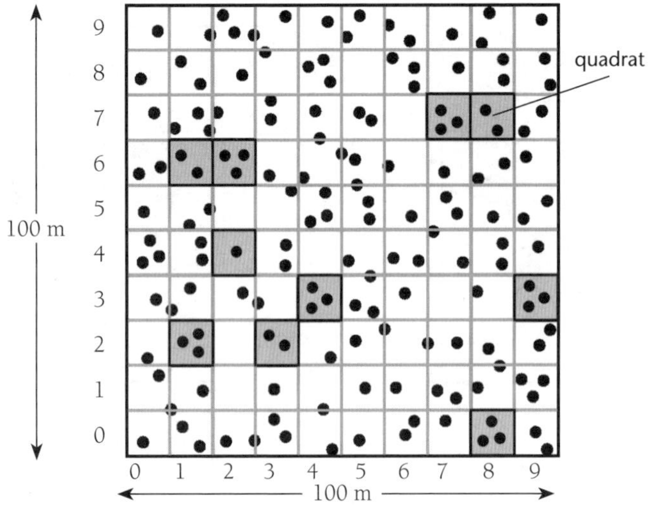

Population characteristics

Population number (size)

The size of a population equals:

$$\frac{\text{Area of habitat}}{\text{Area of sample}} \times \text{Number of organisms in a sample area(s)}$$

Example

Estimating population number

For the thorn population in the 100 m × 100 m field in the preceding *Example*, a quadrat measuring 10 m × 10 m was used at each of the 10 co-ordinates listed.

A total of 25 thorns was counted in the 10 quadrats.

Average thorns per quadrat $= \frac{25}{10} = 2.5$ thorns

To get population count: $\frac{\text{Total area}}{\text{Quadrat area}} \times \text{Average count quadrat}$

$$= \frac{10\,000}{100} \times 2.5$$

$= 250$ thorns

Density

The **density** of a population refers to the number of organisms present in an area.

$$\text{Density of a population} = \frac{\text{Size of population}}{\text{Area of habitat}}$$

Example

Determining population density

Consider the thorn population from the previous *Examples*:

$$\text{Density of thorns} = \frac{250}{10\ 000\ \text{m}^2} \qquad \left[\frac{\text{Population size}}{\text{Area of habitat}}\right]$$

$$= 0.025 \text{ per square metre}$$

Instead of thinking in 'parts of thorns', 0.025 per square metre, we can convert into whole numbers:

0.025 thorns per square metre = 1 thorn per 25 square metres.

Distribution

There are three important types of pattern or distribution of organisms – clumped, uniform and random.

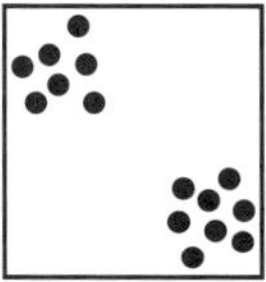

Clumped distribution
Patterns occur when organisms associate together because of environmental or social factors (eg chitons group together in rock crevices to avoid drying out from the sun, feral goats live in herds for protection).

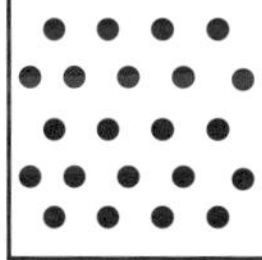

Uniform distribution
Occurs in animals which mark out their habitat into territories for breeding (eg seagulls) or feeding (eg crows). Plants competing for space and light also exhibit uniform distributions.

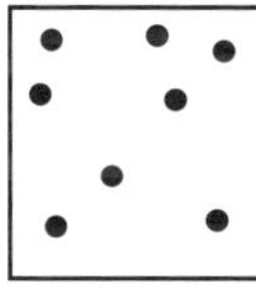

Random distribution
Occurs very rarely in nature. One of the few examples is the distribution of plants in the jungles of Papua New Guinea.

It is important to practise sampling procedures before going on a field trip. A good way to do this is to sample some of the common plants in the grassy areas of the school grounds. The data collected can then be used to practise processing procedures and describing distribution patterns.

Field study survey

Select an area with at least one environmental factor that varies across it.

Example

Lawn survey

The grassy area selected varies in its exposure to direct sunlight.

Select up to six common species that are found in the area. Use a suitable key or reference material to identify them.

Species selected for recording data were broadleaf plantain, ribwort plantain, dandelion, daisy, buttercup and clover. They were initially determined from a *key* of lawn species.

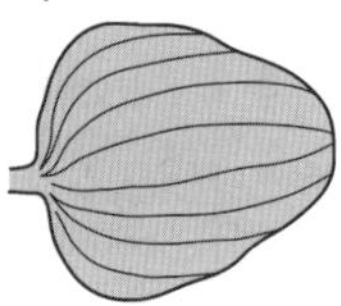

Broadleaf plantain – leaves broad, with ribs distinct.

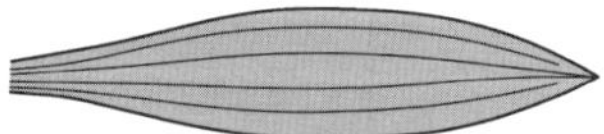

Ribwort plantain – leaves long rather than broad, with *very* distinct ribs, especially on the underside.

Dandelion – leaves long, with distinct lobes or 'teeth' on each side.

Daisy – leaves comparatively short, with small lobes or 'teeth' on each side.

Buttercup – leaf is made of three leaflets, with middle leaflet stalked; comparatively large.

Clover – small leaflets of three lobes; either rounded lobes or blunt lobes.

Work in *groups* (three is a good size). On the field trip it is expected that students will work in groups to collect the data, but every student *must* have a copy of the data because 'processing and interpreting' is an *individual* activity.

Determine how to sample the area to get a representation of the distribution of each species. Do *not* sample *atypical* (ie unrepresentative) parts of the area.

Do not sample the area of lawn that people used as a shortcut, because it is not representative of the rest of the lawn.

Try putting a transect line across the area – consider:

- How *long* must the transect line be (eg 5 m, 10 m, 15 m…)?
- How *many* transects are needed to get an accurate sampling of the whole area (eg 1, 2, 3, 4…)?
- How *far apart* will the samples/quadrats be along the transect (eg 0.5 m, 1.0 m, 2.0 m…)?

It can be helpful to have long ropes (eg 10 m, 20 m, 30 m…) which have been marked off in 1 m intervals to use as transect lines or long measuring tapes (eg 30 m, 50 m).

It is essential to sample the whole area, even if the rope used as a transect line is not sufficiently long to stretch across the whole area.

Try using quadrats in conjunction with transects or just quadrats or on their own. Which gives the most accurate distribution pattern? Try using quadrats of different sizes (eg 20 cm × 20 cm, 25 cm × 25 cm, 50 cm × 50 cm, 1 m × 1 m). Which gives the required accuracy?

The point on a transect line where a quadrat/sample is taken is often called a *station*.

Quadrats of varying sizes can be made quite easily. Most biology departments have ready-made quadrats of about 20 cm × 20 cm or 25 cm × 25 cm. Other sizes can be made from such material as:

- Old wire coat hangers bent into a square.
- Acetate transparency paper (lay it on the ground and count the organisms through it).
- Wooden metre rulers which can be taped together to make squares of varying sizes.
- Orange safety mesh can be obtained and cut to various sizes.

Determine whether the species will be *counted* or a *percentage cover* taken.

Draw up a suitable recording table. This must contain *all the information so that another person could use the data or repeat the collection*, eg:

- Transect number and location.
- Distance apart on the transect line that the samples/quadrats were taken.
- Size of quadrats.
- Whether counts or % cover were recorded.
- Names of the species sampled.

A separate chart was used for each species.

Daisy			**Transects** (10 m apart)					
			(East)				(West)	Av. per quadrat
			1	2	3	4	5	
Stations (2 m apart, 0.5 m × 0.5 m quadrats)	(North)	1						
		2						
		3						
		4						
		5						
		6						
		7						
		8						
		9						
	(South)	10						

It can be very helpful to have a *labelled diagram* of the sampled area showing where the samples (transects/quadrats) were taken, clearly showing the location of the area (eg north, south; for a rocky shore, it is essential to give tidal height/location (eg high tide and low tide)). Record important environmental information (eg shady/sunny; wet/dry soil; rock/sand substrate; air/water temperature), either on the diagram or with the table – *these factors will be important in explaining the distribution of the species.*

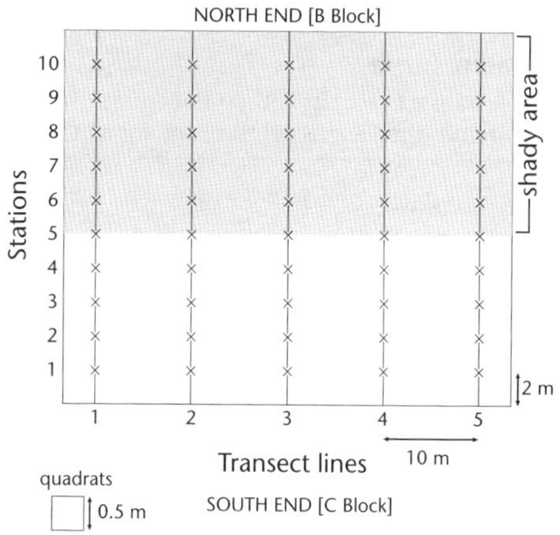

5 transect lines, 10 m apart, were drawn from south (C Block) to north (B Block) across the area. The northern half of the area was in the shade all day (shadowed by B Block), while the southern half was in the sun all day. The shady half had much wetter soil than the sunny half. There were few plants and a lot of bare ground at the beginning of each transect, as people often walk over these areas going to and from class.

Ten stations were marked off at 2 m intervals along each transect, and at each station, a 0.5 m × 0.5 m quadrat was taken. Counts were made of buttercups, daisies, plantains (ribwort and broadleaf), dandelions; for clover, % cover were taken.

For a field trip, you should be provided with a range of sampling equipment. From this you will *select* and use that which *is most appropriate for the community and organisms being investigated.* It is useful to have a *clipboard* and *pencil* for recording purposes (a pencil will write on wet paper).

Processing data

The recorded data must be processed so that *any trends or patterns present will be clearly shown.* This requires the collection of *sufficient* data – *large samples* (eg a 1 m × 1 m quadrat at stations along a transect line) or *multiple samples* (eg several transect lines with equidistant stations for quadrat sampling along each). The second method is preferable, as it allows a good coverage of the whole area to be sampled. From the recordings made for this, the results from the stations in the same zone will be averaged.

Example

Lawn survey

If 5 transects are used, then the counts made in all the station one quadrats will be added together then averaged; similarly for all the station two quadrats, etc.

Averaging is a form of data processing. The processing can be extended by converting the quadrat averages into numbers per m^2.

Example

Quadrat size	Quadrat area	For result per m^2, multiply by:
20 cm × 20 cm	400 cm^2 (0.04 m^2)	25
25 cm × 25 cm	625 cm^2 (0.0625 m^2)	16

Tabulating data

Results should be *tabulated*.

Example

Lawn survey

Table of processed data to show number per m^2 of each species for each station

Station	Species (average number per m^2)					
	Daisy	Broadleaf plantain	Ribwort plantain	Buttercup	Dandelion	Clover (% cover)
1	0	10	10	0	0	0
2	0	10	30	0	0	0
3	20	20	35	0	5	0
4	15	0	0	15	10	0
5	40	0	0	15	10	20
6	45	0	0	20	15	20
7	15	0	0	30	20	50
8	5	5	10	40	10	60
9	0	15	30	55	5	75
10	0	15	25	55	0	80

Graphing data

The averages should show any pattern or trend present. This can be supported by using a line graph such as a *distribution graph*. Always use graph paper when drawing graphs.

Points are plotted on the graph and connected with a smooth line. Each line needs to be labelled with the name of the species, or a *key* given (this is preferable). The graph needs a title to indicate what the plotted information is referring to. The independent variable appears along the x-axis, the **dependent variable** along the y-axis.

Example

Lawn survey

The distance along the transect, in metres, is given on the horizontal (x) axis, along with a reference point (eg from south to north). The numbers per square metre of each species is given on the vertical (y) axis.

(Only three of the six species have been shown for the sake of clarity.)

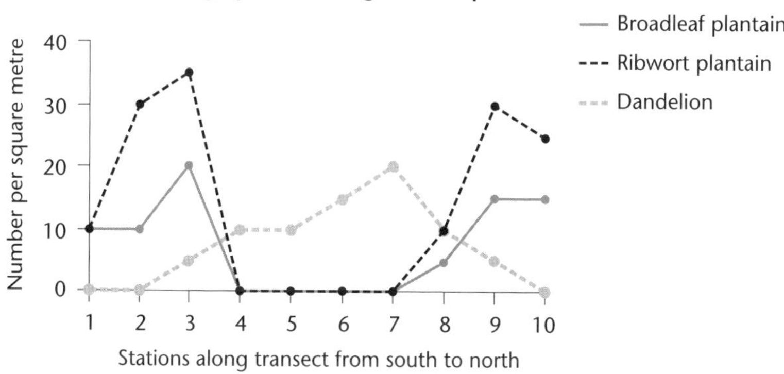

Kite diagrams

When several different species are put on a distribution graph, the overlapping lines can make the trends difficult to see. Therefore, **kite diagrams** are a more common way of showing distribution. Kite diagrams need graph paper (or a grid). One species may be put on the paper, or several species *side by side* (this allows distributions and densities to be easily compared). The vertical axis shows the stations or distance along the transect; the horizontal axis shows the number of organisms for each species (or a scale may be used).

Example

Lawn survey

(Only three of the six species have been shown for the sake of clarity.)

Kite diagram for lawn survey

Broadleaf plantain
Ribwort plantain
Dandelion
Stations
(NORTH)
(SOUTH)
10
9
8
7
6
5
4
3
2
1
10 5 0 5 10 15 10 5 0 5 10 15 10 5 0 5 10
Numbers/m2

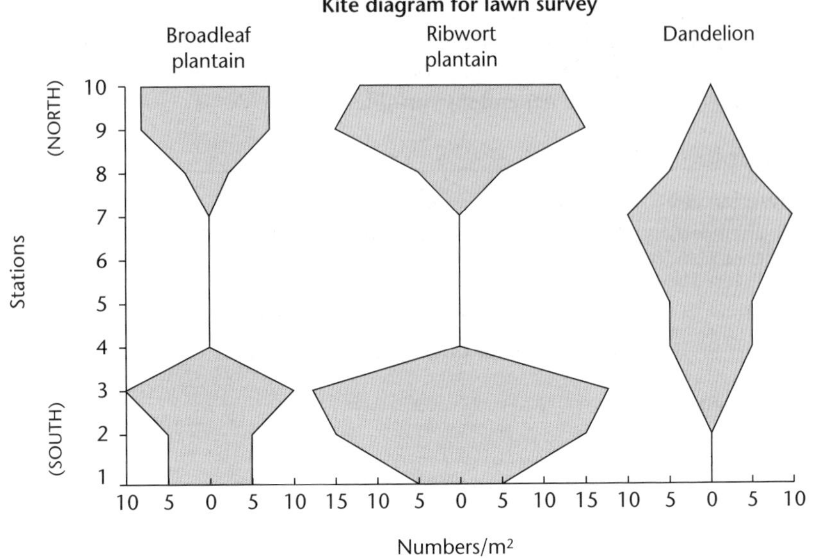

Example

Rocky shore

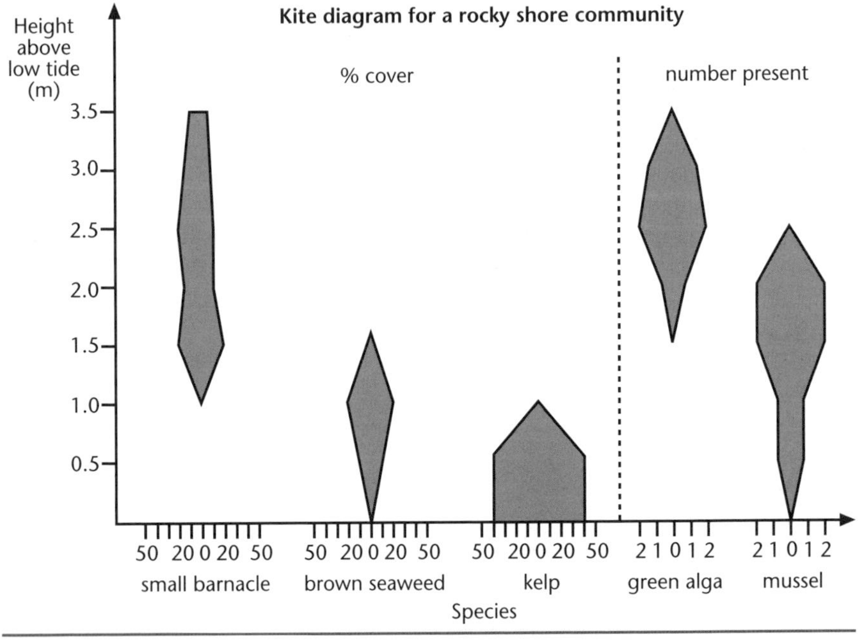

Height above low tide (m)
Kite diagram for a rocky shore community
% cover
number present
3.5
3.0
2.5
2.0
1.5
1.0
0.5
50 20 0 20 50
small barnacle
50 20 0 20 50
brown seaweed
50 20 0 20 50
kelp
2 1 0 1 2
green alga
2 1 0 1 2
mussel
Species

Describing the patterns

The distribution pattern shown in the processed data needs to be *described* with reference to an environmental factor or the biology of the organism

Example

Rocky shore

The following relates to the processed data from the kite diagrams of the rocky shore in the preceding ***Example***.

'The brown seaweed is found attached to the rocks from low tide to 1.5 m above low tide. It is most common at 1.0 m above low tide with approx 40 individuals found per quadrat.'

> This description of the distribution of the brown seaweed describes where the species is found and relates it to an environmental factor (tidal height).

'The brown seaweed has a holdfast with which it attaches to the rocks so that it can live there.'

> This has described an adaptation (the holdfast), which is an aspect of the biology of the organism.

'The seaweed is found in the low-tide zone because is has a low tolerance to drying out/**dehydration**; therefore it needs to be covered with seawater for most of the tidal cycle.'

> The *explanation* of the distribution of the brown seaweed includes a reason relating the seaweed's distribution to an environmental factor (tidal height) and its biology (low tolerance to drying out).

'Living in the low tide zone means that the seaweed is constantly exposed to wave action; this is why it has a strong holdfast which cements it to rocks. It also has a tough, leathery outer layer/epidermis, which protects it from the scouring action of waves. The body of the seaweed has air spaces that give buoyancy; this allows it to float in the water. This will assist protection, but also increases the exposure of the seaweed to gases (O_2 and CO_2) and to nutrients (eg NO_3^-) in the water, which are essential for respiration, photosynthesis and growth. By living in the low tidal zone, the brown seaweed facilitates its ability to carry out these vital processes and its growth is promoted.'

> A *discussion* is more in-depth than a simple explanation and needs to *link ideas*. It must include reference to *both* an environmental factor and the biology of the organism.

The kite diagram includes five species. Although only *one* needs to be described/explained/discussed, if the distribution for one is given incorrectly, the true interaction between species is not realised. It is thus recommended that the distributions of *all five* species are described, explained or discussed because any interactions between the species (eg **competition**, predation, herbivory) offers evidence of the linking of ideas needed for discussion.

'Both the brown seaweed and kelp (a seaweed too) are found together in the tidal zone, so are likely to be competing for space on the rocks. Large numbers of kelp are found between low tide and 0.5 m, while only small numbers of brown seaweed occur here. The maximum number of brown seaweed occurs at 1.0 m above low tide, where kelp no longer occurs. These figures strongly suggest competition between the species, with kelp limiting the lower distribution of brown seaweed.

This sort of consideration is indicative of excellent discussion.

Report

A *report* should take about 2–3 hours to complete. The report uses the data recorded in the field and any resource material about the community and organisms investigated. The report must include:

- The field data.
- A description, with diagram, of how the sampling was done.
- The processed data (tables and/or calculations and/or graphs and/or kite diagrams).
- A description, explanation or discussion of the trend or pattern shown by the processed data.

When the report has been finished, it should be carefully *proofread* before being handed in for assessment. When proofreading, you should make sure that everything required is present, complete, and in the correct order. Proofreading should pick up any errors and omissions and make corrections. A quality report is indicative of a *quality investigation*.

Unit 12.2 Activity 2A: Sampling techniques and distribution patterns

Students are not expected to have seen or studied the organisms and/or the communities presented in the questions; however, students are expected to apply their knowledge of sampling techniques, data processing, environmental factors (biotic and abiotic), and adaptations, in answering the questions.

Any diagrams, tables, and graphs given here may be used as examples of 'best practice' for the report on the investigation.

1. **a.** Define the following terms:
   i. Distribution.
   ii. Density.

   b. Explain the difference between uniform, clumped and random distribution patterns, and suggest reasons why these different patterns occur.

2. Green-lipped mussels (*Perna canaliculus*) are shellfish that may be found on intertidal rocks.

Green-lipped mussel (Perna canaliculus)

A typical rocky shore

Part 1

a. Describe a sampling technique that could be used to collect data on the distribution of the green-lipped mussels between high and low tide. The data collected needs to be *sufficient* to clearly show the distribution pattern.

b. Give the evidence that would say whether the distribution was random, uniform or clumped.

c. Describe how the collected data could be processed to find the density of the mussels.

d. Describe one graphical or diagrammatic method that could be used in processing the collected data.

e. Describe any **i.** biotic **ii.** abiotic factors that may be important in determining the distribution of the mussels.

f. Suggest adaptations of mussels that may be important in determining their distribution pattern.

Part 2

Another study of the same rocky shore sampled a range of species living there. The processed data is shown in the kite diagrams following.

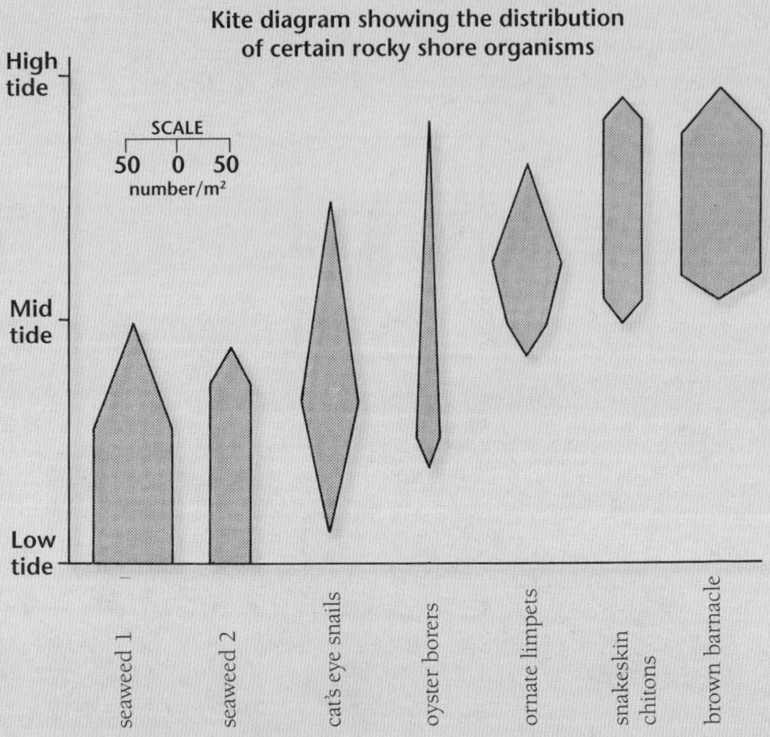

a. Name the community pattern illustrated by these results.

b. Name the main environmental factor determining this pattern.

c. Name two species that probably have a low tolerance to dehydration.

d. Name two species that are probably tolerant of high temperatures and dehydration.

e. Suggest an adaptation that both chitons and limpets have to prevent them being washed off the rocks by wave action.

f. Suggest an adaptation of the cat's eye snails, limpets and oyster borers to prevent damage by wave action.

g. Describe how **i.** interspecific competition **ii.** predation may affect the distribution of species on the shore.

3. *Paphies subtriangulata* (tuatua) and *Paphies australis* (pipi) are two New Zealand shellfish that burrow in the sand of beaches. A study seeks to find if these two species compete with each other, or whether they co-exist.

a. Describe a sampling investigation that could be used to determine the distribution of the two species.

b. Describe how the recorded data would provide evidence for competition or co-existence.

4. A group of students making a study of plants on coastal sand dunes prepared a transect across the dunes. They recorded a number of environmental factors at five stations along the transect, and presented the data in the table below.

Station number	Distance along transect (in metres)	Form of dominant plant species	Average height of plants (in metres)	Factor		
				% cover of surface litter	Relative amount of moisture in surface soil, taking 'Station **V**' as 100%	Relative light intensity at surface of soil taking 'Station **I**' as 100%
I	1	Bare sand dune	nil	0	1	100
II	10	Dune grasses	0.2	1	5	90
III	20	Low shrubs, some with root nodules	1.2	70	30	20
IV	30	Tall shrubs	4.5	85	60	15
V	40	Dune forest	7.0	98	100	10

a. Identify other biotic and abiotic environmental factors (apart from those given on the table) that are likely to change from station **I** to station **V**.

b. Marram grass is a typical sand dune grass that could be found at station **II**. Suggest possible adaptations shown by marram grass to living in this area of the dune.

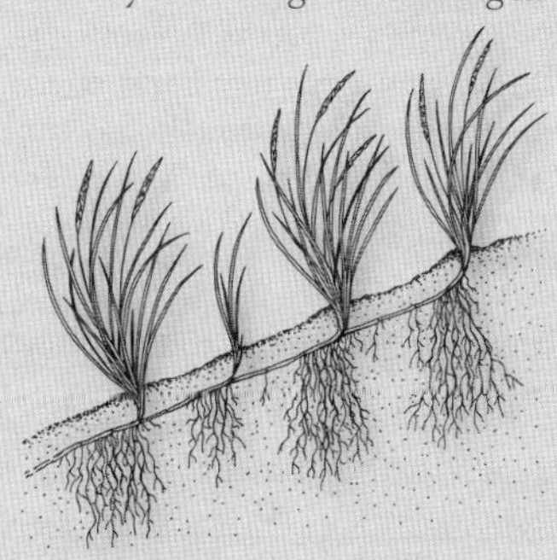

Marram grass

c. Draw a graph showing the average height of the plants in relation to distance along the transect. Describe the pattern shown in the graph, and suggest reasons for it.

5. The picture below shows a species of native tussock grass, the silver tussock (*Poa laevis*), which inhabits high country areas of temperate regions.

Invertebrates, such as grasshoppers, spiders and mites, can be found on the leaves of the tussock, while springtails, amphipods, and other spiders can be found in the leaf litter around the base of the tussock. Numbers of these organisms may be very high.

a. Describe a sampling technique that could be used to collect data on the distribution of the silver tussock in a designated area.

b. Suggest abiotic environmental factors that may determine the distribution of the silver tussock

c. Suggest biotic and abiotic environmental factors that may account for the presence of the invertebrate species.

6. A student observed that the dominant types of plant species appeared to change from the edge of a stream across the field alongside. To obtain more information, she laid a transect line from the stream across the field and counted the numbers of the most common species of plants present in a 1 m × 1 m quadrat placed on the line at 3 m intervals. The data she recorded is given in the table below:

Plant species	Metres from the stream boundary							
	3	6	9	12	15	18	21	24
	Number of plants per m²							
A	50	20	0	0	0	0	0	0
B	0	0	2	12	36	30	12	0
C	0	0	0	0	0	8	30	40
D	18	30	45	20	22	16	8	0

a. Draw a graph to show the distribution pattern of the four plant species.

b. Name the species that is likely to be the most tolerant of a range of environmental conditions, and give a reason for your choice.

c. Name any species that may be in competition for resources, and give a reason for your answer.

7. Organisms that live on rocky shores can be found in distinct zones. The picture below is of a typical rocky shore at low tide.

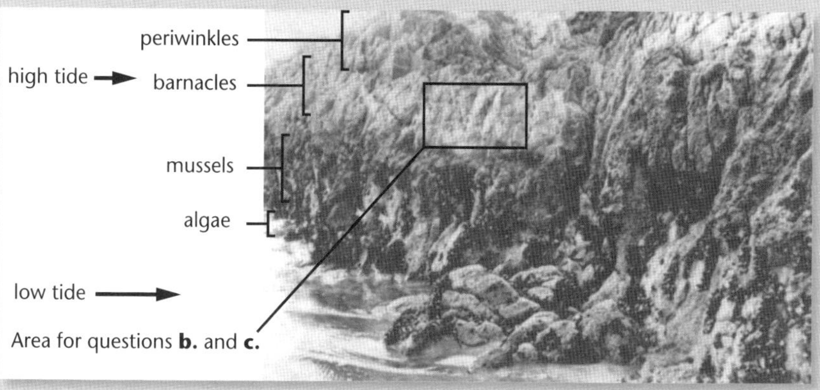

   a. Name four physical factors that could affect these organisms, and give the probable effect of each of the factors in this situation.

   b. Barnacles are firmly cemented to the rocks. Often found with them is their predator, the oyster borer. These are shown in the picture below:

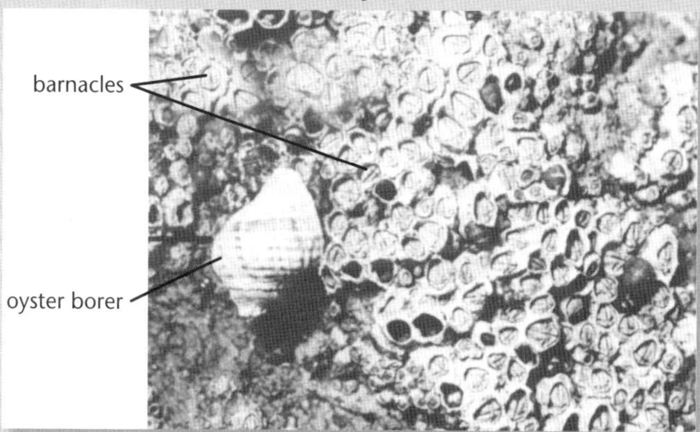

Oyster borer and barnacles.

   i. Describe a suitable method for estimating the number of barnacles in this area.

   ii. State whether this sampling method would also be suitable for estimating the number of oyster borers. If not, describe a suitable one.

   iii. Suggest the two main factors that are likely to determine the zone that the oyster borer occupies.

   iv. Using the picture as a guide, describe a likely factor for intraspecific competition between the barnacles.

c. Two species of periwinkle are shown in the picture below; both live in the high-tide zone.

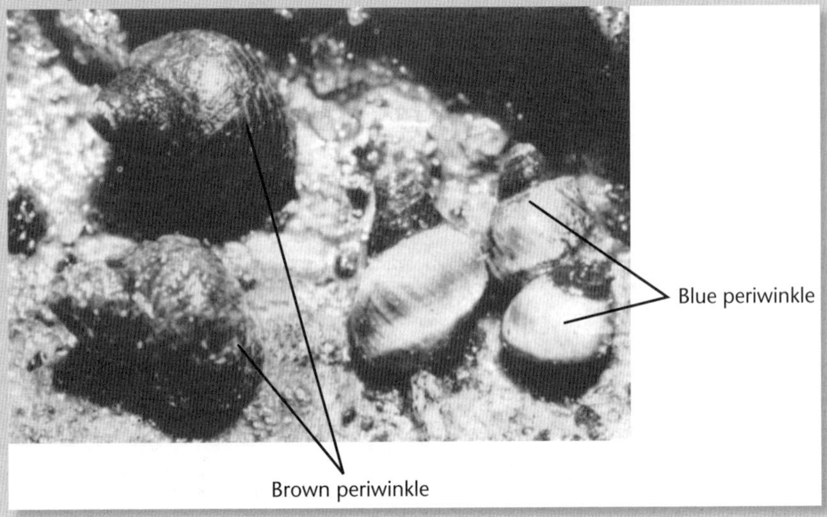

Two periwinkles.

Describe adaptations that both these species are likely to have to living in this habitat and zone.

Unit 12.2 Population

Topic 3: Human population growth in Papua New Guinea

In Topic 3 we move on to consider human population and the relationship between population growth and the impact it has on a country's resources. We look at the growth of population in Papua New Guinea and the reasons for the population increase and think about the impact of PNG's increased population in terms of the environment and species endangerment, and the health and stability of the nation as a whole.

Note: this section is taken from Bourke, R.M. and Harwood T. (eds) (2009). *Food and Agriculture in Papua New Guinea.* ANU E Press, The Australian National University, Canberra. The text is largely based on Section 1.1 Total Population, written by R. Michael Bourke and Bryant Allen. We are grateful to the authors and publisher for permission to reproduce their text in this book. The original book is available online at: http://epress.anu.edu.au/food_agriculture_citation.html

Human population in Papua New Guinea

The number of people in a country and the rate at which they are increasing is a critical issue in any discussion of environment and the production of food and the endangerment of species.

The 2000 National Census put the total population of Papua New Guinea at 5.2 million. Around 81%, or about 4.2 million, of these people live in rural villages. Around 5% live in the National Capital District city of Port Moresby, 8% in other urban areas and 6% in small stations, missions, schools, logging camps and mines, known as 'rural non-village' locations.

Of all the countries around the world, only Rwanda, Bhutan, Nepal and Uganda have a greater proportion of people living in rural areas than PNG.

By 2015 PNG's total population was estimated to be about 7.5 million, and except perhaps for an increase in Port Moresby and some larger towns, the proportions of people living in rural and urban areas has probably not altered significantly.

Population growth

Population growth rates are conventionally determined on the basis of the difference between birth rates and death rates. However, almost all births and deaths in PNG are not registered, so estimates of population growth rates must be calculated from census totals. If censuses are flawed for any reason, estimates of growth will also be inaccurate.

The total population of PNG has been increasing at between 2.1% and 3.2% per year since 1966, with an average rate of growth over this period of about 2.5% per year. At this rate of increase, the total population will double around every 30 years. Every year approximately 105 000 people are added to the population. These people must be fed, clothed, housed, educated and provided with access to health care. If the 2000 population continues to increase at 2.5% per year, it will reach 10.7 million by 2030.

Population growth rates between 1966 and 2000 have been highest in the Islands Region at 2.7% per year, with the Highlands and Momase regions growing at around 2.5% per year. Growth rates in the Southern Region have fluctuated between 1966 and 2000, but overall have been the lowest.

Between the 1990 and 2000 censuses, growth rates in the Highlands and Momase regions appear to have increased significantly to 3.6% per year and 3.3% per year respectively, while in the Islands and Southern regions growth rates have fallen, with almost no growth in the Southern Region.

Province	Rural	%	Rural non-village	%	Urban	%	Total
Western	107837	70	12445	8	33022	22	153304
Gulf	92265	86	3620	3	11013	10	106898
Central	157058	85	21165	12	5760	3	183983
National Capital District	0	0	0	0	254158	100	254158
Milne Bay	188334	90	9327	4	12751	6	210412
Oro	106288	80	15406	12	11371	9	133065
Southern Highlands	526398	96	8813	2	11054	2	546265
Enga	283498	96	4014		7519	3	295031
Western Highlands	371014	84	39094	9	29917	7	440025
Simbu	242748	93	7201	3	9754	4	259703
Eastern Highlands	393418	91	13243	3	26311	6	432972
Morobe	356100	77	46869	10	57743	13	460712
Lae City	0	0	0	0	78692	100	78692
Madang	308135	84	18626	5	38345	11	365106
East Sepik	303706	88	7492	2	31983	9	343181
Sandaun	166919	90	4508	2	14314	8	185741
Manus	34899	80	1276	3	7212	17	43387
New Ireland	103259	87	4346	4	10745	9	118350
East New Britain	174230	79	35613	16	10290	5	220133
West New Britain	109299	59	54969	30	20240	11	184508
Bougainville	167156	95	3897	2	4107	2	175160
Papua New Guinea	4192561	81	311924	6	686301	13	5190786

Rural, rural non-village, and urban populations of PNG, 2000. The 2000 National Census put the total population of Papua New Guinea at 5.2 million. Around 81%, or about 4.2 million of these people, live in rural villages. Around 5% live in the National Capital District city of Port Moresby, 8% in other urban areas and 6% in small stations, missions, schools, logging camps and mines, known as 'rural non-village' locations.

The volcanic eruption in September 1994 at Rabaul, the civil war on Bougainville (1989–1997), and continued growth in the National Capital District will have influenced this pattern, but under-numeration in the 1990 census has probably inflated the apparent rates of increase in the Highlands Region, as has probable overestimates in the population of Southern Highlands Province in the 2000 census. The 1990 to 2000 increase in the Southern Highlands population is demographically impossible, particularly in light of the known out-migration which is occurring from this province. The increases are restricted mainly to the central part of the province and suggest errors in the 2000 census data.

The National Statistical Office Population Projections Task Force predicted in 1999 that the Highlands Region population will reach 3.9 million and the Momase Region population 2.9 million by around 2030.

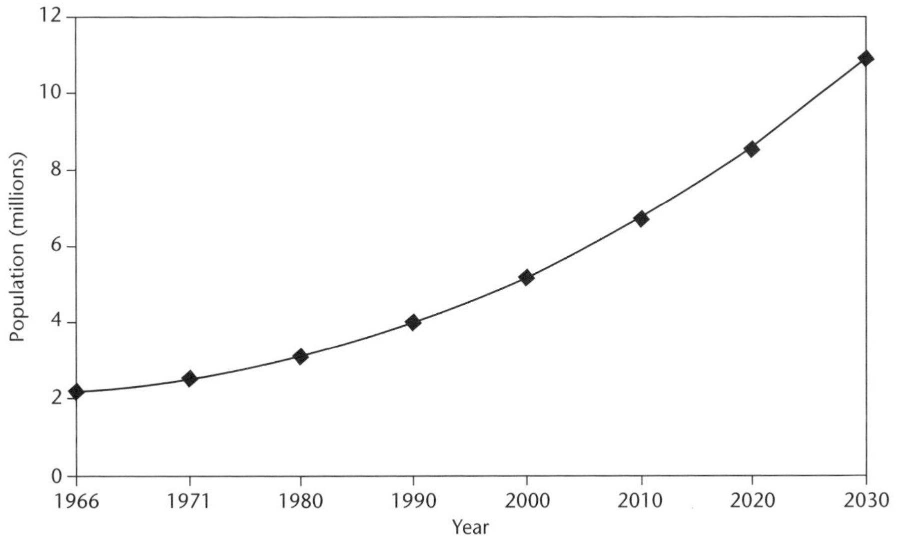

Population growth 1966–2000 and projected growth 2000–2030 at a constant 2.4% per year.

Population distribution

The total population of PNG is not spread evenly throughout PNG. Half of the population lives in six provinces: Southern Highlands, Morobe, Western Highlands, Eastern Highlands, Madang and East Sepik. Another six provinces contain only 14% of the total population: Western, Bougainville, Oro, New Ireland, Gulf and Manus.

In 2000, 38% of the population lived in the five provinces that make up the Highlands Region, 28% lived in the Momase Region, 15% in the Southern Region, 14% in the Islands Region and 5% in the National Capital District. These proportions have remained reasonably stable over the 34-year period 1966 to 2000.

Of the total population, 686 301 (13%) are located in urban areas, notably Port Moresby (254 000), Lae (78 700), Mount Hagen (28 500), Madang (27 900), Wewak (20 250) and a number of much smaller towns and administrative centres. There are 73 urban areas, 40 of which have populations of more than 1000 people. A further 311 924 (6%) live in settlements that are classified as 'rural non-village' census units. These are boarding schools, mission stations, sawmills, logging camps and similar settlements located in rural areas, but which are not villages. The rural village population is 4 192 561 (81%).

Unit 12.2 Activity 3: Human population growth in PNG

Consider the current growth rate in the population of PNG, especially in relation to the area in which you live. As a class, discuss the pressure that population growth will put on the resources of the country and the environment. Think about the policies that should be implemented to ensure that the environment is protected and that living standards improve.

Unit 12.3 Genetics

Topic 1: Mendel's pea experiment and inheritance

In line with the syllabus, Unit 12.3 looks at genetics and inheritance. Students learn that detailed information about plants and animals is contained in each nucleus of every living cell. Studying variation and biotechnological techniques gives us an understanding of how genetics can be used to modify both plants and animals

Topic 1 serves as an introduction to genetics. The content relates to Gregor Mendel (1822–1884), known as the 'father of genetics', and his experiments with the garden pea. The transfer of genetic information is described by looking at:

- Mendel's experiments.
- Monohybrid inheritance.
- Test cross.
- Dihybrid inheritance.
- Inheritance and pedigree.

Key terms

Monohybrid cross	A cross involving one pair of contrasting characters
Phenotype	The physical characteristics of an organism
Genotype	The hereditary makeup of an organism
Gene	A unit of heredity
Locus	The specific position occupied by a gene on a chromosome (plural: loci)
Allele	One of two or more forms that a gene at a given locus can take
Homozygote	An organism carrying two copies of the same allele
Heterozygote	An organism carrying different allelic forms of a gene
Dominant	An allele that is expressed in heterozygotes
Recessive	An allele that is only expressed in homozygotes
Segregation	The separation of alleles during **meiosis**
F_1	A hybrid formed by crossing differing pure-breeding parents
F_2	The offspring produced by inbreeding the F_1
True-breeding	Producing only offspring like the parents when matcd with its own kind
Test cross	A cross carried out to determine an organism's genotype, by mating it with a homozygous recessive organism
Character	A characteristic of an organism, such as the appearance of seeds, pods, flowers, or stems
Dihybrid	A cross involving two pairs of contrasting characters
Genome	Gene complement of an organism
Inheritance	The acquisition of traits by their transmission from parent to offspring
Trait	An identifiable characteristic

The garden pea.

The work of Gregor Mendel

Gregor Mendel was a monk who discovered the basic mechanism of **heredity**. When Mendel did his work in the mid-19th century, **chromosomes** had not been discovered. Mendel performed breeding experiments with varieties of the garden pea. Mendel crossed varieties with clearly contrasting traits (*characters*) – such as red versus white flowers, tall versus short stems, yellow versus green seeds.

Gregor Mendel.

Red versus white flowers and tall versus short stems are examples of **discontinuous variation**, ie each individual can be put into one of a number of distinct groups.

In **continuous variation**, there are no sharp boundaries – eg in humans there are all grades of heights between the tallest people and the shortest; people cannot be classified as either tall or short. Other examples of continuously varying characters in humans are body weight, skin colour, and blood pressure.

All Mendel's plants were **true-breeding** – ie when mated with their own kind, all the offspring resembled the parents.

In one experiment, he crossed a true-breeding red-flowered variety with a true-breeding white-flowered variety. He called the offspring the **first filial generation** or $\mathbf{F_1}$. The F_1 were all just like the red-flowered parent.

Mendel then allowed the F_1 to self-fertilise to produce a **second filial generation** or $\mathbf{F_2}$. Of the 929 F_2 plants, approximately three-quarters (705) were red-flowered and one-quarter (224) had white flowers. This is a ratio of approximately of 3 : 1.

Although the F_1 generation (all red) *looked* like one of the parents (ie the red parent), they were not true-breeding, because some of the F_2 plants had white flowers. Although the F_1 red-flowered plants had the same physical characteristics or **phenotype** as their red-flowered parent, they had a different genetic makeup or **genotype**.

Mendel concluded that although the factor for whiteness was not expressed (ie did not 'show itself') in the F_1, it must nevertheless have been present. Mendel said that red was **dominant** over white, which was **recessive**.

The hereditary factors Mendel studied were later called **genes**. The gene for flower colour exists as two alternative forms or **alleles**.

- **Dominant alleles** are represented by upper-case letters – eg *R* for red-flowered.
- Recessive alleles are represented by lower-case letters – eg *r* for white-flowered.

Mendel deduced that each characteristic is controlled by a pair of 'factors' (ie genes), but only one of each pair is present in each gamete. Thus, during gamete formation, the two members of each pair must *segregate* so that each gamete receives only one gene/factor.

- A true-breeding red-flowered plant can be represented by *RR*, and a gamete carrying the allele for red is represented by *R*.
- A white-flowered plant has the genotype *rr*, and a gamete carrying the allele for white is represented by *r*; an F_1 plant (red-flowered but not true-breeding) is represented by *Rr*.

True-breeding plants are **homozygous** – they have two of the same allele (*RR* or *rr*).

Non-true-breeding plants are **heterozygous** – they have different alleles for a given characteristic (*Rr*).

- A homozygous red-flowered parent is thus *RR*.
- A white-flowered parent is thus *rr*.
- The F_1 of a cross between a homozygous red-flowered parent and a white-flowered plant is thus *Rr*.

Mendel's garden pea experiment explained

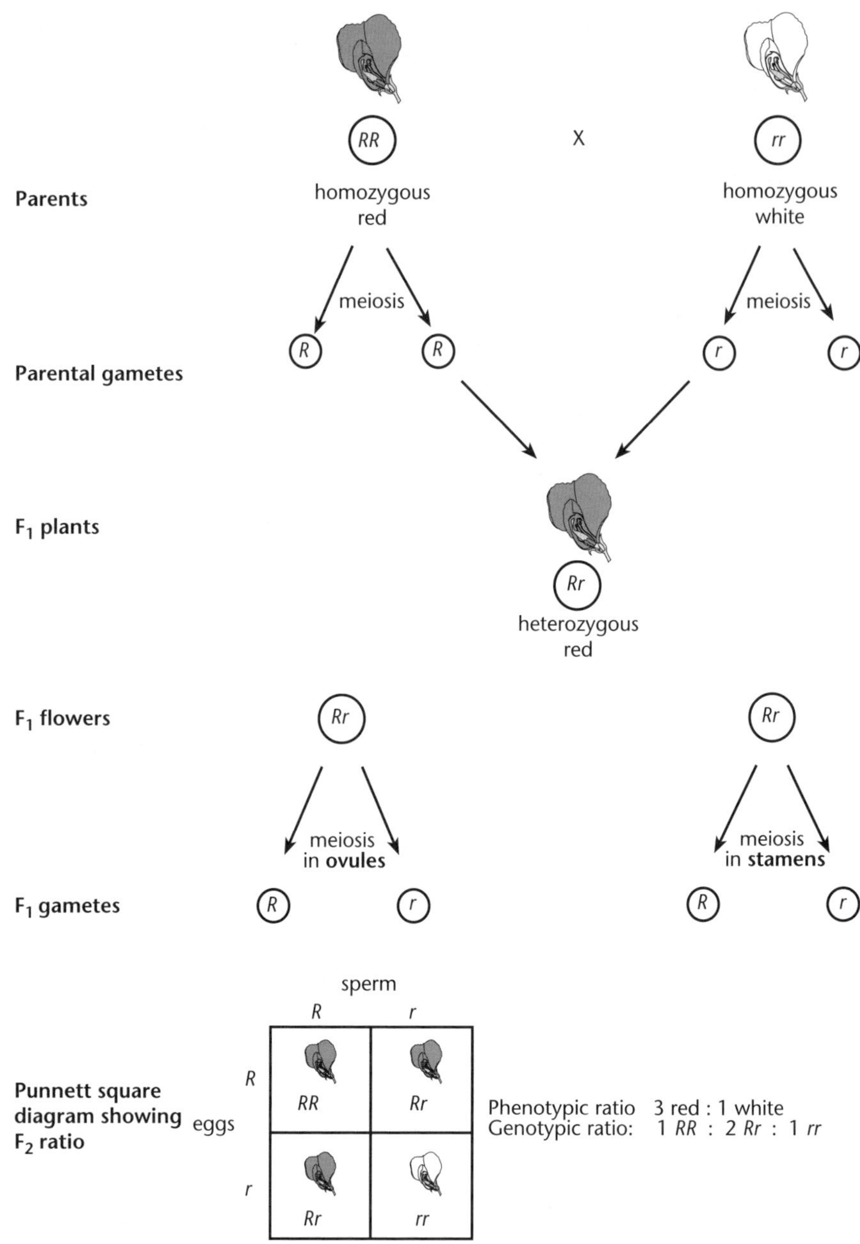

Mendel's experiment explained.

There are four different ways in which two kinds of egg and two kinds of sperm can join. Each is represented by one of the compartments in the Punnett square diagram on the previous page. Two of these give the same result (*Rr*), so this genotype is produced twice as often.

Mendel studied six other pairs of contrasting characters besides red- and white-flowered plants – in every case, he obtained a ratio of approximately 3 : 1 in the F_2 generation.

Test crosses

Plants homozygous for a **dominant trait** look phenotypically the same as plants heterozygous for that trait. They can be distinguished from each other by carrying out a **test cross**.

A test cross involves crossing with the recessive phenotype.

- If the plant under test is homozygous, then all the test cross offspring will show the dominant trait.
- If the plant under test is heterozygous, then the offspring will consist of both phenotypes, in an approximate 1 : 1 ratio, ie about half of each kind.

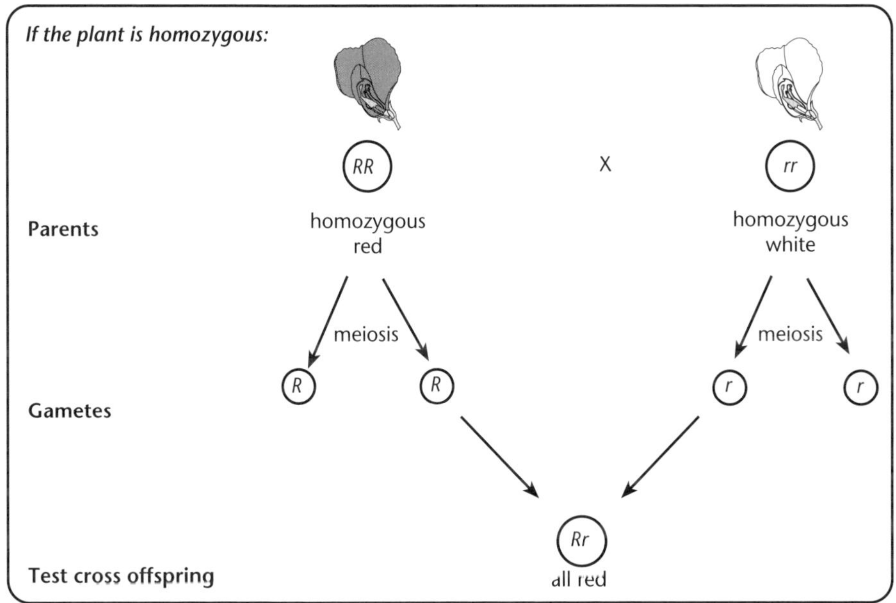

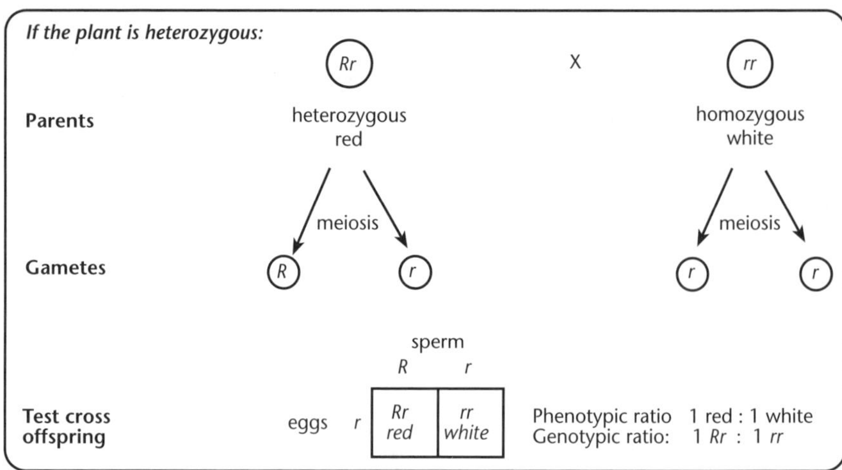

The test cross shows if a red-flowered pea plant is homozygous or heterozygous. Note: there are only two compartments in the Punnett square because there are only two different ways the gametes can combine.

A test cross.

Unit 12.3 Activity 1A: First principles

1. Match the phrases **1–10** to the terms **A–J**.

Phrase	Terms
1. Alternative term for homozygous	**A.** Allele
2. An allele expressed in heterozygotes	**B.** Dominant
3. An allele only expressed in homozygotes	**C.** F_1
4. Having two different alleles of a given gene	**D.** F_2
5. Having two of the same allele	**E.** Genotype
6. One of a number of forms a gene can take	**F.** Heterozygous
7. Produced by inbreeding the F_1 generation	**G.** Homozygous
8. Result of a cross between two pure-breeding varieties	**H.** Phenotype
9. The genetic makeup of an organism	**I.** Recessive
10. The physical characteristics of an organism	**J.** True-breeding

2. Peas can have two types of seed coat, round or wrinkled. The wrinkled seed coat is recessive and is given the symbol '*r*'. Complete the table below to show the phenotypes and genotypes for the seed coats of pure-breeding seeds.

Phenotype	**a.**	**b.**
Genotype	**c.**	**d.**

3. A particular flower can be red or white. Red flowers are due to the dominant allele (*R*), and white flowers are due to the recessive allele (*r*).

a. There are three possible genotypes of the flowers, even though there are only two phenotypes. Describe the three possible genotypes, and include the symbols (*R*) and (*r*) for each genotype.

b. A red-flowered plant is to be crossed with a white-flowered plant. There are two possible combinations of parents. Complete the Punnett squares to show these crosses and the fraction of white flowers in each case:

i.	*R*	*R*
r		
r		

ii.		

iii. The fraction of offspring with white flowers in cross **i.** is:

iv. The fraction of offspring with white flowers in cross **ii**. is:

c. A cross between two red-flowered plants resulted in offspring with a genotype ratio of 1 : 2 : 1, and a phenotype ratio of 3 : 1. Explain why there was a difference between the two ratios.

d. Discuss how it is possible to determine the genotype of a red-flowered plant using simple crossing techniques.

4. In pea plants, yellow seeds (*E*) are dominant to green seeds (*e*), which are recessive.

a. Describe what is meant by the terms:

i. Dominant.

ii. Recessive.

b. i. A homozygous yellow-seeded pea plant is crossed with a heterozygous yellow-seeded pea plant. Complete the Punnett square below to show the genotypes of the offspring. (See page 172.)

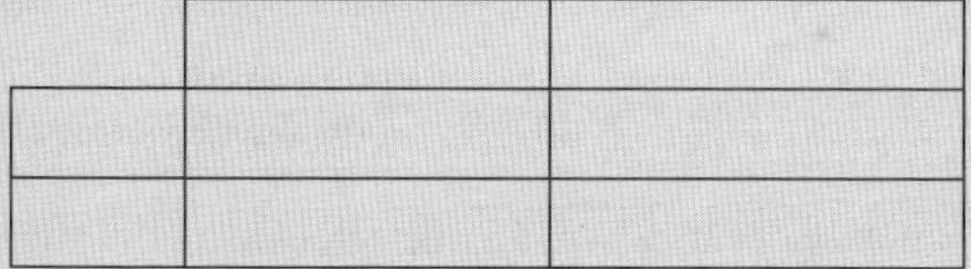

ii. Referring to your answer to part **b.i.**, explain the difference between genotype and phenotype.

c. Two students were given a number of yellow pea-seeds and asked to find out if the seeds were pure-breeding or not. Discuss how the students could solve this problem without the need for specialist equipment or techniques

5. Gregor Mendel is often referred to as the 'father of modern genetics'. His work in the 19th century involved many thousands of breeding experiments using the garden pea. In pea plants, round seeds (*R*) are dominant to wrinkled seeds (*r*), which are recessive. Yellow seed colour is dominant to green seed colour.

a. Describe what is meant by the terms dominant and recessive.

b. In a genetic cross of two pea plants that are heterozygous for seed shape, what fraction of the offspring should have round seeds? (Use the Punnett square to help you with your answer.)

c. Use the example of pea plant seed shape to explain the difference between phenotype and genotype.

d. Yellow seed colour is dominant to green seed colour in the garden pea. Marika has a yellow pea. Discuss how she could determine whether the pea is heterozygous for the dominant allele or not.

Monohybrid inheritance – single character inheritance

Monohybrid inheritance is the inheritance of one feature controlled by *one gene*, which may have two or more different alleles located on **homologous chromosomes**.

Alleles are the different forms of a gene that provide information for a feature/trait. Alleles are typically represented by *letters*, with:

- An upper case/capital letter for a **dominant allele**.
- A lower case/small letter for a recessive allele.

Alleles occur in pairs (as chromosomes are paired).

Examples

D represents dimpled cheeks.

d represents smooth cheeks.

Genotype	Phenotype
DD	dimpled cheeks
Dd	dimpled cheeks
dd	smooth cheeks

(*Dd* and *dD* are the same – convention puts the dominant allele first.)

Complete dominance inheritance occurs when one allele dominates the other. A **dominant allele** is one that is always expressed in the phenotype when present in the genotype. Its presence masks the presence of the **recessive allele**. A recessive allele is one that is expressed in the phenotype when both alleles in the genotype are recessive.

When both alleles in the genotype are the same, the individual is **homozygous** for that feature and is said to be a **pure/true breeder** (as it can pass on only that allele to its offspring).

Examples

T means can roll tongue, *t* cannot roll tongue.

TT or *tt*.

F means has free ear lobes, *f* has attached ear lobes.

FF or *ff*.

When both alleles in the genotype are different, the individual is said to be **heterozygous** for the character and is not a pure/true breeder as it can pass either allele onto its offspring.

Examples

Tt is heterozygous as it can pass either the *T* or the *t* allele onto its offspring.

Ff is heterozygous as it can pass either the *F* or the *f* allele onto its offspring.

Genotype	Phenotype		
DD or *Dd* *dd*	dimpled cheeks smooth cheeks	dimpled cheeks	smooth cheeks
TT or *Tt* *tt*	can roll tongue cannot roll tongue	can roll tongue	cannot roll tongue
FF or *Ff* *ff*	free ear lobes attached ear lobes	free ear lobes	attached ear lobes

Example

Inheritance of cheek types

Problem: What are the chances of two heterozygous dimpled people having a smooth-cheeked child?

Let ***D*** = dimpled, ***d*** = smooth.

The parents are thus: ***Dd*** (Father ♂) and ***Dd*** (Mother ♀). ♂ signifies male, ♀ female.

Both parents have the same chance of producing either ***D*** or ***d*** gametes. The gametes from both the parents are thus ***D*** or ***d***. A sperm can fuse with an egg in the following ways:

- ***D*** sperm with a ***D*** egg – resulting **zygote *DD*.**
- ***D*** sperm with a ***d*** egg – resulting zygote ***Dd*.**
- ***d*** sperm with a ***D*** egg – resulting zygote ***dD*.**
- ***d*** sperm with a ***d*** egg – resulting zygote ***dd*.**

The ratio of the *genotypes* is thus:

1 ***DD*** : 2 ***Dd*** : 1 ***dd*** (Since ***Dd*** is the same as ***dD***, these two are added together as 2 ***Dd***.)

The ratio of the *phenotypes* is:

3 dimpled : 1 smooth

(3 dimpled comes from 1 ***DD*** + 2 ***Dd***. The 1 smooth comes from the ***dd***.)

The chance of a dimpled child is thus $\frac{1}{4}$ (1 out of the 4 possible outcomes). A chance of $\frac{1}{4}$ (ie a probability of 0.25) is just that – a statistical likelihood of an event. Two heterozygous people (***Dd*** and ***Dd***) need not have four children, and even if they do, one does not *have* to have dimpled cheeks.

The preceding problem is best dealt with using a Punnett square.

P: (Parents) ***Dd*** × ***Dd***

G: (Gametes) ***D*** or ***d*** ; ***D*** or ***d***

The phenotype ratio is 3 dimpled : 1 smooth.

♂ (sperm)

♀ (eggs)	*D*	*d*
D	*DD*	*Dd*
d	*Dd*	*dd*

The test cross

An individual that displays a dominant characteristic (**trait**) can either be homozygous dominant or heterozygous.

Example

Not having horns (called polled) is dominant to having horns in cattle.

Let ***P*** = polled, ***p*** = horned. A bull is polled (ie has no horns). To tell if the bull is homozygous dominant (***PP***) or heterozygous (***Pp***), he would be crossed with a herd of horned cows (***pp***).

If the bull is homozygous dominant	If the bull is heterozygous
P: (Parents) ***PP*** × ***pp***	***Pp*** × ***pp***
G: (Gametes) ***P*** ; ***p***	***P*** or ***p*** ; ***p***
♂ (sperm): *P*; ♀ (eggs): *p* → *Pp*	♂ (sperm): *P*, *p*; ♀ (eggs): *p* → *Pp*, *pp*
No calves have horns (ie all polled)	$\frac{1}{2}$ calves polled, $\frac{1}{2}$ have horns

To be sure of the genotype of the individual tested, it must be:

Examples

- Crossed with a *large number* of recessive individuals.

A polled bull (***PP*** or ***Pp***) would be crossed with a herd of horned cows (***pp***); all calves would be polled if the bull was ***PP***.

Or

- Bred a number of times; this is to ensure that the masked recessive allele will show in the offspring if present.

A polled cow (***PP*** or ***Pp***) would be crossed with a horned bull (***pp***) over a number of years; each calf would be polled if the cow was ***PP***.

The original generation/parents are known as the **P** (for parental) generation.

Offspring ('children') from a cross are know as the **F_1** (the **first filial generation**). If these offspring were in turn mated with each other, their offspring (the 'grandchildren') are known as the **F_2** (the **second filial generation**).

A **test cross** involves mating F_1 plants with a homozygous recessive plant (such as a parent called a *back-cross*). A homozygous recessive plant only produces gametes carrying the recessive allele, *so any variation in the offspring must be due to variation in the gametes produced by the F_1 plants.*

Example

Mendel produced F_1 (*Tt*) pure-breeding long-stemmed (*TT*) and pure-breeding short-stemmed parents (*tt*). The test cross was *Tt* × *tt*. As expected, about half the offspring of the test cross were tall and half short.

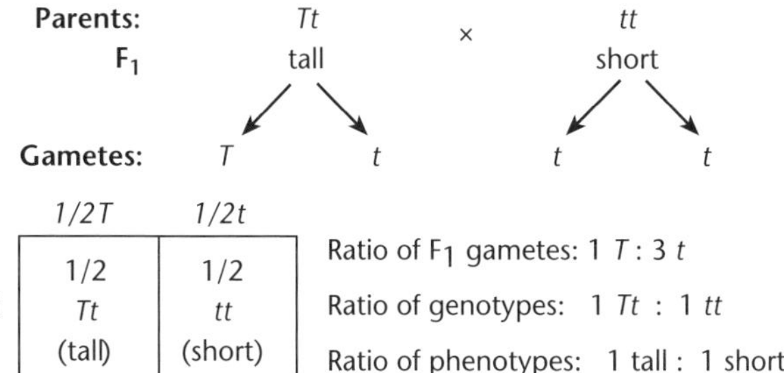

Using a test cross to show Mendel's F_1 plants produced two kinds of gamete in approximately equal numbers.

The 1 : 1 test cross ratio is a consequence of the 1 : 1 ratio of two kinds of gamete produced by F_1 plants.

This segregation of alleles later became known as **Mendel's First Law**, or the **Law of Segregation**:

'Of the two genes controlling each characteristic, only one is present in each gamete.'

Nearly four decades after Mendel did his experiments, the behaviour of chromosomes at meiosis was understood. It was realised that the segregation of alleles as deduced by Mendel paralleled the segregation of chromosomes as observed by microscopists. Only one of each pair of alleles can exist in each gamete, as can only one of each pair of chromosomes. This and other similarities between the behaviour of genes and chromosomes *suggested that genes are carried on the chromosomes*. It was later shown that each gene occupies a characteristic position in a chromosome, called a **locus**.

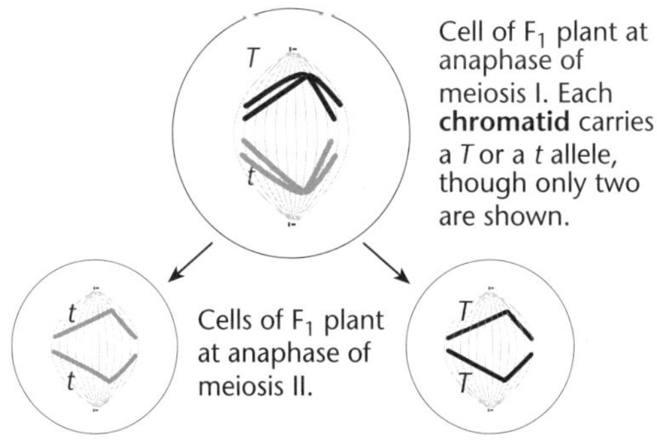

Segregation of genes and chromosomes in Mendel's F_1 plants.

Unit 12.3 Activity 1B: Monohybrid cross

1. Distinguish between the following pairs of terms:
   a. Gene and allele.
   b. Genotype and phenotype.
   c. Homozygous and heterozygous.
   d. Dominant and recessive.
2. A man with dimpled cheeks (all of whose immediate ancestors had dimples) marries a woman without dimples. All of their three children have dimples.
   a. Explain which allele is dominant (dimpled or smooth cheeks).
   b. Each child marries a partner whose cheeks are not dimpled. Explain with the help of a Punnett square what fraction of the grandchildren of the original parents will have dimples.
3. In many plant species, the inability to produce the green pigment chlorophyll is caused by a recessive allele (a) (this is an example of albinism).
   a. If a tobacco plant known to be heterozygous for albinism is self-pollinated and produces 600 seedlings:
      i. Explain *how many* of these seedling would be expected to be albino.
      ii. Explain *how many* would be expected to have the parental genotype.
   b. The albino allele is often known as a *lethal* allele in plants. Explain why this might be.
4. Gregor Mendel is often referred to as the 'father of modern genetics'. His work in the 19th century involved many thousands of breeding experiments using the garden pea.
   a. In pea plants, round seeds (R) are dominant to wrinkled seeds (r), which are recessive.
      i. In a genetic cross of two pea plants that are heterozygous for seed shape, state what fraction of the offspring should have round seeds.
      ii. Use the example of pea plant seed shape to explain the difference between phenotype and genotype.
   b. Yellow seed colour is dominant to green seed colour in the garden pea. Marika has a yellow pea. Discuss how she could determine whether the pea is heterozygous for the dominant allele or not.

Dihybrid inheritance – two character inheritance

The principles of the monohybrid patterns for complete (*see page 170*), incomplete/**co-dominance** (*see Topic 2, page 185*) and multiple alleles (*see Topic 2, page 185*) all apply to dihybrid inheritance.

Dihybrid inheritance is the inheritance of *two* genes controlling *two* different features. The genes may have two or more different alleles.

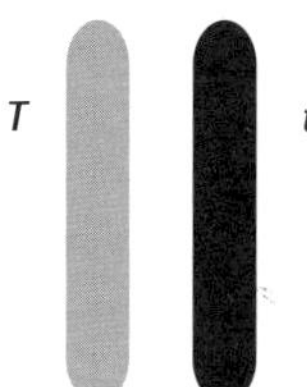

Homologous chromosomes with gene for stem length (heterozygous *Tt*)

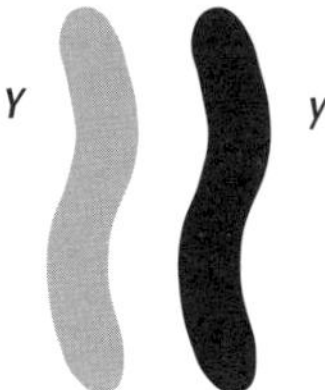

Homologous chromosomes with gene for seed colour (heterozygous *Yy*)

Example

In pea plants, tall stem (*T*) is dominant to short stem (*t*), and yellow seeds (*Y*) is dominant to green seeds (*y*). The two genes are on different homologous chromosomes.

Crossing a pure-breeding tall yellow-seeded plant (*TTYY*) with a pure-breeding short green-seeded plant (*ttyy*)

The cross between the two parents only involves two types of gametes:
TY from ***TTYY***, ***ty*** from ***ttyy***.

P: *TTYY* × *ttyy*
G: *TY* : *ty*
F_1

	TY
ty	*TtYy*

> This can be explained by meiosis as follows:
> - During meiosis for the ***TTYY*** parent, the chromosomes with the ***T*** alleles must go to the opposite ends of the cell during anaphase II; the chromosomes with the ***Y*** alleles must also do the same – each gamete must thus contain one chromosome with a ***T*** allele and one with a ***Y*** allele. Genotypes of gametes are all ***TY***.
> - Similarly, for the recessive plants ***ttyy***, all the gametes must be ***ty***.

Crossing the F_1 gives the F_2.
P: *TtYy* × *TtYy*
G: ***TY*** or ***Ty*** or ***tY*** or ***ty*** : ***TY*** or ***Ty*** or ***tY*** or ***ty***

The easiest way to show the cross is by using a 4 × 4 (four by four) Punnett square:

Male gametes

Female gametes	*TY*	*Ty*	*tY*	*ty*
TY	① *TTYY*	② *TTYy*	③ *TtYY*	④ *TtYy*
Ty	⑤ *TTYy*	⑥ *TTyy*	⑦ *TtYy*	⑧ *Ttyy*
tY	⑨ *TtYY*	⑩ *TtYy*	⑪ *ttYY*	⑫ *ttYy*
ty	⑬ *TtYy*	⑭ *Ttyy*	⑮ *ttYy*	⑯ *ttyy*

16 types of fertilisation, but 9 kinds of zygote (as 9 different genotypes possible), since some yield the same result.

The numbers ① to ⑯ represent 'types' of fertilisation/gametes, resulting in 9 different genotypes in the zygotes.

Genotypes	Number	Phenotypes
TTYY (zygote ①)	1	Tall yellow
TTYy (zygotes ②, ⑤)	2	Tall yellow
TtYY (zygotes ③, ⑨)	2	Tall yellow
TtYy (zygotes ④, ⑦, ⑩, ⑬)	4	Tall yellow
	Total = 9	
TTyy (zygote ⑥)	1	Tall green
Ttyy (zygotes ⑧, ⑭)	2	Tall green
	Total = 3	
ttYY (zygote ⑪)	1	Short yellow
ttYy (zygotes ⑫, ⑮)	2	Short yellow
	Total = 3	
ttyy (zygote ⑯)	1	Short green
	Total = 1	

The overall phenotype ratio is thus: 9 tall yellow : 3 tall green : 3 short yellow : 1 short green.

The 9 : 3 : 3 : 1 ratio is always the probability of getting each phenotype when two heterozygote dihybrids with complete dominance (eg *TtYy* × *TtYy*) are crossed.

Once a dominant allele is present in the genotype, it does not matter what the other allele is, because the dominant will determine the phenotype. A shorthand can be used for genotypes like this.

Example

TTYY, ***TtYY***, ***TTYy*** and ***TtYy*** are all effectively represented as ***T_Y_*** as they all produce the same phenotype (tall, yellow). Similarly, ***TTyy*** and ***Ttyy*** would be ***T_yy***.

For breeding purposes, homozygous individuals are typically needed. To determine whether an individual is homozygous or heterozygous dominant, we need to do a **test cross** (as for **monohybrids**). It does not matter whether the test is for one of the genes or both, the same principles apply:

- Breed the unknown individual with a (homozygous) recessive individual.
- Presence of the recessive feature in the offspring indicates the tested individual is heterozygous and so not suitable for breeding.
- If no recessive features occur in the offspring *after multiple matings*, the tested individual can be safely assumed to be homozygous and so can be used for breeding.

Example

For pea plants, a plant is tall with yellow seeds. To test whether it is homozygous or heterozygous for one or both features (ie ***TTYY*** or ***TTYy*** or ***TtYY*** or ***TtYy***), cross the plant with many short, green-seeded plants. If the offspring consist of:

- All tall with yellow seeds – the parent was homozygous for both genes (***TTYY***).
- Some tall and some short but all with yellow seeds – the parent is only homozygous for seed colour (***TtYY***).
- All tall but some with green seeds and some with yellow – the parent is only homozygous for height (***TTYy***).
- Some tall and some short, some green-seeded and some yellow-seeded, then the parent is heterozygous for both height and seeds (***TtYy***).

The cross that produces some tall and some short, some green-seeded and some yellow-seeded plants is ***TtYy*** x ***ttyy***:

TtYy \ *ttyy*	*ty*
TY	*TtYy*
Ty	*Ttyy*
tY	*ttYy*
ty	*ttyy*

The four different genotypes and their phenotypes are in an expected ratio of:
1 tall yellow seeded : 1 tall green seeded : 1 short yellow seeded : 1 short green seeded.

Using the garden pea (*Pisum sativa*) experiment, Mendel followed inheritance in two characters (**dihybrid cross**) over many generations. Results collected (a ratio of 9 : 3 : 3 : 1) for each pair of characters investigated led to his **Second law of inheritance**, also known as the **Law of Independent Assortment:**

'Two different genes will randomly assort (separate) their alleles during the formation of gametes'.

Unit 12.3 Activity 1C: Dihybrid cross

1. When sweetpea plants pure breeding for purple flowers and long pollen grains were crossed with sweetpeas that are pure breeding for red flowers and round pollen grains, all the F_1 offspring had purple flowers and long pollen grains.

a. Identify which features are dominant; and, using suitable letters, give the genotypes of both parents.

b. i. Give the genotype of the F_1 offspring.

ii. Two of the F_1 offspring were crossed to produce the following offspring:

- 4381 purple flowered with long pollen grains.
- 1470 purple flowered with round pollen grains.
- 1455 red flowered with long pollen grains.
- 480 red flowered with round pollen grains.

Explain these results with the help of a Punnett square.

2. There is an inherited disease in humans called phenylketonuria, which is caused by a recessive mutated allele (*p*). A man and a woman are both heterozygous for this condition (*Pp*). The woman is left-handed (*rr*), while the man is right-handed (*Rr*). The alleles for handedness are not on the same chromosome as the alleles for phenylketonuria.
   a. Give the genotypes for both parents.
   b. Using a Punnett square, work out the possible genotypes of their children. Explain the probability of any child being:
      i. Left-handed.
      ii. Left-handed with phenylketonuria.
3. In mice, the allele for black hair colour (*B*) is dominant to that for brown (*b*); the allele for short hair (*H*) is dominant to that for long hair (*h*). The genes for hair colour and length are not linked (ie they are on separate chromosomes).
   One colony (colony A) of mice was pure breeding for short black hair; a second colony (colony B) was pure breeding for long brown hair.
   a. A male from colony A was mated with a female from colony B.
      i. Give the genotype of the male from colony A and the genotype of the female from colony B.
      ii. These two mice were crossed and produced a large number of offspring (the F_1). Give the genotypes and phenotypes of this F_1 generation.
      iii. These F_1 offspring were then crossed several times and produced the following combinations in the F_2 generation:
         - 117 black long haired.
         - 372 black short haired.
         - 41 brown long haired.
         - 120 brown short haired.

         Give the expected ratio for this cross, and explain why this exact ratio rarely occurs.
      iv. Give all the possible genotypes for the brown short-haired mice.
   b. A brown long-haired male mouse was crossed with a black short-haired female mouse that was heterozygous for both these features. Assuming large numbers of offspring were produced, give the phenotypes of the offspring and the expected ratio in which they occur. Use a Punnett square to help with your answer.

Inheritance and pedigrees

Much can be learned about inheritance by studying family trees, or *pedigrees*. Usually:

- Males are represented by squares and females are represented by circles.
- Successive generations are indicated by Roman numerals.
- Shaded shapes represent individuals showing a particular trait.

Example

The ability to roll the tongue is hereditary.

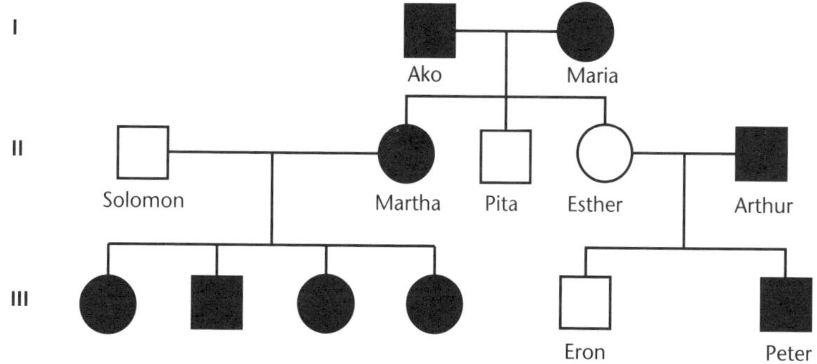

Black shapes indicate an ability to tongue roll, white shapes indicate no ability to tongue roll.

A pedigree showing inheritance of tongue-rolling ability.

Ako and Maria both can roll their tongue but have children who cannot. The allele for inability to roll the tongue must be recessive and Ako and Maria must both be heterozygous (*Aa*). Pita and Esther cannot roll their tongues. They must both be *aa*, having received an *a* allete from each parent. Esther's husband Arthur can roll his tongue and must be heterozygous. Martha's genotype cannot be determined with certainty. If she is homozygous, then all her children would be able to roll their tongues. If she is heterozygous, then she would be expected to have some of each kind. However, she could have all tongue-rolling children even if she is heterozygous.

Pedigree studies

Quite apart from the long generation time, human genetics is difficult to study because we prefer to choose our own mating partners. Human inheritance is thus best studied through family trees or pedigrees.

Example

The following pedigrees illustrate the inheritance of albinism (inability to make the skin pigment melanin) and polydactyly (extra fingers and toes).

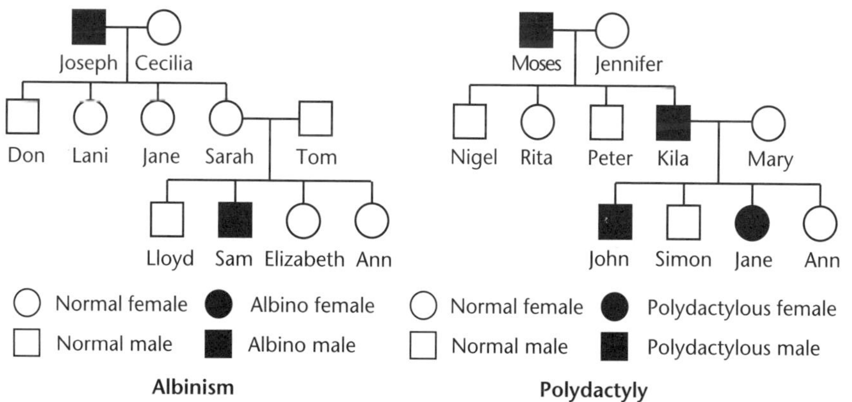

Pedigrees showing inheritance of albinism and polydactyly.

When a child differs from both parents, the child must show the recessive condition (since the allele must have been present in the parents but was not expressed).

Example

In the pedigree for albinism, Sarah and Tom do not have albinism, yet one of their children (Sam) has albinism. Sam must therefore have the genotype *aa* and must have received an allele from each of his parents, both of whom must therefore be *Aa*.

A recessive trait can 'skip' a generation; **dominant traits** cannot.

If an inherited condition is common, it does not necessarily mean that the allele responsible is dominant.

Example

Five digits is normal, but it is recessive to polydactyly, which is dominant.

Unit 12.3 Activity 1D: Pedigrees

1.

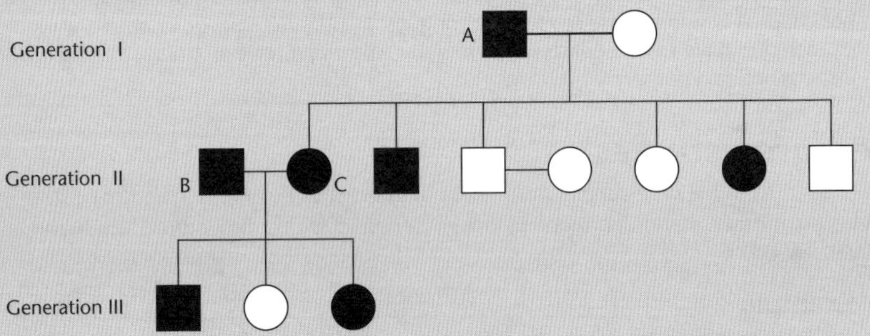

Key: Black = at least 1 dominant allele (normal or carrier)
White = no dominant alleles (CF sufferer)
Squares = males
Circles = females

Cystic fibrosis (CF) is a disease that affects the **lungs** and the digestive system. CF is controlled by the CFTR gene. Everyone has two copies of the CFTR gene; one inherited from each parent. A person with two recessive forms of the CFTR gene is affected by CF. A person with only one recessive form of the gene is unaffected, but is a carrier.

The pedigree diagram above shows three generations of a family with CF sufferers. Use the information in the diagram to answer the following questions.

a. Male A has six generation II children. What fraction of his male children have CF?

b. Explain the genotype of female C.

c. Discuss the phenotype of female C, and the phenotype of any children she has with:

i. Male B.

ii. A normal male.

2. In tigers, a recessive allele codes for white fur. Most tigers have orange-brown fur that is coded by the dominant allele.
The pedigree diagram below shows three generations of tigers. The black shapes indicate dominant alleles for coat colour. The white shapes indicate recessive alleles. Male tigers are shown by squares, females by circles.

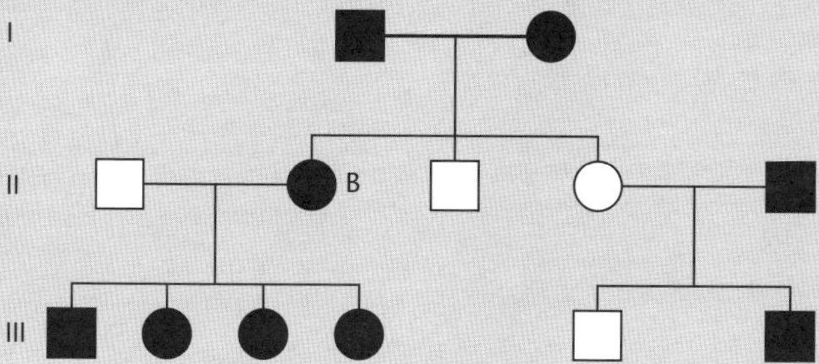

Use the information in the pedigree chart to answer the following questions:
   a. Describe the gender and fur colour in the third generation of tigers.
   b. Explain why, if both of the original tigers have orange-brown fur, some of their cubs have white fur.
   c. Discuss the possible genotype of tiger B, stating why, from the information given, it cannot be determined with certainty.

3. The following pedigree shows the result of crossing two animals, both white in colour. They produced six offspring, five were white and one was black.
In pedigrees like this, males are represented by squares and female by circles.

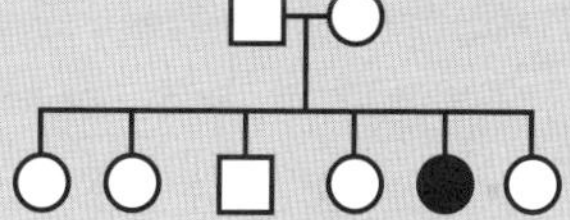

   a. Explain which allele is dominant, white or black.
   b. Using appropriate letters, give the genotypes of the two parents.
   c. Give the expected ratio of white to black offspring if the cross was repeated many times.

Unit 12.3 Genetics

Topic 2: Monohybrid inheritance – an extension

Authors: Martin Hanson with Takis Solulu

In Topic 2, the role of DNA is described in relation to gene expression/inheritance, by:

- Discussing biological concepts and processes relating to the determination of phenotype, through allele interactions involving dominance, incomplete dominance, co-dominance, multiple alleles and lethal alleles; and gene–gene interactions involving polygenes.

Mendel's experiments

In the middle of the 19th century, Gregor Mendel performed a series of breeding experiments using the garden pea, *Pisum sativum*. Although he performed his experiments more than 20 years before chromosomes were discovered, he was able to deduce some of the important properties of genetic material. His success was due to the careful design of his experiments:

- He used **true-** or **pure-breeding** varieties, which, when mated with their own type, produced offspring that all resembled the parents.
- He used *discontinuously* varying characters, ie characters which differ sharply from each other, eg flowers red or white; stems tall (2 m) or short (0.5 m); seeds round or wrinkled; seeds yellow or green.

  This enabled him to sort the offspring into groups, count the number in each group, and calculate the *ratios* between them. (Discontinuously varying characters are relatively unaffected by environment, thus differences he observed were mainly due to hereditary factors.)
- He kept his experiments simple by crossing varieties that differed in a small number of ways.

The flowers of the garden pea are **hermaphrodite** (have both male and female parts). To use a flower as a 'female', Mendel first removed the stamens from the flower before it opened, so it could not pollinate itself. He then used a small brush to transfer pollen from the stamens of a 'male' plant to the **stigma**, and covered the flower with a bag to prevent insects from bringing pollen from other flowers. When he wanted a flower to pollinate itself, all he had to do was to 'leave it alone' since the garden pea is normally self-pollinating.

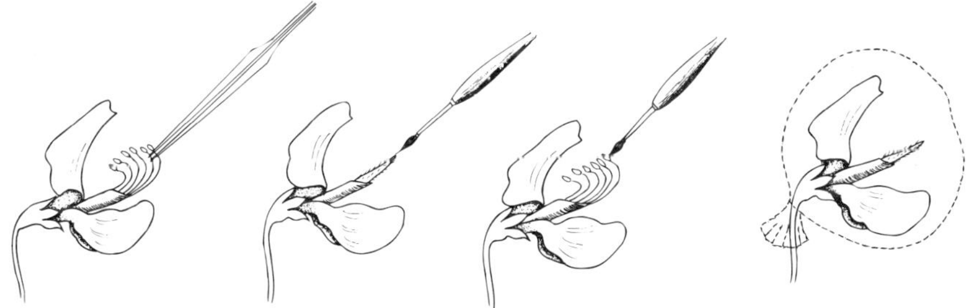

Method of artificial **cross-pollination.**

In **monohybrid** crosses, Mendel considered only one pair of contrasting characters at a time. In **dihybrid crosses**, he considered two pairs of contrasting characters at the same time.

In one monohybrid experiment, he crossed a plant true-breeding for tall stem with a plant true-breeding for short stem. All the offspring, which he called the **first filial generation** or $\mathbf{F_1}$, were as tall as the tall parents.

Next, he allowed the F_1 plants to fertilise themselves, and obtained a **second filial generation** or $\mathbf{F_2}$. Of these 'grandchildren', 787 grew tall and 277 grew short, a ratio of 2.84 : 1 (obtained by dividing 787 by 277). Within the limits of chance variation this is reasonably close to a ratio of 3 : 1 or $\frac{3}{4}$ tall, and $\frac{1}{4}$ short. These results can be summarised as follows:

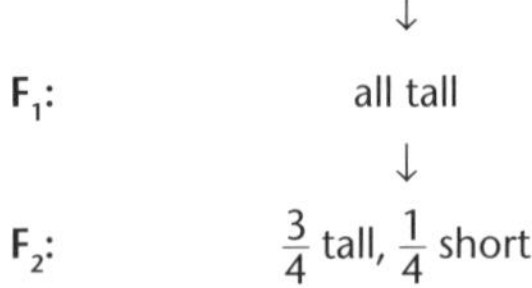

Mendel also found that reciprocal crosses gave the same result, tall female × short male giving the same result as short female × tall male. From further breeding experiments, he found that all the F_2 short plants and about $\frac{1}{3}$ of the F_2 tall plants were true-breeding. These results show that:

- Although the F_1 plants looked like their tall parents (ie had the same **phenotype** or bodily characteristics), they had a different **genotype** or hereditary makeup because they were not true-breeding.
- Although the F_1 plants were all tall, they must have carried information for shortness, since some of the F_2 plants were short. The factor for shortness is **recessive** to the factor for tallness, which is said to be **dominant**.
- Since reciprocal crosses gave the same result, both parents (and hence both male and female gametes) must make an equal genetic contribution to the offspring. Since the male gamete is almost entirely nucleus, the genetic material is probably in the nucleus.
- Because both contrasting phenotypes were present in the F_2, the factors responsible must have retained their identity in the F_1. Since neither factor 'diluted' the other, the genetic material must behave like particles rather than a fluid, as was thought by many at the time. These hereditary particles were later called **genes**.
- Since the zygote receives genes from both parents, stem height must be controlled by *pairs* of genes. Although the genes for tallness and shortness had come together in the F_1, at least some of the F_2 plants must have carried only one kind of factor because they were true-breeding. Therefore, when F_1 plants produced pollen and eggs, the genes for tallness and shortness must have separated or segregated.

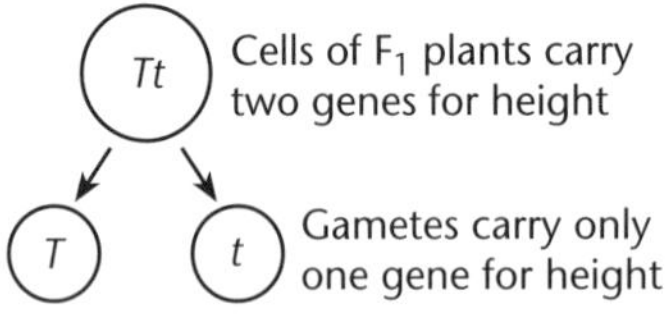

Segregation of alleles.

> An organism's phenotype is often defined as its *appearance*, but in fact most phenotypic characters are not visible, and many are not even structural, such as blood-**clotting** time, blood group, red-green colour-blindness, intelligence, and so on.

Incomplete dominance

When a heterozygote is phenotypically intermediate between the two homozygotes, the alleles are said to show **incomplete dominance**.

Example

When true-breeding, black Andalusian chickens (*AA*) are crossed with true-breeding white ones (*aa*), the F_1 are grey-blue. When the F_1 birds are inbred (brothers and sisters mated together), the F_2 generation consists of black, grey-blue and white birds in a ratio of 1 black : 2 grey-blue : 1 white. Since black and white appear 'undiluted' in the F_2, the alleles have not blended like fluids (as might appear from the F_1), but have remained independent, despite the fact that they interact in their effects in a heterozygote.

	1/2 A	*1/2 a*
1/2 A	1/4 *AA* (black)	1/4 *Aa* (blue)
1/2 a	1/4 *Aa* (blue)	1/4 *aa* (white)

Ratio of F_1 gametes: 1*A* : 1*a*

Ratio of F_2 genotypes: 1 *AA* : 2 *Aa* : 1 *aa*

Ratio of F_2 phenotypes: 1 black : 2 blue : 1 white

Incomplete dominance.

Multiple alleles and co-dominance

In some cases, a gene may exist as more than two alleles, but the rules of inheritance are the same. Each individual cannot carry more than two genes for a given locus; gametes can carry only one.

The ABO blood groups

Blood groups are the result of complex substances on the surface of **red blood cells**. These substances are called **antigens** because they provoke the production of defensive proteins called **antibodies** when introduced into another person. The antigens are under the control of three alleles – I^A, I^B and I^O. I^A and I^B are both dominant to I^O. People who are blood group O are homozygous recessive and have the genotype I^O I^O. A person of group A can be homozygous (I^A I^A) or heterozygous (I^A I^O). Similarly, a person of group B can be either I^B I^B or I^B I^O. A person of group AB has the genotype I^AI^B. The alleles I^A and I^B are said to be **co-dominant** because each of the two alleles has a different kind of effect, but both are expressed in the heterozygote. (In incomplete dominance, the same kind of character is expressed, but to an extent intermediate between the two homozygotes.)

Until the development of **DNA profiling**, blood groups were important in paternity disputes (to determine the father of a child).

Example

A group O child cannot have a group AB father, no matter what the mother's blood group.

Multiple alleles increase the number of possible genotypes. With two alleles there are three possible genotypes, but with three alleles there are six (two for each of groups A and B and one for each of groups O and AB). Since many genes exists as more than two alleles, there is considerable scope for variation in a population.

Lethal alleles

Some genes exert lethal effects when they exist as one of the homozygous genotypes – this results in a 2 : 1 F_2 ratio.

Examples

Coat colour in a particular strain of mouse can be either yellow or grey. Through inbreeding, it has been established that the allele for yellow fur is dominant to that for grey fur. When heterozygous yellow mice are mated, the resulting progeny are found in the ratio of 2 yellow : 1 grey instead of the expected 3 yellow : 1 grey. Further testing shows pure-breeding yellow mice never occur – all yellow mice are heterozygous. In addition, heterozygous yellow females abort homozygous yellow foetuses. The yellow allele has two effects – as a single 'dose' in the heterozygote, it results in yellow fur, but as a double 'dose' in the homozygote it has lethal effects.

In plants there are genes that code for the production of chlorophyll. One allele exists that does not produce sufficient chlorophyll. Usually when homozygous recessive, these alleles cause death of seedlings as they cannot manufacture sufficient chlorophyll for photosynthesis. A 2 : 1 F_2 ratio of leaf colour is seen.

Unit 12.3 Activity 2A: Monohybrid inheritance

1. Match each phrase **1–26** to a term **A–Z**.

1.	The part of a cell which contains chromosomes	**A**	Gene
2.	One of a pair of chromosomes	**B**	Genotype
3.	A threadlike body consisting of many genes	**C**	Heterozygous
4.	The physical or bodily characteristics of an organism	**D**	Recessive
5.	Hybrid formed by crossing two differing pure-breeding varieties	**E**	Dominant
6.	A cell that can only develop further by joining with another such cell	**F**	Autosome
7.	Mating between close relatives	**G**	Segregation
8.	Having the same allele twice	**H**	Chromatid
9.	A kind of nuclear division in which the chromosome number is halved	**I**	Zygote
10.	The genes carried by an organism	**J**	Spindle
11.	An allele that is expressed in heterozygotes	**K**	Nucleus
12.	A fertilised egg	**L**	Centromere
13.	The fundamental unit of inheritance	**M**	Inbreeding
14.	One of a number of alternative forms a gene may take	**N**	Test cross
15.	Produced by inbreeding first-generation hybrids	**O**	Homologue
16.	A chromosome that is not concerned with sex determination	**P**	Allele

17. One of two identical chromosome strands, united by a centromere
18. Having different alleles of the same gene
19. Separation of alleles at meiosis
20. Nuclear division in which the products are genetically identical
21. Having two sets of chromosomes
22. System of fibres concerned with chromosome movement in cell division
23. Characteristics are distinct enough to put individuals into distinct groups
24. The part of a chromosome which attaches to the spindle
25. An allele that is only expressed in homozygotes
26. Mating with a homozygous recessive in order to determine an organism's genotype

Q Diploid
R Meiosis
S Chromosome
T Phenotype
U F_1
V F_2
W Gamete
X Mitosis
Y Homozygous
Z Discontinuous variation

2. Most pea plants have tendrils, which help to support the plant, but plants with no tendrils also occur. The results of four crosses between different plants are shown in the table below:

Cross	Parents	Offspring
1	Plant A (tendrils) x Plant B (tendrils)	All with tendrils
2	Plant B (tendrils) x Plant C (tendrils)	73 with tendrils : 27 no tendrils
3	Plant A (tendrils) x Plant D (no tendrils)	All with tendrils
4	Plant B (tendrils) x Plant D (no tendrils)	52 with tendrils : 48 no tendrils

a. Explain which of the two phenotypes (tendrils or no tendrils) is due to a recessive allele.
b. Using the symbols *T* and *t* if presence of tendrils is dominant, or *N* and *n* if no tendrils is dominant, account for the results of:
   i. Cross 2. ii. Cross 4.
c. How many gene loci control the presence or absence of tendrils in these plants? Explain your answer.

3. The four groups of the ABO blood group system are determined by a series of multiple alleles.

The allele I^O is recessive, and the alleles I^A and I^B are co-dominant. The table below shows the blood group phenotypes and possible genotypes.

Blood group phenotypes	Possible genotypes
Group O	$I^o I^o$
Group AB	$I^A I^B$
Group A	$I^A I^o$ or $I^A I^A$
Group B	$I^B I^o$ or $I^B I^B$

a. Describe the possible blood groups of children of the following parents:

i. Group O mother and group A father.

ii. Group B mother and group A father.

b. In a case of disputed paternity (fatherhood), a woman claims that a man, whose blood is group AB, is the father of her group O child. Could her claim be correct? Explain.

4. In mice, black coat (*B*) is dominant to brown (*b*). Explain how you would find out the genotype of a black mouse.

5. For each of the pedigrees below, explain whether the allele governing the characteristic indicated by the solid symbol individuals is definitely dominant, probably dominant, definitely recessive, probably recessive, or impossible to say.

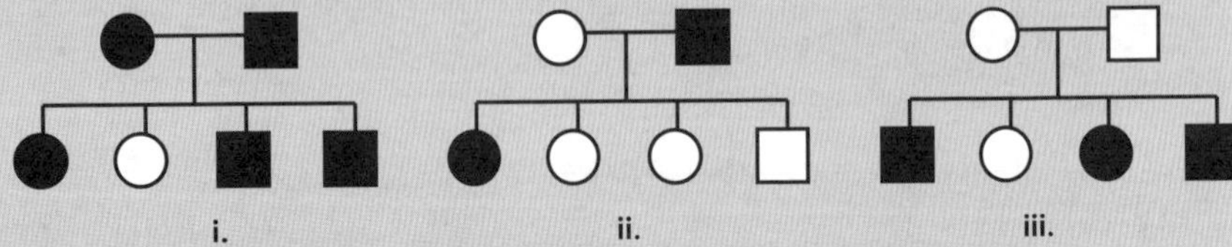

6. A pure-breeding red-flowered antirrhinum was crossed with a pure-breeding white variety, and all the F_1 had pink flowers.

a. Describe the genetic makeup of pollen made by a:

i. Red flower. **ii.** Pink flower.

b. Using the symbols *R* and *r* for red and white respectively, describe what would be the offspring if the F_1 plants were inbred.

c. Describe the expression of the alleles *R* and *r*.

7. Identify what is incorrect about the following statement:

'When two people heterozygous for brown/blue eye colour have children, the result is three brown-eyed children and one blue-eyed child.'

8. The diagram shows the inheritance of brachydactyly, a condition in which the fingers and toes are abnormally short. Affected individuals are shown in solid shapes.

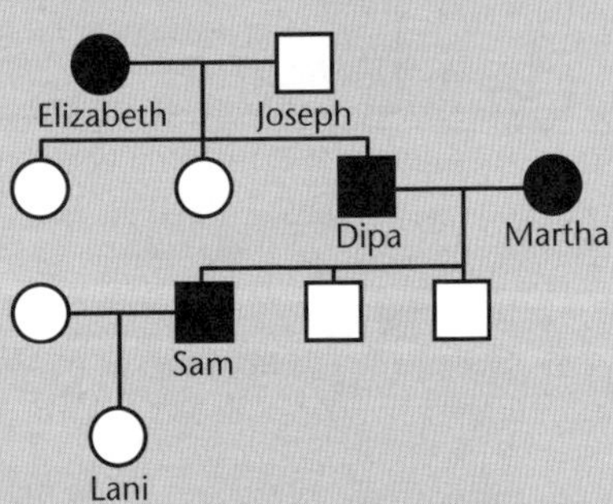

Which of the following statements is justified from the information in the pedigree? Explain your answer.

A Brachydactyly is definitely dominant.

B Brachydactyly is probably dominant.

C Brachydactyly is definitely recessive.

D Brachydactyly is probably recessive.

E No firm conclusion can be drawn.

9. There is a mutation in foxes that causes a platinum coat colour instead of the usual red. Foxes were bred in some countries for their fur and the platinum colour was very popular, so much so that breeders tried to develop pure-breeding strains of platinum foxes. However, no matter how many times they tried, when breeding platinum foxes, some red progeny were always produced. For example, one pair of platinum foxes was bred to produce 120 offspring of which 84 were platinum and 36 were red. Similar ratios were obtained for other matings. Carefully analyse these results to discuss the inheritance pattern of this mutation.

10. Maple syrup urine disease (named after the distinctive smell of the urine) is a rare metabolic disorder that is inherited. The disease appears throughout generations of a family, but the parents of an affected child are not affected themselves. Explain the inheritance pattern of the disease.

11. Pedigrees for the inheritance of three human disorders are shown below:

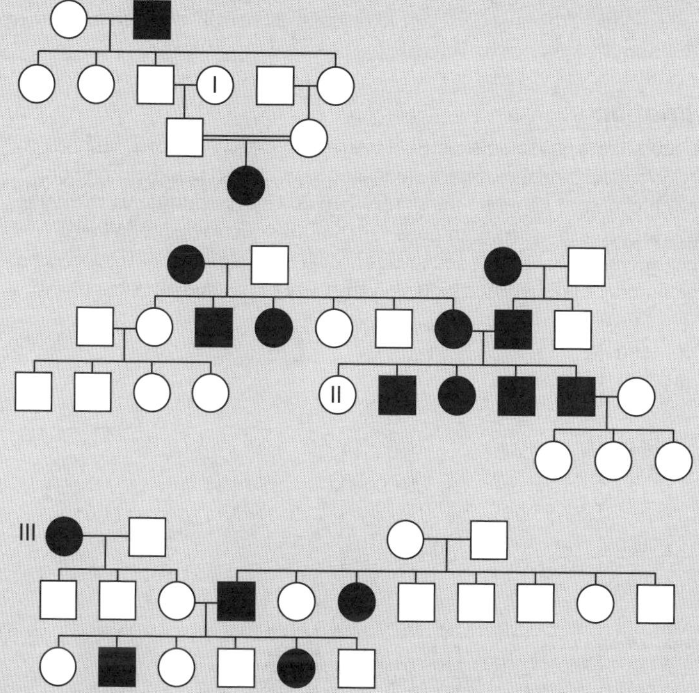

a. Explain the inheritance pattern for each of the three disorders.

b. Describe the genotypes for the three individuals (I, II and III) indicated on each pedigree.

12. In many plant species, the inability to produce the green pigment chlorophyll is caused by a recessive allele (*a*) (this is an example of albinism).

a. If a tobacco plant known to be heterozygous for albinism is self-pollinated and produces 600 seedlings:

i. Explain *how many* of these seedling would be expected to be albino.

ii. Explain *how many* would be expected to have the parental genotype.

b. The albino allele is often known as a *lethal* allele in plants. Explain why this might be.

Complex/polygenic inheritance

Types of variation

Discontinuous variation

This involves characters that are of the 'either-or' kind – eg blood groups, tongue-rolling ability, the ability to taste phenylthiocarbamide (PTC), or to distinguish visually between red and green. Mendel used discontinuously varying characters in his breeding experiments.

Continuous variation

In this kind of variation, there is a smooth series of intermediates between two extremes, so there are no distinct groups. Human examples are height, skin colour, blood-clotting time, blood pressure, and thickness of fat under the skin. Whereas with discontinuously varying characters organisms are *counted*, with continuous variation they are *measured*. When a large number of measurements have been obtained, the data can be plotted in the form of a histogram, with the number of individuals plotted against size of the measurement.

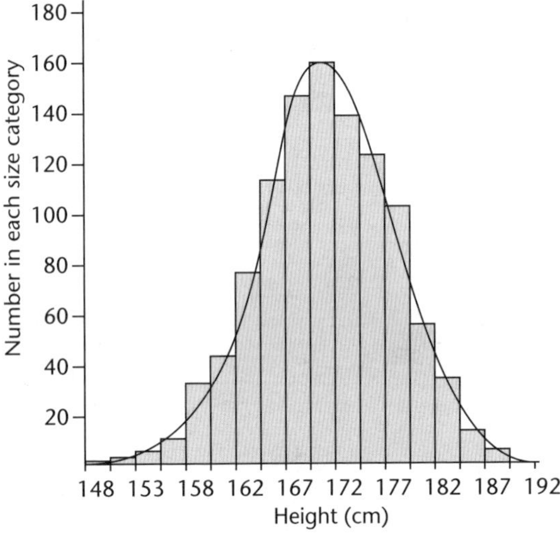

Variation in height amongst 1083 adult men.

Characteristics such as those mentioned above are controlled by many genes – called **polygenic inheritance**. Each individual gene has a small effect on the phenotype, and changing one allele for another at a particular locus for one of the genes has an even smaller effect. Genes in a polygenic system are inherited according to Mendelian principles.

Inheritance and polygenes

Polygenes resulting in continuous variation can be explained by conventional genetics if we assume that:

- There are many loci involved, giving scope for a large number of genotypes. The addition of many individually small gene effects would give many phenotypes differing by small degrees.
- Many, perhaps most, continuously varying characters are sensitive to environmental variations (eg skin colour, body fat, blood pressure). Since no two organisms have identical environments, even individuals with the same genotype will show phenotypic variation.

To see how these two factors could act to produce a smooth distribution of phenotypes, suppose that the intensity of human skin pigmentation is under the influence of two pairs of alleles, and the following assumptions are made:

- Each of the two loci exists as two alleles *A/a* and *B/b* and there is no dominance; each allele of each gene is incompletely dominant to the other alleles.
- Each 'upper case' allele contributes one unit to the darkness of skin pigmentation, and each 'lower case' allele contributes none.
- Both loci have the same effect and the individual gene effects add together, so that in the same environment *AAbb* gives the same pigmentation as *aaBB* or *AaBb*.

The expected results of matings between *AaBb* × *AaBb* people could produce five shadings of skin colour:

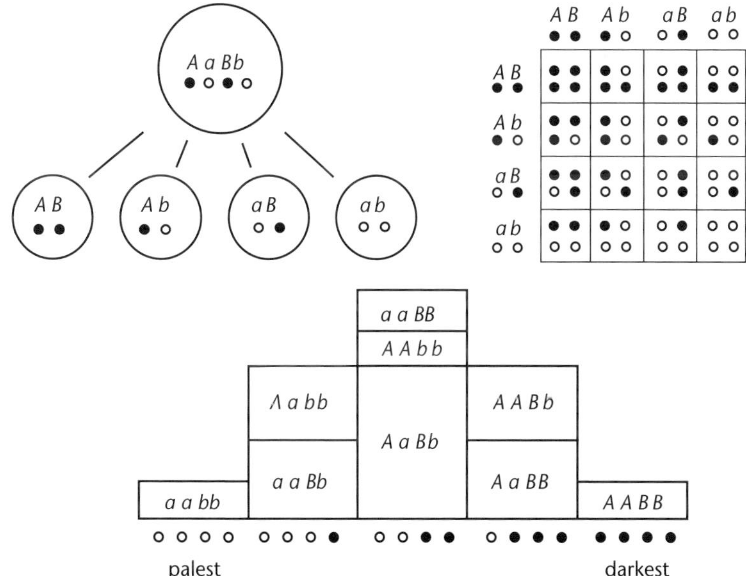

Incomplete dominance at two loci for skin colour.

The situation is probably more complicated than that outlined, with as many as five gene loci probably being involved in skin colour. The more loci (and therefore alleles) involved, the greater the number of phenotypic classes.

Complex characters

Polygenic inheritance is usually dealt with in connection with characters that are simple in the sense that they are *measurable*, such as height, body mass, and so on. But of course an anatomical feature such as an eye is an extremely complicated assemblage of smaller structures, each of which may be itself made up of smaller parts. It is therefore hardly surprising that an eye, a stomach or a heart is the product of the action of many, probably thousands, of genes, and at the moment it is impossible to disentangle their individual effects. However, *each* contributing gene is inherited according to normal Mendelian principles.

Unit 12.3 Activity 2B: Polygenic inheritance

1. **a.** Distinguish between continuous and discontinuous variation, including an example of each.

b. Which is easier to use in the study of inheritance, continuous or discontinuous variation, and why?

2. Human reflex time is probably controlled by many genes. Describe why.

3. The table shows the results of an experiment involving a cross between two pure-breeding varieties of corn – one short-eared (P_1) and the other long-eared (P_2). The figures below show the frequency distribution of ear length in the parental, F_1 and F_2 generations.

Length of ear (cm)

	6	7	8	9	10	11	12	13	14	15	16	17	18	19	20	21
P_1	8	22	26	9	1											
P_2							2	10	13	17	27	18	11	7	2	
F_1					1	11	13	16	15	9	3					
F_2			1	10	1825	42	76	6761	41	28	14	8	1			

a. What evidence is there in the data to suggest incomplete dominance in these genes?

b. Is the variation in the parental generations likely to be due mainly to genetic or environmental variation? Justify your answer.

Unit 12.3 Genetics
Topic 3: Chromosomes, cell division and sex-linked inheritance

Topic 3 explores the role of DNA in relation to gene expression/inheritance. The material covered in this Topic includes:

- Chromosomes.
- How cells divide – mitosis and meiosis.
- Sex chromosomes and karyotype.
- Sex determination.
- Biological concepts and processes relating to the determination of phenotype, through allele interactions involving sex chromosomes.

Genetics is the study of hereditary information. It fall into three parts:

- How the information is carried, and how, in **sexual reproduction**, it is transmitted to the offspring in such a way that no two offspring are exactly alike.
- What the genetic material consists of, and how it is copied, and how the information is used to build a body. This is the study of molecular genetics.
- How the genetic material undergoes changes.

In recent years, techniques have been developed for modifying the genetic material artificially and in a purposeful manner. This is the field of genetic engineering.

The **nucleus** of every cell of every individual contains **chromosomes**. Chromosomes are made of DNA (the genetic material) and protein. Different species may have different *numbers* of chromosomes; the chromosomes usually also differ in *size* and *shape*. Humans have 46 chromosomes (= 23 pairs) in the nucleus of every body cell.

Chromosomes

If parents and offspring ('children') of any species are compared, two things are apparent:

- Offspring closely resemble their parents.
- There are always slight differences, both among offspring and between parents and offspring.

Genetics is the study of how these similarities and differences come about.

Most living things begin life as a fertilised **egg** or **zygote**, which originates by joining together a male **gamete** (**sperm**) and a female gamete or **ovum** (plural **ova**). Each gamete carries information for building the new organism. Half the information in a fertilised egg comes from the male parent (via the sperm) and the other from the female parent (via the egg).

In a cell that is not dividing, the chromosomes are very long (several centimetres), thin and tangled, and cannot be distinguished. Before a cell divides, each chromosome is replicated (copied), forming two identical **chromatids**. These are held together at a point called a **centromere**.

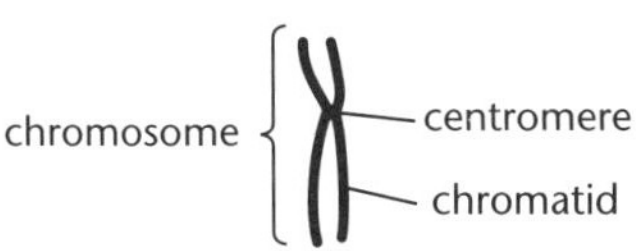

Diagram of a chromosome.

Each chromatid contains a single DNA molecule. In cells that are not going to divide, the chromosomes remain single-stranded. Before a cell divides, each chromosome shortens by coiling and becomes fatter and is visible when stained.

In any particular species, all the cells of the body have the same number of chromosomes – eg human cells have 46 chromosomes, mice have 40, corn has 20, and peas have 14. The gametes have only half these numbers – human gametes contain 23 chromosomes, mouse gametes have 20, pea gametes have 7. Because they have only one set of chromosomes, gametes are said to be **haploid**. Body cells have two chromosome sets, with two of each kind, and are said to be **diploid**. In humans the diploid number ($2n$) is 46 and the haploid number (n) is 23.

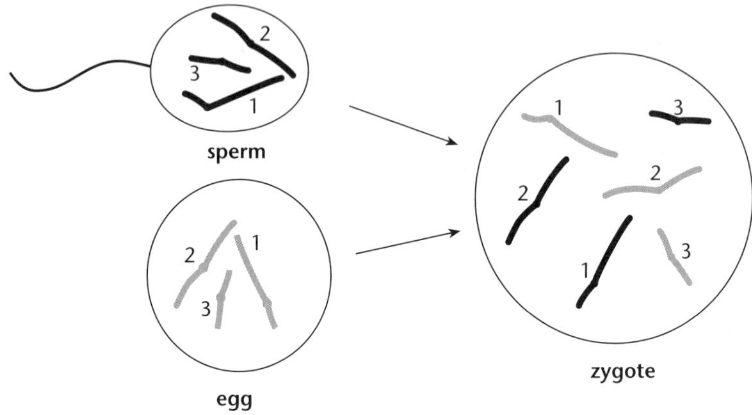

At fertilisation, two haploid cells join to form a diploid cell.

The chromosomes of a gamete differ in their length, centromere position, and banding pattern (revealed by special staining).

When a sperm joins with an egg, the resulting zygote has two chromosome sets, one from each parent. *Maternal* chromosomes are inherited from the mother, *paternal* chromosomes from the father. Each maternal chromosome looks just like one of the paternal chromosomes, and vice versa. Chromosomes are thus said to be in **homologous pairs** (eg the 14 chromosomes in each cell of a pea plant are of seven different kinds, two of each kind).

Staining shows each chromosome has a pattern of banding different from every other chromosome, except for its **homologue** or 'partner'. The chromosomes of a species are characteristic – when photographed and arranged in an orderly pattern, they are the **karyotype** for that species.

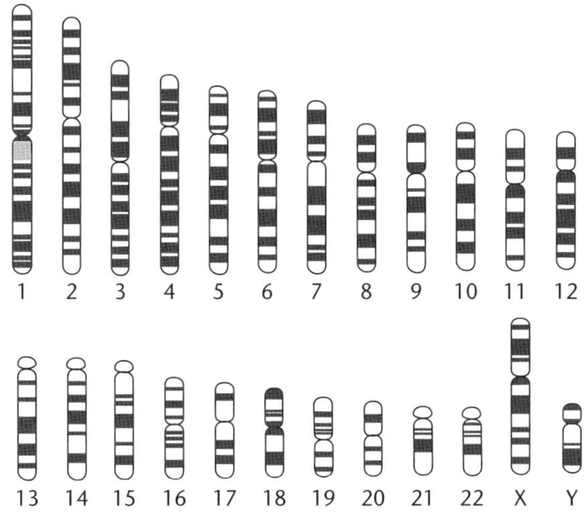

Only one of each chromosome of each homologous pair is shown.

The human karyotype.

Karyotype

The chromosomes present in an individual may be seen on a **karyotype** – a photograph of the chromosomes after they have been stained, matched for size and shape, then numbered and grouped in order (from 1 to 22 in humans).

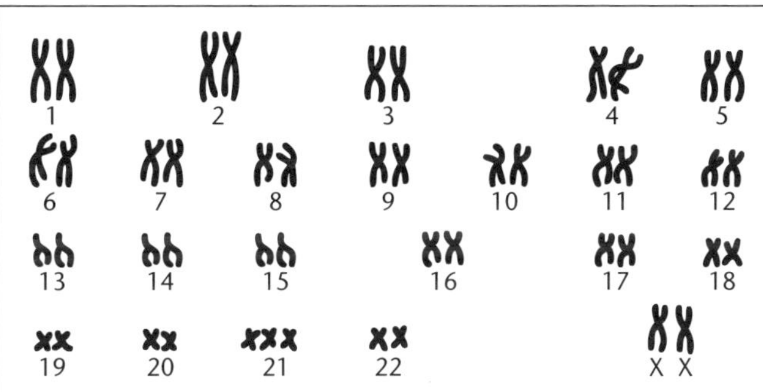

This person is a female with **Down's syndrome**, as she has 2 **X chromosomes** and 3 chromosomes of number 21.

Cell division

Cells divide in either of two ways – by **mitosis** or by **meiosis**.

- Mitosis occurs as part of *growth* and is simply concerned with increasing the number of genetically identical cells. It results in no change in chromosome number.
- Meiosis only occurs in the sex organs and results in a change from the diploid to the haploid number of chromosomes.

At fertilisation the diploid number of chromosomes is restored.

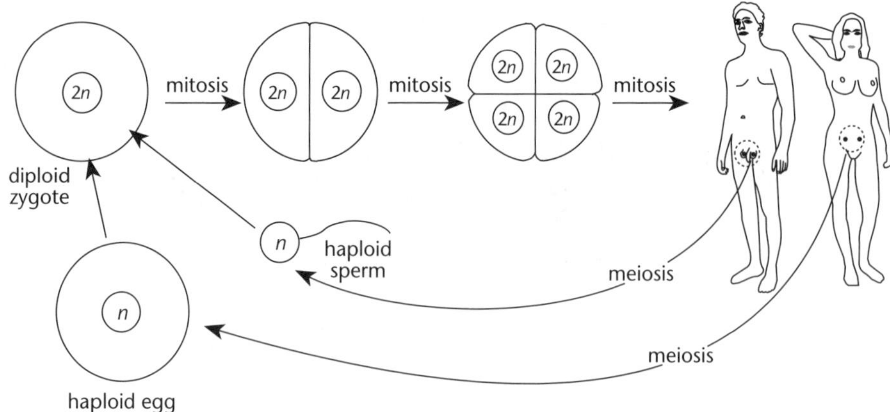

Relationship between meiosis, fertilisation and mitosis.

Mitosis

In very early human development (in the **embryo**), mitosis occurs throughout the body, but eventually becomes restricted to places like the **bone marrow**, skin, and base of the hairs and nails. It also occurs in wound healing. Mitosis occurs in haploid organisms as well as diploid organisms (eg mosses and male bees are haploid, and they grow by mitotic division). In single-celled organisms, mitosis results in **asexual** reproduction.

Before a cell divides, DNA is copied or *replicated* and each chromosome forms two identical chromatids. Until the cell divides, the chromatids remain attached at the centromere.

The events of mitosis

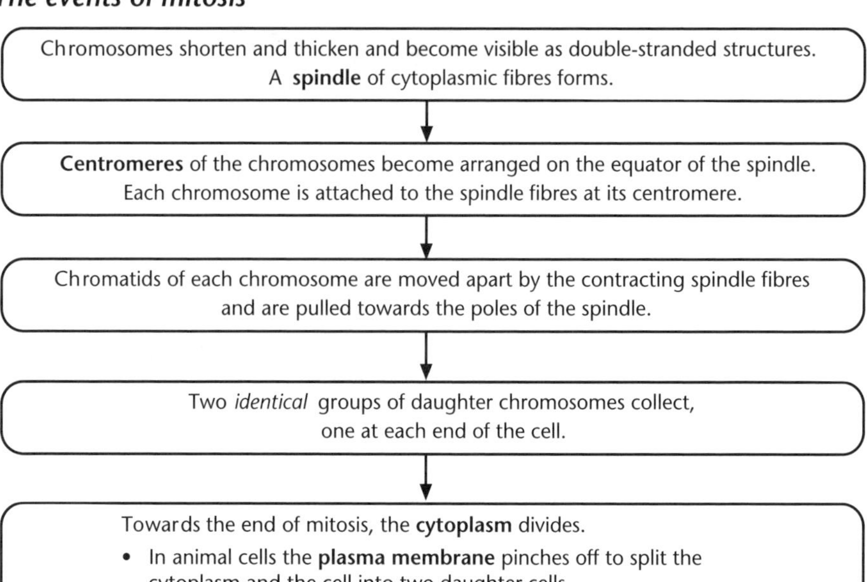

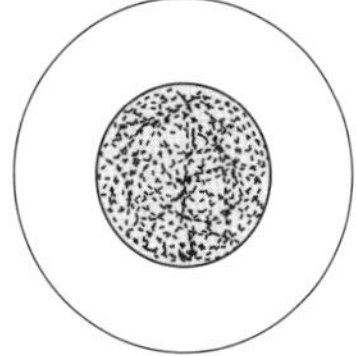

1. Between divisions: chromosomes replicating but not visible as separate structures

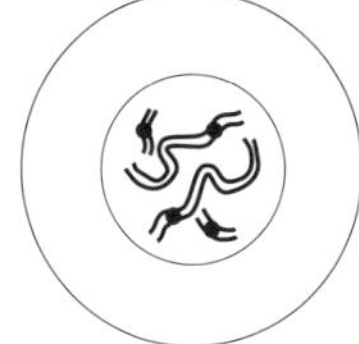

2. Chromosomes shorten, become visible and replicate

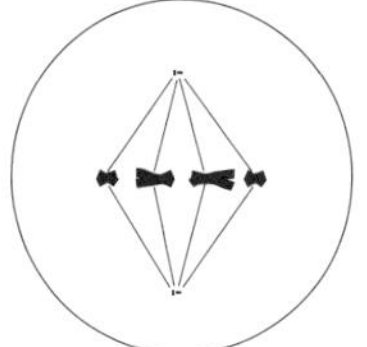

3. Chromosomes become arranged on equator of spindle

4. Chromatids move apart towards opposite poles of spindle

5. Daughter chromosomes lengthen and become indistinct

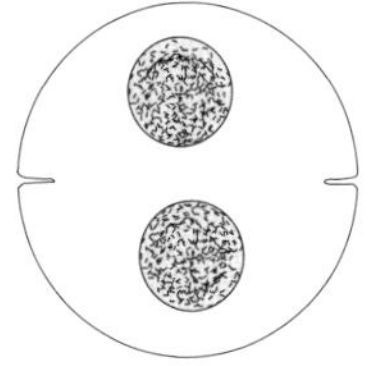

6. Cytoplasmic cleavage

The main events of mitosis in an animal cell, shown in more detail.
Though not part of mitosis, interphase and cytoplasmic cleavage are shown for convenience.

Mitosis.

Meiosis

Meiosis differs from mitosis:

- In meiosis, the daughter cells are genetically *different* from each other; in mitosis the daughter cells are genetically identical.
- The products of meiosis have only half as many chromosomes as the parent cell.
- There are *two* divisions; mitosis consists of only one division.
- Meiosis only occurs in mature organisms – in mammals, it occurs in the **ovaries** and **testes**, in flowering plants, it occurs in the stamens and **ovules**; mitosis occurs throughout life.

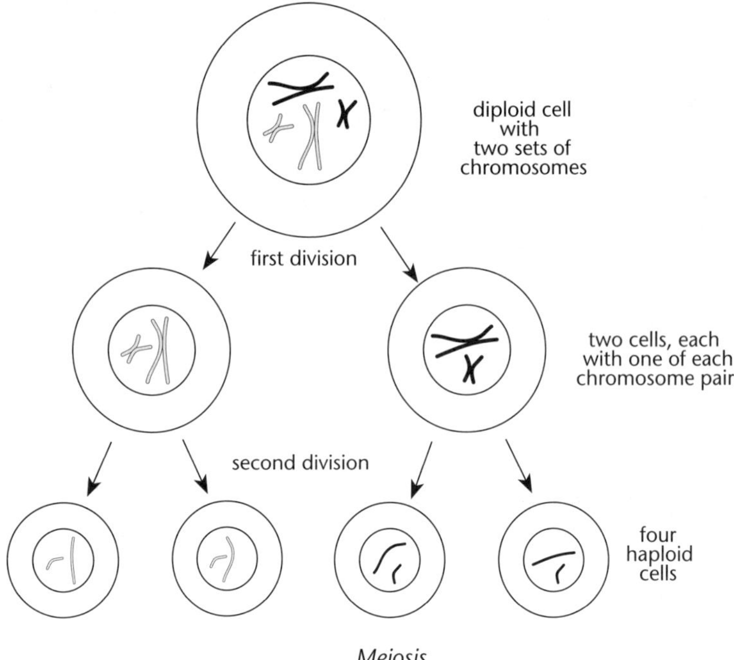

Meiosis.

First division

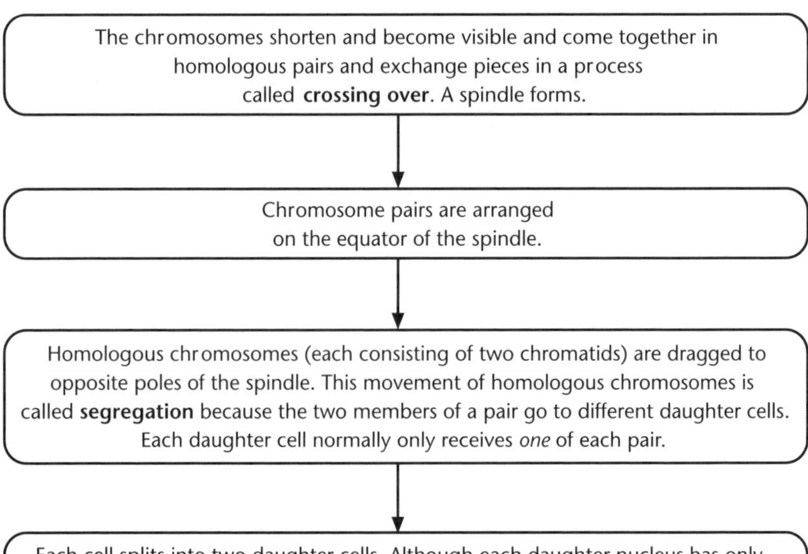

Second division

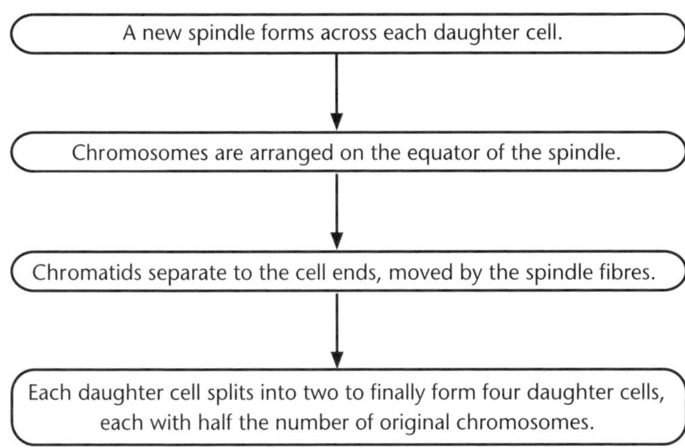

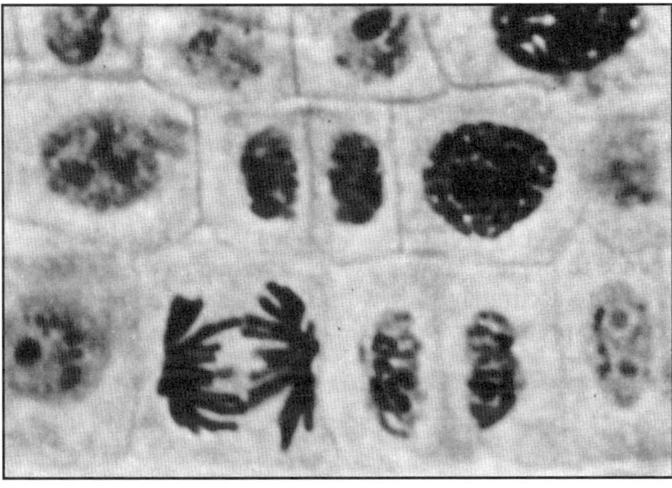

Cells from a growing root tip showing several stages in mitosis. Magnification 400x.

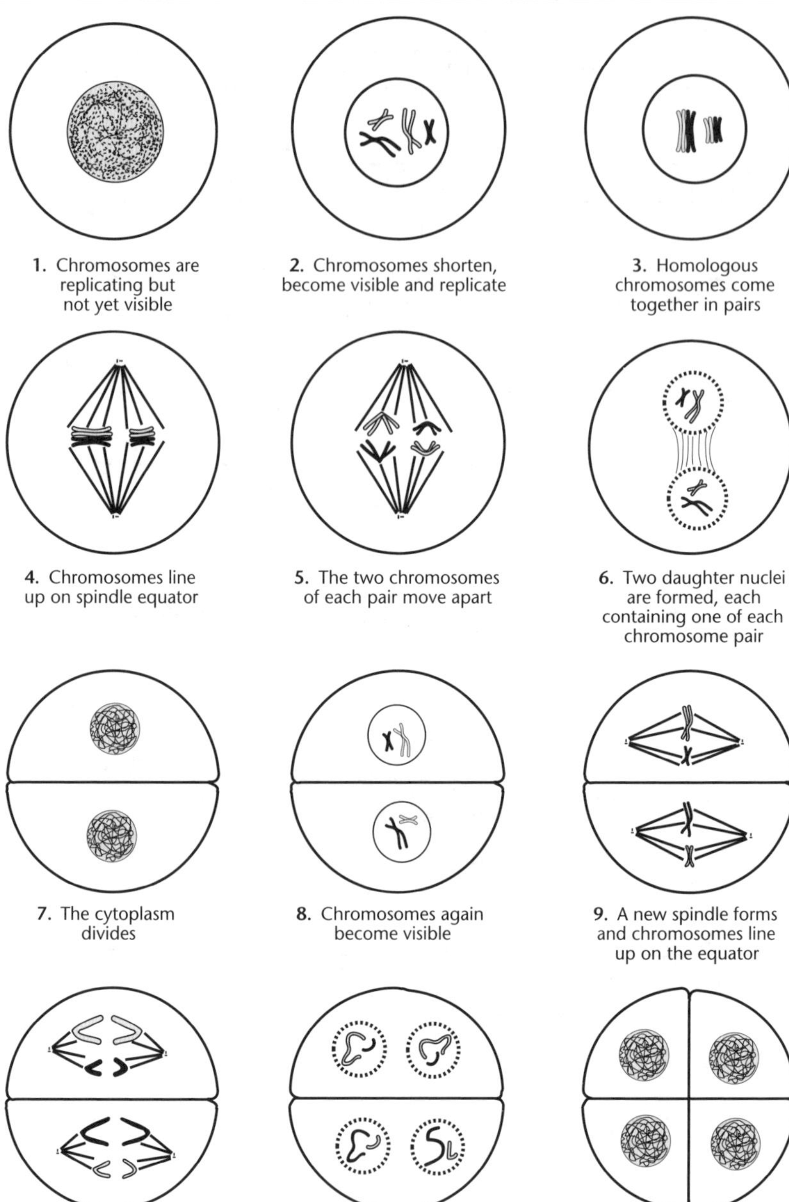

Meiosis.

Meiosis produces genetic variation

The chromosomes of a cell undergoing meiosis separate in different ways. If crossing over is ignored, a cell with two chromosomes can produce four kinds of gametes; with each additional pair of chromosomes, the number of possibilities doubles.

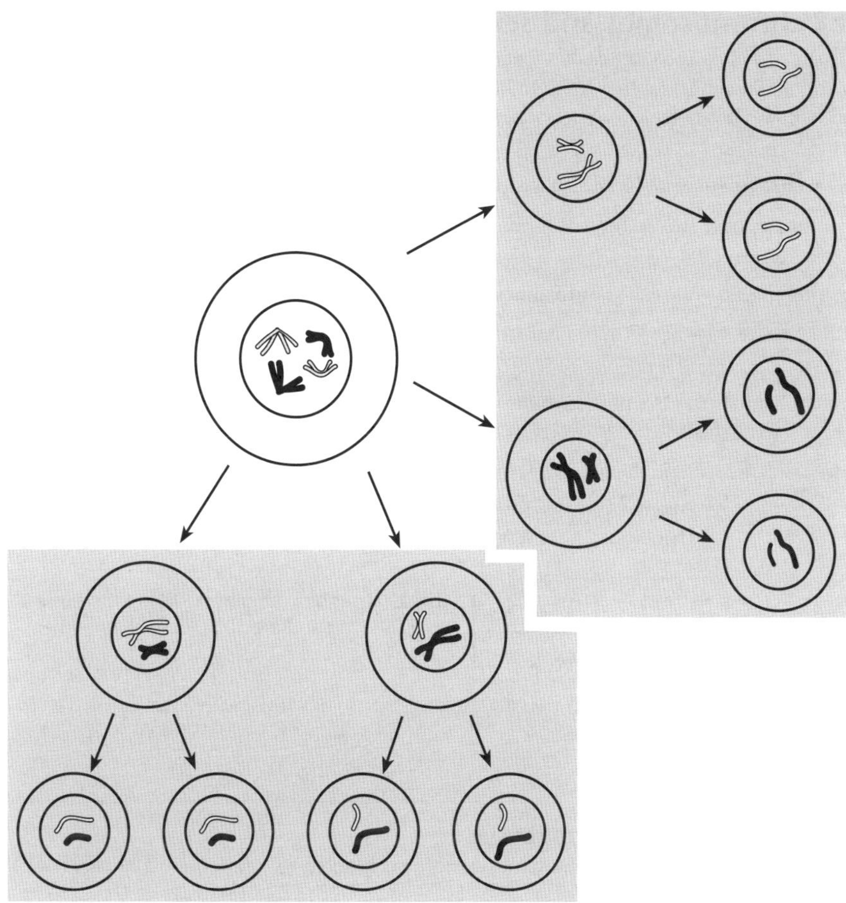

A cell with two pairs of chromosomes can undergo meiosis in two different ways. Each results in *two* kinds of gametes, so *four* kinds of gamete could be produced.

How meiosis produces variation.

Mitosis and meiosis compared	
Mitosis	**Meiosis**
Chromosome number does not change	Chromosome number is reduced from diploid to haploid
Products are genetically identical	Products are genetically different
One division of the nucleus	Two divisions of the nucleus
Homologous chromosomes do not pair up	Homologous chromosomes pair up in first division
Can occur in diploid or haploid cells	Only occurs in diploid cells

Sex chromosomes and sex determination

The sex of a child is determined at the moment of **fertilisation**. It depends entirely on the sperm, and therefore on the father. Half the sperm produced by a man give rise to males and half give rise to females. Sex is determined by two **sex chromosomes**.

- In males the sex chromosomes are different – a large **X chromosome** and a small **Y chromosome**.
- Females have two X chromosomes.

Sex chromosomes in a male and female.

The non-sex chromosomes – ie those not concerned with determining sex – are called **autosomes**.

Humans have 46 – 2 = 44 autosomes.

Females have two X chromosomes (these are a true pair); males have an X and a Y (Y is a much shorter chromosome than X, so has far fewer genes).

Since the father can produce X or Y sperm, it is the father that determines the sex of children:

Parents:	XY	×	XX
Gametes:	X or Y ;		X

	♂ X	♂ Y
♀ X	XX	XY

The chances of a daughter (XX) are the same as a son (XY) and are 50:50 or 50% for either.

It is the presence or absence of Y that determines sex; an individual with a Y is a male, an individual without a Y is a female.

Example

Sex chromosome aberrations

An individual with XXY chromosomes – known as Kleinfelter's syndrome – is a male, not a female.

An individual with only one X chromosome (XO) – known as **Turner's syndrome** – is a female (O means 'no chromosome').

Even though X and Y chromosomes are unlike, they are still homologous. This is because during the first division of meiosis they *segregate* (go to different daughter cells). Each gamete thus normally receives *either* an X chromosome *or* a Y chromosome, but not both. This explains why half the sperm receive the X chromosome and half the Y chromosome. X and Y chromosomes are homologous because *they never normally occur in the same gamete*.

When a man produces sperm, half receive the X chromosome and half the Y chromosome. All a woman's eggs receive *only* an X chromosome. If an X-carrying sperm fertilises the egg, the child is female, and if a Y-carrying sperm fertilises the egg, the result is a boy.

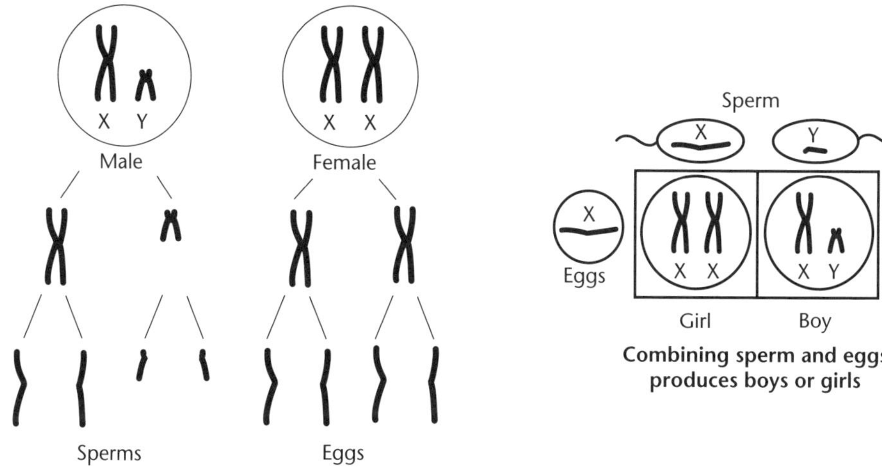

Gamete production and sex determination of a child.

Unit 12.3 Activity 3A: Chromosomes and cell division

1. Match the phrases **1–12** with the terms **A–L**.

Phrase	Terms
1. Being members of a pair of chromosomes	A. Autosome
2. Chromosome not concerned with sex determination	B. Chromatid
3. Division that gives rise to genetically different nuclei	C. Diploid
4. Division that produces genetically identical nuclei	D. Haploid
5. Fertilised egg	E. Homologous
6. Having one set of chromosomes	F. Karyotype
7. Having two sets of chromosomes	G. Meiosis
8. Male gamete	H. Mitosis
9. May be X or Y	I. Sex
10. One of two strands of a duplicated chromosome	J. Chromosome
11. The chromosomal characteristics of an organism	K. Sperm
12. Unfertilised egg	L. Zygote

2. a. The diagram shows the sex chromosomes from two different people, A and B.

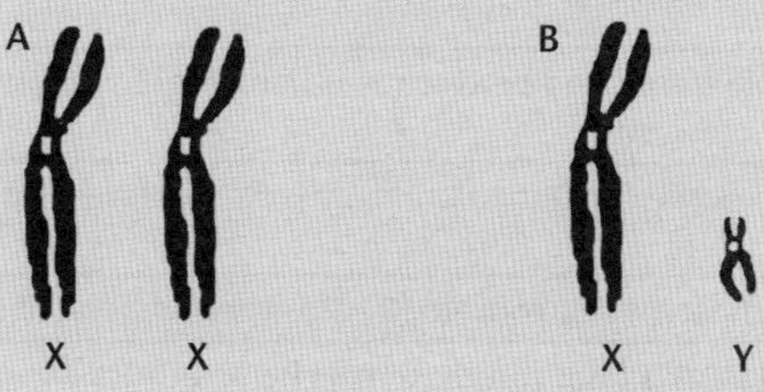

In the diagram, identify the sex of:

i. Person A.

ii. Person B.

b. The sex of a child is set at the moment of fertilisation. Explain how the type of chromosome results in some children being male, and others being female.

3. Different types of cell division take place in the human body.

a. Describe the purpose of mitosis.

b. Explain why mitosis would not be able to produce functional sperm or egg cells.

4. Cell division by meiosis produces genetic variation in the male and female gametes. Discuss how meiosis produces this variation, and why this is important.

5. Cells divide in one of two ways.

a. Choose one of these types of division and describe exactly where this type of division would take place in a named plant or animal.

b. Explain the purpose of this type of cell division in the plant or animal you mentioned in part **a**.

c. Discuss the importance of the number of chromosomes and the number of cells that are produced by your chosen type of cell division. In your answer, state clearly if genetic variation in the daughter cells is important or not.

Chromosomes and cell division

Key terms

Nucleus	The part of a cell containing the chromosomes.
Chromosome	A threadlike body containing the genetic material.
Chromatid	One of the two products of replication of a chromosome.
Centromere	The part of a chromosome that attaches to the spindle during cell division.
Haploid	Having one set of chromosomes.
Diploid	Having two sets of chromosomes.
Mitosis	Division of a nucleus into two genetically identical nuclei.
Meiosis	Division of a diploid nucleus into four haploid, genetically different nuclei.
Spindle	System of protein fibres responsible for chromosome movement in cell division.
Centriole	Structure involved in the organisation of the spindle during division of animal cells.
Homologue	One of a pair of chromosomes, only one of which can exist in each gamete.
Chiasma	Point of interlocking between two homologous chromosomes, resulting from a cross-over.
Sex chromosome	Chromosome concerned primarily with the development into male or female.
Autosome	A chromosome not concerned with determination of sex.
Karyotype	An organism's total chromosomal characteristics.

Eukaryotes are organisms other than bacteria.

Cells are the basic units of life.

- The cell is the smallest structure capable of carrying out all the processes of life, such as movement, growth, respiration and reproduction.
- All sexually reproducing organisms begin life as a single cell, the fertilised **egg** or **zygote**, which is formed by joining together two cells called **gametes** (egg and sperm).
- A zygote develops into a multicellular organism by repeated cell divisions.
- Since each fertilised egg must contain a similar (though not identical) set of information as the previous generation, the genetic message must be copied each time the cell divides.

The **genetic material** is located in threadlike structures called **chromosomes**. In all organisms except bacteria, these are contained in a clearly defined part of the cell called the **nucleus**. Chromosomes are only visible during cell division, and when using ordinary microscopes the chromosomes must first be stained.

Chromosomes have the following important characteristics:

- The number per cell is characteristic of a species.
- They have individuality, differing not only in their length but also in the position of the **centromere** (the part of the chromosome that attaches to the spindle during cell division). They also differ in the pattern of banding when treated with certain stains. Because they can be individually distinguished, they can be numbered. In humans, the non-sex chromosomes or **autosomes** are numbered from 1–22 (in decreasing order of size). The chromosomal characteristics of an organism are its **karyotype**.
- In any gamete there is only *one* of each kind of chromosome. Gametes are said to have the **haploid** number of chromosomes.
- In a body cell there are *two* of each kind of chromosome. Each chromosome thus has one other that looks just like it (the sex chromosomes are an exception). Chromosomes are thus said to be in **homologous pairs**, each member of a pair being called a **homologue**. One of each pair is of *maternal* origin (inherited from the female parent), the other is of *paternal* origin (inherited from the male parent). Body cells are said to have the **diploid** number of chromosomes (two of each kind).

Organism	Diploid chromosome number
Kauri	26
Garden pea	14
Fruitfly	8
Dog	78
Cat	38
Human	46
Chimpanzee	48

Diploid chromosome number refers to the chromosome number found in **somatic** (body) cells.

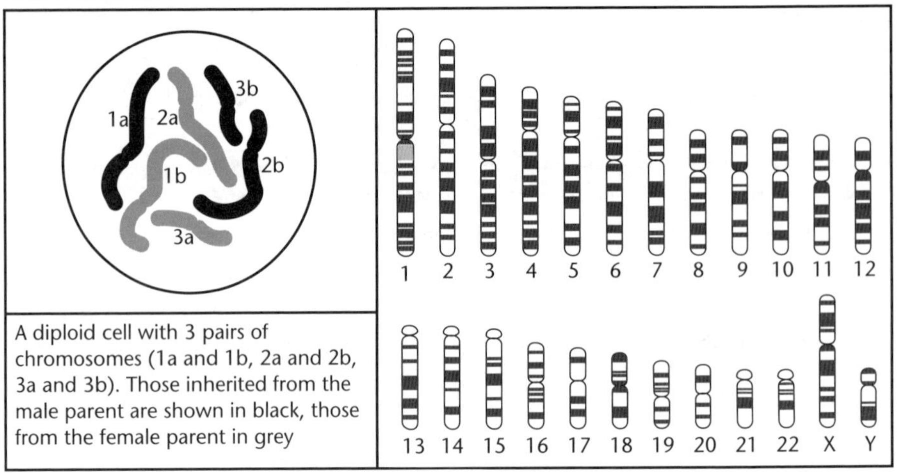

Homologous chromosomes.

Sex chromosomes and sex-linked inheritance

Key terms

Sex chromosome	Chromosome involved in the determination of sex
Autosome	Chromosome not concerned with sex determination
Heterogametic sex	The parental sex which determines the sex of the offspring
Homogametic sex	The sex that plays no part in sex determination

Review: Sex determination

Some organisms (eg most flowering plants) are *hermaphrodite* (ie have both male and female **organs**). Others are **unisexual** (one sex). In many organisms, sex is under genetic control and is fixed at fertilisation, but in some others it may be determined by age or environment.

Genetic determination of sex

This occurs in many animals and in some plants. In many cases, sex is associated with one pair of *sex chromosomes*. The other chromosomes are not concerned with sex and are called **autosomes**. There are four kinds of chromosomal sex determination.

The XY mechanism

This occurs in humans and other mammals, and in some insects such as fruit flies. In the male, one chromosome pair consists of two dissimilar chromosomes called X and Y, the X chromosome being longer than the Y. A male is thus XY and the female, with two X chromosomes, is XX.

When the male produces gametes, half the sperm receive a Y chromosome and half receive an X. Since the male produces two different kinds of gametes (so far as sex chromosomes are concerned), the male is said to be the *heterogametic* sex. The male therefore determines the sex of the offspring. Since the female is XX, all her eggs receive an X chromosome, so the female is the *homogametic* sex.

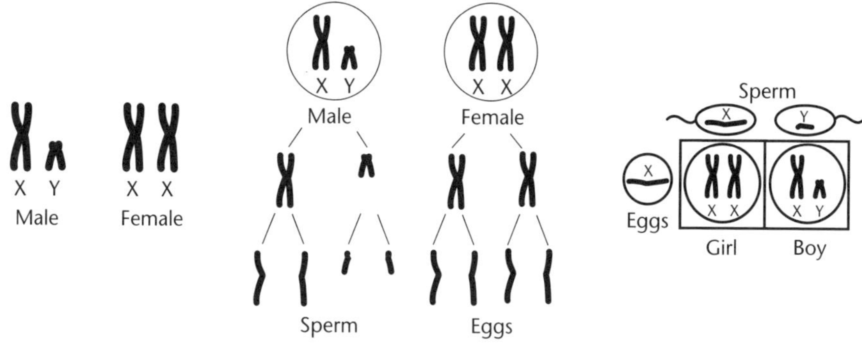

Sex determination in humans (and mammals).

- In the female the two X chromosomes are shown segregating, but only one egg is shown joining with a sperm in the Punnett square. This is because in respect of having an X chromosome, all eggs are the same. So far as X and Y chromosomes are concerned, there are only two kinds of **fertilisation**, one resulting in a male zygote and the other in a female.
- The X and Y chromosomes are an exception to the rule in that the two members of an homologous pair look identical. Their homology is due to their behaviour at meiosis I – the X and Y chromosomes *segregate*, so that each daughter cell receives an X or a Y chromosome.

Extension: Other sex-determining mechanisms

The XO mechanism

In wetas, locusts and other grasshoppers, the male has only one sex chromosome and is XO, the female being XX.

Example

In the tree weta *Hemideina crassidens*, the male has 15 chromosomes and the female has 16.

The WZ mechanism

In birds, butterflies and moths, the female is heterogametic, with two unlike chromosomes (W and Z). The male is ZZ.

Haplodiploidy

In ants, bees and wasps, there are no sex chromosomes. Males are **haploid** because they develop from unfertilised eggs, and females are diploid because they develop from fertilised eggs.

Sex-linked traits

Sex-linked traits are caused by genes on the sex chromosomes (almost always the X chromosome, in which case the condition is said to be X-linked). The Y chromosome is much shorter than the X, so it cannot carry as many genes. There must therefore be many loci on the X chromosome that are not represented on the Y chromosome. Males have only one X chromosome, inherited from the mother. Males therefore cannot be heterozygous for X linked genes, which are always expressed (males are said to be *hemizygous* for X-linked genes). Since females have two X chromosomes, a sex-linked recessive allele will only be expressed if she is homozygous. For this to happen, *both* her parents must carry an X chromosome with the defective allele. This is why sex-linked recessive alleles are much more rarely expressed in females than in males (in animals in which the male is the homogametic sex, the reverse is true).

An important difference between X-linked and autosomal traits is that with the X-linked alleles, reciprocal matings give different results. The reason is that the mother contributes an X chromosome to all her children, but the father contributes them only to his daughters.

Example

Red-green colour-blindness in humans is a sex-linked trait.

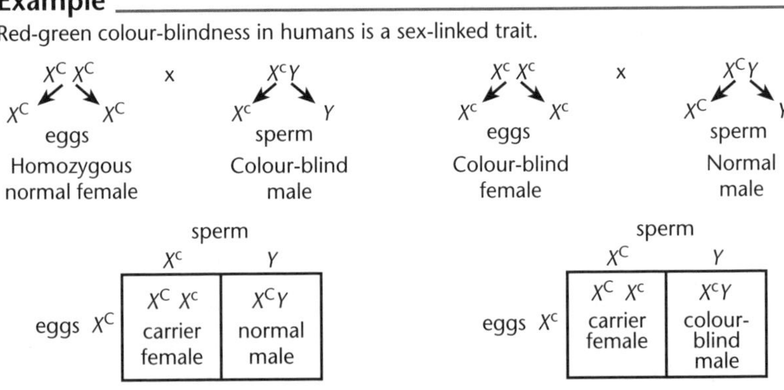

With X-linked traits, reciprocal matings produce different results.

Haemophilia is another sex-linked trait. The potential seriousness of haemophilia is such that it is considered a *genetic defect*.

Genetic defects

The number of possible defects is as great as the number of human genes, estimated to be about 31 000. Although there are a great many genetic abnormalities, many of them are very rare. Some of the better-known ones are listed below.

Autosomal dominant traits	
Huntingdon's disease	Progressive mental deterioration, beginning in middle age.
Polydactyly	Presence of extra digits.
Brachydactyly	Shortened fingers.

Autosomal recessive traits	
Albinism	White hair, pink skin, pink eyes.
Phenylketonuria	Darkening of cartilage, urine turns black on exposure to air.
Cystic fibrosis	Excessive secretion of **mucus** by glands of the breathing passages and gut.
X-linked recessive traits	
Red-green colour-blindness	Inability to distinguish red from green.
Haemophilia A	Blood does not clot.
Duchenne muscular dystrophy	Progressive weakening of muscles, usually fatal by adulthood.
X-linked dominant traits	
Vitamin D-resistant rickets	Mechanical weakness and deformation of bones, not alleviated by vitamin D

Y-linked inheritance

The Y chromosome contains only about 60 genes. The X and Y chromosomes are homologous nevertheless because they pair and subsequently segregate during meiosis.

The most important gene is the sex-determining region of the Y chromosome, the SRY gene, which triggers the development of male characters during development.

This gene is found on a segment of the Y chromosome that is not found on the X chromosome, so is said to be *Y-linked*. No crossing over occurs between the X and Y chromosome at this point. A Y-linked gene is only found in males (unless as a result of **translocation** mutation it shifts to the X chromosome and is inherited by female offspring and must therefore be passed on to any male offspring). There are very few Y-linked genes known for certain to be functional.

The X and Y chromosomes do undergo crossing over and recombination, but only in the short regions near the ends. These regions contain the 20 or so genes that are equally inherited by both sexes and are said to be pseudo-autosomal. They show the normal diploid inheritance patterns that are seen for other genes found on autosomes.

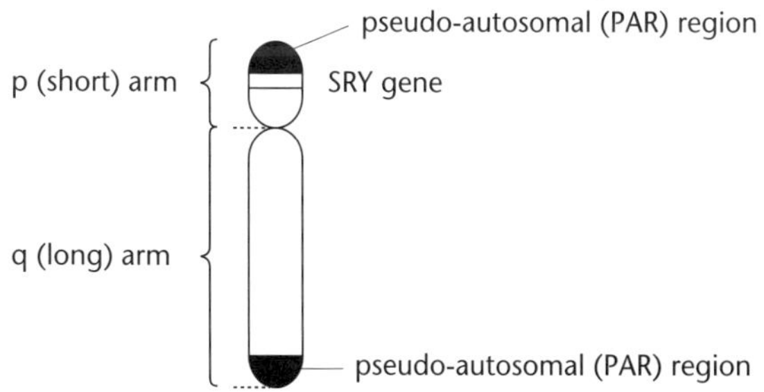

Some important regions of the Y chromosome.

Pedigrees and sex-linked inheritance

It is often possible to tell from a family tree whether an allele is sex-linked. The following are the important points to bear in mind:

- In autosomal recessive and **dominant traits**, males are affected as often as females.
- X-linked recessive traits are more common in males than in females, since a male only has to have one parent (his mother) with the allele, whereas for a female to develop the trait she must have both parents with the allele.
- A male showing an X-linked condition receives the allele from his mother and passes it on to his daughters; he cannot pass it to his sons.
- X-linked dominant traits are more common among females than males. This is because, statistically speaking, an affected male passes on the allele to all his daughters, but an affected female passes it on to half her sons and half her daughters.

Terms used in sex-linked inheritance

Autosomal dominant traits	Genetic characteristics always expressed phenotypically, present on autosomal chromosome.
Autosomal recessive traits	Genetic characteristics only expressed phenotypically in homozygote, present on autosomal chromosome.
Sex-linked traits	Traits controlled by genes on the sex chromosoms.
X-linked dominant traits	Genetic characteristics always expressed phenotypically, present on X sex chromosome.
X-linked recessive traits	Genetic characteristics present on X sex chromosome, expressed phenotypically in affected males (eg X^cY), only expressed phenotypically in homozygous females (eg X^cX^c).
Y-linked inheritance	Gene present on a Y sex chromosome.

Unit 12.3 Activity 3B: Sex linkage

1. **a.** Describe the meaning of the term 'sex-linked'.

b. Name an example of a sex-linked condition in humans.

c. In cats, a pair of alleles (*B* and *b*) are sex-linked. In females, the homozygous condition $X^B X^B$ gives black coat, $X^b X^b$ gives a ginger coat colour, and $X^B X^b$ gives a patchwork of ginger and black called tortoiseshell. A tortoiseshell female is mated with a black male. Give the possible genotypes of the parents and the offspring. Explain your answer.

2. The frequencies of some genetically determined recessive traits are far higher among males than among females in human populations. Examples are red-green colour-blindness and haemophilia. Outline the reasons for this in a paragraph.

3. In humans, there are rare cases of females with only one X chromosome. These females are sterile and this condition is known as Turner's syndrome.

Gene and its function	Allele
Gene controlling colour vision	X^N normal colour vision X^n colour-blind

Describe the genotype of a Turner's syndrome woman who is colour-blind. Explain your answer.

4. A man with normal vision is married to a woman with normal vision whose father is colour-blind. Determine the probability that their first child will be a colour-blind boy, explaining your reasoning.

5. Haemophilia is a hereditary inability to make one of the blood-clotting proteins, due to an allele on the X chromosome. Determine if each of the following is true or false, giving your reasoning:

a. The condition only occurs in males.

b. A male can inherit the allele from either of his mother's parents.

c. When both parents carry the defective allele, all the sons are affected.

6. In the following pedigrees, the gene for the condition shown by solid shapes is rare in the general population. For each pedigree, say whether the condition represented by the solid character is autosomal recessive, autosomal dominant, X-linked recessive, or X-linked dominant, justifying your answer in each case.

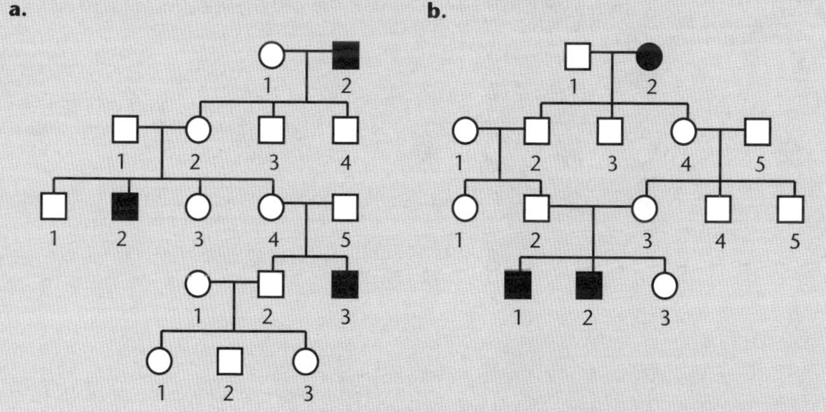

c.

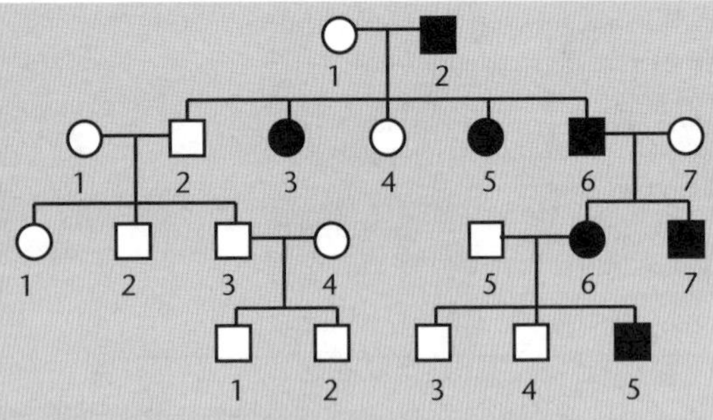

d.

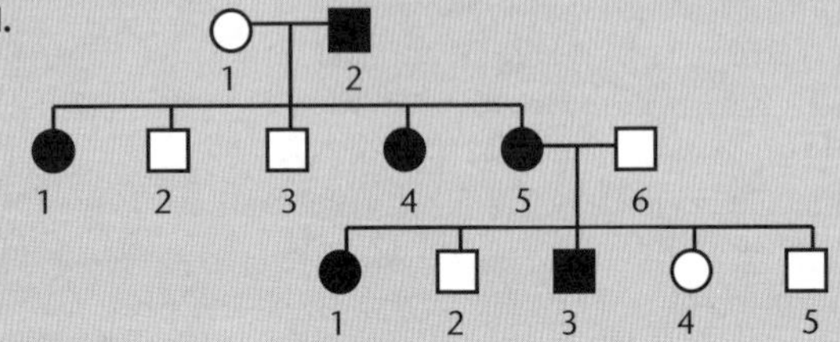

7. State your reason(s) for deciding whether the following are true or false.
   a. Human X-linked recessive traits only occur in males.
   b. In any particular population, a given X-linked recesive allele is carried by more females than males.
   c. In a given population, a given X-linked dominant trait is twice as common in females as in males.
8. The following pedigree shows the inheritance of a certain kind of muscular dystrophy, a fatal condition characterised by progressive muscular weakness. Affected individuals are shown in dark shapes.

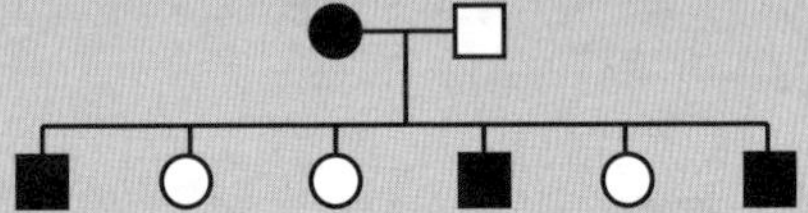

Using *autosomal* and/or *sex-linked* and/or *dominant* and/or *recessive*, define the condition.

Unit 12.3 Genetics

Topic 4: Biology of DNA – biochemistry of the genetic material

Biology material is covered in this Topic 'Describe the structure and function of the genetic material', by looking at:

- Structure of DNA.
- Replication of DNA, its occurrence and importance.
- Describing gene expression through protein synthesis (transcription and translation).
- Discussing the role of DNA in determining the structure of a protein.

The genetic message – DNA

Just as a plan and raw materials are needed to build a house, there must be *information* to build an organism. Living things all use the same kind of information, stored in a chemical called **deoxyribonucleic acid**, or **DNA**. DNA contains the information we inherit from our parents.

DNA is the *genetic material* and is located in threadlike bodies called **chromosomes**, which are in the nucleus of every cell. DNA:

- *Stores information* for building a new organism.
- Can be copied or *replicated*, so it can be handed down to future generations. ('Like begets like', ie offspring always resemble their parents (though not usually exactly) – the information for building a new organism must therefore have been replicated (copied) in the parents before being handed down to the next generation.)

Chromosomes are divided into smaller segments called **genes**. A gene consists of thousands of smaller units called **nucleotides**. Each nucleotide consists of three parts:

- A *phosphate* group.
- A *sugar*, **deoxyribose**.
- A *base* – which can be of four kinds – **adenine** (A), guanine (G), **thymine** (T), **cytosine** (C).

A DNA molecule consists of two chains of nucleotides arranged in a spiral shape called a *double helix*.

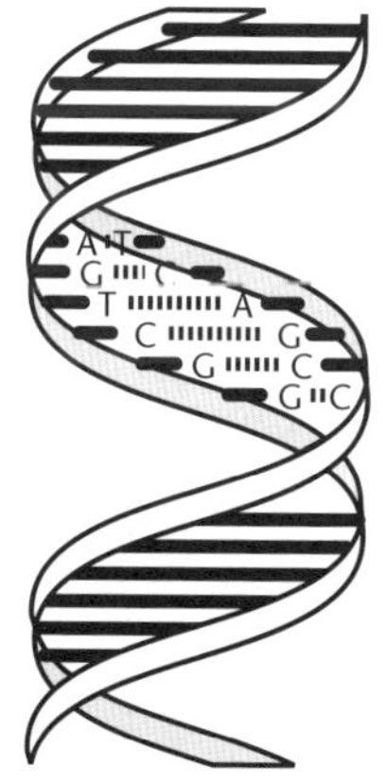

A simple representation of the DNA double helix

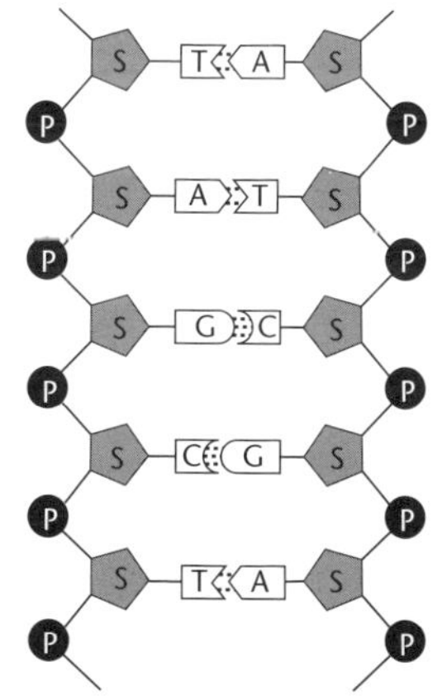

The helix straightened out to show base pairing between the two chains

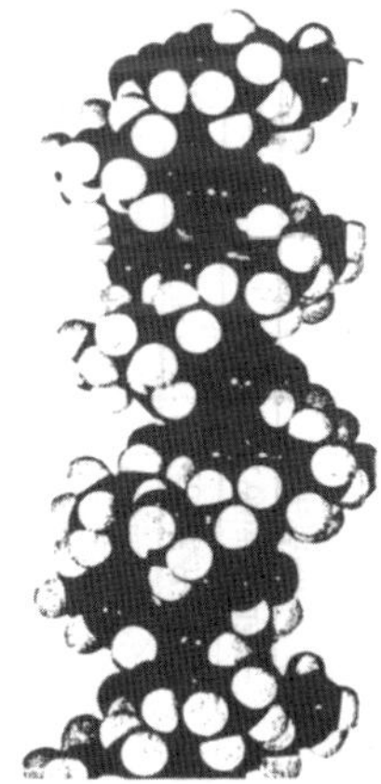

A DNA model.

How DNA structure is represented.

The chains are linked together by the **bases**:

- Adenine pairs with thymine
- Guanine always pairs with cytosine.

Because of this *complementary base pairing*, the order of the bases in one strand determines the order of the bases in the other strand. Thus, if the order of bases in one strand is known, so is the order in the other. This is the key to how DNA is copied, or *replicated*.

DNA replicates itself *between* cell divisions. The two strands of the double helix separate. New nucleotides then pair up with the exposed bases. The two strands are 'zipped up' by an **enzyme**, resulting in two new DNA molecules, each consisting of one 'old' strand and one 'new' one.

Replication of DNA requires:

- A supply of the raw materials – the nucleotides.
- Energy from respiration (this is one reason why dividing cells use a lot of energy).
- Enzymes to catalyse the process.
- Information in the existing DNA.

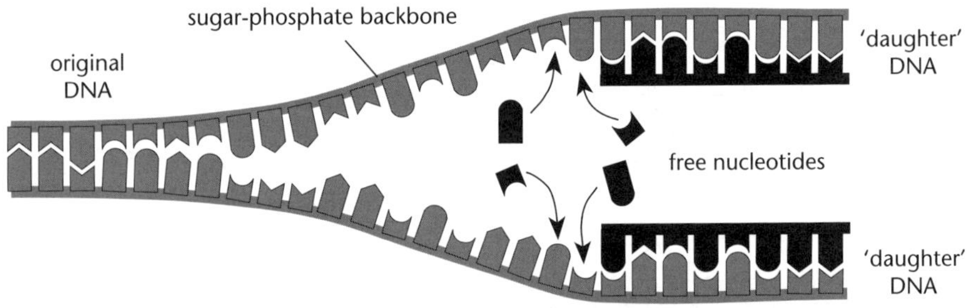

A simplified overview of how DNA replicates itself.

Although DNA was first extracted in 1868 (from **pus** cells, by F Meischer), it was not until 1953 that its detailed structure was worked out by James Watson and Francis Crick:

DNA structure

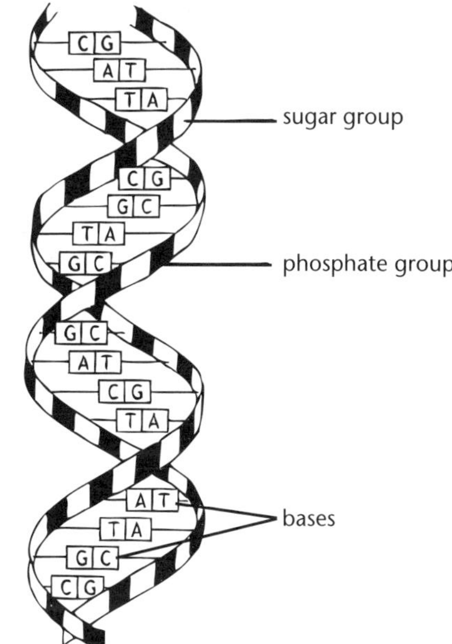

DNA is the very large molecule found in chromosomes which carries the genetic code. Its shape is a *double helix* – two cross-connected strands twisted together.

The side strands are made of alternating sugar and phosphate groups; cross strands are the paired bases.

Each strand of DNA is made up of many repeating units called **nucleotides**. A nucleotide is made up of three parts:

- A deoxyribose sugar.
- A phosphate group.
- A nitrogen base.

The only difference between nucleotides is the base that they contain:

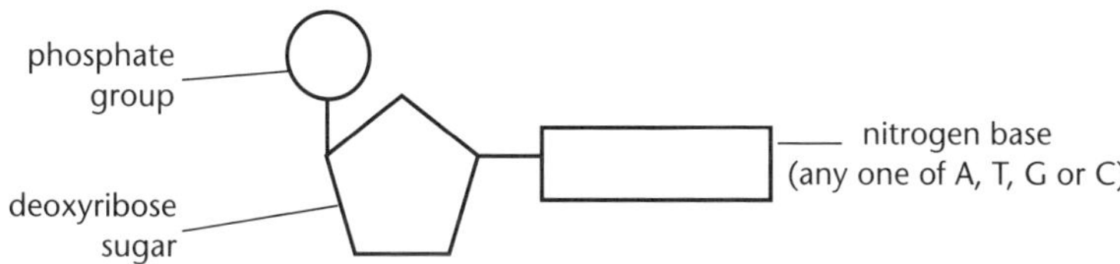

There are four different bases – adenine (A), cytosine (C), guanine (G), thymine (T). The bases always pair according to the **base pairing rule**:

- **Adenine** with thymine (A – T).
- **Cytosine** with **guanine** (C – G).

The pairs are held together by weak **hydrogen bonds** and these hold the two strands of DNA together.

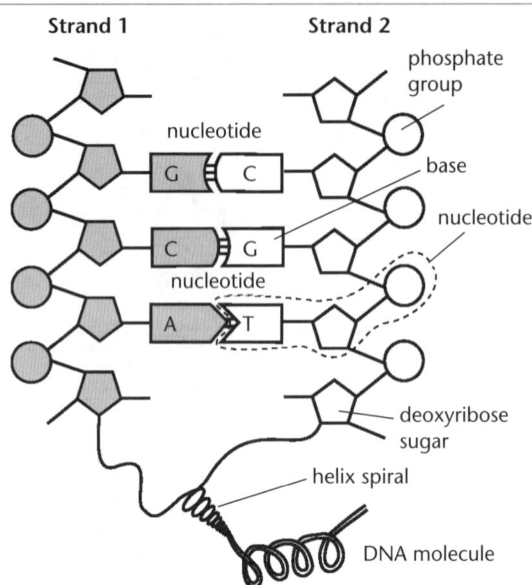

Structure of DNA.

It is the *sequence of bases* along one strand (the **template strand**) that is the genetic code. A sequence of three bases (a **triplet**) codes for an **amino acid**; amino acids are the building blocks of **proteins** (eg the triplet GCA on strand 1 shown above codes for the amino acid arginine).

All the triplets that code for one complete protein make up a **gene**. A molecule of DNA may have thousands of genes. Different sequences of bases along a gene may produce proteins that are slightly different (eg one base sequence in humans codes for blue eye colour while another codes for brown eye colour) – different forms of a gene are called **alleles**.

DNA replication

It is essential that DNA can *replicate* itself, so that chromosomes can be copied to give every new cell made the same genetic code. It is the base pairing mechanism that allows DNA to replicate.

Replication occurs in a series of steps, each controlled by enzymes, with energy supplied from the molecule **ATP**.

Two identical DNA molecules result from replication, each having one original strand and one new strand (known as *semi-conservative replication*).

DNA replication occurs *prior* to cell division (mitosis and meiosis), in what is known as **interphase** in the cell cycle. In this phase, the chromosomes are long unravelled threads of chromatin.

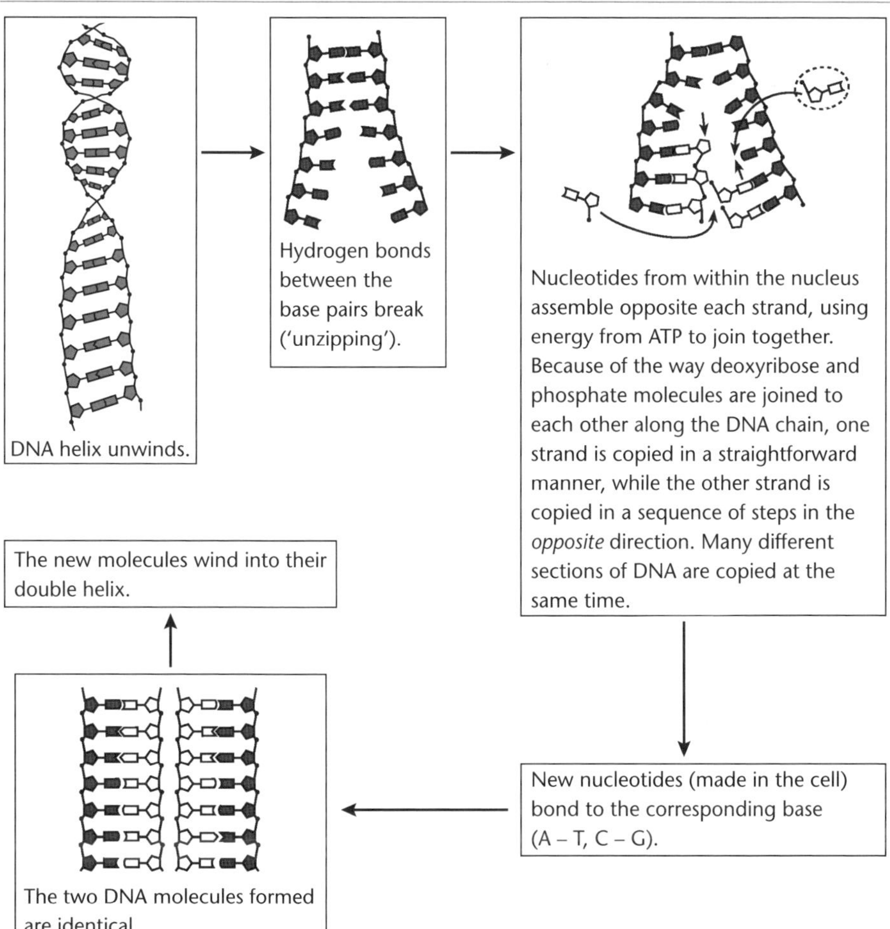

DNA replication.

Mitosis

The process of mitosis is necessary to allow cells to divide to enable growth to occur and to replace worn or injured cells. Simple cell division (or **mitosis**) produces two *daughter cells genetically identical to their parent cell*, so that the same cell functions and processes continue. (The other form of cell division, called **meiosis**, results in the formation of sex cells.)

Although mitosis is a continuous process, certain stages have been recognised and given specific names, such as *interphase* and *prophase*.

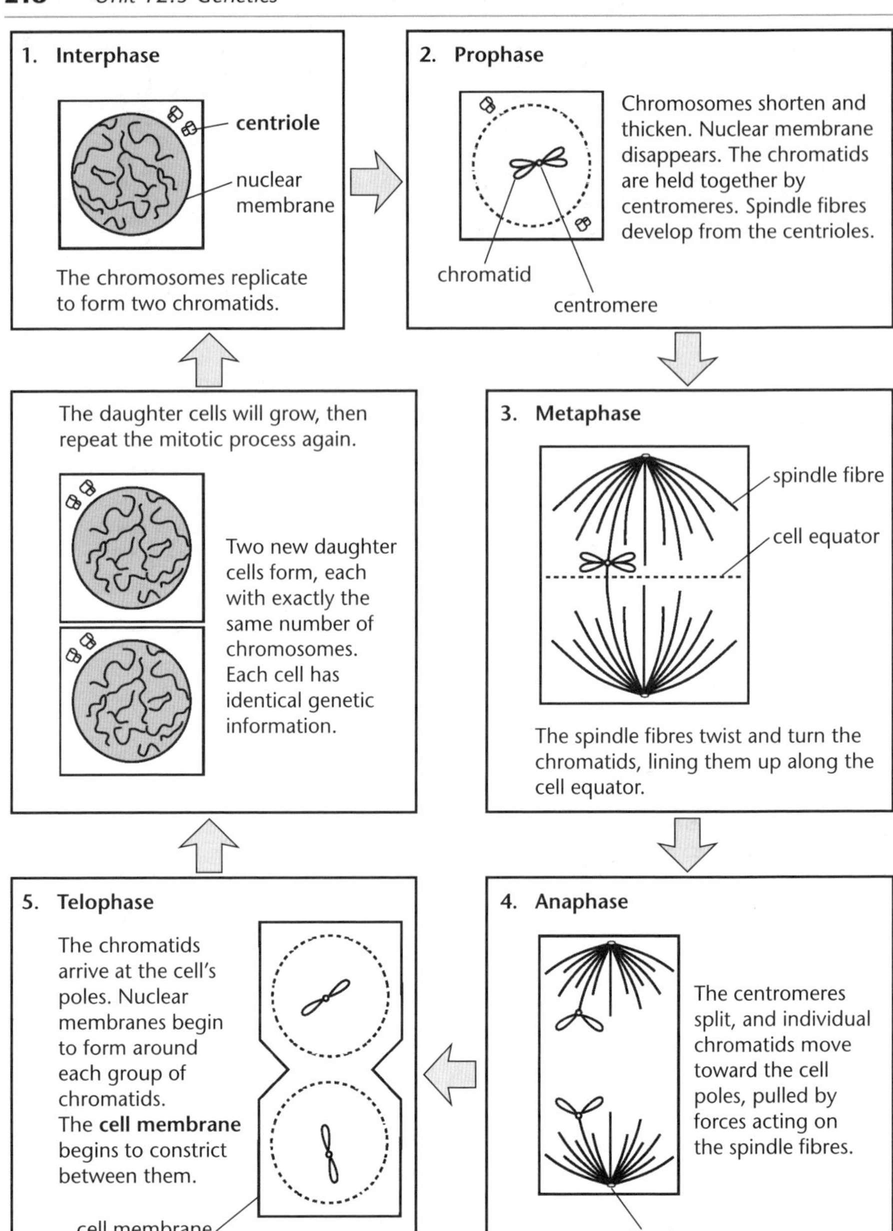

Mitotic cell division.

Unit 12.3 Activity 4A: DNA

1. Match the phrases **1–10** with the terms **A–J**.

Phrase	Term
1. A long chain of amino acids	A. Adenine
2. Building unit of proteins	B. Amino acid
3. Codes for the production of a protein	C. Base
4. In DNA, always pairs with cytosine	D. Chromosome
5. In DNA, always pairs with thymine	E. Gene
6. In DNA, links sugars together	F. Genome
7. One of the building units of DNA	G. Guanine
8. The entire genetic information of an organism	H. Nucleotide
9. The variable part of a nucleotide	I. Phosphate
10. Threadlike structure consisting of many genes	J. Protein

2. a. Each DNA nucleotide contains a phosphate, a sugar and a base. The diagram shows the arrangement of these components on one side of a DNA strand.

i. Complete the matching bases in the diagram.

ii. Clearly label each of the following parts in the diagram:

- Nucleotide.
- Phosphate.
- Sugar.

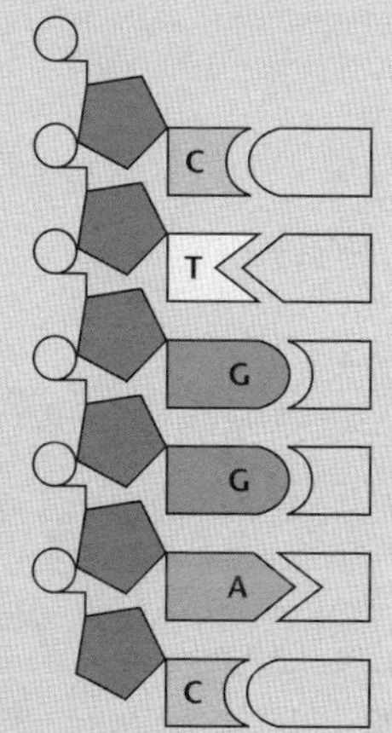

b. Explain why the base-pairing rules are an important part of DNA replication.

c. A particular flower can be red or white. Red flowers are due to the dominant allele (*R*), and white flowers are due to the recessive allele (*r*). Discuss how the information carried by DNA can result in the red and white colours of the plants.

3. a. Describe the three main components of a DNA molecule. You may use the diagram to help you.

b. The sequence of DNA shown in the diagram was repeated in a larger section of DNA, and it was found that adenine (A) occurred 300 times. State how many times each of the following components would occur:

i. Cytosine (C).

ii. Thymine (T).

iii. Guanine (G).

c. Explain how you worked out your answer to part **b**.

d. When DNA replicates, one of the two strands in each 'daughter' DNA molecule is new.

i. Discuss what is required for this type of replication.

ii. Explain why it is important that the DNA be copied exactly.

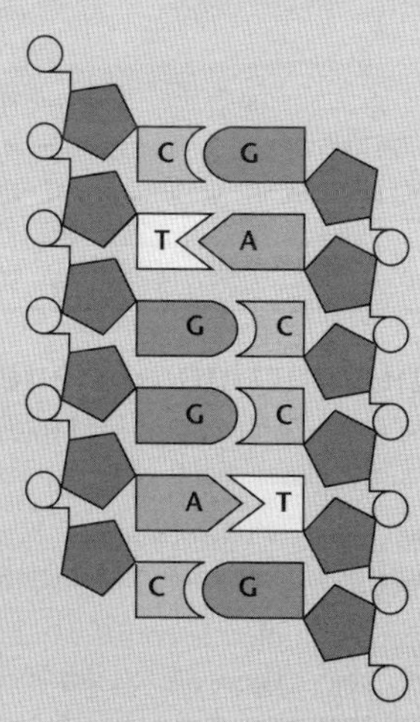

Unit 12.3 Activity 4B: DNA structure and replication

1. Describe the structure of DNA.
2. Describe the structure of a nucleotide.
3. The diagram below is of a model of part of a DNA molecule. Give the names of the parts labelled A, B, C.

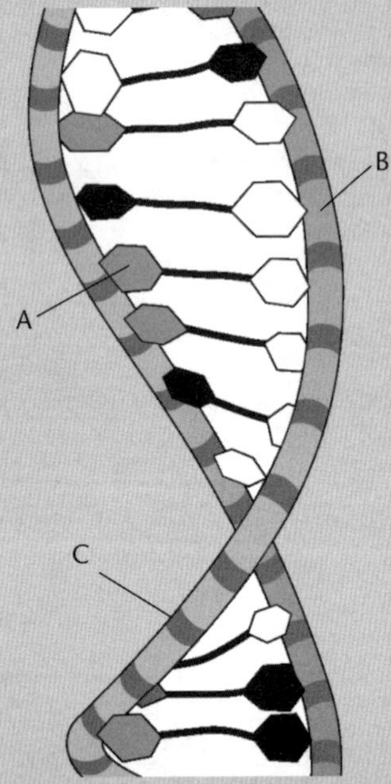

4. Explain the significance of the base pairing rule.
5. If the amount of thymine in a cell was 28% and the amount of guanine was 22%, give the percentages of cytosine and adenine in the cell.
6. If the base sequence on the template strand of a DNA molecule is ATC GGA TCT AGC, give the base sequence on the non-template strand.
7. Discuss the process of DNA replication. In your answer, you need to describe the process – its occurrence, and explain its importance.
8. Describe the similarities and differences between the DNA in the cells of a cat and the DNA in the cells of a rat.

The function of genes

The entire information for building an organism is called the **genome**, broken up into smaller bits, the genes. Each gene carries coded information for building a protein.

Proteins are vital substances, because every organic chemical in a cell is either a protein or is made by an **enzyme**, and enzymes are proteins.

A protein molecule consists of up to 1000 or more **amino acids**, strung together like beads of a necklace. There are 20 kinds of amino acid in proteins. It is the *sequence* or order of the amino acids that determines the properties of the protein.

Collagen is a tough, fibrous *protein* in **tendons** and **ligaments**. **Keratin** is another fibrous *protein*, and is found in hair, nails, claws and feathers.

- **Haemoglobin** is the red, oxygen-carrying *protein* in the blood.
- **Insulin** is a hormone (chemical messenger) made of *protein* that regulates the concentration of glucose in the blood.

It is the sequence of bases in the DNA that 'tells' the cell the order in which the amino acids must be joined together to make a protein. Each amino acid is coded for by a sequence of three bases – the code is therefore a *triplet* code.

In making a protein, only one of the two strands is used as a source of information.

How proteins are made – protein synthesis

All the organic constituents of a cell are either proteins or substances synthesised by protein **catalysts** (enzymes). Information specifying the amino acid sequence of a **polypeptide** is encoded in the base sequence of a segment of DNA called a **gene**.

Except for the small amount of DNA in **mitochondria** and plastids, all the DNA in eukaryotes is confined to the nucleus. Experiments with radioactive tracers have shown that proteins are made by the **ribosomes** in the **cytoplasm**. Before a polypeptide chain can be produced, the necessary information must be transferred from nucleus to cytoplasm. The transfer of this coded information and its subsequent decoding involve **ribonucleic acid (RNA)**. DNA and RNA are collectively called **nucleic acids** and are very similar, except that in RNA:

- The sugar is **ribose** instead of deoxyribose.
- Thymine is replaced by another **pyrimidine, uracil**.

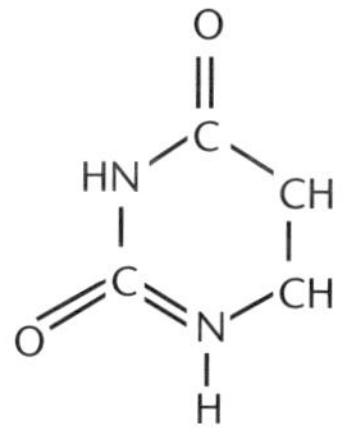

Uracil.

- RNA is *single-stranded* (although it may be bent back on itself and so *appear* double stranded).

Making a protein involves the integrated action of three kinds of RNA:

- **Messenger RNA (mRNA)**. This acts as the working copy of a gene.
- **Transfer RNA (tRNA)**. This brings each amino acid into association with the mRNA.
- **Ribosomal RNA (rRNA)**. Combined with proteins, this is a constituent of ribosomes. Ribosomes 'read' the coded message in the mRNA and use it to link amino acids in the correct order.

Although the essential mechanisms of protein synthesis are similar in **prokaryotes** and eukaryotes, they differ in some important details. The following description applies to eukaryotes. In eukaryotes, protein synthesis involves three stages, the first two of which occur in the nucleus:

- **Transcription**, which involves making an RNA copy of the gene, called the **primary transcript**.
- **RNA processing**, in which the primary transcript is 'edited' by removal of certain non-coding portions, to produce mRNA.
- **Translation**, which occurs in the endoplasmic **reticulum** of the cytoplasm. In this process the ribosomes and tRNA operate to decode the information in the RNA copy of the gene, linking the amino acids in the order encoded in the mRNA.

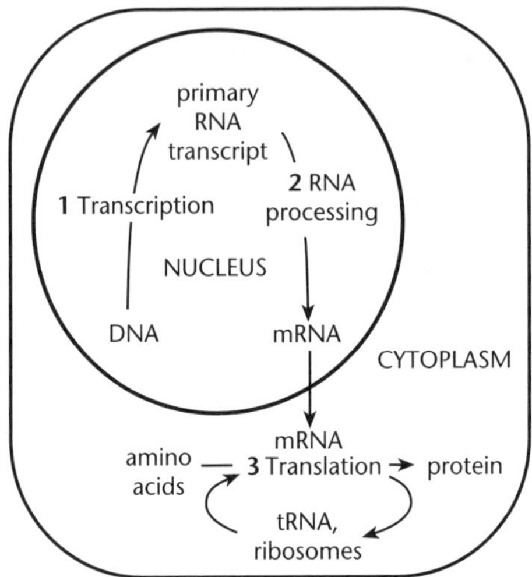

The three stages of protein sysnthesis in a eukaryote.

Transcription

In this process, one of the two strands of the DNA is copied (transcribed) into RNA. It is similar to DNA replication except that RNA is produced instead of DNA. Under the influence of the enzyme **RNA polymerase**, one of the two DNA strands is used to make a complementary sequence of ribonucleotides.

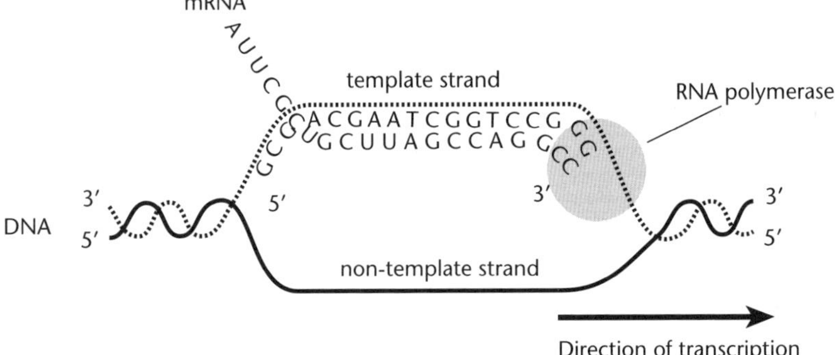

How DNA is transcribed into RNA.

The **base pairing rules** are the same as when DNA replicates except that adenine of DNA pairs with uracil instead of thymine. The DNA strand involved in base pairing with the ribonucleotides acts as a kind of template or 'pattern', and is therefore called the **template strand**.

As with all polynucleotides, the RNA grows in the 5' → 3' direction by adding new nucleotides to the 3' end. As with DNA, the two polynucleotide strands are anti-parallel (head to tail), so the RNA polymerase moves along the template strand in the 3' → 5' direction.

A specific nucleotide sequence on the template strand called the **promoter** sequence determines which of the two DNA strands is used as a template and where to bind to the DNA.

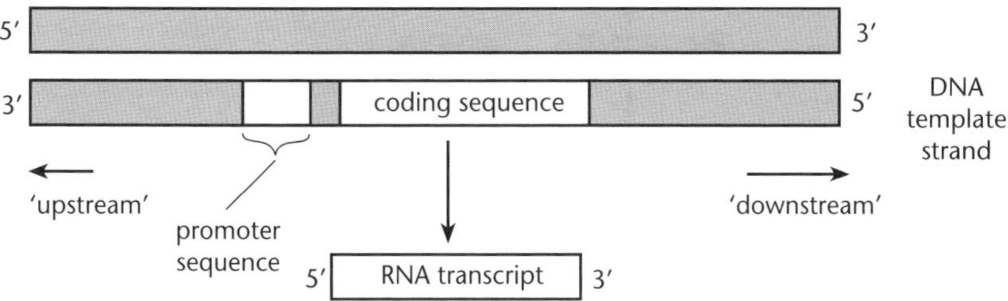

Promoter and coding regions of a eukaryotic gene.

Once the RNA polymerase has 'recognised' and bound to the promoter (called *initiation*) it begins to move along in the 3' → 5' direction, making RNA as it goes (the promoter sequence is not copied, it just provides the 'starting signal' for the RNA polymerase). The synthesis of the RNA is called *elongation* and is the second phase of transcription. In the third phase, the RNA polymerase reaches a sequence of bases on the DNA called the terminator sequence. At this point, the RNA polymerase enzyme stops transcription and drops off the DNA–RNA complex. Although each DNA strand extends the entire length of a chromosome, in different genes either strand can act as the template.

Transcription thus requires three things:

- A source of *raw materials*, the ribonucleotides.
- A *catalyst*, RNA polymerase.
- A source of *information*, the template strand of the DNA.

RNA processing

The information in the RNA transcript is not yet ready to be translated because in a eukaryotic gene, coding regions are separated by sections of DNA that are *not translated into proteins*. The segments carrying information that is translated are called **exons** because they are **ex**pressed. The **in**tervening sequences seem to have no function and are called **introns**.

Most genes have many introns and exons.

A eukaryotic gene (the gene for human myoglobin).

Initially, the RNA copy of a gene includes both introns and exons. Before the message can be used to make a protein, the introns are removed and the exons then joined or *spliced* together to make a continuous strand of mRNA.

In some cases, the introns catalyse their own removal. RNA molecules that can act as such catalysts are called **ribozymes**. In addition, it has been discovered that, depending on which exons are spliced together, more than one type of mRNA molecule can be made. It would appear that, in some cases, one gene can code for more than one polypeptide.

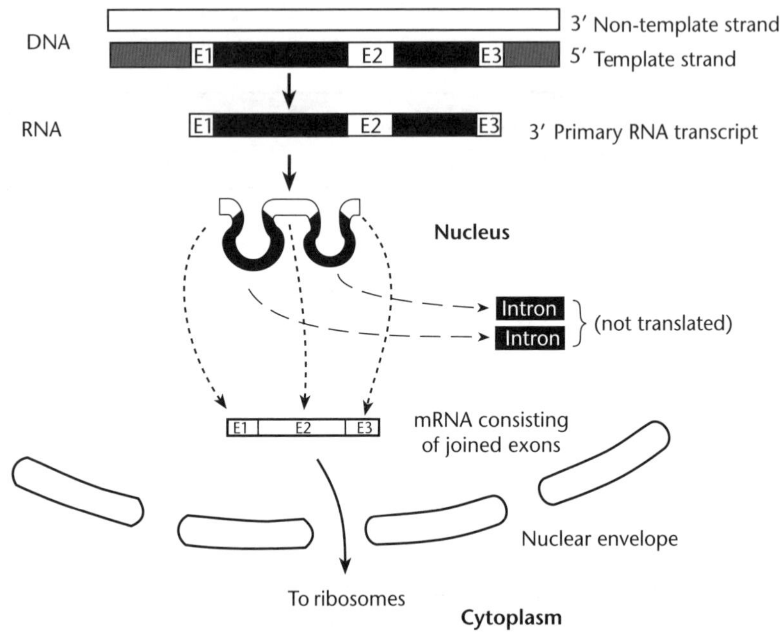

Editing of primary RNA transcript by removal of introns.

In many genes, there are a dozen or more introns, which may be of considerably greater total length than the exons. It has been proposed that by increasing the length of the DNA through the presence of introns, there may be more places where the crossing over of DNA can occur and hence an increase in genetic diversity.

After removal of the introns, the RNA leaves the nucleus via the pores in the **nuclear envelope** and passes to the cytoplasm. The pores in the nuclear envelope have a complex structure, and in some way the mRNA is actively 'fed' through them and passes to the endoplasmic reticulum in the cytoplasm.

Translation

Translation is the linking of amino acids in the sequence specified by the base sequence of mRNA. Since an mRNA molecule is a copy of a gene, the information it contains is still in code form, the sequence of bases representing the order of amino acids in the polypeptide. The four-letter 'alphabet' of DNA must therefore be translated into the 20-letter 'alphabet' of proteins.

Each base cannot represent an amino acid, since only four amino acids could be coded for. Two bases (a doublet code) would still be insufficient since it would give only $4 \times 4 = 16$ sequences:

AA	AC	AG	AT
CA	CC	CG	CT
GA	GC	GG	GT
TA	TC	TG	TT

At the minimum therefore it must be a triplet code, with three bases representing each amino acid. This gives $4 \times 4 \times 4 = 64$ triplets – more than enough.

In protein synthesis, the bases in an mRNA molecule are 'read' in groups of three, called **codons**. Each codon specifies a particular amino acid in the polypeptide. It is the function of the tRNA and the ribosomes to decode the mRNA and translate it into the correct amino acid sequence of a polypeptide.

Transfer RNA (tRNA)

The function of tRNA is to act as an '*adaptor*' molecule, bringing each amino acid to the mRNA codon that specifies it. tRNA is a relatively small molecule, consisting of a chain of 70–90 nucleotides. Since there is a different kind of tRNA for each amino acid, there are at least 20 kinds of tRNA. A tRNA molecule is folded back on itself giving it a complex 3-dimensional shape. When flattened out, it has the shape of a clover leaf.

The molecule is held in shape by hydrogen-bonded base pairs, so bases along one part of the molecule must be complementary to bases further along.

Each tRNA molecule must be able to combine with:

- One of the 20 kinds of amino acid.
- The mRNA codon that specifies the amino acid.

Having combined with its amino acid, how does a tRNA find the mRNA codon that specifies that amino acid? This depends on three unpaired bases on the tRNA. Since they are complementary to a codon on the mRNA and can pair with them, these unpaired tRNA bases together form an **anticodon**.

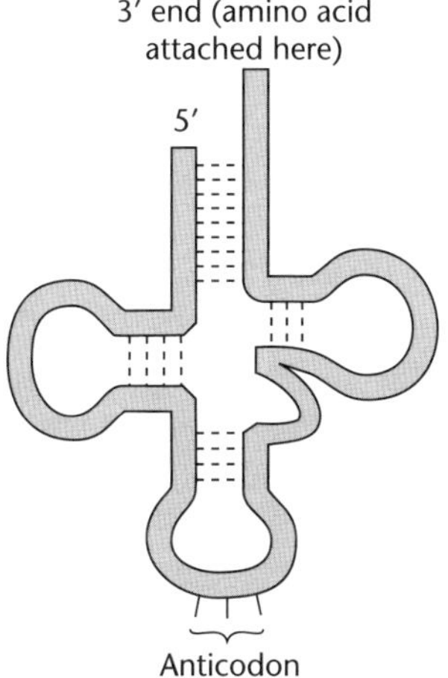

A transfer RNA molecule shown flattened out.

Ribosomal RNA (rRNA)

A problem with decoding mRNA is that although the bases have to be read as three-letter words, *they are not grouped in threes*. Consider the following message:

BIGFATSAMSATALLDAY

There are three 'frames' in which this message could be read, depending on whether one starts at the first, second or third letter.

Starting at the first we read: BIG FAT SAM SAT ALL DAY

Starting at the second we read: IGF ATS AMS ATA LLD AY

Starting at the third we read: GFA TSA MSA TAL LDA Y

If tRNA anticodons were to combine randomly with the codons of mRNA, there would be no consistent reading frame and the result would be chaos. This is prevented by **ribosomes** – tiny granules about 25 nm across, consisting of rRNA and protein. Each ribosome consists of two subunits, which only come together to form a ribosome when the mRNA is being translated.

Like other forms of RNA, rRNA is synthesised on a DNA template. The sections of chromosome involved form one or more darkly staining bodies called *nucleoli*. Although the protein component of ribosomes is synthesised in the cytoplasm, the ribosomes are assembled in the nucleus. They then pass out through the pores in the nuclear envelope to the endoplasmic reticulum.

How ribosomes read the message

The ability of a ribosome to translate the coded information in an mRNA molecule depends on two essential properties:

- It can 'recognise' the 5' end of an mRNA molecule and move along it one codon (three bases) at a time.
- It allows codon–anticodon pairing to occur. Because a ribosome moves along the mRNA one codon at a time, and because codon–anticodon pairing can only occur in the presence of a ribosome, the places that tRNA links up with mRNA also move along one codon at a time.

A ribosome has two binding sites for tRNA. One (the P-site) is occupied by the tRNA attached to the growing polypeptide chain. The other (the A-site), binds the incoming tRNA that holds the next amino acid to be added.

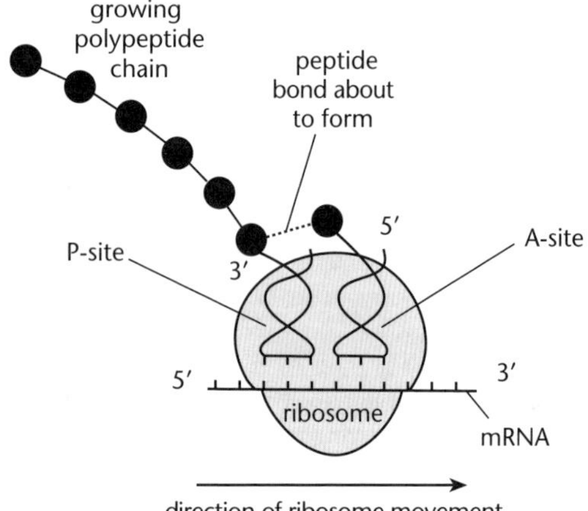

Relationship between mRNA, ribosome and tRNA.

In outline, what happens is as follows:

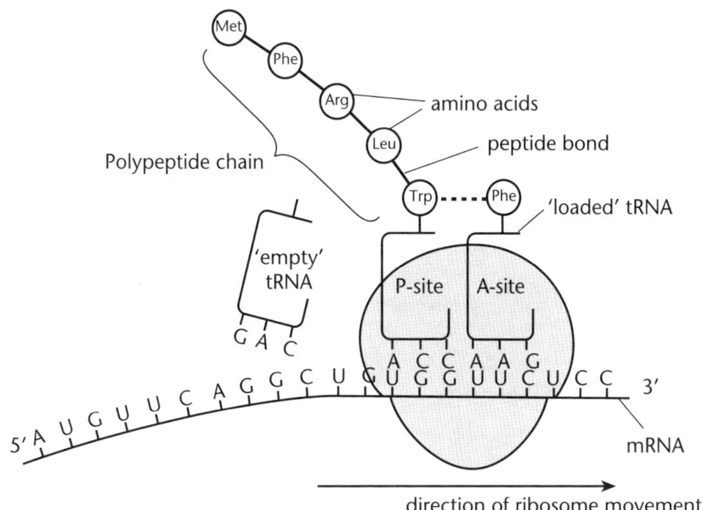

Translation of mRNA by ribosome.

A ribosome joins with the 5' end of an mRNA molecule. The codon AUG acts as a 'begin here' symbol (as well as coding for the amino acid methionine, which is removed after the polypeptide chain has been completed). The ribosome moves along the mRNA one codon at a time. Each time it does so, the tRNA at the A-site is displaced to the P-site. At the same time a peptide bond is formed between the newly arrived amino acid and the end of the peptide chain. The tRNA that had previously occupied the P-site then breaks free, and can pick up another amino acid. Every time the ribosome moves along one codon, another amino acid is added to the polypeptide chain. When the ribosome reaches a 'stop' codon, it breaks free and can then begin 'reading' another mRNA molecule to make another protein molecule. Normally, several ribosomes move along an mRNA at any one time, the whole complex being called a *polyribosome*.

Several ribosomes 'read' an mRNA molecule simultaneously making several polypeptides.

Breakdown of mRNA

Each **mRNA** molecule is 'read' many times before being eventually broken down by an RNase enzyme. A cell can therefore only make the same protein if new mRNA molecules continue to pass out from the nucleus.

To decrease the rate at which a cell makes a protein, fewer of the mRNA molecules coding for it are made in a given time. This enables the cell to adjust the rate at which it makes a protein in response to changing circumstances. (The other two kinds of RNA are not destroyed, but continue to be recycled.)

Protein synthesis in prokaryotes

Bacteria make proteins in an essentially similar way to eukaryotes, except:

- Since there are no introns, there is no RNA processing. There are thus only two stages, transcription and translation.
- Since there is no nuclear envelope to separate the sites of transcription and translation, the ribosomes don't have to 'wait' for mRNA to be completed. The ribosomes begin moving along the mRNA before transcription has finished. Transcription and translation thus occur simultaneously.

Characteristics of the code

Because there are *two* DNA strands, the code is usually represented in terms of mRNA codons.

First base of codon (5′ end)		Second base of codon: U	C	A	G		Third base of codon (3′ end)
	U	Phe	Ser	Tyr	Cys	U	
	U	Phe	Ser	Tyr	Cys	C	
	U	Leu	Ser	STOP	STOP	A	
	U	Leu	Ser	STOP	Trp	G	
	C	Leu	Pro	His	Arg	U	
	C	Leu	Pro	His	Arg	C	
	C	Leu	Pro	GluN	Arg	A	
	C	Leu	Pro	GluN	Arg	G	
	A	Ile	Thr	AspN	Ser	U	
	A	Ile	Thr	AspN	Ser	C	
	A	Ile	Thr	Lys	Arg	A	
	A	Met + START	Thr	Lys	Arg	G	
	G	Val	Ala	Asp	Gly	U	
	G	Val	Ala	Asp	Gly	C	
	G	Val	Ala	Glu	Gly	A	
	G	Val	Ala	Glu	Gly	G	

Each amino acid is indicated by its standard abbreviation, eg 'Pro' stands for proline, 'Lys' for lysine.

The genetic code.

The first base of each codon is indicated by the letters U, C, A and G down the left-hand column. The second base is indicated by the row of letters across the top, and the third base is indicated by the letters down the right-hand side.

- It is a *triplet code*, three bases specifying each amino acid.
- It is *degenerate*, meaning that some (in fact most), of the 20 amino acids are represented by more than one codon. Only methionine (Met) and tryptophan (Trp) are represented by single codons.

- The degeneracy is not random – where several codons represent the same amino acid the alternative codons are nearly always in the same cell of the table, indicating that only the last base is different.
- One codon (AUG) is both a 'start' signal and codes for methionine (Met).
- Three codons indicate 'end of message' and are called 'stop' codons. Unlike the 'start' codon, these do not code for an amino acid.
- Because of the polarity of polynucleotide chains, opposite sequences within a codon are read differently, eg CGA is different from AGC.
- The code is *non-overlapping* – each block of three bases is read separately (rather than the last base of one codon acting as the first base of the next).
- The code is *universal* – with very minor exceptions, the codons specify the same amino acids in all living things so far studied. This is powerful evidence for a **common ancestor** for all present-day organisms.

Extension: Cracking the code

Breaking the code posed two problems:

- How many bases specify each amino acid?
- What sequence of bases specifies each amino acid?

How many bases?

It was Francis Crick who showed that the code is a triplet code, the 'words' having three letters. He used a **bacteriophage** that parasitises *E. coli*. Each time a bacterium is attacked, about 100 new virus particles are produced, which then go on to infect other bacteria in a chain reaction. The result is a clear area or **plaque** on a bacterial culture. Different mutant strains of virus produce plaques with different appearances. When two kinds of virus infect a bacterium *simultaneously*, 'mating' and recombination of the two viral DNA strands may occur. Crick found that in certain cases, two different mutant strains could, when mated together, produce the wild-type virus. Somehow the one mutation had *suppressed* the other.

Crick explained this by suggesting that one class of mutants had been produced by **insertion** of an extra base in the DNA (a 'plus' mutation), whilst the other had been produced by the deletion of a base (a 'minus' mutation). The insertion or deletion of a base would alter the 'reading frame', causing the bases to be read in the wrong groups of three. As a result of such a *frame shift* mutation, the whole of the gene 'downstream' of the mutation would become nonsense.

Example

Inserting the letter Q at position 3 in the message

THE BIG FAT BUG ATE ALL DAY

results in gibberish:

THQ EBI GFA TBU GAT EAL LDA Y

However, if a 'plus' mutation were to be combined with a 'minus' one, the second would cancel out the effect of the first, and sense would be restored. If, say, G were removed from position 7 we would have:

THQ EBI FAT BUG ATE ALL DAY

As a result of a 'plus' and a 'minus' mutation, only the amino acids coded for by the bases between the two mutations are affected.

The very fact that triplets between the two mutations were translated at all shows that a triplet can be changed and yet still 'mean' an amino acid.

Crick discovered that by mating two 'plus' mutants, he could (as a result of recombination) produce 'double plus' and 'double minus' mutants, in which two bases had been inserted or deleted. The result was invariably a mutant. However, when he produced 'triple plus' or 'triple minus' mutations, *some of them had reverted to wild type*. This is exactly what would be expected if the code were a triplet one, since the insertion (or deletion) of three bases would result in no change in the reading frame after the third added or deleted base.

Which triplets code for which amino acids?

Showing that the code is a triplet code was only the first step. The next stage was to establish which triplets code for each amino acid. This task was accomplished in 1961 by Marshall Nirenberg and others.

In breaking the code, two techniques were crucial. The first involved making proteins *in vitro* ('in glass'), in a test tube containing mRNA, tRNA, ATP and certain enzymes. Secondly, it was discovered how to make artificial RNA with known base sequences without the need for a DNA template.

Example

A synthetic mRNA consisting of only uracil resulted in the production of a polypeptide containing only phenylalanine. Hence UUU 'means' phenylalanine. Likewise, it was found that AAA codes for lysine and CCC codes for proline.

Nirenberg then found that an artificial *trinucleotide* of known base sequence could act as a miniature mRNA molecule, and bind to tRNA and ribosomes. By using a mixture of all 20 tRNA–amino acid complexes, he found that the ribosome and trinucleotide combined with just *one* of the tRNA–amino acid combinations.

Example

UUG would combine with ribosomes and leucine–tRNA, showing that UUG 'stands for' leucine.

Extension: The central dogma

In outline, the mechanism by which the information encoded in DNA is used to make proteins can be summed up as 'DNA makes RNA and RNA makes proteins', or

$$\text{DNA} \xrightarrow{\text{information}} \text{RNA} \xrightarrow{\text{information}} \text{protein}$$

The DNA constitutes the genotype and the proteins make up the phenotype. This relationship between DNA, RNA and protein was called by Crick the **Central Dogma** of molecular biology.

By stating that information can flow from DNA to protein but not in the reverse direction, Crick was saying that genes provide the information to build the body but that the body cannot directly influence the plan that helped to build it. This assertion goes to the heart of the way modern biologists envisage the mechanism of **evolution**. The neo-Darwinian view is that although changes in DNA are random, a small proportion of such changes are, *by chance*, advantageous. These 'better' genotypes help build bodies that are more successful because they produce more offspring, and therefore, more copies of the changed DNA.

A modern follower of Lamarck would say that when the body adjusts to changing conditions, these changes are inherited – ie changes in the body result in changes in the genes. This would be equivalent to saying that changes in the amino acid sequences of proteins can lead to changes in the base sequence in DNA. Not only is there no evidence of this, it is difficult to imagine how it could happen.

Since Crick stated the Central Dogma, it has had to be modified as a result of two discoveries:

- In some viruses, RNA is used as a template for making more RNA, so the RNA can contain the information for its own replication, just like DNA.
- In **retroviruses**, viral RNA is injected into the host cell and is used as a template to make DNA. Later, this viral DNA is used to make viral RNA. In these viruses, information in RNA is used to make DNA.

The possible pathways of information flow can thus be written as:

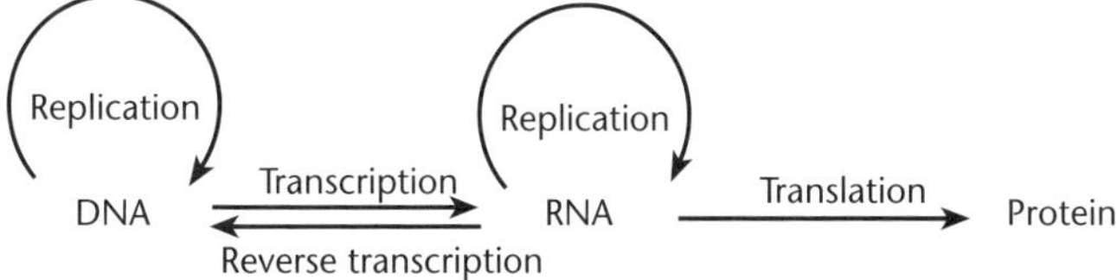

Revised summary of the Central Dogma.

The essential feature of the central dogma can thus be restated:

'Information can flow from nucleic acid to protein, or from one kind of nucleic acid to another, but not from protein to nucleic acid or from protein to protein'.

Extension: Which came first – DNA or protein?

Proteins cannot be synthesised without the information in nucleic acids. In both prokaryotes and eukaryotes, this is DNA. DNA cannot synthesise itself, because it needs enzymes to link the nucleotides together. Thus DNA needs enzymes for its production, and enzymes need DNA for their production.

Until recently it was believed that all enzymes are proteins. This leads to a classic 'chicken and egg' problem – how could either of two interdependent chemicals have originated first? This seemingly insurmountable problem of how life originated has been solved – at least in principle, with two discoveries:

- In some viruses, the genetic material, though consisting of RNA, can replicate itself.
- Some RNA molecules, called **ribozymes**, can act as enzymes. Examples are the catalysis of peptide bond formation by the RNA of ribosomes, and the removal of exons from mRNA by the mRNA itself.

RNA can thus carry out the two key functions of life – catalysis and self-replication. The first bio-molecule may thus have been neither DNA nor protein, but RNA.

Unit 12.3 Activity 4C: Making proteins

1. The diagram shows the flow of genetic information resulting in the formation of a polypeptide chain.

   a. Give the names of steps 1, 2 and 3 and describe where each one occurs in a eukaryotic cell.

   b. How is the end of a polypeptide chain signalled by mRNA?

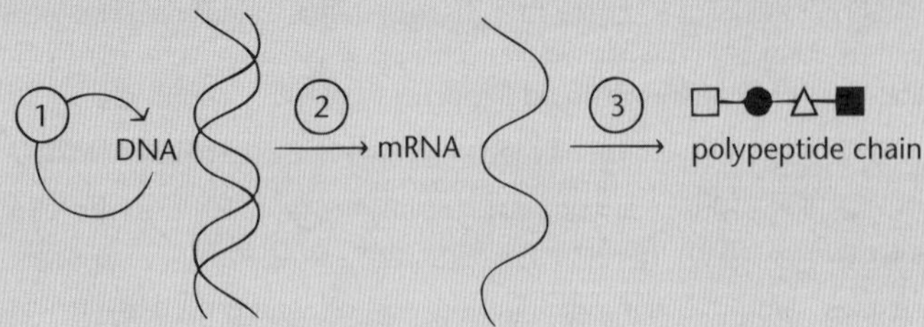

2. a. Select one item from the list below to name each nucleic acid to which the following statements refer.

DNA, mDNA, tDNA, mRNA, tRNA, rRNA

   i. ...is a double helical molecule.

   ii. ...transports amino acids.

   iii. ...is a component of ribosomes.

   b. i. Complete the following:

```
G T _ A C _ T _ A G C A _ _ C  }
_ _ T _ _ G _ C _ _ _ _ A A _  } DNA
G _ _ _ _ _ _ _ _ _ _ _ _ _ _    mRNA
_ _ _ _ _ _ _ _ _ _ _ _ _ _ _    tRNA
```

   Use the genetic code to answer the following questions:

   ii. Suppose the first base G from the left is deleted from the mRNA segment. What will be the amino acid sequence produced in the organism?

   iii. Why would it be incorrect to say that such an organism is a mutant?

   iv. Why can a codon not consist of fewer than three bases?

3. Protein synthesis is a complex process in which the genetic information stored in the sequence of bases along the DNA molecule is used to determine the sequence of amino acids in a protein molecule. The scheme following shows a diagrammatic representation of protein synthesis.

   a. Name the two types of RNA molecules which are involved in translating the genetic message in the DNA into the amino acid sequence of the protein.

   b. Briefly describe what is occurring at the stages labelled **i.**, **ii.** and **iii.** in the diagram. Include names of the types of molecules involved in your explanations.

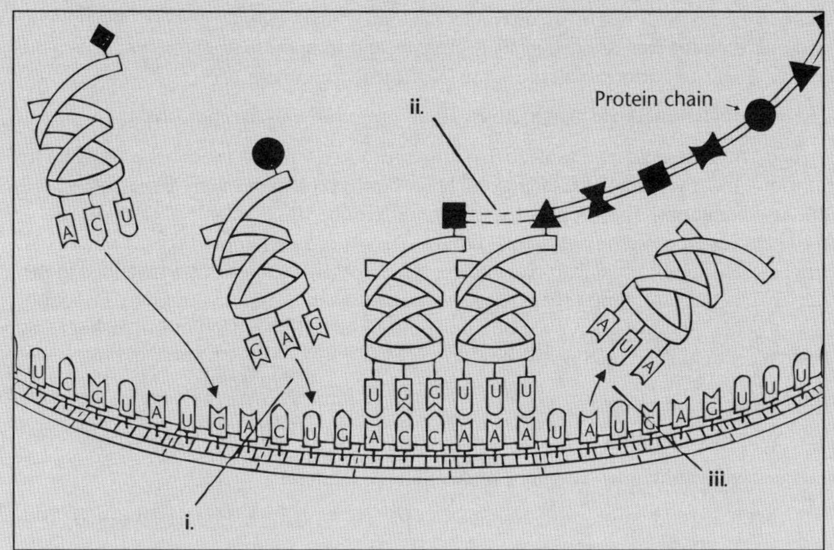

4. True or false? Explain each answer.
   a. A mutation is a change in the genetic code.
   b. A change in a triplet may not result in a change in protein structure.
5. Each of **a.–f.** is a property of one or more of DNA, tRNA, rRNA and mRNA. Excluding bacteria, name the nucleic acid(s) to which each of the following applies:
   a. Single-stranded.
   b. Contains paired bases.
   c. Contains uracil.
   d. Has a short lifespan.
   e. Combined with protein.
   f. Produced in the nucleus.
6. What is meant by 'the genetic code is universal'?

Proteins: products of genes

The function of a gene is to code for the production of a protein (or strictly speaking, a **polypeptide**). This is known as the 'one gene, one polypeptide' hypothesis. Proteins differ from one another in the sequence of their **amino acid** subunits. A gene must therefore *specify the order in which amino acids are linked together*.

Extension: Evidence for the chemical action of genes

The first experiments to suggest a gene directs production of an enzyme were done by two American scientists, George Beadle and Edward Tatum in the 1940s. They used *Neurospora*, a reddish mould often found growing on bread. Besides being easy to keep and having a life cycle of only a few days, this fungus has two advantages over more complex organisms:

- Since it is *haploid* for most of its life cycle, all its genes are expressed – there is no dominance.
- Besides reproducing sexually, it also produces vast numbers of spores **asexually**. Since these are genetically identical to the parent, many copies of a given strain can be quickly produced.

Neurospora can be grown on an **agar** jelly medium containing only glucose, mineral salts and one vitamin. This is called *minimal medium*, and from these simple raw materials the fungus is able to synthesise all the substances it uses for growth.

The ability of *Neurospora* to manufacture its body chemicals must be under the control of genes.

By damaging these genes, Beadle and Tatum hoped to learn how the genes exerted their control. First, they irradiated large numbers of *Neurospora* spores with X-rays, known to cause mutations. They hoped that this would produce fungi unable to carry out all the chemical processes of the normal fungus. They succeeded in producing a number of strains of the fungus unable to grow unless certain organic substances were added to the minimal medium. Amongst these *nutritional mutants* were strains that needed the amino acid *arginine*. Amino acids are the building units of proteins, so if arginine is lacking, any protein containing it could not be produced.

Beadle and Tatum found that not all these 'arginine mutants' were the same:

- Strain 1 could grow provided that arginine or citrulline or ornithine were provided.
- Strain 2 could grow if arginine or citrulline were present.
- Strain 3 could grow only if arginine was present – no other amino acid would do.

The simplest explanation for these observations is that arginine is produced in a sequence of reactions:

Beadle and Tatum suggested Strain 1 lacked the enzyme necessary to convert the precursor substance to ornithine. The fungus would still be able to make arginine so long as ornithine or citrulline were provided. Strain 2 presumably lacked the enzyme that converted ornithine to citrulline, in which case arginine could be made from citrulline if this were available. Strain 3 would have lacked the enzyme that converted citrulline to arginine, so only arginine would support growth.

It appeared therefore each of the arginine mutants lacked a different enzyme. The next step was to determine the nature of the genetic differences between the three strains. They performed two kinds of breeding experiment:

- Each mutant strain was mated with the normal wild type. In each cross, wild type and mutant offspring were in a 1 : 1 ratio. This showed each mutant differed from the wild type by a *single gene locus*.
- When different mutant strains were mated together, some offspring were wild type. This showed different strains *complemented* each other – ie each parent provided what the other lacked. The genes must therefore have occupied different loci, ie they were *non-allelic*.

Since then, our view of the role of genes has been modified in several ways:

- Many proteins are not enzymes (eg haemoglobin, keratin).
- Many enzymes consist of more than one polypeptide (chain of amino acids). Each polypeptide is coded for by a different gene. It is thus now more appropriate to speak of '*one gene, one polypeptide*'.
- Some genes code for the production of ribosomal RNA and transfer RNA.

Proteins

All the **organic** materials that make up an organism are proteins or are substances made by protein catalysts (enzymes). Since proteins are produced under the control of genes, they are the link between the genes and the body they help to build. The effect of a gene mutation may be the production of a protein with a slightly different structure. This often affects the functioning of the protein.

Proteins have many functions:

- **Enzymes** (biological catalysts) are proteins (some forms of RNA are also catalysts).
- Proteins form an essential constituent of all **cell membranes**. Many membrane proteins are important in the **active transport** of substances into and out of cells.
- They may have a *structural* function, eg in vertebrates, collagen (in tendon, ligament and bone), keratin (in hair, feathers, claws, nails and epidermis).
- Some *hormones* are proteins, eg insulin.
- The **antibodies** that help defend more complex animals against disease are all proteins.
- Some proteins have a *transport* function, eg haemoglobin.
- Some proteins are *contractile* and are responsible for movement, such as the **actin** and *myosin* of muscle, and *tubulin* that makes up spindle fibres in cell division.

A protein molecule consists of one or more chains of up to several thousand **amino acids** linked together *in a particular order*. Each chain is called a **polypeptide**. The information for the sequence of amino acids in a polypeptide is stored in a **gene**. A change in a gene is a **mutation**, and usually results in an altered protein. The effect of a mutation can vary from none at all to a completely non-functional protein.

Amino acids

There are over 100 kinds of amino acid, only 20 of which are used to make proteins.

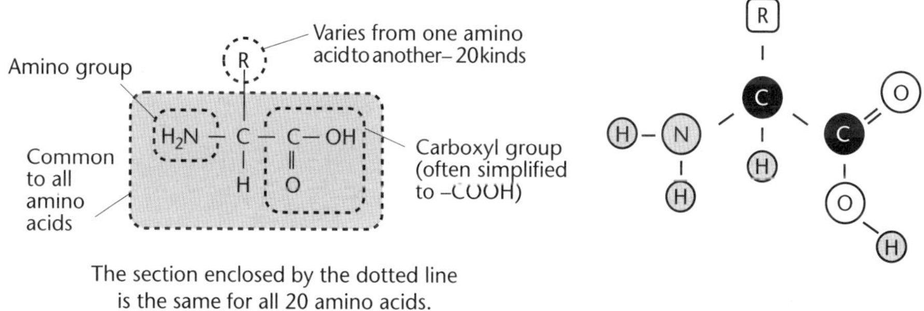

Structure of an amino acid.

The $-NH_2$ group is called an *amino* group and is basic. The –COOH group is called a *carboxyl* group and is acidic. In a protein, these two groups do not exist as such because they are 'used up' in linking amino acids together. The R– group is different for each amino acid and gives each amino acid its distinctive properties.

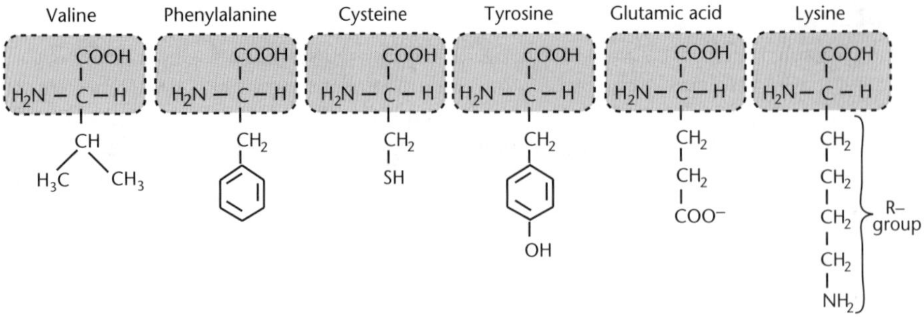

Some amino acids.

One of the most important ways in which R– groups affect the properties of a protein depends on whether or not the R– group is attracted to water. R– groups that are attracted to water are said to be *hydrophilic*; R– groups not attracted to water are said to be *hydrophobic*.

A water molecule, H_2O, is said to be polar. Polarity results from the unequal distribution of electrons between atoms of covalent chemical bonds – in H_2O, oxygen has a slight negative (–) charge and the hydrogens have a slight positive (+) charge. Water molecules are strongly attracted to each other and to polar groups of other molecules.

Examples

The R– group of glutamic acid is hydrophilic because the COO– portion of the R– group is negative and is attracted to the slightly positive hydrogens of water.

The R– group of valine is hydrophobic because the R– group is a neutral hydrocarbon, CH_3CHCH_3, which is not attracted to polar water molecules.

The polarity of amino acids is important in determining the final shape of a protein.

How amino acids are linked together

Amino acids join together by the amino group of one reacting with the carboxyl group of another. The result is a *dipeptide*, the two amino acids being held together by a **peptide bond**.

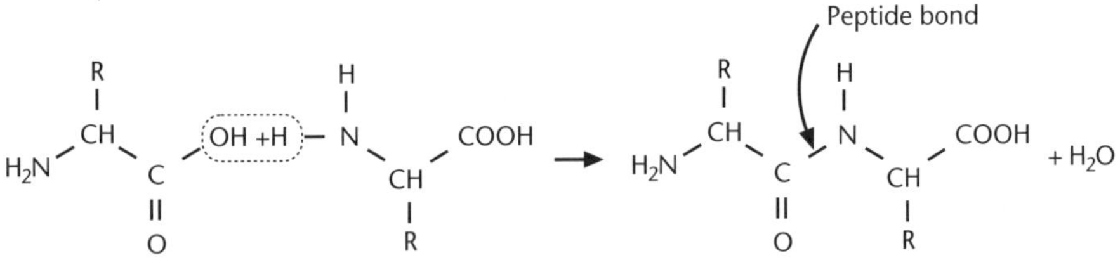

Formation of a dipeptide from two amino acids.

Formation of a dipeptide is an example of a *condensation* reaction, in which two molecules combine to form a larger one with the formation of *water*. If another amino acid is added, a tripeptide is formed, and with more amino acids, a *polypeptide* is produced.

Amino acids in a polypeptide are no longer amino acid molecules but are called amino acid *residues*. The backbone of all polypeptide chains consists of the same repeating sequence of –NH, –CH, and C=O groups (the R– groups provide the variety).

```
        R           H           O           R           H           O
        |           |           ||          |           |           ||
etc     CH          N           C           CH          N           C
   \  /    \     /    \     /     \     /      \     /     \     /     \
    N        C           CH          N            C           CH          etc
    |        ||          |           |            ||          |
    H        O           R           H            O           R
```

Section of a polypeptide.

Fibrous and globular proteins

Proteins come in two main kinds, according to the shape of the molecule.

- *Fibrous proteins* – many polypeptide chains form rope-like bundles (eg **collagen** in tendon, ligament and bone, and **keratin** in the epidermis of the skin, and structures derived from it such as nails, hair and feathers).
- **Globular proteins** – the polypeptide chain is irregularly folded into a ball-shaped molecule, eg enzymes.

Generally speaking, fibrous proteins have a mechanical function (eg protection, support), globular proteins take part in chemical reactions.

Structure of proteins

The structure of proteins can be described at four different levels of complexity – primary, secondary, tertiary and quaternary.

Primary structure

This is the sequence or order of amino acids along the polypeptide chain. Each kind of protein has a primary structure that differs from that of every other protein. It is the amino acid sequence that determines the properties of the protein.

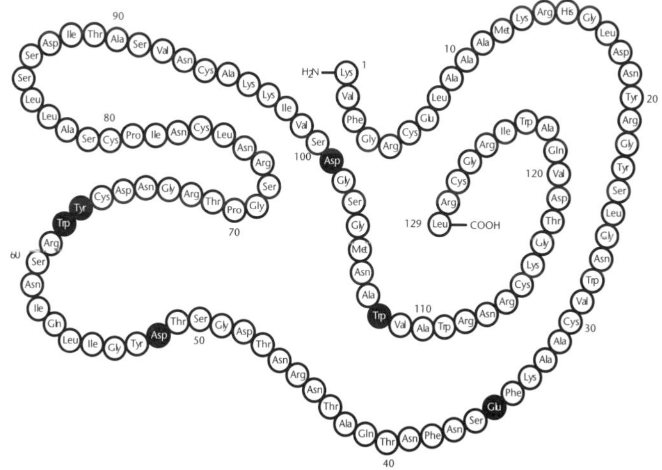

Lysozyme is an enzyme that attacks the outer coat of many bacteria.
The name of each amino acid is abbreviated. Amino acids at the **active site** are shown in black, and are actually much closer than shown, as a result of folding of the chain.

Primary structure of lysozyme.

The number of possible sequences is vast beyond imagination.

There are 20 × 20 = 400 possible dipeptides, 20 × 20 × 20 = 8000 possible tripeptides, and 160 000 possible tetrapeptides. The number of possible sequences for a modestly sized protein of 100 amino acids is 20^{100}, or 10^{130}. Most of the possible sequences would be biologically useless – a change in only one amino acid is, in some cases, enough to cause the protein to be inactive.

Secondary structure

In both fibrous and globular proteins, the polypeptide chain is bent or twisted to form a geometrically regular **secondary structure**. The two most common kinds are the α *helix* (alpha helix) and the β (beta) arrangement or *pleated sheet*. In the α helix, the peptide chain forms a spiral or helix, held in shape by **hydrogen bonds** linking the C=O and N–H groups of adjacent turns of the peptide backbone.

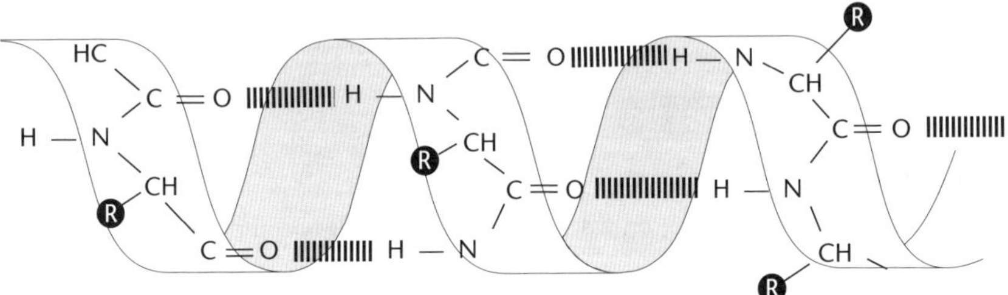

Only atoms on the near side of the helix are shown. The thick dotted lines are hydrogen bonds, the black circles are R– groups.

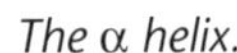

The α helix.

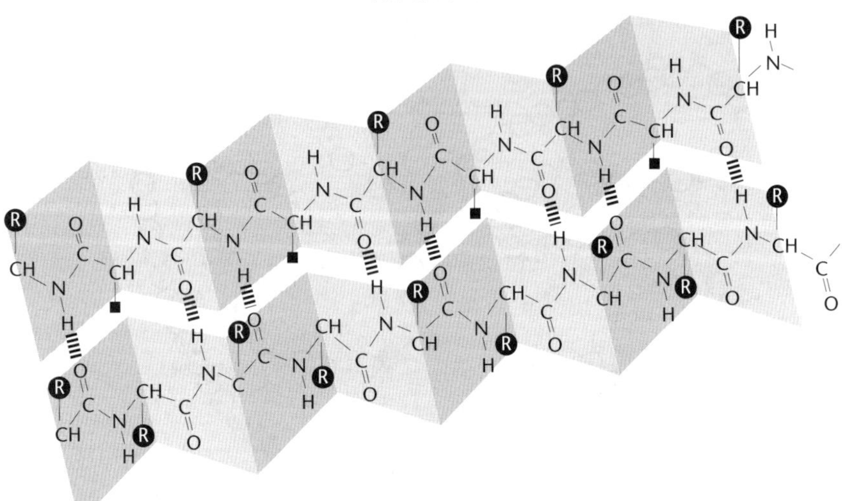

Two or more sections of polypeptide chain run next to each other, linked by hydrogen bonds (between the C=O and N–H groups).

The β pleated sheet.

Both α helices and β pleated sheets can be found within a protein.

Tertiary structure

In **globular proteins**, the polypeptide chain is folded irregularly to form a **tertiary structure**. Tertiary structures vary greatly from one protein to another.

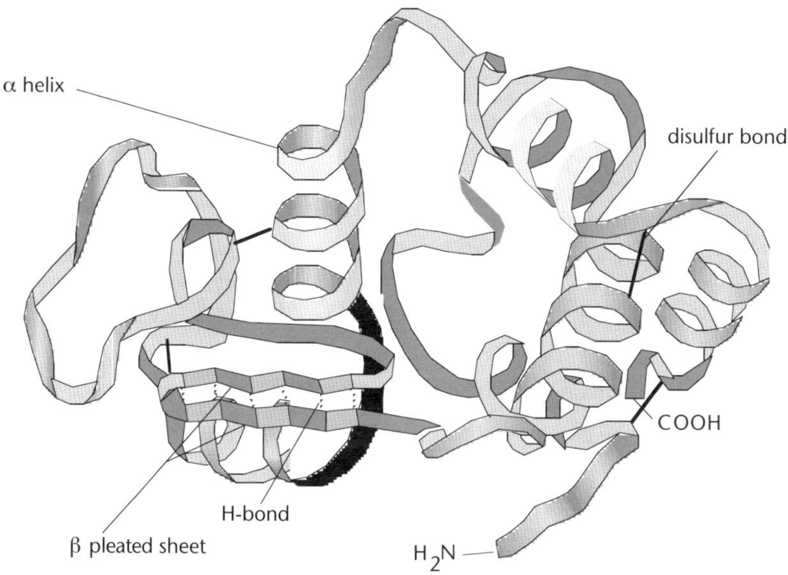

Tertiary structure of lysozyme.

Though globular proteins are large molecules, their chemical activity usually resides in a small number of R– groups located on a small part of the protein surface – in enzymes this is called the **active site**. The R–groups responsible for this chemical activity are a long way apart along the peptide chain but are brought close together by secondary and tertiary folding of the peptide chain.

Although the peptide chain is 'floppy', it is held in its tertiary structure by several kinds of forces. The most important are hydrogen bonds (H-bonds) between polar R– groups and surrounding water molecules.

hydrogen bond

Tyrosine

Hydrogen bonding between water molecules and a polar R– group.

Non-polar R– groups are buried in the interior between adjacent folds of a chain since they are not attracted to water – this helps give a protein its characteristic shape. (Other links are *disulfur* bonds between cysteine residues.)

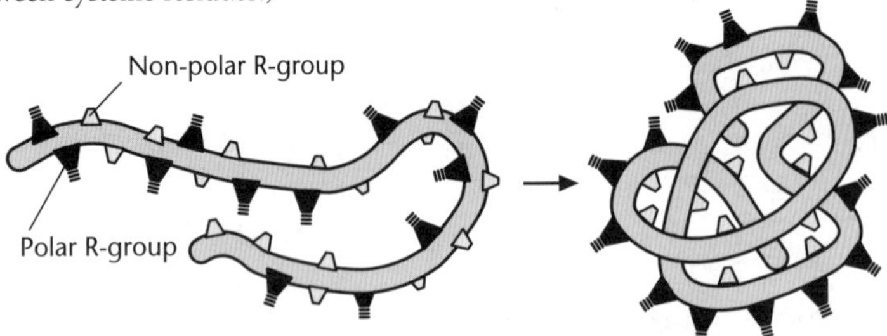

Polar R-groups (black circles) face outwards into the water, non-polar R-groups (grey) are buried in the interior.

How water helps to hold a globular protein molecule in shape.

In a globular protein, the R– groups thus perform two quite different functions – a small cluster carries out the chemical activity of the protein, the remainder help retain the molecule's shape, bringing the 'active' R– groups close together.

Hydrogen bonds, being weak, are easily broken by moderate heat – the molecules collide and vibrate strongly enough to destroy the secondary and tertiary structure. This *denaturation* results in a protein being inactivated, ceasing to function as normal, and usually precipitating.

Quaternary structure

Many globular proteins consist of two or more polypeptide chains held loosely together to form a **quaternary structure**.

Example

Haemoglobin is a *tetramer*, consisting of four peptide chains – two α chains and two β chains. Each chain has an iron-containing *haem* group that binds an oxygen molecule. The forces holding the chains together are mainly hydrogen bonds between polar R– groups and water. Polar R– groups are on the outside, facing the water; non-polar R– groups lie buried in the clefts between the individual polypeptide units.

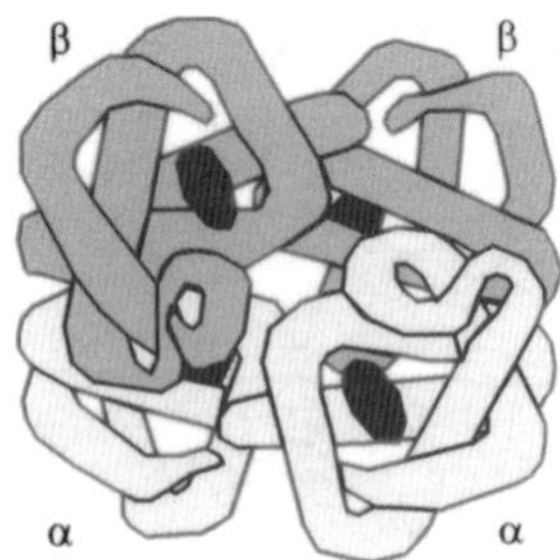

The black discs represent the *haem* groups.

Haemoglobin – a globular protein with quaternary structure.

Membrane proteins

Globular proteins are a major constituent of cell membranes. Some transport substances into and out of the cell, others play a part in detecting chemical signals from other cells. Membrane proteins float in the lipid bilayer, held by hydrogen bonding between polar R– groups and water.

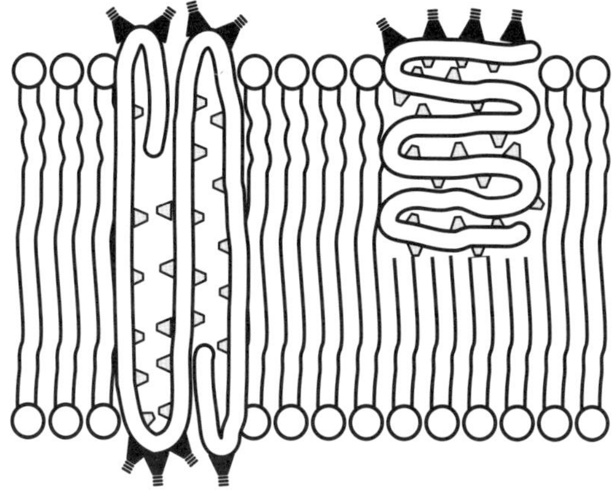

Hydrogen bonds holding membrane proteins in cell membranes.

Enzymes

Enzymes are the catalysts that allow the chemical reactions of life to occur. Apart from some RNA catalysts ('ribozymes'), all biological catalysts are globular proteins.

Enzymes take part in the reactions they catalyse by temporarily combining with the substance(s) they act on (the substrate), forming an *enzyme-substrate complex*. This occurs at the active site. The active site changes shape slightly to 'clasp' the substrate. This *induced fit* mechanism is possible because most of the forces holding the enzyme in shape are sufficiently weak to allow some movement. (The instability of enzymes to heat is thus the price paid for enzymes to work as fast as they do.)

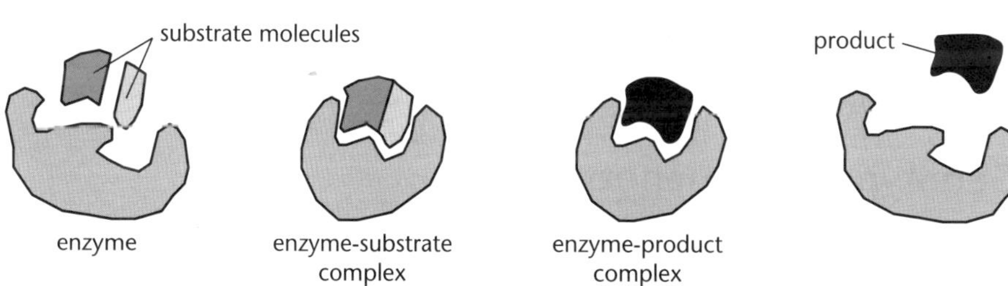

The induced fit mechanism of enzyme action.

Metabolic pathways

Most of the chemical processes in cells (collectively called **metabolism**) occur in sequences called *metabolic pathways*, in which the product of one enzymatic reaction becomes the substrate for the next.

The speed at which a metabolic pathway may be adjusted to meet changing circumstances can result from a change in the:

- Activity of an enzyme; *or*
- Rate of its production.

Enzyme activity can be adjusted if it is inhibited by the end product of the pathway. If the end product accumulates because it is being made more quickly than it is needed, it automatically slows down its own production.

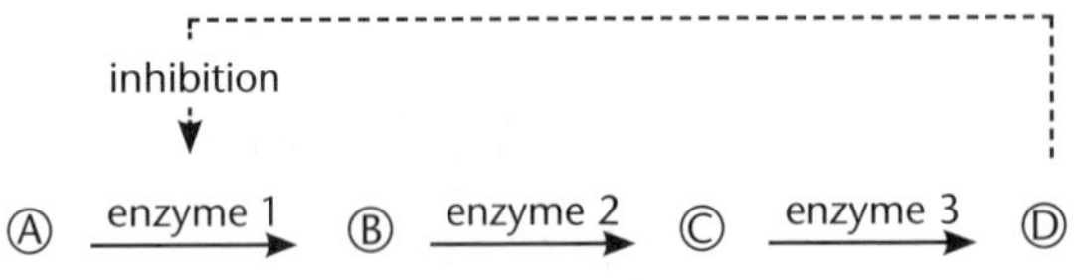

End-product inhibition of an enzyme.

This kind of inhibition (called **negative feedback**) depends on the end product combining with the enzyme at a site other than the active site. This causes the enzyme to change shape slightly so that the active site can no longer combine with the substrate.

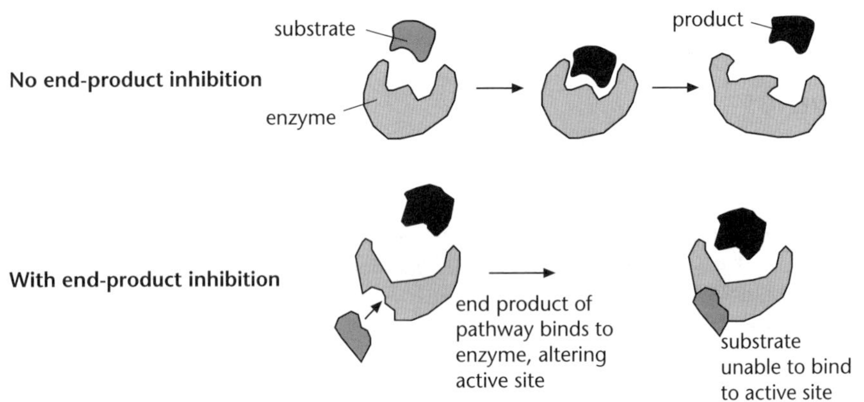

Mechanism of end-product inhibition.

Enzyme activity can also be affected by mutation of the gene controlling its production.

Genotype and phenotype

The 'one gene, one polypeptide' hypothesis did much to demystify the workings of cells. However, an organism is far more complex than a system of individual chemical processes. Proteins are only the first link in a chain of effects that a gene has on the total phenotype. Multicellular organisms are so complex that even when the entire base sequence of an organism's DNA is known, it is not possible to predict all the details of the body that develops. There are a number of reasons for this:

- A phenotypic character may be affected by many genes.
- A gene may have more than one phenotypic effect (**pleiotropy**).

- Differences in the environment may give rise to differences in phenotype.
- The activity of **structural genes** may be affected by protein products of **regulator genes**.

Unit 12.3 Activity 4D: Proteins – products of genes

1. Describe two roles of proteins common to *all* cells.

2. Describe how it is possible for two proteins to consist of identical amino acid sub-units and yet have very different properties.

3. Suppose a mutation results in an enzyme in which a single amino acid is replaced by another. Give two quite different ways in which the activity of the enzyme could be affected.

Unit 12.3 Genetics

Topic 5: Mutations – the origin of new alleles

Author: Martin Hanson

Topic 5 examines the role of DNA in relation to gene expression by:

- Investigating and describing the determination of phenotype through gene mutations.
- Investigating and describing the determination of phenotype through chromosomal mutations.

Mutations

A **mutation** *is a sudden, relatively permanent change in the genetic material.*

Recombination is also a change in the genetic material, but since it occurs every generation it is not stable and is thus distinct from mutation.

> A mutation is *not* a change in the genetic code. Code and message are quite distinct (there is only one Morse code, but it can be used to transmit an infinite number of messages). The genetic code does not change.

Types of mutation:

- *Gene or point mutations* – changes in individual genes that may result in the production of new alleles.
- **Chromosomal mutations** – block and ploidy – involve the structure of chromosomes, and thus affect many genes.
- **Polyploidy** – affects chromosome number, but the structure of each chromosome is unaltered.

Characteristics of gene mutations:

- Like recombination, they are *random* – they do *not* occur in response to *need*.
- Most are *harmful* under most environmental conditions. Most existing alleles have passed the test of natural selection, so almost any random change is likely to be harmful (though in a changing environment this may not be so).
- In diploid organisms, most gene mutations are *recessive*. So long as a mutant recessive allele is rare, it will be present in heterozygotes and will not be expressed. Recessive mutant alleles are therefore shielded from natural selection by dominant alleles. As a result, all diploid organisms carry a *genetic load* of harmful recessive alleles.
- They are *rare* – eg in fruit flies, a new mutation is typically present in about one in 30 000–50 000 gametes.
- They can occur in any cell at any time. Mutations that occur in body cells are called *somatic* mutations and cannot be handed on to future generations (although they can cause cancer).
- Mutations are *reversible*.

Mutation is essential for evolution

Though the number of genotypes that can be produced by recombination (independent **assortment** and crossing over) is vast, it can only explain variation *within* a species. For new species to arise, *new* alleles must be produced by mutation.

Types of gene mutation

A **gene mutation** is a change in the sequence of bases in the DNA. There are two kinds – base substitutions and inversions, and frame shifts.

Base substitutions and inversions

One base is replaced by another (*substitution* mutation) or two bases exchange places (*inversion* mutation). The effects range from zero to lethal, depending on various factors.

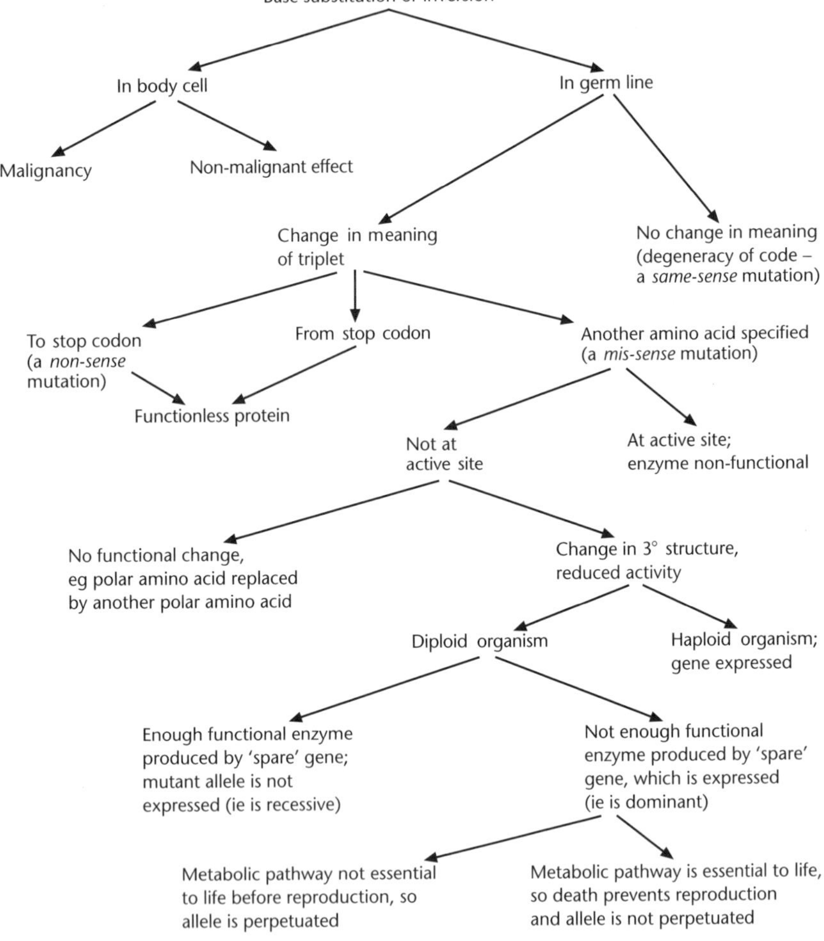

Possible effects of a base substitution or inversion in a gene coding for an enzyme.

Frame shifts

Frame shifts are caused by the *insertion* or *deletion* of a base changing the 'reading frame'. As a result, every triplet 'downstream' of the alteration is changed.

Example

Consider the following sentence (with spaces inserted between triplets to make easier reading):

THE BIG FAT RAT SAT ALL DAY

Adding an 'S' after RAT completely destroys the sense thereafter:

THE BIG FAT RAT SSA TAL LDA Y

The deletion of a base would be equally catastrophic:

THB IGF ATR ATS ATA LLD AY

Unless a frame shift mutation occurs very near the end of a gene, it will almost certainly lead to the production of a completely functionless protein. However, a second frame shift mutation of the *opposite* kind to the first will restore the reading frame.

Example

Combining the two 'mutations' in the previous Example gives:

THB IGF ATR ATS SAT ALL DAY

The greater the distance between an addition and a deletion, the greater the chance that the change in the reading frame will produce a 'stop' codon. Since 3 out of the 64 triplets signal 'stop', there is a $\frac{3}{64}$ (about 1 in 20) chance that any given triplet will be transformed into a 'stop' codon. If an addition and a deletion are close enough together, the number of changed amino acids may be small enough for the protein to still be functional. Most cases of back-mutation are probably of this kind (though note that an exact reversal of the original mutation will be extremely rare, as will be the exact reversal of a base substitution).

How gene mutations are produced

Mutations that occur during DNA replication or recombination, or that are due to an unknown cause, are said to be *spontaneous*. Most, however, are **induced** by external agents called **mutagens**. There are three classes of mutagen:

- *Radiation* – such as X-rays, γ-rays, α-rays, β-rays and neutrons. These are called *ionising* radiation because they break chemical bonds, producing ions or other particles that are so reactive that they combine with almost any molecule with which they come into contact.

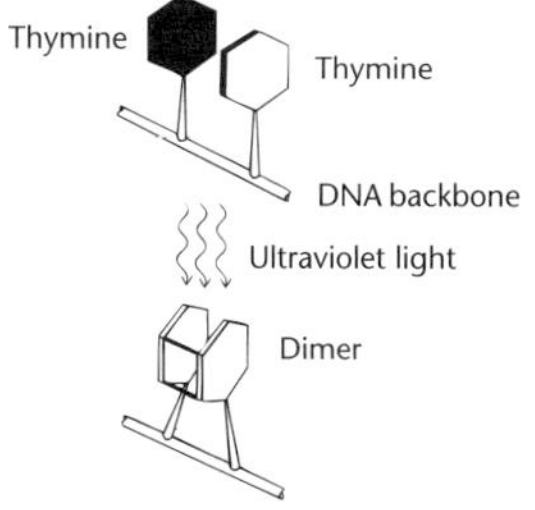

Formation of thymine dimer by UV radiation.

If this happens to be DNA, then one base may be changed into another. Ionising radiation therefore acts *indirectly* on DNA.

Ultraviolet light on the other hand specifically damages DNA. Where two thymine bases occur next to each other along one of the two DNA strands, they can become linked to form a *thymine dimer*. The result of this is to interfere with the replication of the DNA at the next division, and a mutation possibly resulting in skin cancer.

- *Chemicals* – there are hundreds of chemical mutagens, which act in different ways.

  Some change one base into another, eg nitrous acid changes cytosine into uracil. Since this pairs with adenine, the effect is to change a C–G pair into an A–T pair.

  Substances such as 5-bromouracil and 2-aminopurine are called *base analogues* because they are sufficiently like the normal DNA bases to be incorporated into replicating DNA, but have slightly different pairing properties. As a result, the next generation of DNA contains the wrong base.

  Acridine dyes, which become lodged in the DNA helix and interfere with replication, causing an extra base to be inserted, ie a frame-shift mutation.

- *Viruses* – some viruses insert their DNA into that of the host cell. In doing so they may disrupt host genes, causing mutation. The hepatitis B virus produces liver cancer in some people many years after the original infection.

No safe dose

The probability of a mutation occurring is directly proportional to the size of the dose of mutagen. Even at 'background' or natural radiation levels, some mutations occur. A medical X-ray is 'safe' only because the risks are trivial in comparison with the risks of not diagnosing what could be a serious condition.

DNA repair

The actual rate of mutation is much higher than it appears because DNA is constantly being 'checked' and repaired by enzymes. Since each DNA strand carries the information for building the other half, a chemically altered base can be removed and the correct one substituted.

The existence of DNA repair mechanisms becomes apparent when they break down.

Example

In **xeroderma pigmentosum**, the afflicted person develops many skin cancers after even small exposures to UV radiation. The cause is a mutation in the gene coding for the DNA repair enzyme that normally recognises and repairs the thymine dimers produced by the action of UV.

Gene mutation and protein function

The amino acid sequence of a polypeptide is determined by the gene coding for it. A mutation in a gene may change the properties of an enzyme in two different ways:

- One of the amino acids at the active site may be replaced by another.
- A change in an amino acid distant from the active site may change the three-dimensional structure of the protein, or it could render it less stable to heat.

In some cases, the effects of a mutation can be explained in terms of a defective protein.

Examples

Sickle-cell anaemia

Sickle-cell anaemia is a genetic disease in which the haemoglobin is abnormal. It is particularly common in West Africa and amongst people of West African descent. In homozygotes, the haemoglobin becomes crystallised at low oxygen concentrations, causing the red cells to become sickle-shaped. As a result, **capillaries** may become blocked, and in addition the defective red cells are attacked by white corpuscles, causing severe **anaemia**. Heterozygotes are hardly affected, and are said to have the *sickle-cell trait*.

The cause is a single amino acid difference in the β-chain, valine replacing the normal glutamic acid at position 6. Glutamic acid is highly polar and thus strongly attracted to water, but valine is non-polar. As a result, the haemoglobin molecules are more attracted to each other than to water, causing them to aggregate into crystals. This effect is greatest at the low oxygen concentrations that occur in capillaries.

Himalayan albinism

Some mutations are characteriesed by abnormal temperature-sensitivity of certain body processes. Himalayan rabbits are white except for the ears, snout, feet and tail, which are black. The cause is the abnormal temperature-sensitivity of an enzyme involved in the synthesis of the black pigment melanin. Even at normal body temperature the enzyme is **denatured**, but at the extremities of the body where the temperature is several degrees lower, the enzyme remains active and pigment is produced. Siamese cats are another example.

Normally only the extremities are black. In this animal the hair was shaved from an area on the back, and kept cool with an ice pack.

Himalyan albinism in rabbits.

Gene mutation and metabolic pathways

A number of genetic diseases in humans are caused by the inability to produce a particular enzyme due to a mutation in the gene that codes for the enzyme. The following are some of the best known of these 'metabolic blocks' – all are due to autosomal recessive alleles failing to produce one of the necessary enzymes that are involved in the metabolism of the amino acid phenylalanine.

- *Albinism*

  The body cannot make **melanin**, the pigment that gives skin and hair their colour.

- *Alkaptonuria*

  Homogentisic acid accumulates in the blood and is excreted in the urine, which slowly turns black on exposure to air.

- *Phenylketonuria (PKU)*

  Results from the inability to convert the amino acid phenylalanine to tyrosine. Phenylalanine accumulates in the blood, and in young children causes mental retardation. Another result is a shortage of tyrosine (from which melanin is made), so phenylketonurics tend to be very blond. If diagnosed early enough, the condition can be treated with a special diet containing proteins low in phenylalanine.

When a metabolic block occurs, phenylalanine or one of its derivatives fails to be broken down. Block 4, for example, occurs when the body cannot make the enzyme that converts DOPA to melanin, causing albinism.

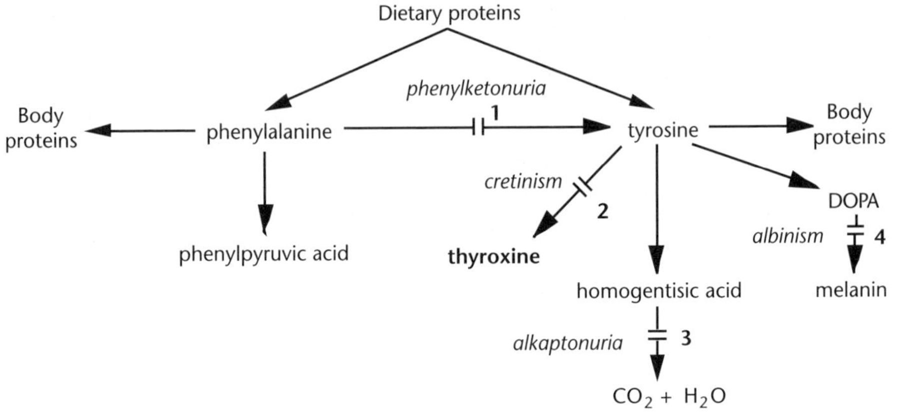

High levels of phenylalanine are harmful and the normal pathway involves its sequential breakdown to CO_2 and H_2O, which are breathed out.

The metabolic pathway of phenylalanine.

Extension: Why uracil in RNA?

Very occasionally, **cytosine** spontaneously changes to uracil. In RNA this does not matter so much because each RNA molecule is only one of many copies. Such a change would be much more serious in DNA since there is only one copy (two in homozygotes). This kind of mutation is prevented by an enzyme that regularly moves along the DNA 'checking' for the presence of uracil. Any uracil encountered is removed and cytosine is substituted. If uracil were naturally present in DNA, the cell would not be able to distinguish between 'bona-fide' and 'alien' uracil.

Cancer as mutation

A special case of mutation is cancer. Cancer is a group of diseases in which cells multiply in unrestrained fashion and spread to other parts of the body. Cell division and the tendency of cells to maintain their position in the body are under genetic control. A cell becomes a cancer cell when these controls break down. A key event in the formation of a cancer cell is the mutation of a normal gene to form an *oncogene*. Over 20 different oncogenes have been discovered, each one derived by mutation from a *proto-oncogene*. Proto-oncogenes are normal genes that play an essential part in regulating cell behaviour such as division.

Unit 12.3 Activity 5A: Gene mutation

1. Describe what is meant by each of the following terms:
   a. Mutation.
   b. Frame shift.
   c. Missense mutation.
   d. Mutagen.
   e. Metabolic block.
2. The primary structure of a protein molecule is determined by the sequence of its amino acids. The following diagram shows a small piece of a protein molecule, together with the sequence of bases on the corresponding coding strand of the DNA molecule that codes for that part of the protein.

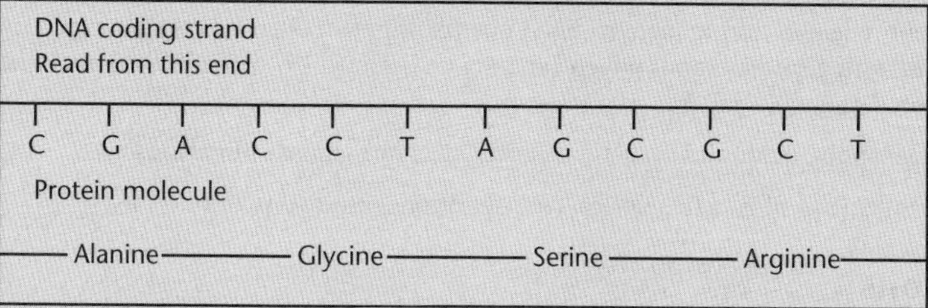

   a. As a result of a mutation the DNA coding strand became altered to:

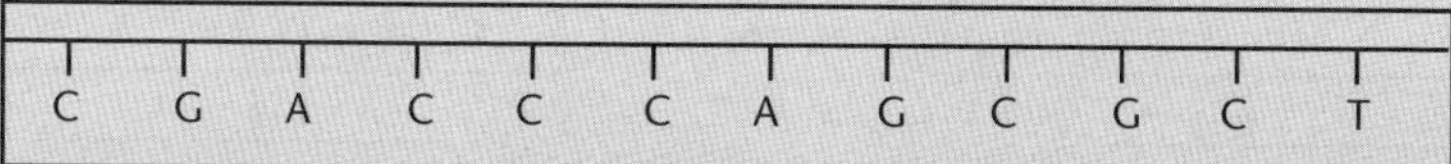

   However, the amino acid sequence remained unaltered as:

Alanine	Glycine	Serine	Arginine

   Explain why the amino acid sequence of the protein is unaffected by this mutation.

   b. A different mutation altered the original sequence of bases on the DNA coding strand as shown below:

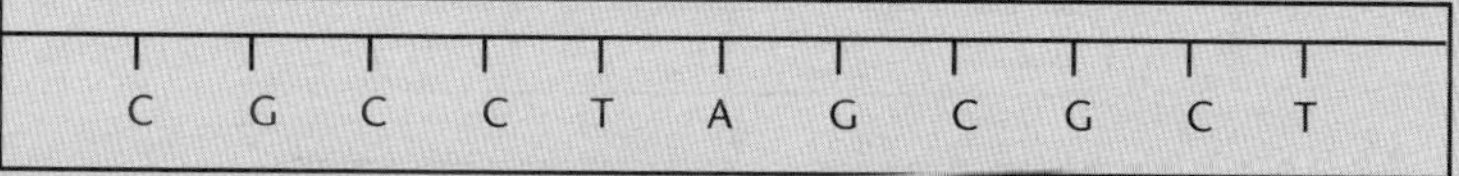

   Describe the effect that this second mutation is likely to have on the sequence of amino acids in the protein, and explain why this would occur.
3. What is meant by saying that mutations are *random*?
4. 'Most mutations are harmful.' Rewrite this statement in a more acceptable form, giving an example to illustrate why the original statement should be criticised.
5. What are the three major classes of mutation, and how do they differ?
6. Why are most mutant alleles rare in natural populations?
7. What are somatic mutations, and can they have evolutionary significance?
8. a. What is a mutagen?
   b. Name two fundamentally different categories of mutagen (exclude viruses) and give an example of each.

Chromosomal mutation

Closely related species often differ in both number and structure of their chromosomes; clearly then, evolution must involve changes in both of these.

Changes in chromosome structure – block mutations

As a result of radiation and chemicals such as mustard gas, arsenic and asbestos, chromosomes may be broken. Although cells have enzymes that can repair such breaks, chromosomes can still undergo permanent change, for two reasons:

- A break is not always repaired.
- If two breaks do occur, the 'wrong' ends may be rejoined.

As a result of a structural change, a chromosome will no longer be able to pair with its partner along its entire length during meiosis. Such mis-pairings can often be seen through a microscope (in contrast with gene mutations which can only be detected by their effects on the phenotype). Chromosome mutations fall into two categories:

- Changes in the quantity of genetic material (deletions and **duplications**).
- Rearrangements of genetic material (translocations, inversions and fusions).

Deletions

As the name suggests, a section of chromosome becomes deleted:

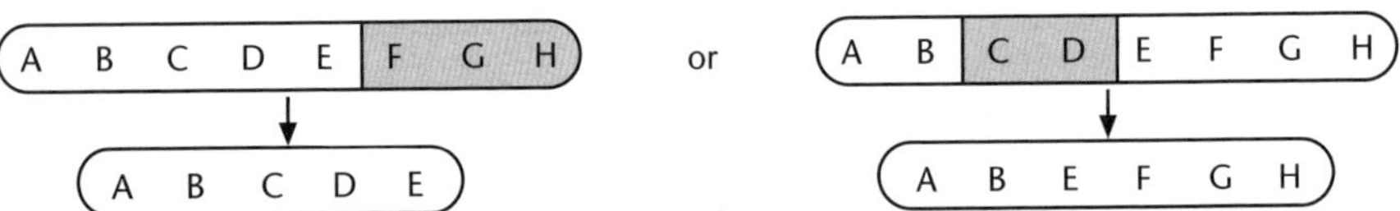

Deletions are often fatal even when an individual is heterozygous, since they usually involve loss of a substantial number of genes.

Duplications

A **duplication** is a repeated section of chromosome:

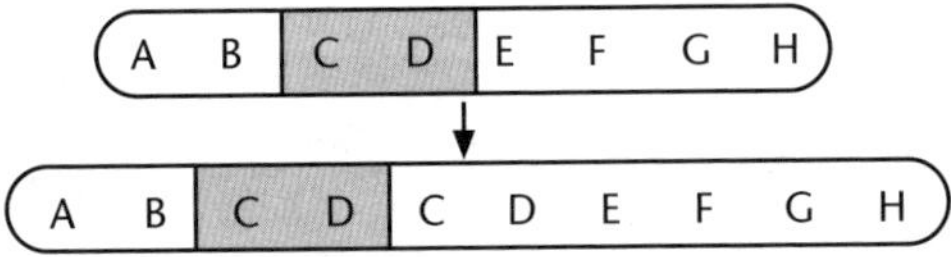

Duplications are generally less serious than deletions since there is no loss of genes, but some duplications do result in abnormal phenotypes.

Example

The condition *bar-eye* in *Drosophila* is associated with a duplicated section of the X chromosome.

The quantitative balance between genes (or rather the relative amounts of their protein products) must therefore be an important factor influencing the phenotype.

Duplications and deletions probably arise as a result of slight inequalities in the sites of the two breaks that occur in crossing over during meiosis I between *homologous* chromosomes:

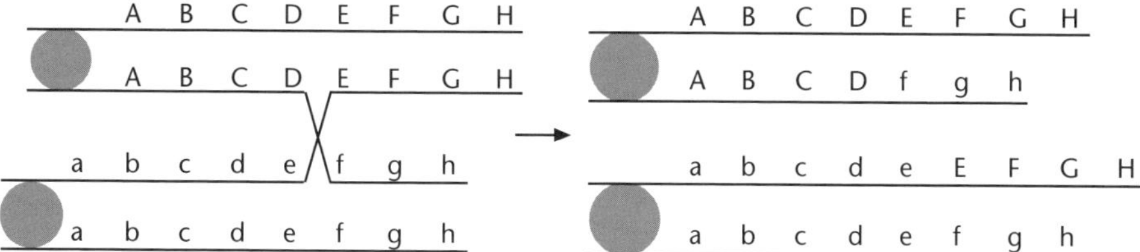

For every duplication produced in this way, there will be a deletion in one of the chromatids of the homologous chromosome.

Duplications are probably important in evolution because they allow the two loci to evolve independently. There are a number of proteins whose amino acid sequences show that they are descended from a common ancestor and have probably originated by duplication.

Example

Myoglobin (muscle haemoglobin) and the α and β chains of haemoglobin can be traced to a common ancestral protein.

Inversions

If a chromosome is broken in two places and repaired 'the wrong way round', an **inversion** occurs:

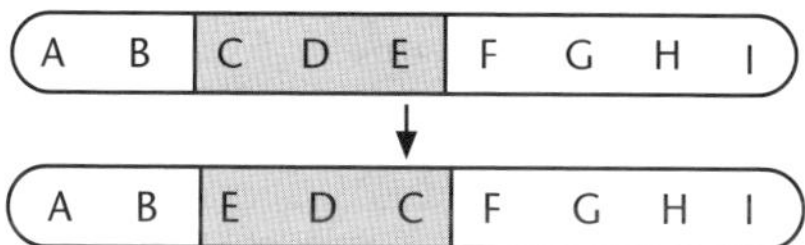

Although there is no loss of DNA, inversions can produce problems during meiosis in organisms heterozygous for the inversion (in which one homologue is normal and the other has an inversion).

Translocations

When breaks occur in *non-homologous chromosomes*, the ends of the non-partner chromosomes may rejoin:

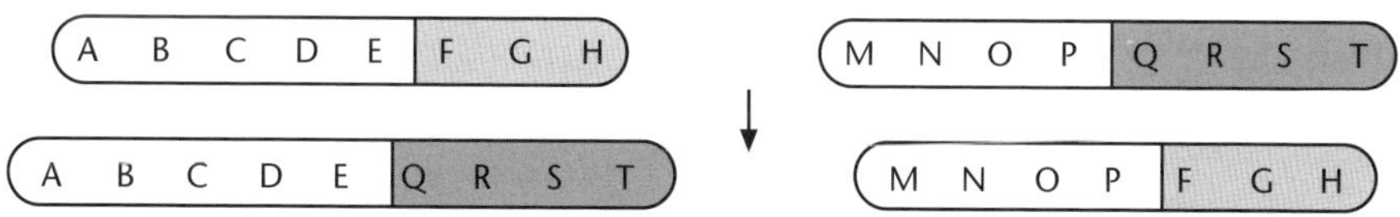

As with inversions, there is no loss of genetic material, but problems arise during chromosome pairing in heterozygotes, and some of the meiotic products are inviable.

Fusions

If two non-homologous chromosomes become broken, they may rejoin to form a single, larger chromosome. If the breaks occur near centromeres that are near the ends of the chromosomes, few genes may be lost:

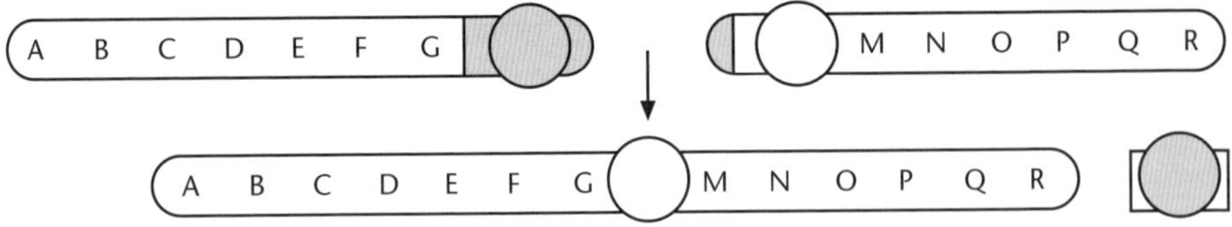

Example

Fusion has occured in human evolution, chromosome 2 having originated by fusion of two ancestral chromosomes.

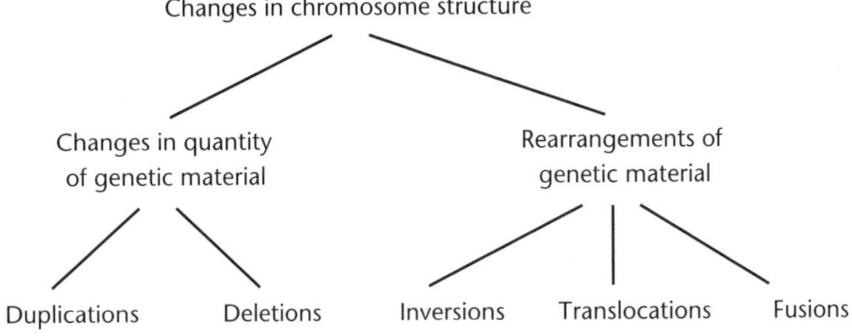

Summary of structural chromosome changes.

Changes in chromosome number – ploidy

Although uncommon in animals, numerical changes in chromosomes have been important in the evolution of over a third of flowering plant species.

Genome – the entire complement of genetic material in an organism.

Haploid – the number of chromosomes in the gametes, abbreviated to n.

Diploid – the number of chromosomes in the zygote, abbreviated to $2n$.

Monoploid – the number of chromosomes in each set, abbreviated to x. In animals and in the majority of plants, 'haploid' and 'monoploid' are interchangeable. But in many plants, the gametes carry two or more different sets of chromosomes, which may be derived from different evolutionary ancestors.

Aneuploidy – only certain chromosomes are represented an unusual number of times.

Euploidy – variation in chromosome number involving the *entire* genome. Many plant species are permanently *polyploid*, with three or more chromosome sets. There are two kinds of polyploidy – **autopolyploidy** and **allopolyploidy**.

Aneuploidy

Aneuploidy is variation in chromosome number involving only part of the chromosome set. The simplest kind of aneuploidy occurs when only one kind of chromosome is involved. The various possibilities are named according to the number of times that chromosome is represented. Bearing in mind that the normal diploid set is $2n$:

Monosomy – only one is present instead of the usual 2, giving a total of $2n - 1$.

Trisomy – three copies ($2n + 1$).

Tetrasomy – four copies ($2n + 2$).

Since monosomy involves the loss of an entire chromosome, monosomy is more serious than trisomy. The severity of the effects also depends on which chromosome is involved; loss of a large chromosome (and therefore more genes) would be expected to be more serious than the loss of a small one. Aneuploidy is more common in plants than in animals (in which the effects are almost always harmful).

Aneuploidy results from the failure of chromosomes to separate or *disjoin* during cell division. This is called **non-disjunction** and can occur during meiosis (either the first or second division) or mitosis. Instead of moving to opposite poles of the spindle, a pair of chromosomes (or chromatids in the case of the second meiotic division) moves to the *same* pole. One daughter cell receives both chromatids and can be represented by ($n + 1$), while the other receives neither and can be represented as ($n - 1$). When a normal gamete (n) joins with an ($n + 1$) gamete, the result is a trisomic. If a normal gamete joins with an ($n - 1$) gamete, a monosomic results. Fertilisation between two ($n - 1$) gametes would produce a nullisomic ($2n - 2$), but not surprisingly these are very rare.

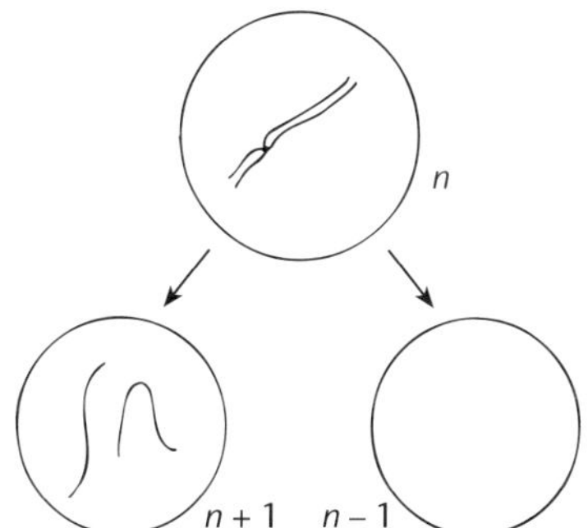

Non-disjunction can also occur in meiosis 1, or in mitosis.

Origin of aneuploidy by non-disjunction in meiosis 2.

Aneuploidy in humans

Aneuploidy in humans is characterised by two peculiarities:

- All cases involve the smaller chromosomes.
- Most involve the sex chromosomes.

The involvement of the smaller chromosomes is probably the result of different survival rates *in utero* rather than a tendency for non-disjunction to occur in certain chromosomes. Aneuploidy for larger chromosomes would be expected to be more serious, since more genes are involved.

The reason why sex chromosomes are more frequently involved in aneuploidy than **autosomes** may be because selection is more intense for disadvantageous X-linked alleles than autosomal ones. This is because all genes on the X chromosome are expressed in males.

Turner's syndrome (X0)

This is monosomy for the X chromosome, and occurs in 1 in about 2500 births. The person is phenotypically female but with undeveloped sex organs, small stature and a characteristically 'webbed' neck. In addition, the afflicted person tends to have a low non-verbal IQ.

Klinefelter's syndrome (XXY)

This occurs in 1 in about 400 births and originates by fusion between an XX egg and a Y sperm, or an X egg and an XY sperm. Affected individuals have male genitalia but tend to have a somewhat female body shape (which can be treated with **testosterone**). IQ and life expectancy are normal.

Other aneuploidies involving sex chromosomes are known, such as XXX (1 in 1000 births) and XYY (1 in 250 births).

Down's syndrome

This is trisomy of chromosome 21, and is associated with certain facial features, reduced IQ and increased susceptibility to disease. One of the most striking features of **Down's syndrome** is that its frequency rises with the age of the mother, particularly after age 40. Evidently the probability of non-disjunction increases with the age of the potential egg. This may be linked with the fact that in mammals meiosis begins at about the time of birth, but then comes to a halt and is not resumed until **ovulation** (which may be 45 or more years later).

Polyploidy

In a polyploid, *every* chromosome is represented three or more times.

Autopolyploidy

Autopolyploidy involves the multiplication of the entire genome *within a single species*. Thus an autotriploid has three sets of chromosomes and an autotetraploid has four sets. Autopolyploidy results from the failure of chromosomes to separate during cell division because the spindle does not function properly. As a result, all the chromosomes finish up in the same nucleus, which has twice as many chromosomes as it should have. This can happen in two ways:

- In the first or second division of meiosis. Either way, the normal halving of the chromosome number fails to occur and two diploid nuclei result, which later give rise to diploid gametes. If a diploid gamete subsequently fuses with a normal haploid gamete, a triploid zygote is produced. If it fuses with another diploid gamete, the resulting zygote is tetraploid.
- In mitosis, resulting in a tetraploid cell. Each replicated chromosome separates, but the cell itself does not divide. The descendants of this cell will also be tetraploid. This is called *somatic doubling*, and if it occurs in a shoot tip it may give rise to a tetraploid shoot. The shoot may later produce flowers, tetraploid cells in the stamens and ovules undergoing meiosis to produce diploid pollen and eggs.

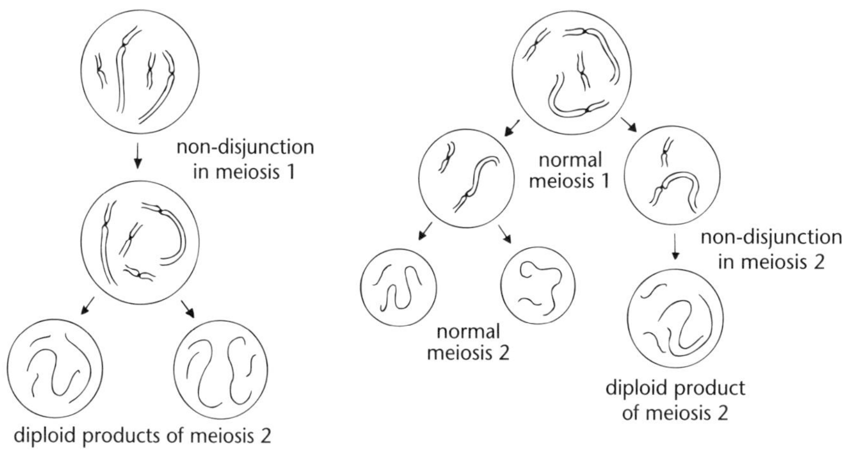

Chromosome doubling as a result of non-disjunction in meiosis.

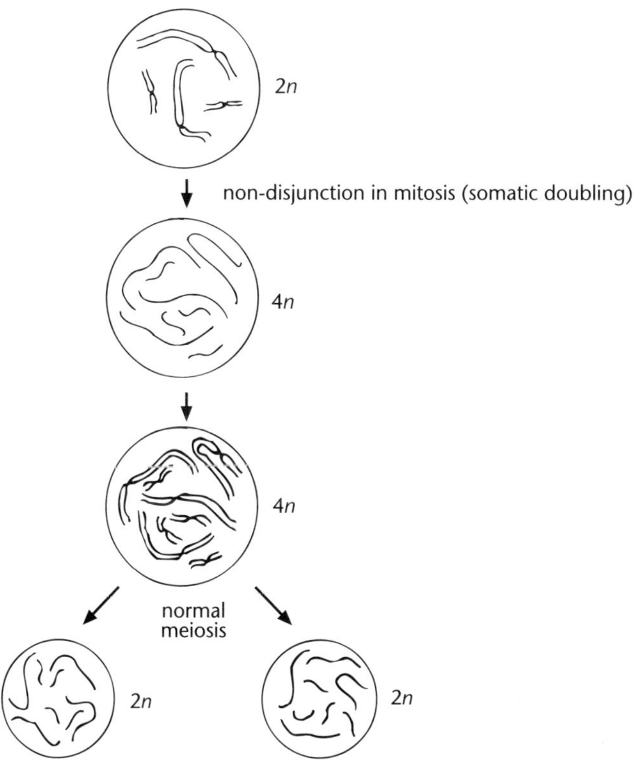

Chromosome doubling as a result of abnormal mitosis.

Although chromosome doubling can occur quite naturally, it can also be induced artificially by the drug *colchicine* (which prevents the formation of the spindle).

Reduced fertility in autopolyploids

Although autopolyploids are often vigorous and may be larger, they are usually **infertile**, especially autotriploids (produced by fertilisation between a diploid and a haploid gamete). Triploids are highly infertile because of the very low probability that meiosis will produce cells with balanced chromosome sets (this is why bananas are seedless). Dandelions are triploid with $3n = 24$. With three homologues, meiosis 1 results in each daughter nucleus receiving either one or two of each chromosome trio. The probability that a daughter nucleus will receive eight chromosomes (one of each kind) is $\left(\frac{1}{2}\right)^8 = \frac{1}{256}$. Similarly, the probability that it will receive 16 (two of each kind) is also $\frac{1}{256}$. Only these two kinds of gamete have balanced sets of chromosomes, so only $\frac{2}{256} = \frac{1}{128}$ of the gametes would be viable. To produce a viable zygote, *both* gametes would have to have balanced chromosome sets. The probability that both gametes would carry eight chromosomes is $\left(\frac{1}{256}\right)^2$, which is 1 in about 65 000. Hardly surprisingly, dandelions are almost completely infertile. (It is just as well that they are *parthenogenetic*, the eggs developing without fertilisation.)

Autotetraploids tend to be more fertile than autotriploids. With four homologues, each daughter nucleus can receive two chromosomes. Such a 2 + 2 segregation often occurs, so a much higher proportion of gametes have balanced chromosome sets.

Allopolyploidy

Allopolyploidy, probably more important in evolution than autopolyploidy, results from hybridisation between species. The first direct proof that polyploids can arise in this way was obtained by Karpechenko in the 1920s. He crossed radish (*Raphanus sativus*, $2n = 18$) and cabbage (*Brassica oleracea*, $2n = 18$). The hybrid was sterile, as expected, since each chromosome had no homologue with which to pair at meiosis.

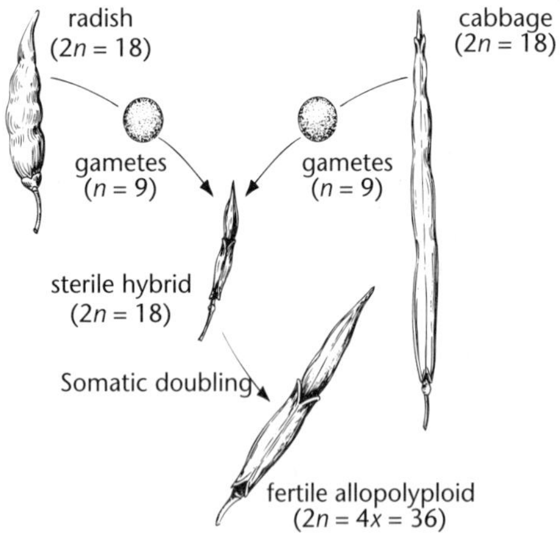

Production of Raphanobrassica, an allopolyploid.

Normal chromosome segregation therefore could not occur and no viable gametes could be formed. Eventually however, some viable seeds were obtained. Chromosome doubling had occurred in the sterile hybrid, producing somatic cells with 36 chromosomes. Each chromosome now had a partner which was not only homologous but was identical since it had arisen by chromosome duplication.

The fully fertile hybrid (called *Raphanobrassica*) had two sets of cabbage chromosomes and two sets of radish chromosomes, and is called an *amphidiploid* (*amphi* = Greek for 'both'). Numerous other amphidiploids have been created artificially, by treating shoots with colchicine to induce chromosome doubling, thus converting the sterile hybrid into a fertile amphidiploid. Many plants are suspected to have had an allopolyploid history, but this often has to be deduced by comparison with present-day species.

Example

Modern wheat, *Triticum aestivum* ($2n = 42$), has the possible ancestry shown below, with *A, B* and *D* representing different chromosome sets.

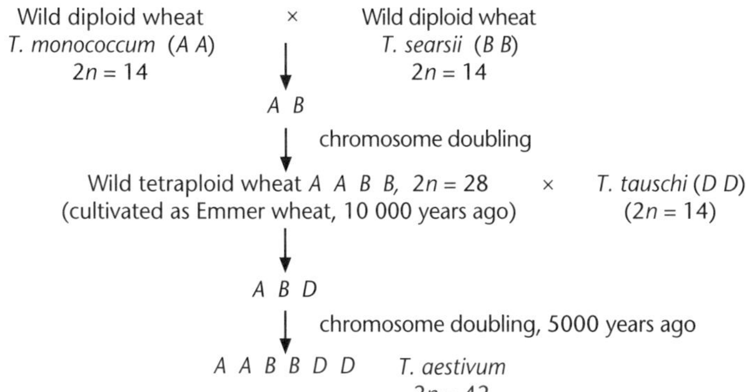

Modern wheat is thus an allohexaploid with three ancestors, each of which has contributed 14 chromosomes. Thus, the haploid number ($n = 21$) is made up of three groups of 7, the monoploid number. Hence $2n = 6x$. In wheat, therefore, *haploid and monoploid are not the same.*

Many other cultivated plants have an allopolyploid ancestry.

Examples

Cotton, $2n = 4x = 52$, an allotetraploid.

Loganberry, $2n = 6x = 42$, an allohexaploid.

Strawberry, $2n = 8x = 56$, an allooctaploid.

The alpine buttercup *Ranunculus nivicola* is a New Zealand buttercup. This has a diploid number of 92 and occurs on Mt Taranaki, the Central Plateau and the Kaimanawa ranges. The only other North Island *Ranunculus* species are *R. insignis* and *R. verticillata*, both with diploid numbers of 46. In a number of respects *R. nivicola* is intermediate between these other two species, and may have been derived by somatic doubling of a hybrid.

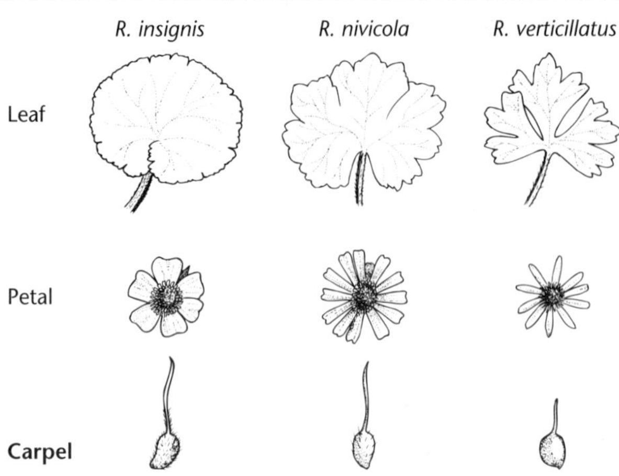

R. nivicola is intermediate between its two parent species, *R. insignis* and *R. verticillatus*.
Allopolyploidy in Ranunculus.

There is some evidence for this view. In the Tararua ranges both species occur close together, and some plants have been found which resemble *R. nivicola* in being intermediate between *R. verticillata* and *R. insignis*. Unlike *R. nivicola*, these plants proved to be sterile, and so may well have been hybrids. The crucial test would have been to examine the chromosomes at meiosis, but cultivated plants died before this could be done.

Other New Zealand plants which may have polyploidy ancestry are certain species of *Hebe*.

Ecological significance of polyploidy in plants

Whether polyploidy could be considered good or bad depends on the environment. This is illustrated by the fact that the frequency of polyploid species increases markedly from the equator towards the poles, and also with altitude.

Example

37% of flowering plant species are polyploid in Sicily, while in Southern Greenland over 70% are polyploid. A similar gradient exists with altitude.

It seems that the lower the temperature, the greater the advantage of polyploidy. It may be that as a result of the advances and retreats of the ice caps during successive glaciations, a more variable environment existed at higher latitudes and altitudes than in warmer regions. Under these conditions, the greater genetic variability of polyploids could have been advantageous. In warmer climates, the advantage of having more genes may be offset by the disadvantage of the increased energy needed for their replication and transcription.

Polyploidy in animals

In the animal kingdom **polyploidy** is rare. It occurs in a number of animals that reproduce parthenogenetically (without fertilisation), such as some crustaceans and some leeches. The evidence for its harmfulness (in humans at any rate) is clear – polyploids being the second most common chromosomal abnormality amongst human miscarriages (about 10% of which are triploids).

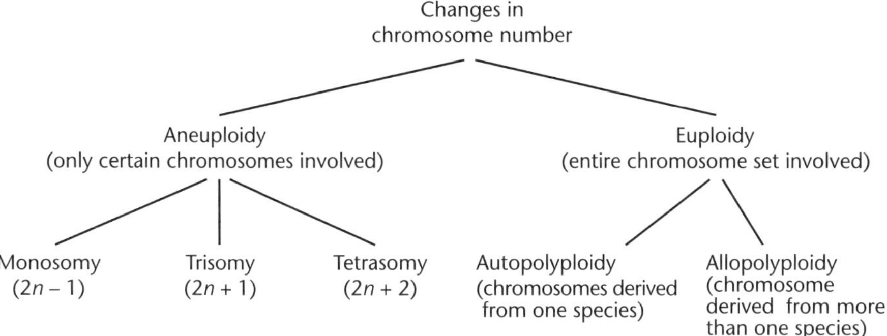

Summary of numerical chromosome changes.

Unit 12.3 Activity 5B: Chromosomal mutation

1. Describe what is meant by each of the following:
   a. Translocation.
   b. Aneuploidy.
   c. Non-disjunction.
   d. Tetraploid.
   e. Amphidiploid.
2. The diagram represents two non-homologous chromosomes. The numbers and letters mark points along the chromosome.

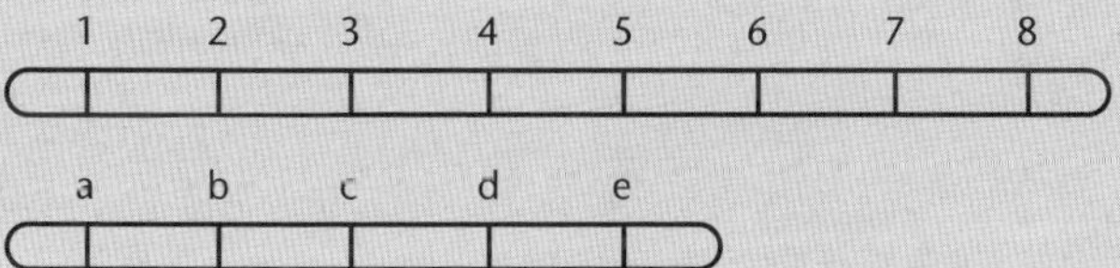

   a. Chromosome mutations can result in changes to the original sequence of genes. Redraw each of the chromosomes to show the result of the mutations named below.
      i. Inversion. ii. Translocation.
   b. Inversion and translocation can be caused by mutagens. Describe what is meant by a mutagen and give an example.

3. The diagram shows human chromosomes arranged in two karyotypes. One person has a chromosomal disorder, the other has a normal phenotype.

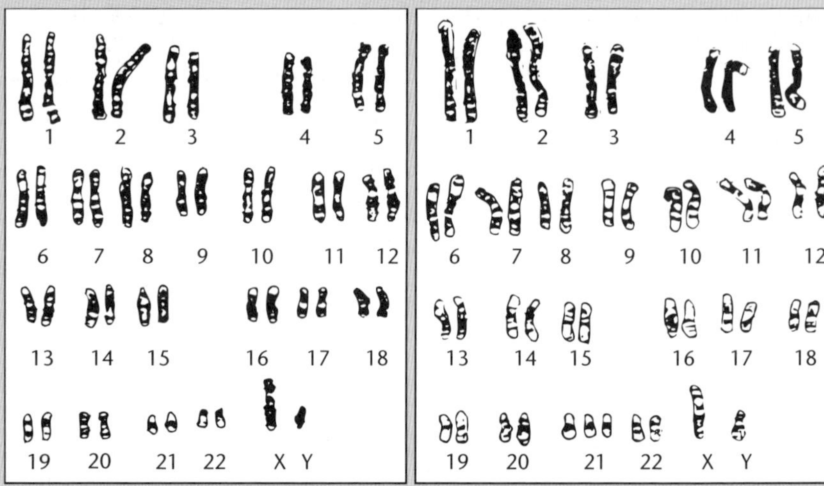

Karotype of Individual A   Karotype of Individual B

a. What evidence indicates that these chromosomes are not those of a gamete?

b. What sex is individual B, and what is the evidence which confirms that a chromosomal disorder is present?

c. Briefly state how this disorder may have occurred.

4. Match each of the terms i–x to one of the descriptions A–J.

i Karyotype	A Produced by increasing chromosome number within one species
ii Homologues	B Variation in chromosome number involving the entire genome
iii Allopolyploid	C Failure of chromosomes to separate during cell division
iv Autopolyploid	D An organism's chromosomal characteristics
v Aneuploidy	E Produced by chromosome doubling of a sterile hybrid
vi Non-disjunction	F Has chromosomes derived from more than one ancestral species
vii Euploidy	G Used to artificially increase chromosome number
viii Amphidiploid	H The entire genetic complement of an organism
ix Genome	I Members of a pair of chromosomes
x Colchicine	J Change in chromosome number involving only part of the chromosome set

5. Two chromosomes have gene sequences A B C D E F G H I J and M N O P Q R S T. Use these sequences to illustrate the distinction between inversions, translocations, duplications and deletions.

6. People with Turner's syndrome are said to be chromosomally 'X0'.

a. Turner's syndrome is an example of what general genetic phenomenon?

b. Explain how an 'X0' zygote could be produced by normal parents.

7. The diagram shows how one kind of polyploidy may originate.

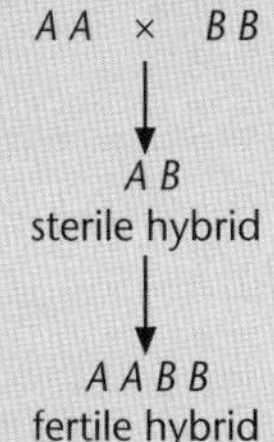

A and *B* each represent a chromosome *set*.

a. Why is the hybrid *A B* sterile?

b. If the two original parent plants have 6 and 8 chromosomes respectively, how many chromosomes would the *A A B B* fertile hybrid have?

Unit 12.3 Genetics

Topic 6: Genetic variation and diversity

Topic 6 looks at genetic variation and change. The Topic deals with:

- Genetic biodiversity – allele frequencies and gene pools.
- Mutation.
- Independent assortment, segregation, recombination.

Genetic biodiversity

Heredity is the passing on of characters/features or **traits** from an individual to its offspring. The basic unit of heredity is the **gene**. A gene is a specific length of DNA on a chromosome which has information for a particular characteristic or trait. Alternative forms of genes are called **alleles**.

The **genotype** is the genetic make-up of an individual for a characteristic. The physical expression of this in the individual is the **phenotype**. The expression of the genotype may be modified by the environment.

Example

The coat colouring of Siamese cats is the result of a heat-sensitive gene which is only expressed at cool temperatures; hence, Siamese cats only have coloured fur at their cooler extremities (ears, paws, nose, tail), while the rest of their body is pale.

This may also apply to other breeds of cat, such as the Birman in the photograph.

Members of a species typically live in **populations** and so share a common **gene pool**. The gene pool is all the *alleles that are present in the population*. The **allele frequency** is the number of times that an allele occurs in a population.

Genetic biodiversity is an expression of the *range* of all the alleles present in the gene pool – the greater the number of *different* alleles present, the greater the genetic biodiversity. This is important in the survival of a species, as the greater the range of alleles, the greater the variation present, which provides more material for evolution to act on. When the environment changes, the chance that favourable alleles are present in the population is high and these favourable alleles can be selected for (natural selection), allowing the individuals with the favourable alleles to cope with the change. Therefore, through their favourable alleles, some individuals will have the **adaptations** necessary for new environmental conditions. Individuals without favourable alleles will tend to be eliminated – the frequency of their alleles in the gene pool will be reduced, maybe to zero.

Unfortunately, biodiversity is contrary to most human plant and animal husbandry practices, where uniformity rather than diversity is selected for.

Example

Exotic pine forests are increasingly being planted out with young pines grown from **clones** of just one or a few pine plants. Though these clones are considered to be 'genetically superior' because they are known to grow quicker and straighter and produce harder wood with fewer knots, etc, the range of alleles that the mature pine forest represents is severely limited. Should a major environmental change occur (such as the introduction of a new pest), it is more likely that these 'genetically superior' trees will not be able to withstand the change because their gene pool may not contain the alleles necessary to cope with the disruption. A 'natural' pine forest grown from seedlings from a variety of trees will have a wider genetic base (wider gene pool) and will have a much greater chance of containing the 'necessary' alleles to combat the environmental change.

As an awareness of the importance of biodiversity increases, steps are being taken to preserve the genetic bases of species. Seeds are being stored in seed banks, captive breeding populations of endangered animals are being maintained in zoos, wildlife reserves are being set aside and attempts are being made to reduce the impact of human activities on ecosystems.

Mutations

A **mutation** is a change in the genetic material of an organism. Mutations are *spontaneous* and *random,* and typically are rare. However, their rate may be increased by exposure to environmental factors such as ionising radiation and **mutagens** (mutation-inducing chemicals such as benzene and formaldehyde).

Block or chromosomal mutations

Mutations may involve *whole genes*. This occurs when parts of a chromosome are deleted, repeated, or translocated to other chromosomes. These mutations are known as **block mutations** or **chromosomal mutations**. They may remove alleles from the gene pool (through deletion from chromosomes) or produce different combinations of alleles (through translocation to other chromosomes).

Other types of mutations may involve the:

- Addition of whole chromosomes.
- Removal of whole chromosomes. The situation in which an individual has more or less chromosomes than the diploid number (ie normal chromosome number in a body cell – 46 in humans) is known as **aneuploidy**.

Examples

Down's syndrome can be caused by individuals having an additional chromosome in the 21st pair. Total number of chromosomes in each cell is 47.

Turner syndrome – an individual has only one sex chromosome (X) instead of two (the person is a female). Total number of chromosomes in each cell is 45.

Point mutations

Point mutations are changes to the bases in DNA, so occur *within a gene*. DNA bases may be deleted, inserted or substituted during DNA replication. This results in a change in the base sequence of the DNA that codes for a gene, *so producing a new allele*.

A sequence of three bases (called a **triplet**) in DNA codes for an amino acid; amino acids are the building blocks of proteins. Any change to the bases in a triplet may change the amino acid coded for.

Example

The triplet **CCA** codes for the amino acid glycine.
If the base **C** (in the second position) is substituted for by **G** to give the triplet **CGA**, the amino acid alanine would be produced instead.

CCA → glycine
CGA → alanine

The substitution of alanine for glycine results in the manufacture of a changed protein, which may then have an altered biological function.

If the protein changed from a point mutation is an enzyme, then the enzyme may no longer be able to act as a catalyst.

Example

A gene in a biochemical pathway enables the production of **melanin** (the protein that causes colouration in the skin):

A mutation in the gene that codes for the enzyme needed to convert substance X to melanin means that the enzyme's structure is changed so that it can no longer catalyse this reaction. Melanin can't be made, so the individual's skin lacks pigment (ie the person is an albino).

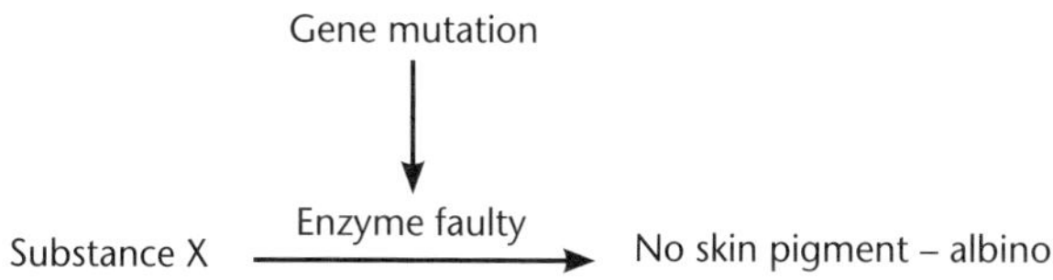

Mutations which produce new alleles are the source of genetic variation

While most mutations are typically harmful, sometimes the altered function will be beneficial to the individual.

If mutations occur in the sex cells or gametes (or the cells that produce them), then the mutations will enter the gene pool of the population and become subject to natural selection. If the new allele is beneficial to the survival of the individuals that have it and the mutation enhances their reproductive success, then the mutation will be selected for and its frequency will increase in the gene pool. The genetic biodiversity of the population has been increased.

Example

There must have been a change in the genes that code for enzymes in the bacteria that inhabit geothermal hot pools, for these bacteria can now operate in temperatures approaching 100°C – temperatures that would kill all other organisms by denaturing their enzymes.

Meiosis

Meiosis is the process of cell **division** that occurs in the sex organs (ovaries and testes), and produces the sex cells or **gametes** (ova and sperm).

In many species, each *somatic* cell (general body cell) has pairs of chromosomes. This is known as the *diploid* number (or 2*n*).

Example

Human cells have 23 pairs of chromosomes to give 46 in total; this is our diploid number.

Pairs of chromosomes are known as **homologous chromosomes** (or homologues). Each chromosome in a pair is identical in length and shape and has the same genes at the same position (or locus). The alleles of these genes may be different (eg ***A*** and ***a***) – this is because one chromosome in each homologous pair comes from the father/paternal parent and the other from the mother/maternal parent.

The gametes produced in meiosis have *half* the diploid number of chromosomes (*one from each pair*) – known as the **haploid** number (or *n*).

Example

In humans, the haploid number is 23 (half of 46).

When the ovum and sperm fuse in **fertilisation**, the diploid number of chromosomes is restored in the **zygote** (the fertilised ovum). Subsequent mitotic divisions of the zygote to form the embryo maintain the diploid number.

Example

The cell drawn is a diploid cell with four chromosomes (two homologous pairs). Therefore, for this cell, $2n = 4$. Between divisions (either meiosis or mitosis) the cell is said to be in **interphase**. It is during this time that the chromosomes replicate themselves through the process of DNA replication – each chromosome is now two **chromatids**.

Prophase I

Step 1: Chromosomes have shortened and thickened, becoming clearly visible as two chromatids held together by a **centromere**. The group of four chromatids is called a **tetrad** or a **bivalent**.

chromatid

centromere

bivalent

Step 2: Parts of adjacent chromatids may **cross-over** and recombine.

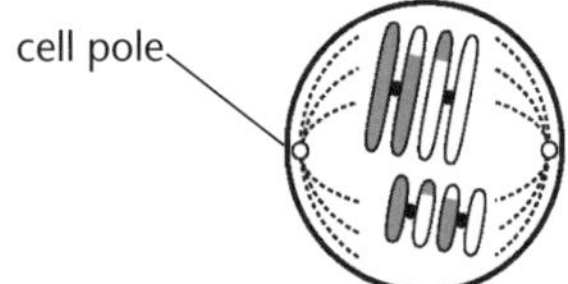

Metaphase I

Step 3: The chromosomes arrange themselves along the cell equator, with each member of a pair of homologous chromosomes orientated towards an opposite pole. **Spindle** fibres begin pulling the homologous chromosomes apart.

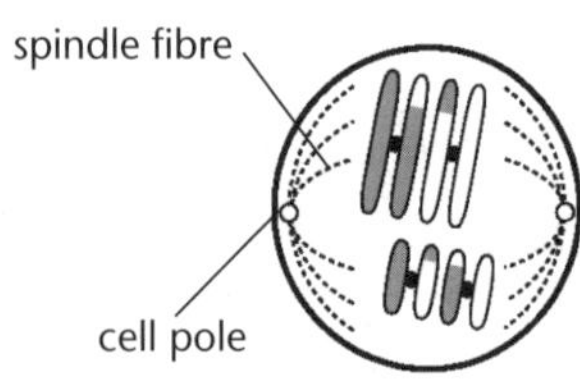

At the next stage of meiosis (anaphase I), any combination of chromosomes from each of the homologous pairs can move towards the cell poles.

Anaphase I

Step 4: The homologous pairs separate and move towards the cell poles. The cell membrane begins to close up between them.

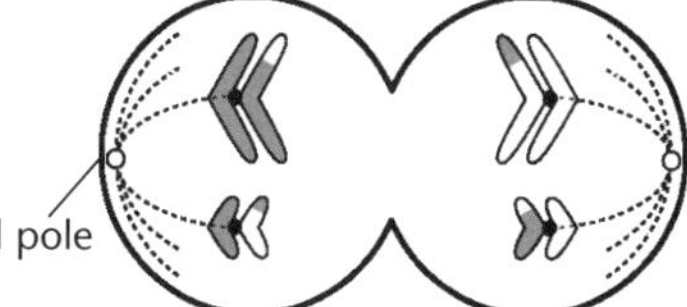

Telophase I

Step 5: Two new cells form, both genetically unique because of the random assortment of alleles on the chromosomes that separated and from crossing over at metaphase I.

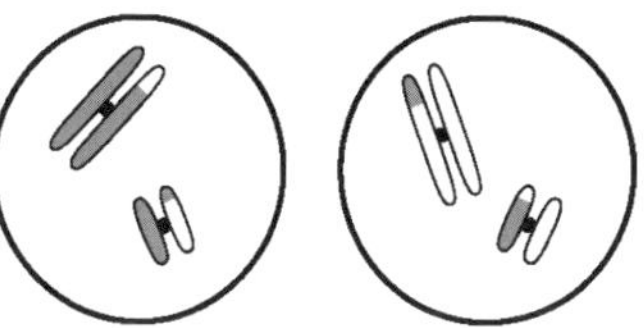

The chromosomes may now undergo a resting time in which they unravel. Shortening and thickening again indicates the start of the second stage of divisions. This is known as prophase II.

Metaphase II

Step 6: Chromosomes again line up on the cell equator of each cell. Spindle fibres form. As in metaphase I, the positioning of chromosomes along the cell equator is random.

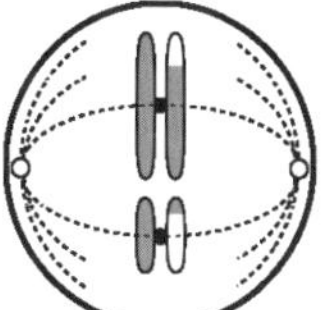
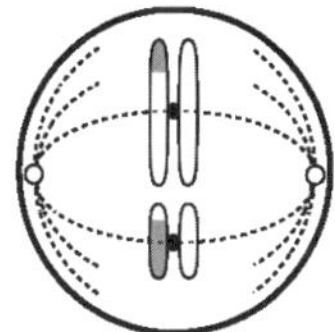

Anaphase II

Step 7: Chromatids separate and move towards the cell poles. Cell membranes constrict between them.

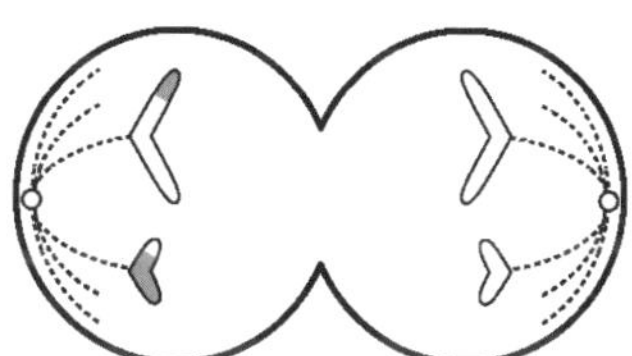

Telophase II

These cells are the gametes, and for them $n = 2$.

Step 8: Four new haploid daughter cells have formed, each a different genetic make-up from the other.

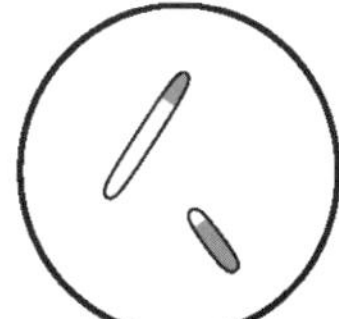

At Step 4, which member of which chromosome pair gets pulled into each new cell is *random*.

At Step 7, which chromatid gets pulled into each new cell is *random*.

This random assortment of chromosomes and chromatids, like the exchange of genetic material during Step 2, and the different alleles present on each chromosome, *increase the genetic variability of the gametes formed.*

Independent assortment, segregation, crossing over and recombination

Independent assortment

Independent assortment occurs when the homologous pairs line up in a random order during metaphase 1 of meiosis.

Example

A cell has three homologous pairs ($2n = 6$). One member of each pair is shown as dark and the other as light. The possible combinations of these pairs are all a result of chance. Possible combinations of three homologous chromosomes at metaphase I are:

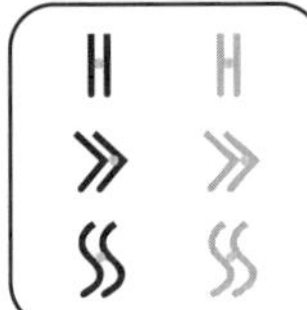
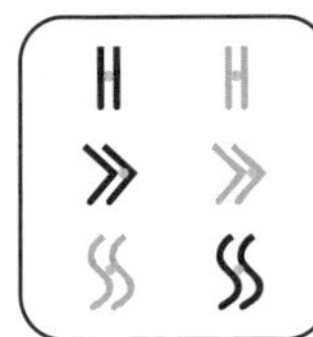
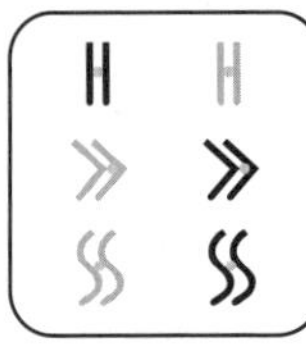

From the independent assortment of the pairs at meiosis I, there are eight possible combinations of the chromosomes at the end of telophase II:

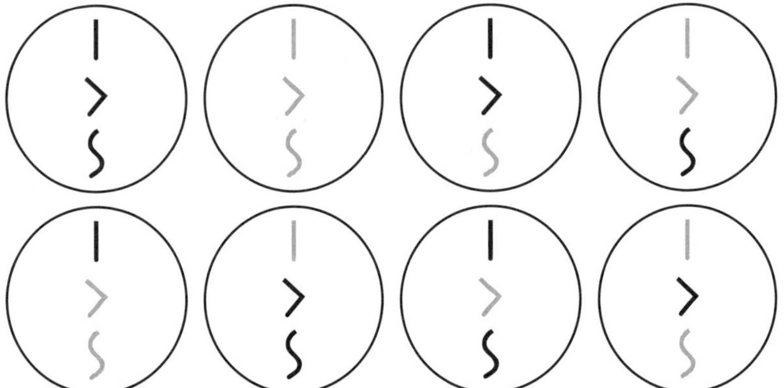

For humans – with 23 pairs of chromosomes – there are over 8 million possible combinations of the chromosomes in the gametes that are produced. It is no wonder that no two humans are identical (apart from identical twins, who are produced from the same fertilised ovum, so have identical genetic material).

Segregation

Segregation refers to the independent separation of alleles on different chromosomes.

Example

Three homologous chromosomes had the following alleles on them for three different genes:

A A a a
B B b b
C C c c

For the three pairs of homologous chromosomes in metaphase I, each is heterozygous. The possible combinations of alleles in the gametes at the end of telophase II as a result of segregation is:

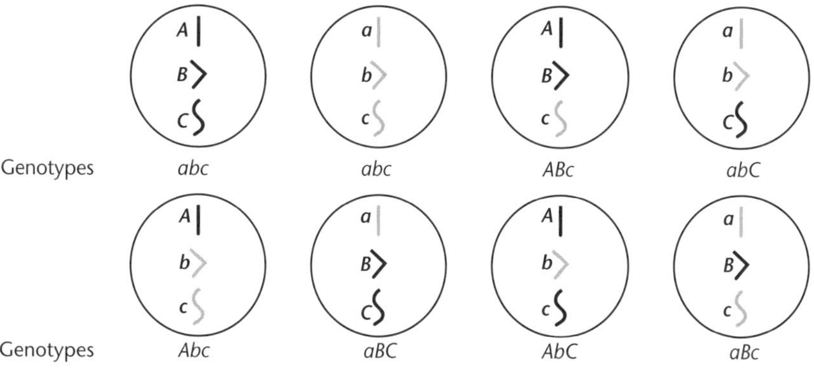

The alleles of the three genes would segregate independently of each other during meiosis and so produce the following possible combinations in the gametes:

ABC, abc, ABc, abC, Abc, aBC, AbC, aBc.

The laws of probability give equal chances of getting each combination.

Crossing over and recombination

Crossing over and **recombination** occur in prophase I of meiosis. Parts of the chromosome of non-sister chromatids may cross over at a point called a **chiasma**, followed by breaking and recombining.

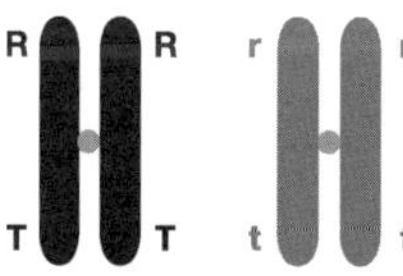

Homologous chromosomes, heterozygous for both alleles (*Rr* and *Tt*)

chiasma

Crossing over occuring between alleles on non-sister chromatids

Parental types

Recombinants

Recombination of the alleles; two chromosomes and their alleles remain the **parental types** while two become the **recombinants.**

RT and *rt* are the parental genotypes, *Rt* and *rT* are the recombinant genotypes.

As a result of crossing over and recombination, different *combinations of alleles* to that of the parents can result. These chromosomes (or the gametes that contain them) are known as **recombinants**; the original combinations of alleles are known as the **parental type**.

Fertilisation

Fertilisation is the fusion of a sperm and ovum – bringing together the maternal and paternal chromosomes so restoring the diploid state ($2n$) in the zygote. Both the ovum and sperm have a unique combination of alleles as a result of independent assortment, segregation and recombination. It is chance which sperm gets to fertilise an ovum. The resulting individual has its own unique set of alleles, which, though making it similar to its parents, also makes it different from them.

It is the process of mutation that creates the new alleles which are the source of genetic variation. But it is the three processes of **independent assortment**, **segregation**, and **recombination** that shuffle the alleles creating new combinations which produce the great variation seen within members of a species.

Unit 12.3 Activity 6A: Genetic variation

1. Define the following terms:
   a. Genetic biodiversity.
   b. Gene pool.
   c. Allele frequency.
2. Distinguish between the following pairs of terms:
   a. Gene and allele.
   b. Genotype and phenotype.
   c. Haploid and diploid.
   d. Gamete and zygote.
   e. Mitosis and meiosis.
   f. Independent assortment and segregation.
   g. Crossing over and recombination.
   h. Point mutation and chromosomal mutation.
3. Explain the significance of mutations in genetics and evolution.
4. a. Describe the main function of meiosis.
   b. An older term for meiosis is *reduction-division*. Explain why this term could be used instead of meiosis.
   c. Describe, with the help of a diagram, homologous chromosomes.
   d. Describe, with the help of a diagram, a tetrad. (Note that some texts use the term *bivalent* for a tetrad.)

5. There are more than one species of fox. The red fox (*Vulpes vulpes*) has 34 chromosomes in each of its body cells, while the arctic fox (*Alopex lagopus*) has 52 chromosomes in each of its body cells. The two species of fox have been known to breed and produce a hybrid.

a. Give the diploid and haploid numbers for each of the two species of fox.

b. Explain how many chromosomes you would expect the hybrid to have in each of its body cells.

c. Discuss whether this hybrid is likely to be a new species.

6. For the gamete-producing cell shown:

a. Give its **i**. diploid **ii**. haploid number.

b. Draw and describe the cell as it would appear in prophase I in meiosis.

c. Draw two or three diagrams of this cell in meiosis I to illustrate independent assortment.

d. Draw two or three diagrams of this cell at the end of telophase II to illustrate segregation.

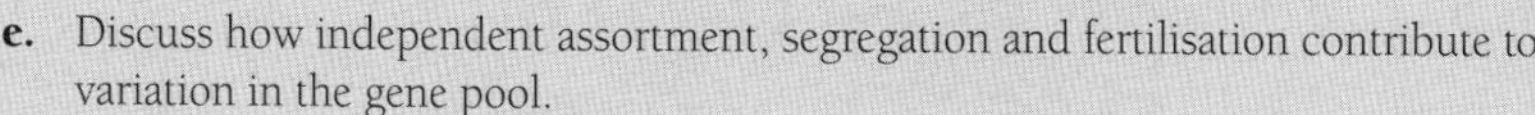

e. Discuss how independent assortment, segregation and fertilisation contribute to variation in the gene pool.

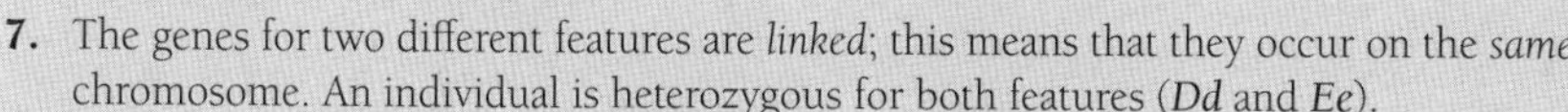

7. The genes for two different features are *linked*; this means that they occur on the *same* chromosome. An individual is heterozygous for both features (*Dd* and *Ee*).

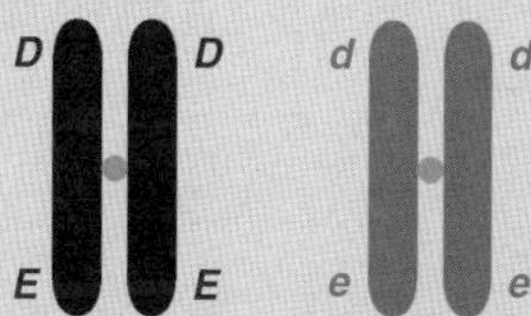

a. Starting with the diagram above, draw and describe the crossing over and recombination of these alleles in prophase I of meiosis.

b. Give the genotypes of the:

i. Parental chromosomes.

ii. Recombinant chromosomes formed in telophase II.

8. Alleles segregate independently of each other during meiosis. An individual is heterozygous for three features. The alleles for each are on separate chromosomes, as shown:

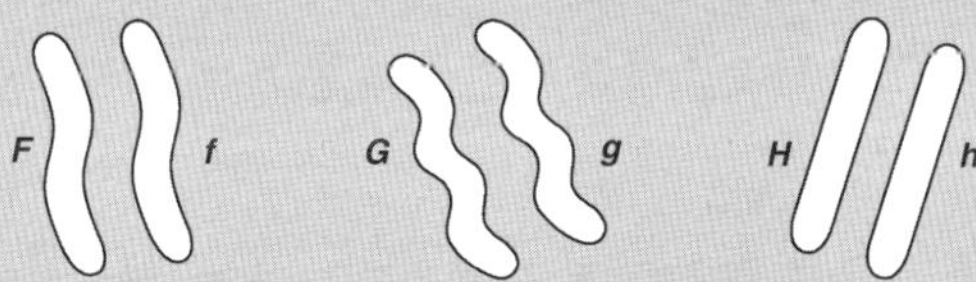

Give all the possible combinations of these alleles in the gametes formed at the end of meiosis.

9. Discuss the reasons for the members of a family being both alike (have a 'family resemblance') yet recognisably different.

10. Compare and contrast mitosis and meiosis.

Unit 12.3 Genetics

Topic 7: How genetic knowledge is applied – an overview

Authors: Martin Hanson with Takis Solulu

Topic 7 briefly introduces the applications of genetics, by looking at:

- Selective breeding.
- Genetic engineering.

Applications of genetics

The current (and future) knowledge about how DNA, genes and chromosomes work to produce physical characteristics can be used to advantage, usually in meeting human needs or demands.

Selective breeding

Selective breeding is the selection of organisms by humans for breeding. The process involves the transferral of thousands of genes from two selected parents (usually of the same species) into an offspring to produce an organism that contains desirable traits.

The purposes of selective breeding include to:

- Produce more organisms with the same, or 'better', traits.
- Consistently produce plants or animals of known physical qualities – important for people who rely on selectively bred organisms for their livelihood.
- Create a 'breeding bank' of organisms with superior traits, and to use these organisms to breed with other organisms on a commercial basis (eg the production of bulls that possess the genes for superior quality meat – the sperm of these bulls can be used to produce similar offspring).

Selective breeding was originally based on phenotype and was fairly hit-and-miss. For example, milk-producing animals that produced the most milk were bred together in the hope of producing more such animals, or the seeds of plants that produced the best crops or were the hardiest were kept and planted. The understanding of the genetic basis of inheritance means selective breeding is now much more precise.

Selective breeders aim to produce organisms that:

- Are homozygous for the desirable trait(s).
- Do not possess (especially recessive) alleles for harmful or undesirable characteristics.

Several breeding techniques have been developed to ensure the best possible results are produced in the shortest amount of time, including:

- Inbreeding in plants and animals.
- Artificial insemination (in animals).
- Artificial pollen transfer (in plants).

Inbreeding

Selective breeders may use **inbreeding** (breeding closely related organisms, eg brother and sister, or father and daughter), which increases the chance desirable genes will come together in the offspring. Inbreeding can be repeated for many generations, particularly in plants, which often have a shorter life cycle than animals and in which 'harmful' (ie dangerous or undesirable) genes are easily eliminated by destroying affected plants.

A disadvantage of inbreeding is that it decreases the genetic diversity of the species, to the point where some genes may be lost from the population (**inbreeding depression**). Genetic diversity within a species offers protection because some organisms may be better able to survive disease or a change in climate. When these organisms reproduce, their genes are passed on to their offspring, assisting the survival of the species. Some commercial crops, eg potatoes and pine trees, could be decimated should a disease these crops are not resistant to, breach national biosecurity. To ensure some genetic diversity remains, breeders **outcross** (breed non-related organisms together).

Genetic engineering

Selective breeding is the manipulation of organisms at the physical level. **Genetic engineering** (also known as GE, genetic modification, recombinant DNA technology) involves manipulation at the genetic level, producing organisms of known genetic make-up. Genetic engineering can also cross the **species barrier**, taking genetic material out of one species and putting it into the DNA of another different species.

Genetic engineering involves four main steps:

1. Finding the desirable **target** gene in the **donor** organism and removing it from its DNA. The target gene could be used to make a substance needed by humans or other animals, or it could be a gene for resistance to a certain disease.
2. Putting the desirable gene into the DNA of a **recipient** organism (which could be of a different species).
3. Testing to see if the process works.
4. If the process works, either make the recipient organism **express** the target gene so a certain substance is made, or make lots of identical recipient organisms, all of which have the target gene.

Genetic engineering processes can occur between many different organisms. For example:

- Genetically engineered potatoes contain a gene that makes them resistant to a certain type of rot.
- Apples and tomatoes can be engineered so they don't ripen in storage, or to make their skin a certain colour.

Example

Production of the enzyme chymosin for use in New Zealand cheese making

The enzyme *chymosin* causes milk to curdle, forming the solid *curds* (used to make cheese) that can be easily separated from the liquid whey (discarded). Chymosin is found in the stomach of calves and, since it is an enzyme, there is a gene in calf DNA for its production.

Chymosin enzyme has two disadvantages:

- It is in short supply (and therefore, is expensive).
- Calves must be killed in order to extract the enzyme from their stomachs – some people (eg vegetarians) want to eat cheese that is manufactured without causing the death of any animals.

Through genetic engineering:

- Large amounts of chymosin can be made from bacteria.
- The chymosin gene can be made artificially.

The (highly simplified) process of genetic engineering of the chymosin gene is outlined in the diagram below:

1. Artificial copies of the chymosin gene are made in the laboratory. These artificial genes are identical to the chymosin genes found in calf DNA.
2. Suitable bacteria are isolated and their DNA in the form of a circular **plasmid** is taken out and cut open with special enzymes.
3. The artificial chymosin gene and the bacterial plasmids are mixed together. Some of the artificial genes 'stick' into the plasmid and become a permanent part of the bacterial plasmid.
4. The plasmids are put back into the bacteria. Tests can be done to find out which bacteria contain plasmids with the chymosin gene in them.
5. The bacteria containing the chymosin gene reproduce and can be **induced** to produce large quantities of chymosin enzyme. This enzyme can be separated from the bacteria and used to make cheese.

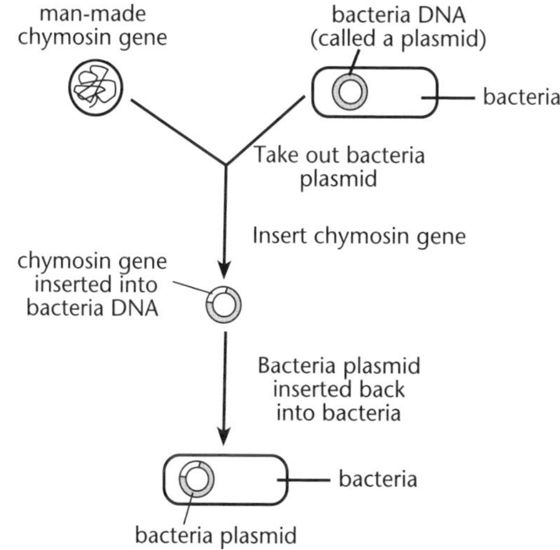

Genetic engineering of chymosin gene.

Genetic engineering is a complicated process and success rates are not high. The major advantage of genetic engineering is its precision – if successful, it can produce organisms that contain only the desired gene. If genetic engineering is used in conjunction with **tissue culture**, thousands of genetically identical plants – all containing the desired gene – can be produced quickly.

Cloning

Cloning refers to the creation of an organism that is genetically identical to another organism. **Clones** occur naturally in the form of identical twins, where the zygote splits and develops into two identical embryos, and then into **foetuses**. The cloning of plants through the use of **cuttings** and, more recently, **tissue culture** has occurred for many years without controversy. However, there is much debate over the ethics of cloning animals (including humans), and research into human cloning is banned in many countries.

Reproductive cloning involves a technique known as SCNT (somatic cell nuclear transfer) – Dolly the sheep was the result of this technique.

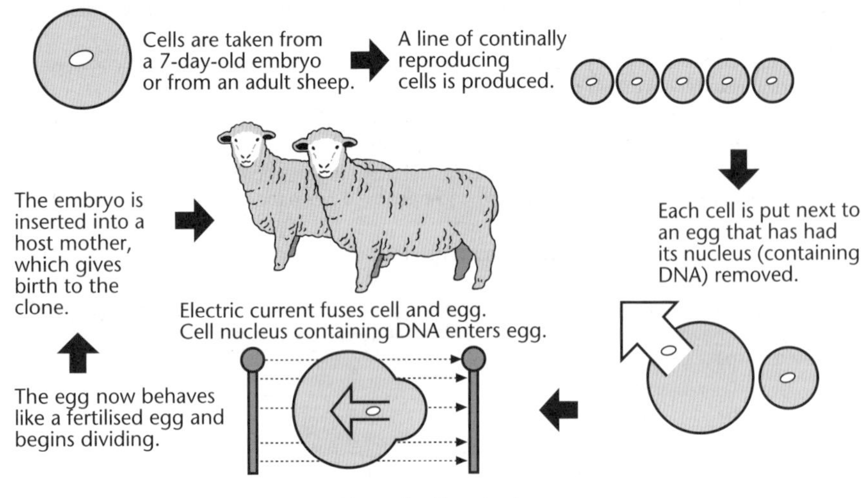

Reproductive cloning.

The main steps of reproductive cloning are as follows:

1. Cells containing two copies of every chromosome (**diploid**) are taken either from an embryo or from adult tissue.
2. These cells are reproduced artificially for many generations to ensure they are genetically normal cells – this is called a **cell line**.
3. The nucleus is removed from the cells.
4. An egg cell is obtained from a female and the nucleus, along with all of its DNA, is removed and disposed of. This egg cell is now called **enucleated**.
5. The nucleus and the enucleated egg are placed next to each other. Chemicals or an electric current is applied and the nucleus and egg fuse together. The nucleus moves into the egg.
6. The egg divides by mitosis like a normally fertilised egg and develops into an embryo.
7. If the embryo survives and normal cell division continues, it is implanted into the **uterus** of a surrogate female, where it develops. The surrogate mother then gives birth to the clone, which should be identical in every way to the original adult.

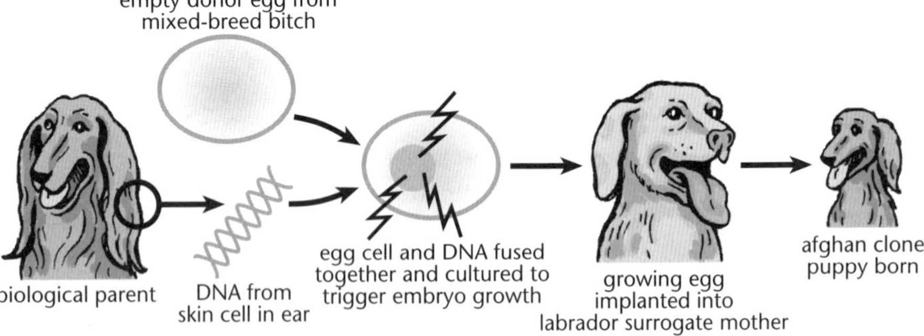

Diagram illustrating the main steps in the cloning of 'Snuppy', an afghan hound.

Unit 12.3 Activity 7A: Applications of genetics

1.

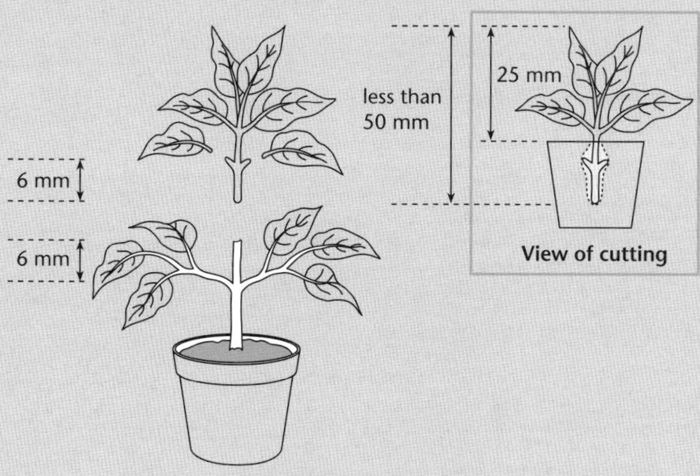

View of parent plant

a. Fruit farmers can grow new plants from a small 'cutting,' which is a piece of an existing plant. Usually, it is a piece of stem with a few leaves. Explain the advantages to a fruit grower of using this method of reproduction compared with growing plants from seeds.

b. Explain how genetic modification could be used to change the genetic make-up and improve the production of a food crop such as apples.

Unit 12.3 Genetics

Topic 8: Manipulation of DNA – techniques in gene technology

Author: Martin Hanson

Topic 8 covers how to describe applications of biotechnological techniques by:

- Examining techniques involving PCR, ligation, use of restriction enzymes, gel electrophoresis, gene sequencing, DNA chips and tissue culture.

Gene technology

Until the origin of agriculture, all genetic change was the result of two processes – the production of random variation by recombination and mutation, and selection of the 'best' genes by the environment. Then, about 10 000 years ago, humans began to bring about genetic change in the animals and plants they domesticated, by *selection*.

At first, this would probably have been unconscious.

Example

Less docile animals would have been more troublesome and so would be more likely to be used for food than kept for breeding purposes. The most commmonly harvested grain would come from those plants in which the seed heads were most strongly attached to the stem. Seeds that became detached easily would have escaped the harvest and so would not have been used for the next year's crop.

At some stage, plants and animals would have been deliberately chosen for breeding because of their desirable qualities. By selective breeding, the genetic make-up of many species has been progressively altered under domestication. Traditional methods of plant and animal breeding had three major limitations:

- The production of 'better' varieties was largely a matter of luck, since recombination and mutation are both random processes.
- Gene pools of different species are normally isolated from one another, so geneticists were limited to crossing varieties of the same species, or in some cases, closely related species.
- It takes much longer to produce new varieties by selective breeding.

Molecular genetics has changed all this. Since it was discovered that the genetic code is universal, the transplantation of genes between unrelated organisms has become a reality.

This technology is called genetic engineering, and is a major branch of **biotechnology**, the industrial use of biological processes. DNA produced by artificially combining genes from different organisms is called **recombinant DNA**. **Genetically modified organisms** (**GMOs**) carrying genes that have been transplanted from other species are said to be *transgenic*.

Gene technology is a very important field in many countries, and is surrounded by controversy and misunderstanding. Gene technologies are used in numerous areas such as the agricultural and food industries, medicine, forensic science, conservation and evolutionary studies. The essential techniques and tools that are used in gene technologies follow.

The genetic engineer's toolkit

Manipulating DNA involves the use of a variety of tools and techniques.

> While there is significant detail covered for each technique that follows, this is more for the purposes of understanding how the techniques are applied, rather than for assessment purposes.

Cutting DNA – restriction endonucleases

Restriction enzymes are used by bacteria against invasion by bacteriophages (viruses that attack bacteria).

Restriction enzymes are named according to the kind of bacterium from which they are obtained.

Example

Eco RI is obtained from *E. coli* (*R* denoting the strain, and *I* indicating that it was the first restriction enzyme to be obtained from this bacterium).

Restriction enzymes act as chemical 'scissors', cutting DNA at specific sites. The ends created by a restriction enzyme are either staggered (*sticky*) or flush (*blunt*).

Examples

Micro-organism	Enzyme abbreviation	Recognition sequence	
Escherichia coli	*Eco* RI	5′ . . . G↑A A T T C . . . 3′ 3′ . . . C T T A A|G . . . 5′	Create sticky ends
Hemophilus influenzae	*Hind* III	5′ . . . A↑A G C T T . . . 3′ 3′ . . . T T C G A|A . . . 5′	
Hemophilus aegyptius	*Hae* III	5′ . . . G G↑C C . . . 3′ 3′ . . . C C|G G . . . 5′	Create blunt ends
Serratia marcescens	*Sma* I	5′ . . . C C C↑G G G . . . 3′ 3′ . . . G G G|C C C . . . 5′	

Base sequences recognised by some commonly used restriction enzymes.

Restriction enzymes that generate sticky ends not only cut at specific sequences, but generate the same kind of ends in any piece of DNA, regardless of the species of origin.

Example

Pieces cut from human DNA have ends complementary to ends cut from any other DNA (regardless of species), provided the same restriction enzyme is used.

The size of fragments created depends on the number of base pairs in the recognition sequence.

Example

Hae III recognises a sequence 4 base pairs in length. This sequence would be encountered by chance on an average of every $4^4 = 256$ base pairs. Hence *Hae* III creates fragments on average 256 base pairs long.

Eco RI recognises 6 base pairs, so would generate fragments $4^6 = 4096$ base pairs long (approx. 4 kilobases).

Joining DNA fragments – DNA ligase

DNA ligase acts as a molecular 'stapler', joining DNA fragments. Its normal role includes the joining of **Okazaki fragments** in DNA replication.

DNA fragments are joined as follows:

- Sticky-ended DNA fragments are created from two DNA sources using the same restriction enzyme.
- DNA from the two sources is mixed and complementary ends allowed to form complementary base pairs. This process of forming a loose and temporary join by hydrogen bonding is called **annealing**.
- DNA ligase is added to form a permanent link.

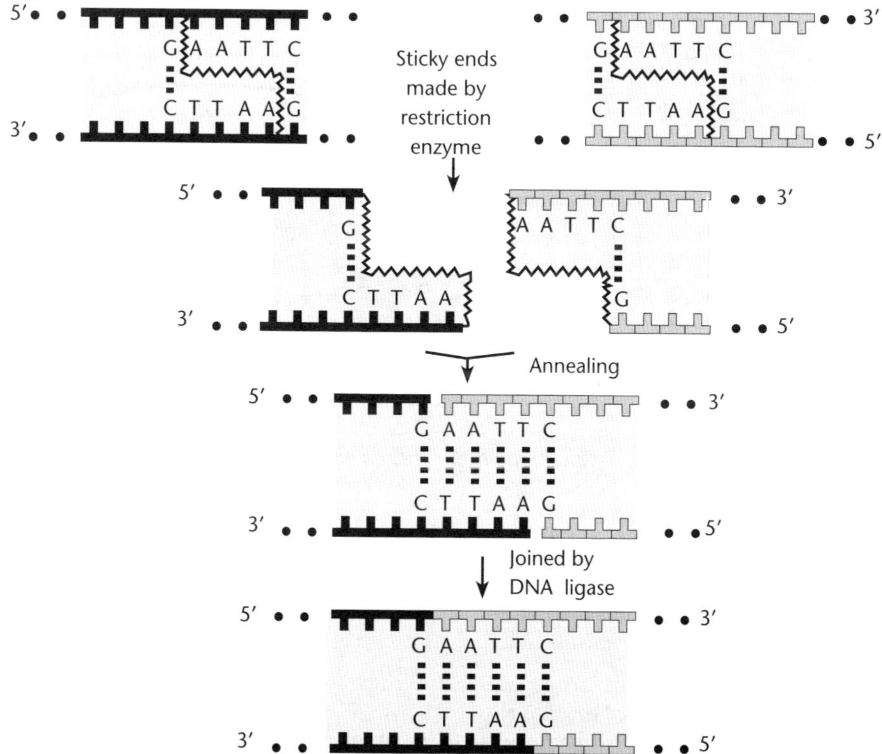

Use of DNA ligase to join sections of DNA with sticky ends.

Separating DNA fragments – gel electrophoresis

Electrophoresis means 'to carry with electricity'. The phosphate groups in DNA are negatively charged, so in an electric field DNA fragments migrate towards the anode (positive electrode). The fragments are placed in a gel, and encounter resistance that depends on their size. The shorter the fragment, the less resistance and the faster it moves through the gel. A mixture of fragments becomes separated into **discrete** bands, each band consisting of millions of DNA fragments of identical length.

The bands are not visible, but the progress of the electrophoresis can be followed by the:

- Inclusion of a 'label' of some sort that attaches itself to the DNA fragments being separated. This label is typically a dye that is made visible by the addition of ethidium bromide, which fluoresces orange under UV light.
- Entire gel being immersed in a solution of methylene blue – methylene blue attaches itself firmly to the DNA and shows the positions of the DNA fragments as dark blue bands.

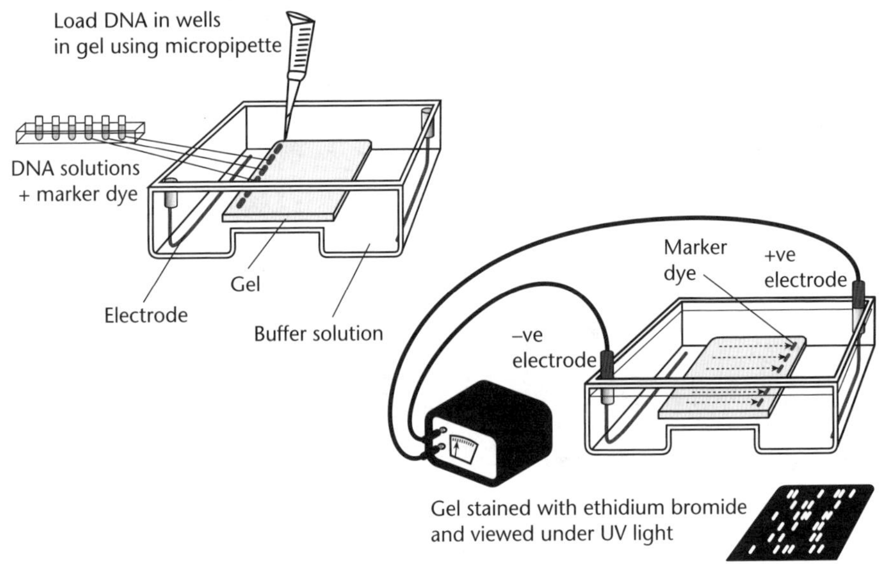

Separating polynucleotide fragments using gel electrophoresis.

The kind of gel used is tailored to suit the sizes of the fragments to be separated.

- For longer fragments, **agarose** (a highly purified form of agar) is used.
- For shorter fragments, **polyacrylamide**, which has smaller pores, is used. Pore size can be more finely tuned by varying the concentration of the gel.

Gel electrophoresis can also be used to measure the length of DNA fragments. The rate of movement of fragments whose length is already known (determined by independent methods) can be used as standards for comparison with fragments of unknown length.

Amplifying a DNA sample – the polymerase chain reaction

Using the **polymerase chain reaction** or **PCR**, a minute quantity of DNA (as little as one molecule) can be increased (*amplified*) a billion-fold or more. In essence, PCR involves the repeated replication of DNA in the test tube, the amount of DNA doubling each replication cycle. After 20–30 replication cycles, the amount has increased between a million (10^6) and a billion (10^9) times. The procedure is very quick, each cycle taking only about five minutes.

In principle, all that is needed to replicate DNA in the test tube is:

- A parent DNA molecule, the two strands of which act as **templates**.
- The enzyme **DNA polymerase**.
- A supply of the four kinds of **nucleotide** (actually **nucleoside** triphosphates).
- Two **oligonucleotide** primers (short nucleotide chains).

The PCR consists of the following steps:

- *The DNA is heated to 95°C to denature it.* Two single strands of DNA are produced. Each of these can be used as a template to build the other strand.
- *DNA polymerase, four nucleoside triphosphates and excess quantities of the oligonucleotide primers are added. The mixture is cooled to allow the primers to anneal to the DNA at either end of the region to be amplified.*

  DNA polymerase can only add new nucleotides at the 3' end of a pre-existing strand. This pre-existing strand is the primer, and its base sequence must be complementary to part of the strand to be sequenced. The primers have to be synthesised artificially, so their nucleotide sequence must be known.

  Determining the **primer** nucleotide is not too difficult because:

  - Something may already be known about the protein coded for by the DNA sample.
  - Much eukaryotic DNA is 'junk', consisting of well-known repeated sequences.
  - A primer can be as short as 10 nucleotides, and if lower temperature annealing is used, complementarity does not have to be exact. A range of primers may have to be tried before success is achieved.

  Each primer is an oligonucleotide (8–20 nucleotides long), and hybridises with a section at either end of the region of DNA to be replicated

  The primers are in excess to ensure that as the mixture is cooled, the sample DNA strands preferentially anneal with the primers rather than with each other.
- *The DNA is replicated, producing two double-stranded molecules.*
- *The DNA is heated again to denature, and the process repeated for 20–30 cycles.* After the second cycle, there are 4 DNA molecules, and after the third cycle there are 8, and so on.

  The heat used denatures most DNA polymerases. DNA polymerase from *E. coli* used to be used, but since this is denatured by the heat, new enzyme had to be added each cycle. Heat-stable DNA polymerase from *Thermus aquaticus*, a thermophilic ('heat-loving') bacterium living in hot springs, is now used. This *Taq* polymerase has an optimum temperature of about 70°C. An even more heat-tolerant polymerase, from the hyperthermophile *Pyrococcus furiosus* (a bacterium from hot vents on the ocean floor) is also now being used.

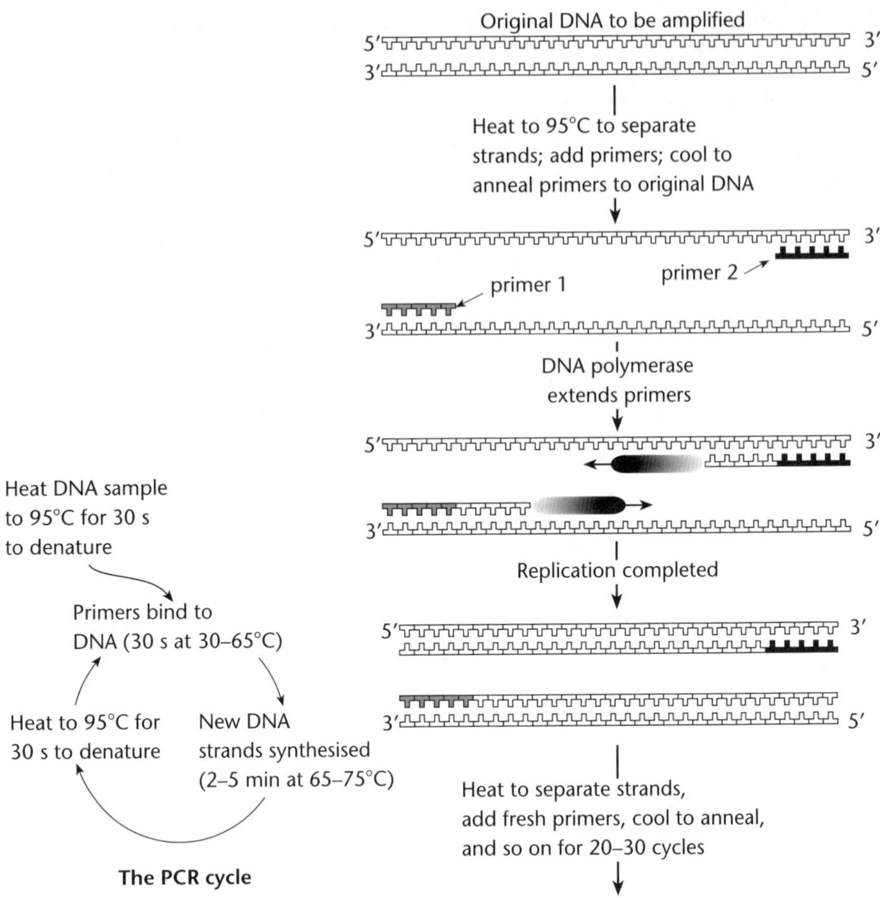

The polymerase chain reaction.

Uses of PCR

- Mitochondrial DNA from remains of extinct animals and plants can be amplified and compared with that of modern species.
- DNA from a tiny blood stain can be amplified and used as forensic evidence.
- Presence of individual micro-organisms of a particular species can be detected by amplifying their DNA.

Danger of contamination

The very power of PCR means that extreme care must be exercised to prevent contamination with extraneous DNA, since this would also be amplified.

DNA/gene sequencing

The most common technique for determining the base sequence of a length of DNA is the **interrupted replication** or Sanger method.

Procedure

1. Break up DNA into shorter fragments using restriction enzymes.
2. Separate by gel electrophoresis.
3. Divide the DNA from each of the separated bands into *four* test tubes, and use each as a template to build a complementary DNA strand using radioactively labelled primer. Each tube contains many identical fragments of the DNA strand to be sequenced, together with DNA polymerase and a supply of the four nucleoside triphosphates and radioactively labelled primer. In addition, each tube contains a *trace* of *one* of four **nucleotide analogues**.

Nucleotide analogues differ from normal nucleotides in lacking an OH group on carbon 3. These can be added to the 3' end of a pre-existing chain but cannot accept any more since their carbon 3 has no OH group.

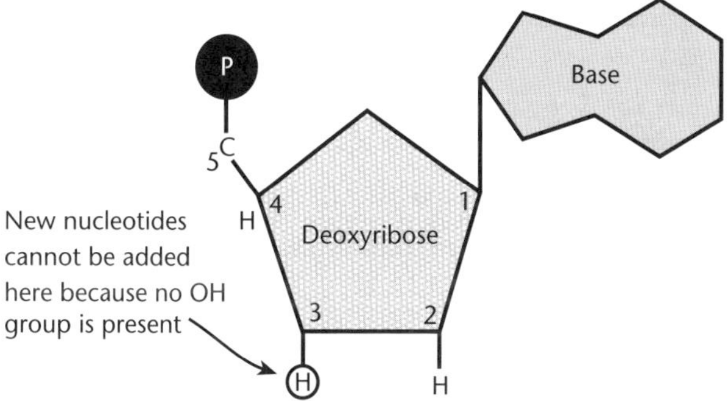

A nucleotide analogue.

The nucleotide analogues 'fool' the DNA polymerase, which adds them to the 3' end of a nucleotide chain. *But*, since a nucleotide analogue cannot accept any more nucleotides, DNA synthesis is halted with the analogue at the 3' end.

Example

Suppose the analogue G* is an ingredient. DNA replication proceeds along the thousands of copies of the DNA until halted by incorporation of the analogue. The concentration of G* is much lower than that of the normal nucleotide, so a normal nucleotide is much more likely to be incorporated than an analogue. Provided concentration of the analogue is not too low, replication of most DNA strands will be interrupted at some point, though precisely where is a matter of chance. The result is a mix of partly replicated DNA fragments, all ending in a nucleotide analogue – some will be short because replication was interrupted early; others are longer because the analogue hadn't been incorporated until later in the replication process.

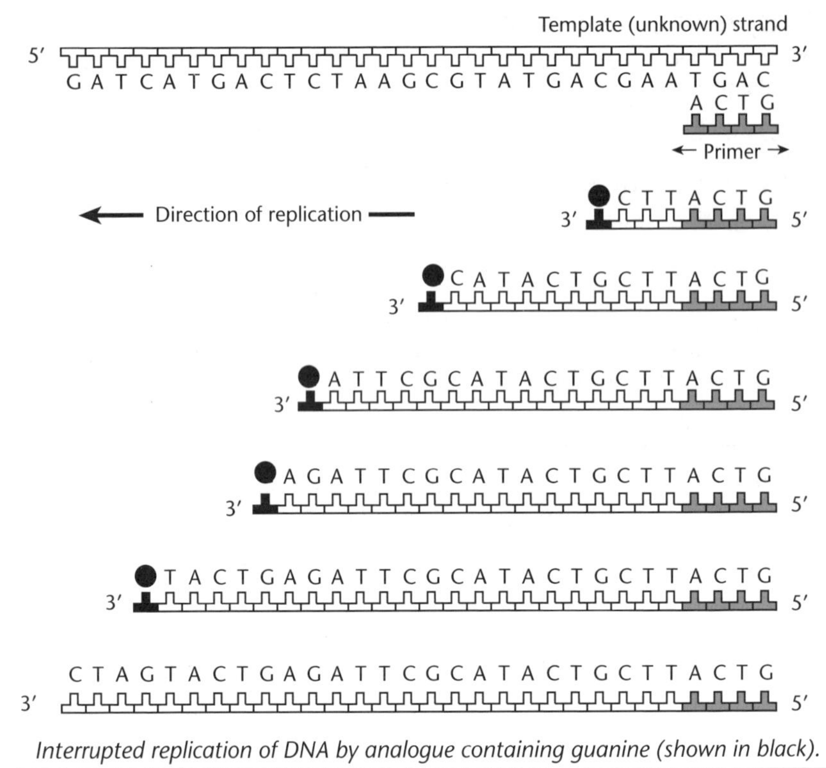

Interrupted replication of DNA by analogue containing guanine (shown in black).

4. Allow DNA replication to proceed as far as it can go.
5. Separate the fragments in each tube by electrophoresis (using a polyacrylamide gel with pore spaces fine enough to resolve differences of a single nucleotide).
6. Read off the sequence. The positions of the DNA fragments are able to be determined by exposing the gel to X-rays; the radioactive primers at the start of each synthesised fragment show up on X-ray film, corresponding to their positions on the gel.

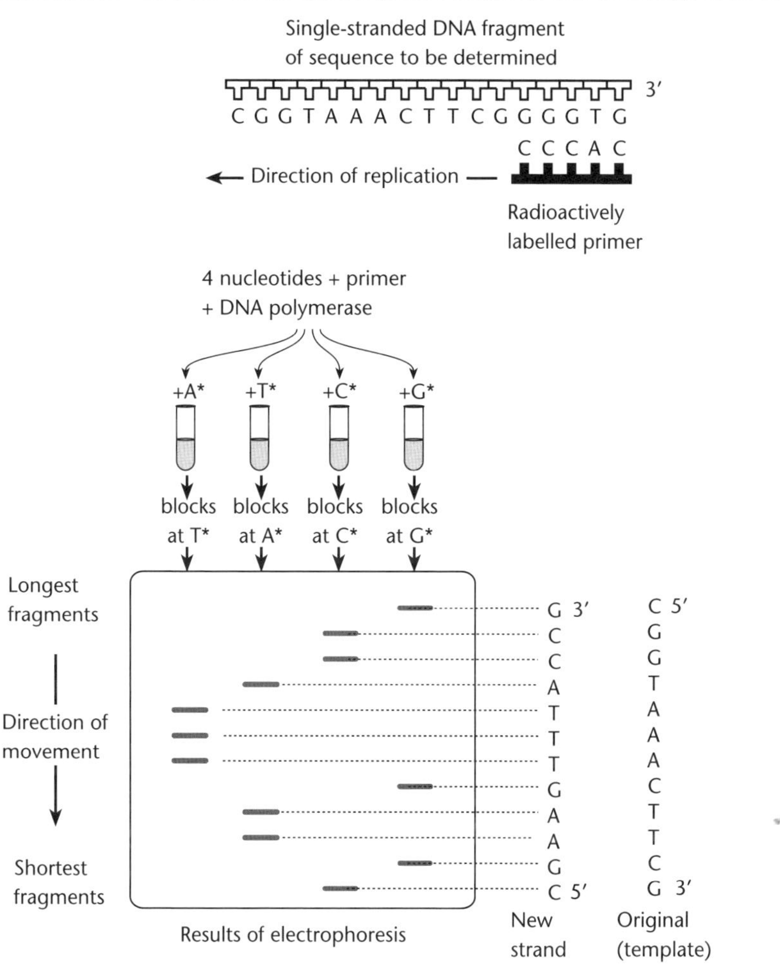

Interrupted replication method for DNA sequencing.

Automated DNA sequencing is used to rapidly sequence up to 600 bases at a time. Instead of separating each DNA fragment into four test tubes, only one tube is used because the nucleotide analogues are labelled with a fluorescent dye, a different colour for each of the four types of bases. Only one lane needs to be used to separate the fragments during gel electrophoresis, as the order of the fluorescent bands corresponds directly to the terminal base of each fragment in each band on the gel.

Example

Nucleotide analogue G* may be labelled with a yellow dye, C* with blue, T* with red, and A* with green. If the DNA fragment was sequenced, the result of the electrophoresis would be that as shown.

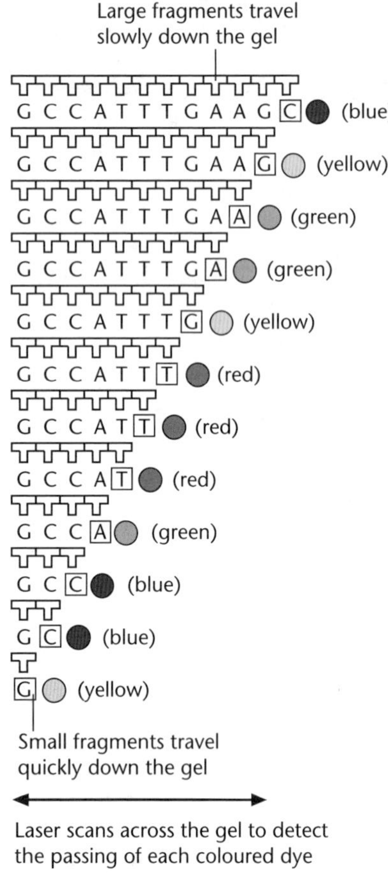

Automated method for DNA sequencing.

DNA chips

Also known as **microarrays**, DNA chips are devices based on the natural base pairing that occurs between complementary DNA sequences that have the potential for a wide variety of uses. First developed in the 1990s, the original DNA chips resembled computer chips. They generally comprised a small square of material, usually glass, nylon or a type of plastic divided up into a grid.

Onto each grid cell is placed (either by hand or robotically) a segment of single-stranded DNA or mRNA that can be as short as two or three nucleotides or as long as a single gene, called the **probe**. The **target**, single-stranded DNA, is then introduced to the grid, and once it locates a complementary probe, base pairing occurs. Because the sequence of each probe is known, once the target is located (usually though the use of fluorescent tags), the sequence of the target is easily established.

Uses of DNA chips

- *Expression analysis*. DNA chips were first developed when a group of scientists who wanted to establish the exact sequence of gene expression in yeast cells that were producing reproductive spores. Every possible mRNA sequence that a yeast cell was known to produce was placed on a chip. Resting yeast cells (those not producing spores) were processed, and the DNA and mRNA being produced at that particular point in time were introduced to the chip. Therefore, only certain spots on the grid fluoresced, as only these genes were being expressed. The scientists then induced the yeast cells into producing spores over many 12-hour periods. By tracking the order in which spots on the grid fluoresced over a 12-hour period, the order of gene expression could be analysed. This has obvious implications for similar studies in humans (eg the order in which genes are expressed during foetal development or that trigger certain genetic diseases, such as certain types of leukemia).
- *Mapping genetic differences*. Human differences – such as skin colour, susceptibility to certain diseases or an individual's clinical response to a drug therapy – are due to genetic differences. Over 99% of human DNA is the same; however, there are many loci in the human genome where only one base may differ between individuals. These differences (approximately 200 000 in total), are called **single nucleotide polymorphisms** (SNPs). Analysis of these SNPs is done by making a DNA chip that contains many variations of just one gene. An individual's DNA is amplified by PCR, introduced to the probe DNA, and the particular version of the gene they have is identified.
- *Disease diagnosis*. SNP analysis could be used to diagnose and treat a genetic disorder in a doctor's surgery, either as a means of administering preventative therapy before the disease starts in high-risk individuals, or through selecting the best therapy for the individual based on a particular type of genetic disorder they have (eg some types of leukemia resemble each other clinically, but their genetic basis is different). It is important to administer correct treatment as soon as possible, so establishing the genetic basis is important for successful treatment.

There are potentially many other applications of this technique, which will only become more apparent as the technology becomes more sophisticated and accessible.

Tissue culture

Tissue culture encompasses a range of techniques for growing plants from tiny portions of plants called **explants**. These may be individual cells or, more frequently, buds, discs of leaf, stem or root tissue. Unlike animal cells, plant cells are **totipotent** (ie an entire plant can be derived from a single cell if treated with the correct chemicals and plant hormones). Thus, if foreign genes are inserted into plant cells, the resultant plant will carry an exact copy of the foreign gene in every single one of its cells.

Foreign genes are frequently introduced into explants using the soil bacteria **Agrobacterium**, which readily and naturally inserts its own genes into plant cells.

Discs are cut out of leaves of the plant to receive the gene and treated with *Agrobacterium* with plasmids containing the gene to be transferred, and also a gene for kanamycin resistance. In response to chemicals produced by the wounded plant tissue, *Agrobacterium* cells infect leaf disc cells. The discs are grown in a medium that stimulates the cells of the wounded tissue to divide, producing an undifferentiated mass of cells called a **callus**. The growth medium contains various essential mineral ions, and organic nutrients such as glycine and certain B vitamins. The medium also contains two antibiotics – *kanamycin* (which kills uninfected plant cells), and an antibiotic such as cefotaxime (which kills *Agrobacterium* without harming plant cells). Also included in the medium is *cytokinin*, a plant growth substance that induces shoot formation. The entire procedure is conducted under **aseptic** conditions. After production of shoots, the medium is replaced by one containing **auxin**, which induces root formation. After several weeks, the plantlets are potted and grow into mature plants.

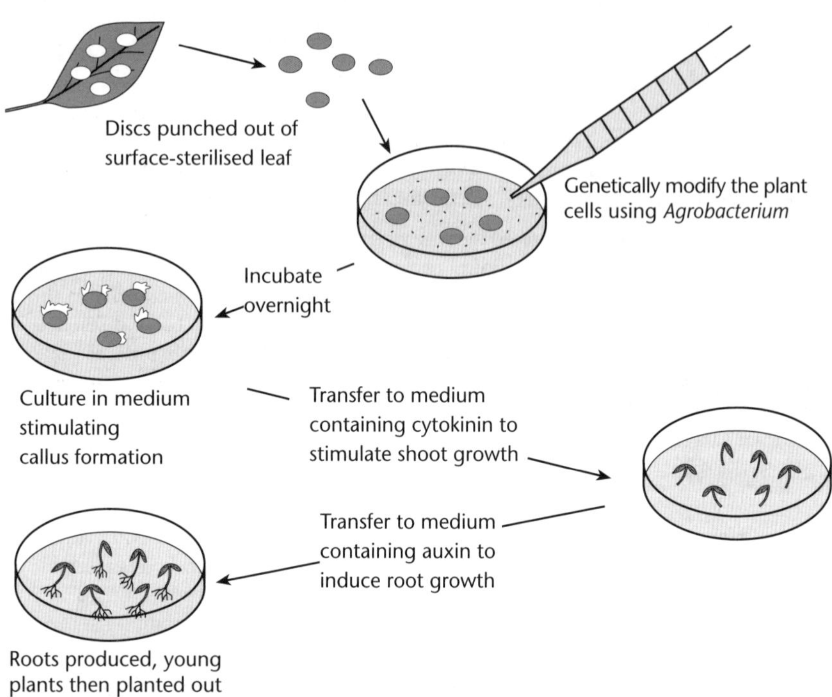

Production of transgenic plants from leaf discs infected by Agrobacterium.

Producing plants from tissue culture has important advantages, such as:

- Large numbers can be produced very quickly, as seed production and germination are not necessary.
- All are of identical genotype to the 'parent', and all will contain the desired foreign gene.
- It can be done at any time of the year.

Unit 12.3 Activity 8A: Putting genes to work – Gene technology

1. Match the items in the list **A–T** with those of the list **1–20**.

1.	Joins DNA fragments	**A**	DNA polymerase
2.	Separates DNA fragments according to size	**B**	Gel electrophoresis
3.	Amplifying the number of copies of a trace of DNA	**C**	Restriction endonuclease
4.	Used in DNA sequencing	**D**	Polymerase chain reaction
5.	Links a deoxyribonucleotide to an existing DNA chain	**E**	DNA ligase
6.	Cuts DNA at specific points	**F**	Agarose
7.	High-energy raw material used in polynucleotide synthesis	**G**	Nucleoside triphosphate
8.	Medium used in electrophoresis	**H**	Nucleotide analogue
9.	Used in plant tissue culture to produce a mature plant	**I**	Annealing
10.	Able to give rise to all cells of an organism	**J**	Polyacrylamide
11.	Oligonucleotide needed to begin DNA synthesis	**K**	*Taq* polymerase
12.	Cuts DNA yielding 'sticky' ends	**L**	*Eco* RI
13.	Mass of undifferentiated plant cells produced by wounding	**M**	Totipotent
14.	Used to detect gene transfer in plants	**N**	Ligation
15.	Stimulates shoot formation in plants	**O**	Primer
16.	Used in PCR	**P**	Explant
17.	Gel used to separate smaller DNA fragments	**Q**	Cytokinin
18.	Stimulates root formation in plants	**R**	Kanamycin
19.	Joining DNA fragments by covalent bond formation	**S**	Callus
20.	Linking DNA fragments by complementary base pairing	**T**	Auxin

2. The diagram shows the results of separation of DNA fragments in determination of the base sequence of the DNA, the bands representing fragments of different sizes. The bands have been stained to make them visible.

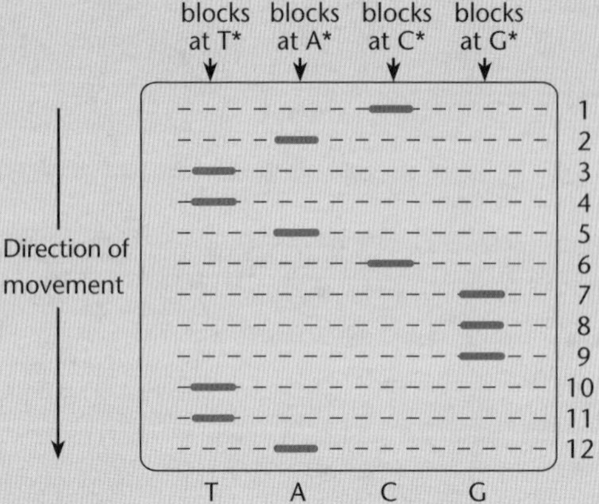

a. What are the coordinates of the longest fragment?

b. What is the first base of the fragment being synthesised?

c. Using the letters A, T, C and G, and beginning at the 3' end, what is the sequence of the first six bases in the original DNA strand (ie the strand being sequenced)?

d. Name the technique that would be used to identify the position of DNA fragments on a gel in circumstances in which a stain is not used.

3. Consider two restriction enzymes: one that recognises a sequence of 4 bases, and one that recognises a sequence of 8 bases.

Which would you expect to cut a given length of DNA into (on average) the greater number of fragments?

4. The following diagram shows the steps in the polymerase chain reaction (PCR). Use it to answer the questions that follow.

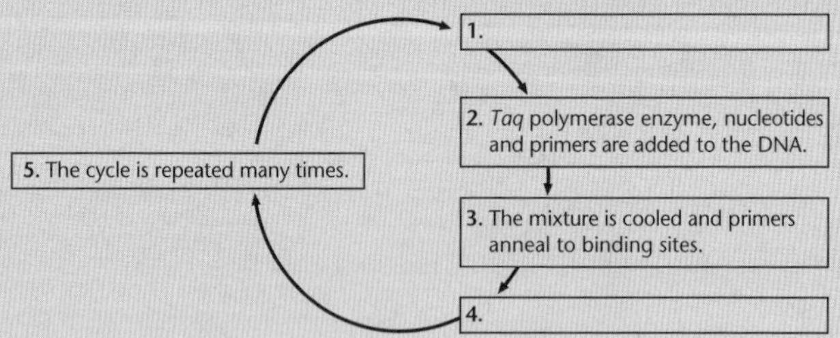

a. Suggest suitable descriptions for steps **1** and **4**.

b. Explain the significance of the *Taq* polymerase enzyme.

PCR is often used in forensic science before DNA profiling is carried out.

c. Discuss the implications of using PCR in forensic science.

Unit 12.3 Genetics

Topic 9: Application of gene technology – biotechnology

Martin Hanson with Takis Solulu

Topic 9 covers how to describe applications of biotechnological techniques to meet human needs and demands by:

- Describing applications involving gene cloning, transgenesis, DNA profiling, genome analysis, stem cell research and xenotransplantation.

Meeting the needs and demands of humans is the driving force behind the development and use of gene technology.

Combinations of the techniques covered in Topic 8 are used in the following applications

Gene cloning

One of the most important requirements in gene technology is to be able to replicate a desired sequence. Although the PCR is immensely powerful and indispensable in many situations, the most convenient method uses the metabolism of a host organism – usually a bacterium.

To get the DNA of interest into a host bacterium it must first be inserted into a larger piece of DNA. This is then replicated inside the bacterium, alongside the gene of interest. After many reproductive cycles, large numbers of identical copies of the gene are produced. A series of identical 'photocopies' is called a **clone**. The bacterial or viral DNA acts as a 'carrier', and is called a **cloning vector**.

Two main kinds of cloning vector are used – plasmids and bacteriophages.

Plasmids

Plasmids are present in most bacteria, and are small circular sections of DNA. They are very small (about 5 kb, compared with 4000 kb for the main chromosome of *E. coli*). Unlike the ordinary bacterial chromosome, plasmids are present as 20 or more copies. They carry genes for resistance to antibiotics. Their small size makes them easily separable from the main chromosome.

Plasmids have two important properties crucial to their role as cloning vectors:

- *They can be taken up by bacteria of the same or even another species*. (The uptake of DNA by bacteria is the mechanism by which resistance to antibiotics can spread from one species of bacterium to another.)
- *Because they carry genes for resistance to antibiotics, they can be used as markers*

Example

The gene that is required to be transferred is incorporated into a plasmid that also carries a gene for resistance to a streptomycin, and the plasmid transplanted into a streptomycin-sensitive strain of bacteria. If the recipient bacteria are treated with streptomycin, any survivors must have received the plasmid, and hence the required or 'wanted' gene.

If DNA comprising or containing a 'wanted' gene has been cut from the donor species by using a restriction enzyme generating sticky ends, it can be joined to DNA in a plasmid that has also been produced using the same restriction enzyme, since the same kinds of sticky ends will be generated.

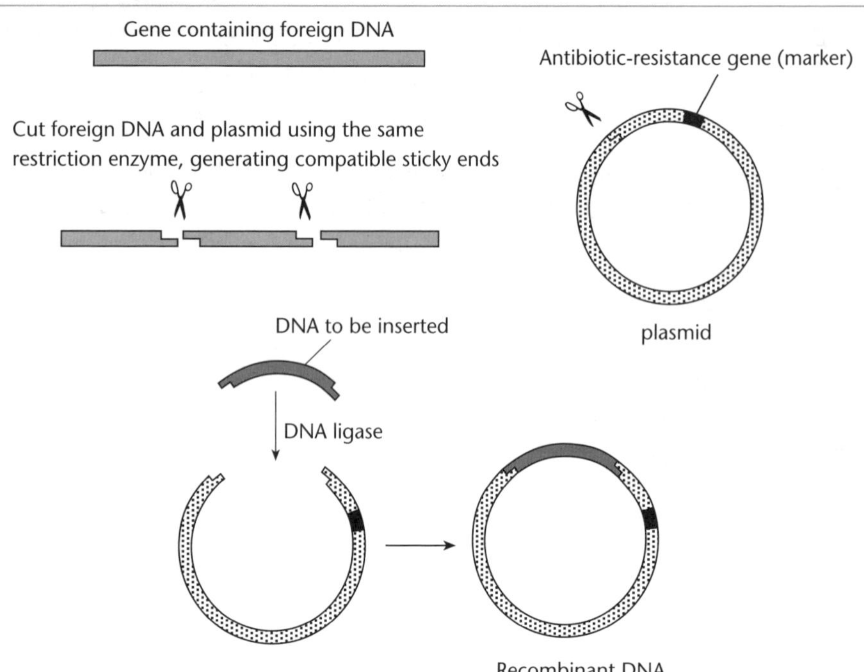

Inserting DNA into a plasmid.

The plasmids are then added to a suspension of the bacteria. Uptake of the plasmids (**transformation**) is promoted by prior chilling in ice-cold **calcium** chloride solution followed by a brief warming to 40°C. A small proportion (about 0.01%) of the bacteria are transformed by taking up the DNA; the plasmids are then maintained in the bacterial cells as self-replicating units.

After treatment with the vector plasmids, the bacteria are allowed to multiply. A suspension of the bacteria is then treated with the appropriate antibiotic and plated out onto agar. Those bacterial colonies that grow are the ones that received the wanted DNA.

Bacteriophage

A **bacteriophage** (or 'phage') is a virus that attacks bacterial cells.

The usual result of phage attack on a bacterium is the lysis (break-up) of the host cell and the release of hundreds of new phage particles. Each of these can then infect neighbouring cells, and so on. Each cycle takes about 20 minutes. When this happens in a dense population of bacteria on an agar plate (called a 'lawn' of bacteria), a clear area or **plaque**, about 3 mm diameter, is produced on the agar. A plaque thus marks the position of millions of phage particles that are destroying bacterial cells.

The phage most commonly used as a vector is **phage λ** (**phage lambda**). To insert a piece of DNA into a phage, the phage particles are first broken up into their protein and DNA components. A non-essential part of the phage DNA is then removed to make room for the desired DNA.

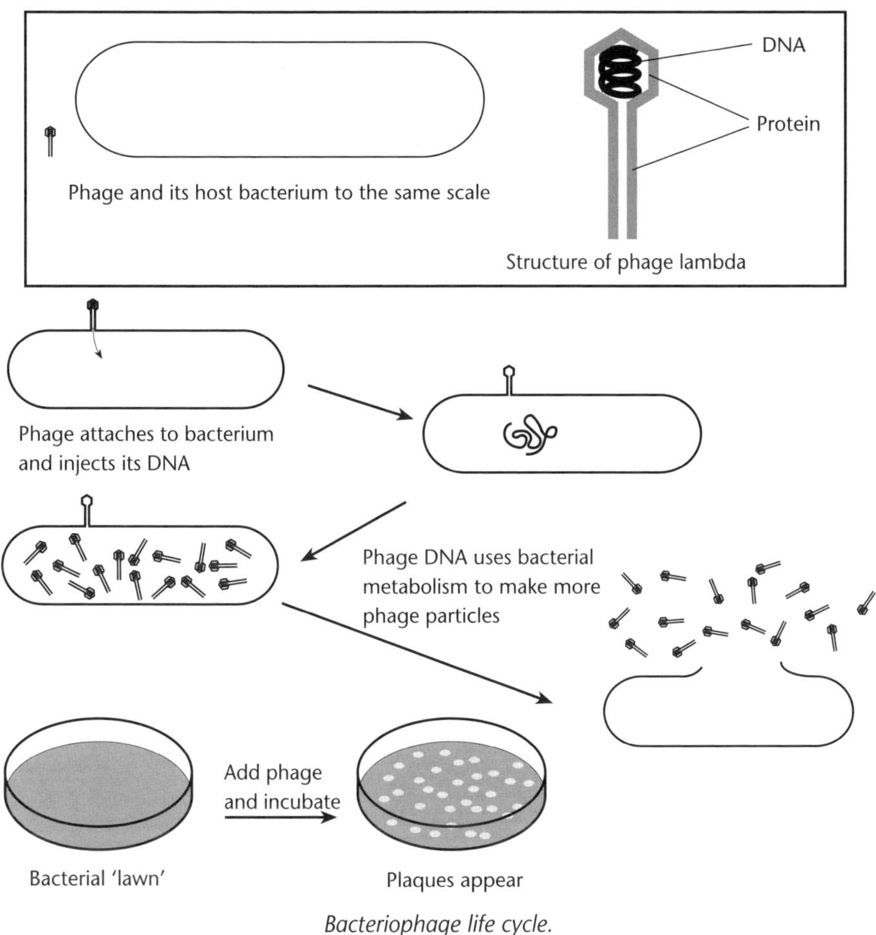

Bacteriophage life cycle.

The DNA of the phage and the desired DNA are then cut to give either blunt or sticky ends. If blunt ends are produced, the two ends, 'tails' of several repeats of the same base that will naturally base pair and anneal when mixed together are joined on, such as AAAAAA (called a 'poly-A' tail) and TTTTTT (called a 'poly-T' tail). The viral and desired DNA are then mixed and after the ends have been annealed, they are covalently joined using DNA ligase. The protein coat is then added back and the phage then used to infect the cells of the recipient bacteria.

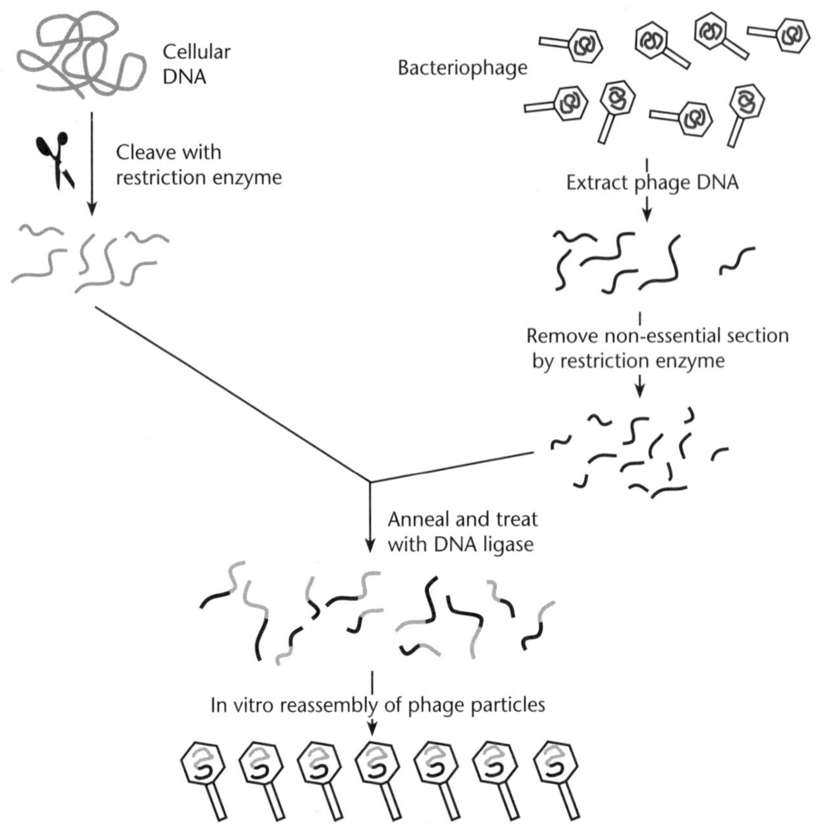

Inserting desired DNA into phage lambda.

Transferring genes from one species to another – transgenesis

Transgenesis involves removing DNA from one organism and splicing it into the genome of a different species. It is possible because of the near-universality of the genetic code and the discovery of restriction endonucleases and other enzymes that enable specific sections of DNA to be cut, joined and replicated.

Although the genetic code is the same in bacteria, fungi, plants and animals, there are several important differences in the way proteins are made, which complicate the effective transfer of genes between eukaryotes and prokaryotes:

- **Prokaryotes** have no introns and lack the means of removing them from RNA. A prokaryote therefore cannot make a functional protein from an entire eukaryotic gene since it would translate the intron sequences as well as the exons. The resulting protein would contain stretches of 'junk' polypeptide corresponding to the introns. To make a functional protein, the introns must first be excised from the primary RNA transcript.
- Although the genetic code is essentially the same in eukaryotes and prokaryotes, the mechanisms regulating gene action are different. In particular, the promoter sequences

that RNA polymerase recognises are not the same in different kinds of eukaryote. To work effectively in another species, a transplanted gene must therefore be accompanied by the necessary promoter and other regulator sequences.

- In eukaryotes, many proteins are modified after their production. Some only become active after addition of a non-protein prosthetic group. Others are 'pruned' to yield the active product.

Example

In insulin, the inactive precursor polypeptide is cut into two chains which remain held together. A bacterial cell therefore cannot produce insulin because it cannot convert the precursor into the active product.

Transgenesis involves several stages.

1. The DNA is obtained from the source organism.

2. The DNA fragments are inserted into a **cloning vector**.

3. The recombinant DNA formed is inserted into a 'host' organism (usually a bacterium) within which the recombinant DNA is replicated. If the cloning vector is a plasmid, it is introduced into the host by bacterial transformation. If the vector is a bacteriophage, the recombinant DNA in the phage is replicated by infecting a host bacterium. Whatever kind of cloning vector is used, the result is a large number of copies of the recombinant DNA in either bacterial colonies or plaques.

4. The clone containing the desired gene is located.

5. The gene is then inserted into the host organism in which the gene is to be exploited.

Obtaining DNA for transgenesis

There are two ways of isolating a gene:

- Indirectly by making the DNA for the gene from its mRNA.
- Directly from the DNA.

Genes from mRNA for transgenesis

The most frequently used method is to extract **mRNA** and produce a DNA copy by **reverse transcriptase**. This method has two important advantages:

- It avoids the problem of introns, since these are not present in the final mRNA product.
- The various kinds of mRNA represent copies of only part of the genome, so the desired gene represents a higher proportion of the **cDNA** than it would of the total genome.

Though mRNA is only a small proportion (about 1%) of the total RNA in a cell, it can be extracted using the fact that most eukaryotic mRNA has a poly-A tail on the 3' end. A poly-T nucleotide is added to the surface of a column of **cellulose** or other inert material. A solution containing the total cellular RNA is then poured through the column. The poly-A tails of the mRNA bind to the poly-T nucleotide, the rest of the RNA passing straight through. The mRNA is then removed by chemical treatment.

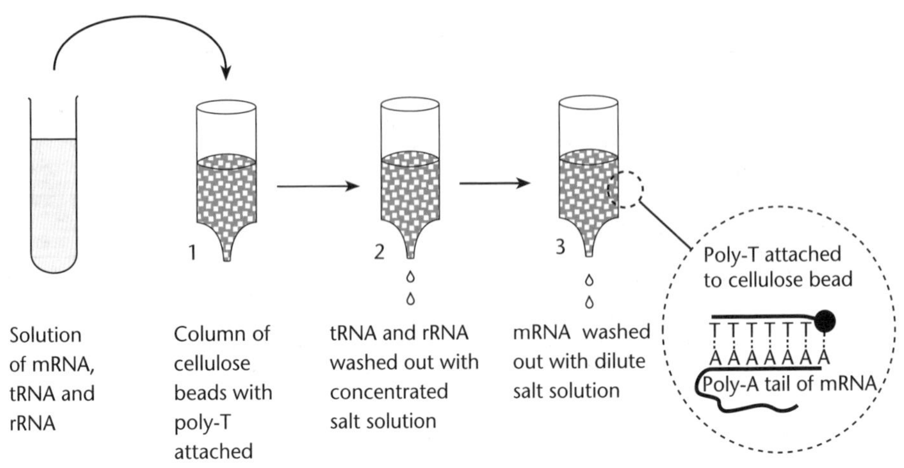

Extracting mRNA using poly-T.

The mRNA is used to make cDNA. The DNA fragments are spliced into a suitable cloning vector and amplified by introducing them into a host bacterium. Many different clones will be produced, only a tiny proportion of which contain the gene of interest. Between them, these clones contain copies of all the genes that were transcription products active in the tissue from which the mRNA had been extracted. Collectively, these clones constitute transcription products of the active genes of the donor tissue and hence represent a cDNA library.

A cDNA library represents only a small part of the total (genomic) DNA. Some genes are transcribed by all the cells of the body. These 'housekeeping' genes are those coding for the enzymes concerned with basic functions such as respiration and protein synthesis. Others are only translated in particular kinds of cell (eg the genes for haemoglobin are only active in developing red blood cells from the bone marrow).

Obtaining a gene directly from the DNA for transgenesis

This involves extracting the entire genome and is thus a more indiscriminate method than the mRNA method. If promoter and other controlling sequences are required, this method must be used since these are not transcribed into mRNA.

The DNA is extracted and digested by restriction enzymes to yield many different fragments. These are then cloned as described for cDNA. The resulting **genomic DNA library** is considerably larger than a cDNA library.

Finding the right gene for transgenesis

Having obtained either a cDNA or a genomic DNA library, the next step in transgenesis is to locate the desired gene.

The result of cloning a gene is an agar plate containing a large number of clones. If a plasmid is used as a cloning vector, the clones take the form of bacterial colonies. If a bacteriophage is used, the clones will be plaques on a 'lawn' of host bacteria.

Only a tiny proportion of these clones will contain the desired gene. The most common way of finding it is to use a gene probe. If a bacteriophage is used as the vector, the essentials of the technique are similar.

A gene probe is a length of nucleic acid with a base sequence complementary to at least part of the gene sought. Each probe can thus base-pair with only DNA containing a complementary base sequence. The probe has to be long enough to contain a unique base sequence that will only pair with the DNA of interest.

A 20-base sequence is sufficient.

There are several ways of preparing a DNA probe:

- Extracting an mRNA copy of the gene and using reverse transcriptase to make a DNA copy. mRNA is most easily extracted from cells specialised for making a very small number of proteins, in which there are correspondingly few types of mRNA.
- Determining the sequence of seven or eight of the amino acids of the protein, and using the genetic code to deduce the DNA base sequence. A short length of DNA is then made artificially (degeneracy of the code means several alternative sequences may have to be tried).

A probe that has found its target (its complementary sequence) must be identifiable. This is made possible by first making the probe radioactive by labelling the 5' end with ^{32}P.

A piece of nitrocellulose filter or nylon is gently pressed on the agar containing the colonies. The membrane is then pierced through to the agar to provide orientation marks for subsequent alignment of positive hybridisation signals. The membrane is then carefully lifted off. Some of the bacteria adhere to the filter, producing an exact replica of the pattern of colonies.

1. The bacteria are lysed (broken open) chemically to free the DNA.

2. The DNA is **denatured** by alkali to make it single-stranded.

3. The probe is added and allowed to hybridise with the DNA on the filter.

4. The filter is washed to remove any unbound probe.

5. The position of the bound probe is revealed by leaving it in contact with X-ray film. After development, the position of the hybridised probe is revealed, and the colonies on the agar carrying the DNA fragment of interest therefore identified.

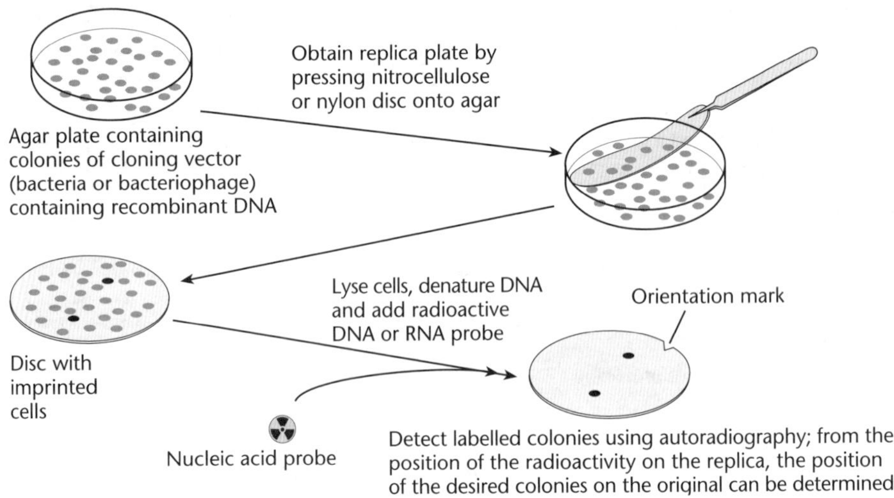

Using a gene probe to find the desired gene.

An alternative to using a gene probe is to use an antibody to the protein product of the gene. The method is similar to that shown for a gene probe except that the nitrocellulose replica plate is treated with radioactively labelled antibody. An autoradiograph enables the colonies producing the protein of interest on the original plate to be identified.

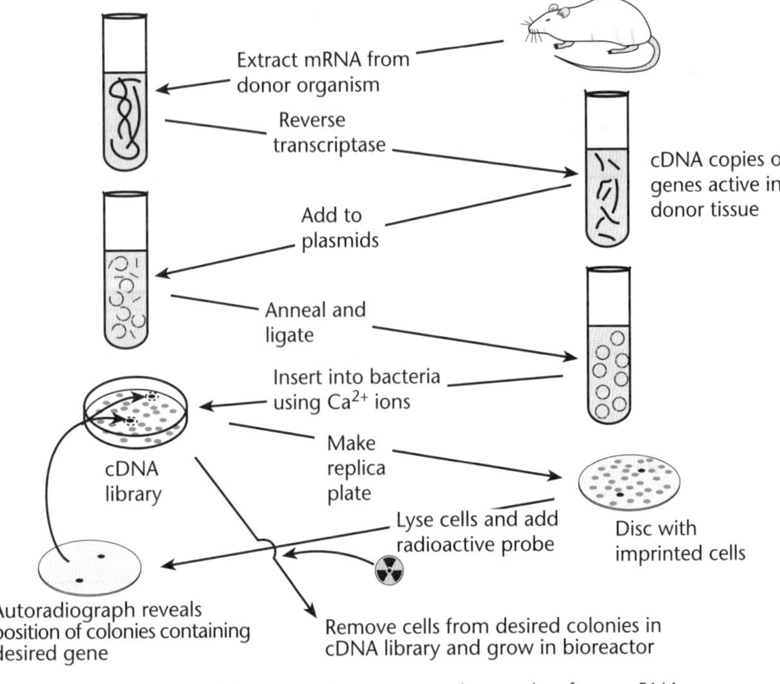

Summary of the stages in transgenesis, starting from mRNA.

Case study: Production of human insulin

Insulin was the first human hormone to be produced commercially by recombinant gene technology. Insulin regulates the concentration of blood **glucose**, and failure to produce enough of the hormone results in *diabetes mellitus*, characterised by raised blood glucose concentrations and presence of glucose in the urine. Until recently the disease was treated by daily injections of beef or pork insulin. Although these are as active as human insulin, the amino acid sequences are not identical and this sometimes resulted in immune reactions.

Insulin is produced as a precursor molecule which is subsequently cleaved into two polypeptide chains, named A and B, which remain held together by disulfide bonds. Thus, although there is a single gene for insulin, the final result is two polypeptides. Insulin is made as follows.

- Two artificial 'genes' coding respectively for the A and B chains are synthesised *in vitro*. These are then separately inserted into the *E. coli* gene coding for β-galactosidase. As a result, each kind of bacterium produces a compound protein consisting of β-galactosidase continuous with either the A or the B insulin chain.
- After purification, each protein is then treated chemically to release the insulin chain from the β-galactosidase.
- The A and B chains are then purified, mixed and treated to link them together to form the active insulin.

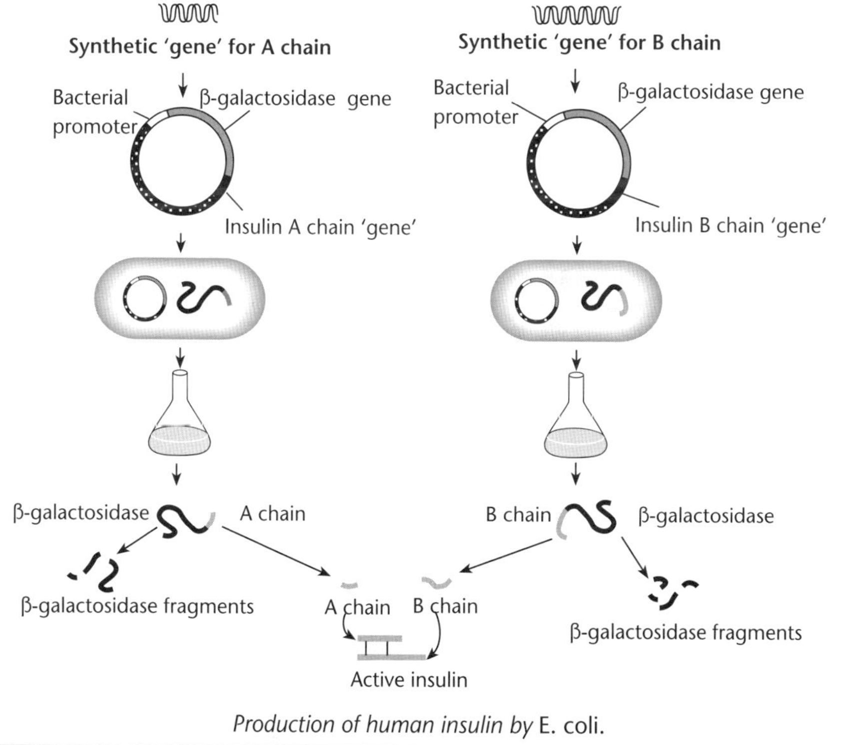

Production of human insulin by E. coli.

Creating the transgenic organism

Introducing genes into animal cells

Genes can be introduced into animals for either of two reasons:

- To produce a transgenic animal, in which all the cells of the body receive the gene, including those ancestral to the gametes. To achieve this, the gene must be introduced into a fertilised egg.
- To rectify genetic defects in particular tissues.

Examples

- Certain inherited immunodeficiencies can be corrected by inserting a gene into bone marrow cells, which are ancestral to cells giving rise to immune cells.
- It is hoped that in the future cystic fibrosis may be treated by introducing a normal gene into the cells lining the lungs and gut.

The introduction of foreign DNA into mammalian cells is called transfection. Four methods are common.

- **Microinjection** – DNA is injected directly into the nucleus of a recipient cell.
- **Electroporation**, in which the cells are subjected to an electric pulse. This causes minute holes to briefly develop in the plasma membrane, allowing DNA to be taken up from the medium.
- Using a virus as a carrier, genetically engineered to make it harmless.
- **Lipofection** – DNA is coated in minute artificial lipid vesicles or *liposomes*, which are taken up by the plasma membrane. The foreign DNA may then be expressed as part of the host cell's genome. It is likely that gene therapy will be done this way.

Example

A transgenic mouse would be produced by first washing newly fertilised eggs out from the oviducts of a female mouse. Before the male and female pronuclei (sperm and egg nuclei) have fused, DNA containing the desired gene would be injected into one of the pronuclei. The eggs are then implanted into an appropriate female mouse. Samples of the offspring are later taken for DNA analysis.

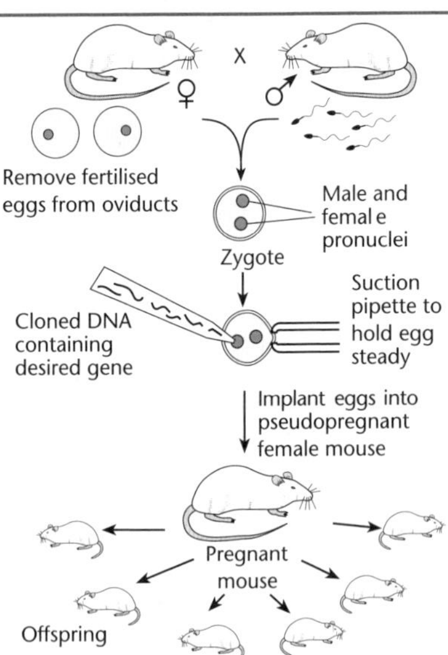

Production of transgenic mice by microinjection.

Inserting genes into plant cells

Although animal cells contain the entire genome, single cells cannot be induced to grow into an entire organism because the process of **differentiation** appears to be irreversible. Plants have a great practical advantage over animals – their cells are **totipotent**, meaning that given the appropriate chemical treatment, a single mature cell can grow into an entire plant. Any gene transferred into such a cell can thus be transmitted to the next generation.

The most common method of transferring genes to plant cells is to use *Agrobacterium tumefaciens*. This common soil bacterium parasitises a variety of **dicotyledonous plants**, causing the production of a tumour-like growth called a *gall* inside which the bacteria live.

When a bacterium enters the plant (through a wound), it inserts part of a *tumour-inducing* (Ti) plasmid into a host cell. The part of the plasmid that is transferred into the host cell is called *T-DNA* and this is then incorporated into a chromosome of the plant. Some of the genes in the plasmid are necessary for the infection of a host cell, while others induce host cells to multiply to produce a gall.

The bacteria used by the genetic engineer are genetically modified in two ways:

- They have had the gene necessary for gall-formation deleted, while retaining the genes needed to infect host cells.
- They have had genes for resistance to certain antibiotics (eg kanamycin) added.

To use *Agrobacterium* as a vector for introducing a gene into a plant, the desired gene is first isolated, spliced into a plasmid of *Agrobacterium* and cloned. Cells of the host plant are grown in culture and treated with the bacterium and also with the antibiotic for which the plasmid carries a resistance gene. This serves to identify the cells into which the plasmid has been successfully inserted, since only these cells can grow in the presence of the antibiotic. From the cultured cells, entire plants are then grown.

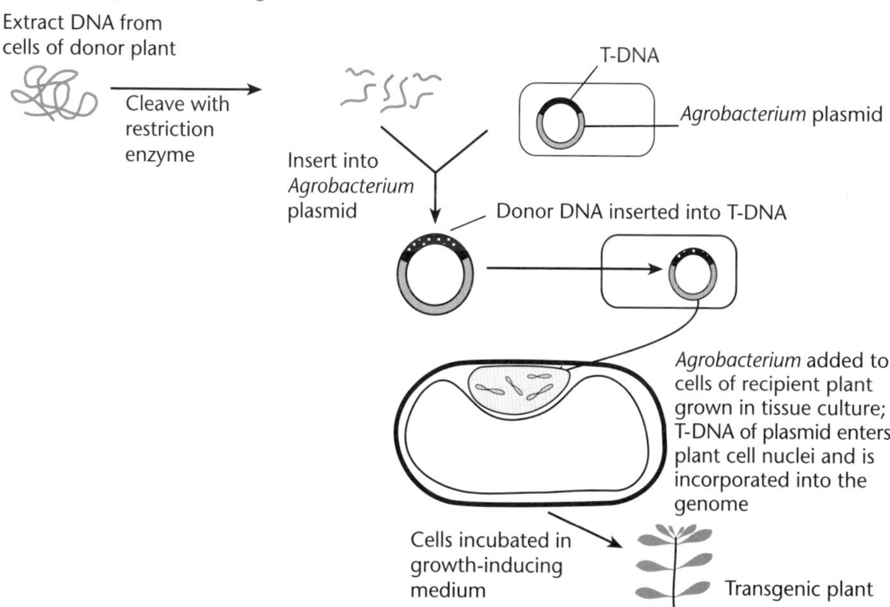

Use of Agrobacterium *to insert foreign DNA into a plant cell.*

Agrobacterium has proved highly successful in producing transgenic plants such as tomato, tobacco, cotton and apple. Unfortunately it does not infect **monocotyledons**, which include some of the most important crops such as the cereals. Alternative methods are used for these plants. One, the 'shotgun' or **biolistic** method, involves firing minute tungsten or gold beads coated with DNA at intact plant cells. Not only does the DNA enter the cells, but a small proportion of them incorporate it into their chromosomes. Electroporation is another method, but the cell walls have to be digested away first to produce naked cells called *protoplasts*. After treatment with polyethylene glycol, the two protoplasts fuse, producing a hybrid cell containing the DNA of two cells.

Expressing the gene – control of transcription

The mechanisms for switching genes on and off in eukaryotes are different from those in prokaryotes. This does not matter if the bacteria into which a gene has been transplanted are only used to clone the gene. If, however, the bacteria are used to produce the *protein* coded for by the gene, the eukaryotic gene must be expressed in the bacterium. To enable this to happen, the gene is usually spliced 'downstream' of the prokaryotic promoter and operator. The gene can then be switched on by addition of the appropriate inducer.

Example

If a gene is spliced into the lactose **operon** of *E. coli*, the gene can be switched on by adding lactose to the medium. In this way the production of the protein can be controlled by the technologist.

DNA profiling

'Genetic fingerprinting', or **DNA profiling**, is a technique that makes it possible to determine relationships between individuals by comparing their DNA. The technique was developed by Alec Jeffreys in 1985, and has been of enormous significance in forensic science (the science of obtaining evidence to be used in court).

Examples

It has been used to establish cases of paternity. In cases of incest, it can be shown that the accused is genetically more similar to the child than if the father were a non-relative.

In rape cases, it has been used to exonerate the innocent and/or confirm guilt.

In Europe it has been used to establish whether a person has relatives in the country he or she is seeking to enter.

DNA fingerprinting is also important in captive breeding of endangered species. When total numbers are low, DNA profiles can help to prevent inbreeding by making it possible to select unrelated individuals for mating, thus increasing the diversity of the gene pool.

DNA profiling has two important advantages over (traditional) blood group tests:

- Since blood groups are not unique, they can only be used to *eliminate* a suspect. A DNA profile is absolutely specific (except in the case of identical twins), and so can give a *positive* identification.
- Any tissue containing cells – skin, hair root, bone – can serve as a DNA source.

About 90% of the human genome has no known function. Much of this non-coding DNA consists of introns, the RNA copies of which are removed before translation.

Exons that code for amino acid sequences in essential proteins vary little, since any departure in the structure of an essential protein is likely to be harmful. Because introns are not translated into protein, they are not subjected to natural selection. *Intron sequences thus vary greatly in a population, and form the basis for DNA profiling.*

Much of this intron DNA consists of regions called **minisatellites**, where there is great variability from one individual to another. A minisatellite consists of short sequences of 20–100 nucleotides, called *variable number tandem repeats (VNTRs)*. Each VNTR consists of a **core sequence**, repeated many (3–30) times. Some core sequences are common to all individuals but the number of times they are *repeated* at a given site varies with the individual.

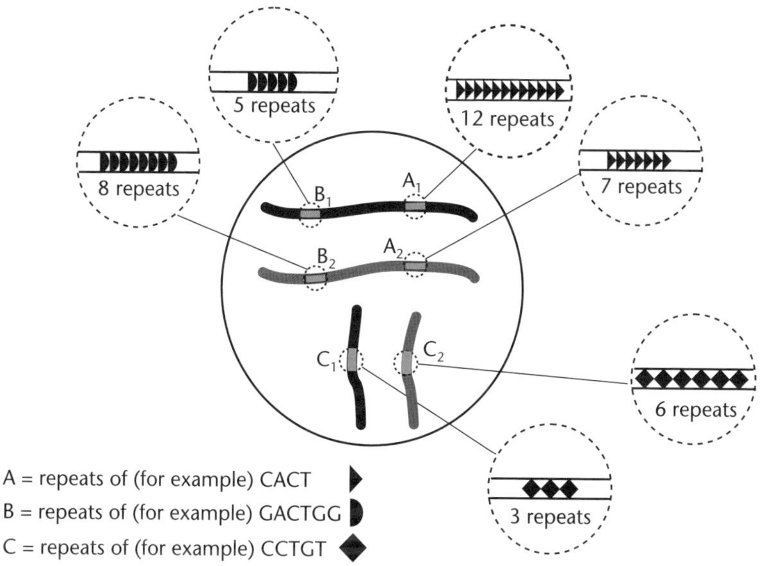

Each VNTR is located at a particular site on a pair of homologous chromosomes. The kind of repeat sequence is characteristic of the site but the number of repeats at the two corresponding sites can vary independently. Although two people may share one repeat, the chance that they will share two repeats is much smaller.

Distribution of three imaginary VNTR loci on two pairs of chromosomes.

Each autosomal VNTR site exists as two copies, occupying the same position on homologous chromosomes. In most cases ('heterozygotes'), the two VNTRs have different numbers of repeats. In **homozygotes** the two sites have identical repeat numbers. Each chromosome has many VNTRs and the number of repeats varies independently at different sites. At any given site the number of repeats varies from person to person, so does the length of each VNTR region. Hence each individual has a unique 'profile' of VNTRs, which can be used to distinguish them from every other person (except for genetically identical twins).

When VNTRs are compared within a family, for a given site, it can be seen that each person inherits one VNTR from each parent.

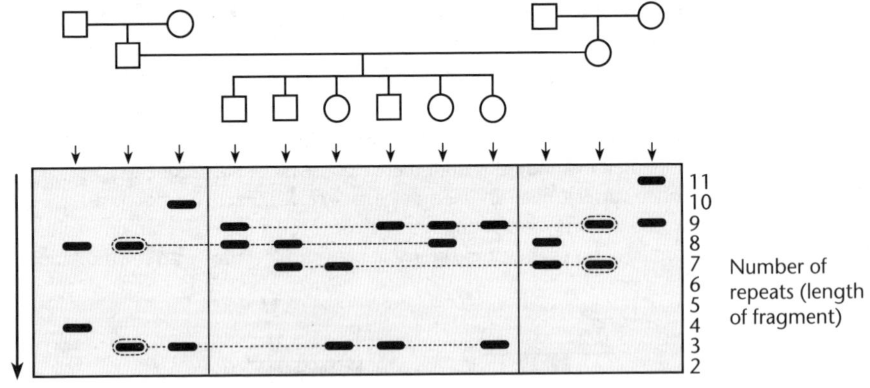

Autoradiogram of DNA from 12 people in a family.

By taking blood samples from hundreds of people, scientists have established the percentage of individuals with a given repeat number at a given VNTR site.

Example

Suppose that the VNTR site shown above has 14 repeats in 10% of the general population. The probability that anyone taken at random from the general population has that number of repeats in this VNTR is thus 0.1. The probability that any two people taken at random will share the same number of this VNTR is
$0.1 \times 0.1 = 0.01$.

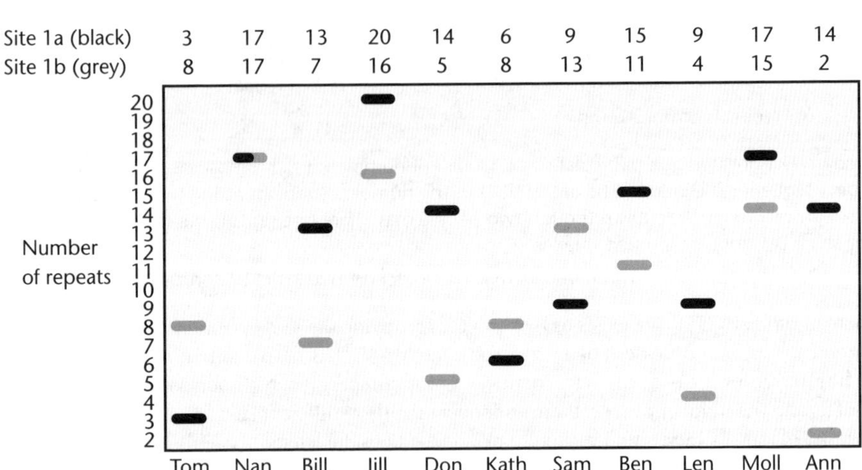

Although some individuals have one site in common, no two share both 'alleles'. ('Alleles' with 10, 12, 18 and 19 repeats are absent from this sample.) Note that Nan is homozygous for the 17-repeat 'allele'–Site 1a (black) on the paternal homologue, Site 1b (grey) on the maternal one.

Occurrence of 16 'alleles' in 11 unrelated people.

Example

Suppose Taku and Gawac are suspected of committing a rape. If the DNA of Taku matches DNA from the crime scene and victim with respect to a particular VNTR, then Gawac along with 90% of the population are excluded as suspects. While Taku may be the main suspect, a 10% chance of guilt is insufficient evidence to convict him. More loci would have to be compared to convict Taku.

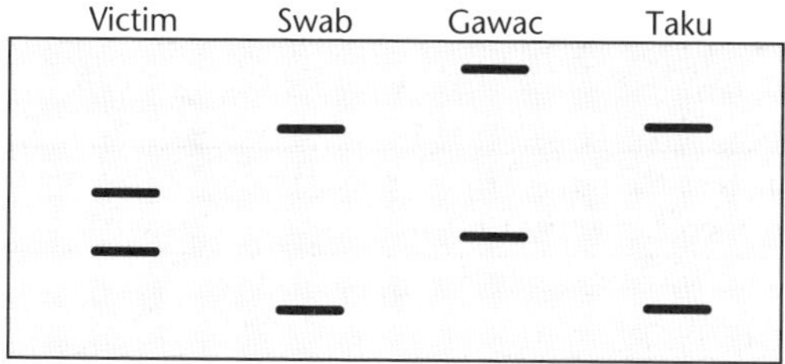

Result of a single locus comparison of DNA from a rape victim with that of two accused men and a vaginal swab obtained from the victim. Evidence from a single locus would exonerate Gawac, but more loci would have to be compared to convict Taku.

DNA profiling is such a powerful tool because if more than one VNTR region is compared, the level of certainty rises greatly.

Example

Suppose that 1 in 20 (0.05) people share the same number of repeats at one site, and 1 in 10 (0.1) share the same number of repeats at a second site. The probability of a chance match for both is 0.05×0.1, or 0.005. Only a few sites need to be compared to reduce the chances to millions to one.

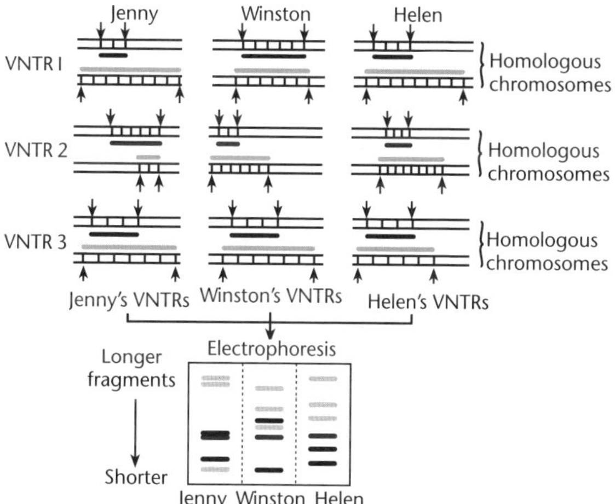

Arrows represent restriction sites at which VNTR fragments were cut. VNTRs from paternal chromosomes shown in black, maternal in grey.

Genetic fingerprinting of three individuals using three VNTR loci.

Notice that variation in fragment length depends on both the number of repeats and on the number of bases in each repeated sequence. Thus CGCGCGCGCGCG (6 repeats of 2 bases) would give the same fragment length as 3 repeats of a 4-base sequence.

Making a genetic profile

1. A sample of the DNA of the individual is digested with a restriction enzyme which does not attack any site within a repeated sequence, so the repeats are left intact.

2. The fragments are separated by gel electrophoresis. At this stage they are invisible, and are greatly outnumbered by other DNA fragments which are not wanted.

3. The gel is then subjected to **southern blotting**.

The gel is immersed in sodium hydroxide solution to denature the DNA, making it single-stranded.

The gel is overlaid with a nitrocellulose or nylon filter, and the filter is then covered with a thick wad of paper towelling. This acts as a wick, drawing up the buffer solution through the filter by **capillary action**, leaving the DNA on the filter.

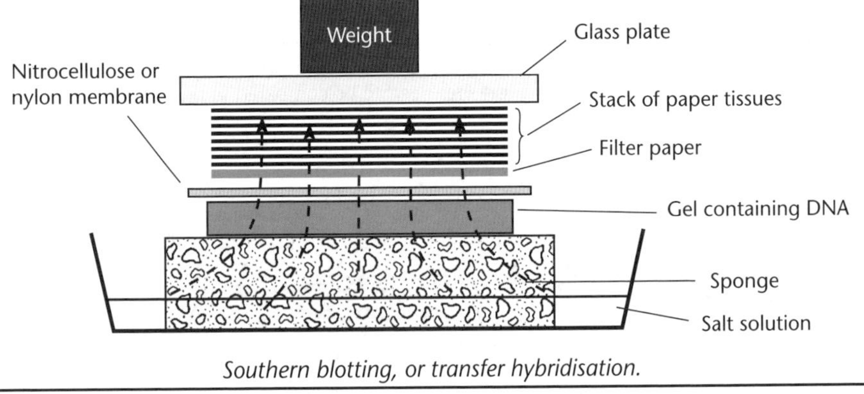

Southern blotting, or transfer hybridisation.

4. The nitrocellulose or nylon paper is baked or UV-irradiated to attach the DNA permanently to the membrane.

5. A DNA probe is used to identify the repeat fragments. This probe consists of single-stranded DNA complementary to the repeat sequence, labelled with radioactive ^{32}P. The probe binds to the repeat fragments, thus labelling them.

6. The non-bound probe is washed away and an X-ray film laid over the filter.

After developing the film, the positions of the repeat sequences are revealed as dark bands on the film. The spacing between the fragments indicates their rates of movement through the gel, and hence their relative sizes.

The overall pattern looks like a supermarket bar code and is unique for any individual except similar twins.

Within a given family, if the VNTR pattern of one parent is known, then the pattern of the other parent can be at least partly deduced through their children. The more children, the more complete can be the reconstruction of the genetic profile of an unknown parent.

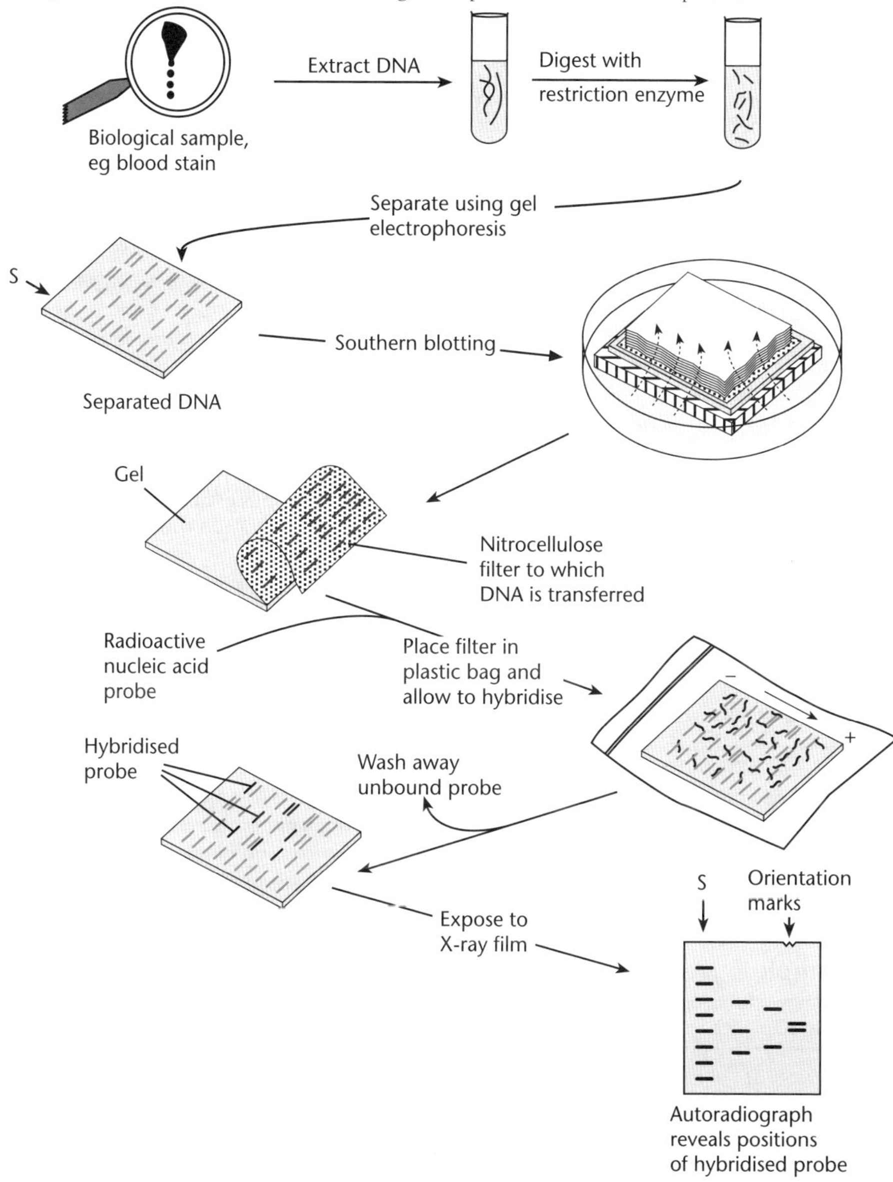

The bands in the lane marked 'S' are fragments of known length, which act as standards for comparison

Summary of procedure for making a DNA profile.

Genome analysis

Genome analysis involves the use of numerous techniques, including traditional breeding methods to analyse the genome of an organism.

Example

The Human Genome Project, started in 1990, has involved some 250 collaborative groups worldwide. The entire human genome, at the chromosomal, gene and nucleotide levels, is being deduced for the following reasons:

- To understand the genetic basis of humans. The implications of this are wide ranging, from a better understanding of genetic triggers for the changes that happen during foetal development to evolutionary relationships between *Homo sapiens* and other hominid species.
- To establish the precise location and base sequences of genes that result in disorders such as Huntington's chorea, high blood pressure, heart disease, allergies, alcoholism and some types of cancer.
- To develop gene therapies that target and correct the affected gene itself, maybe even before birth.

There have been three main lines of work in the Human Genome Project, involving a mixture of 'traditional' genetics and new gene technologies:

- Genetic mapping of genes was done through genetic crosses to establish the order of linked genes on chromosomes and the relative distances between linked genes. This was completed in 1994.
- Physical mapping of the chromosomes, where each chromosome was digested into fragments by restriction enzymes and the order of the fragments established, was completed in 1998. By the end of this stage, it was possible to say on which chromosome and at which locus a particular gene was.
- Sequencing of the genome using a combination of gene technologies such as gel electrophoresis and gene sequencing to establish the exact order of up to 6 billion nucleotide base pairs. In 2003 it was revealed that 99% of the human genome had been sequenced to 99.99% accuracy.

The genomes of other important species are being analysed for the future benefit of, and use by, humans. The DNA of such organisms as *E. coli*, yeast, the fruit fly *Drosophila melanogaster*, several species of roundworm, corn and mice has been sequenced. Understanding the genetic make-up of commercially important crops and plants, and the pests that affect them, is important should the desire to genetically modify them for human use arise. The inheritance patterns of genes and associated segments of DNA such as markers and promoters can be more closely studied. Comparative analysis of the genes of other species may help with the interpretation of human DNA.

Example

The known function of a nucleotide sequence in yeast may yield clues to the function of a similar sequence in humans, particularly if it occurs in intron DNA.

Stem cell research

The ultimate goal of stem cell research is the use of human cells instead of drugs and other pharmaceuticals as therapy for human illnesses – particularly those that are life-threatening or severely compromise the quality of life of an individual.

Stem cells

Stem cells are cells yet to differentiate into the functional cells that characterise multicellular organisms such as humans. All stem cells have three common properties:

- They are able to *divide* and replicate themselves many times. Differentiated cells such as blood cells, skin cells or bone cells function for a certain period of time and then are replaced by new cells. Stem cells can divide many times for long periods to give rise to millions of new stem cells.
- They are *unspecialised* – ie they cannot perform specific functions, such as propagating a **nerve impulse**, as a nerve cell will do. However, given the right conditions – whether naturally (in the body) or artificially (in a laboratory) – stem cells will differentiate to become cells specialised to carry out a specific function.
- They can *differentiate* to become specialised cells – much research has been occurring over the past 20 years to establish the requirements of differentiation. For stem cells to differentiate, certain genes must be expressed in specific sequences. This involves the exchange of chemical stimuli between neighbouring cells.

There are two main types of stem cells – embryonic stem cells and adult stem cells.

- **Embryonic stem cells**. After fertilisation of the female gamete by the male gamete, the resultant zygote undergoes many mitotic cell divisions to become a ball of undifferentiated cells called a **blastocyst**. Within the blastocyst is a group of cells called the inner cell mass. This cell mass is composed of embryonic stem cells that have the potential to differentiate into any of the hundreds of different cell types within the individual – called **pluripotency**. The purpose of embryonic stem cells is to generate the many organs and tissues that make up the individual.
- **Adult (somatic) stem cells**. Small numbers of stem cells have been found in many organs and tissues within the human body.

Example

Adult stem cells are present in the brain, cornea and retina of the eye, heart, fat, skin, bone marrow, blood vessels, skeletal muscle and intestines. Stem cells are also found in the **umbilical cord**, give rise to blood cells and are classified as adult stem cells. Adult stem cells are found in very small numbers within each tissue type or organ. They are dormant until injury or illness stimulates their division and differentiation into new functional cells of the tissue or organ of origin.

Most adult stem cells do not appear to be pluripotent, although there is evidence some have the ability to differentiate into multiple cell types – called **plasticity**.

Example

Haematopoietic stem cells usually give rise to new blood cells, such as red blood cells, B and T **lymphocytes** and other **white blood cells**, and **platelets**. However, haematopoietic stem cells have also been induced into becoming skeletal muscle cells, cardiac muscle cells and liver cells.

Stem cell lines

Embryos used for stem cell research generally originate from *in vitro* fertilisation clinics, from unwanted embryos donated for research purposes. However, scientists are creating embryos by

therapeutic cloning, where the nucleus of a somatic cell is removed and inserted by micro-injection into an enucleated fertilised egg.

Adult stem cells are derived from various tissues, depending on their intended final use, but most adult stem cell research is focused on stem cells found in the heart, brain and bone. Stem cells have been in use therapeutically for 20 years (eg the use of blood-forming stem cells from bone marrow to replace those destroyed by the effects of chemotherapy). The goal of current research is the cure of serious illnesses such as Parkinson's and Alzheimer's, **Type 1 diabetes** and **rheumatoid arthritis**, and to repair damaged tissue as a result of a heart attack or damage to the spinal cord or brain.

The first step in such research is to create a **stem cell line** – a culture of stem cells that has replicated many times for several months on **Petri dishes** to produce millions of genetically normal, pluripotent, undifferentiated stem cells. Embryonic stem cell lines are derived from the stem cells of a 5-day-old blastocyst that has been created by *in vitro* fertilisation or by therapeutic cloning. These cells are removed and placed on a Petri dish, the surface of which is covered by embryonic mouse cells. This is called the **feeder layer**, and is needed by the stem cells as a surface upon which to stick, and as a source of nutrients for cell division and growth. The cells proliferate, and within a few days must be **subcultured** onto fresh Petri dishes or they will die. Batches of cells from a stem cell line can be frozen and transported to other laboratories for research work.

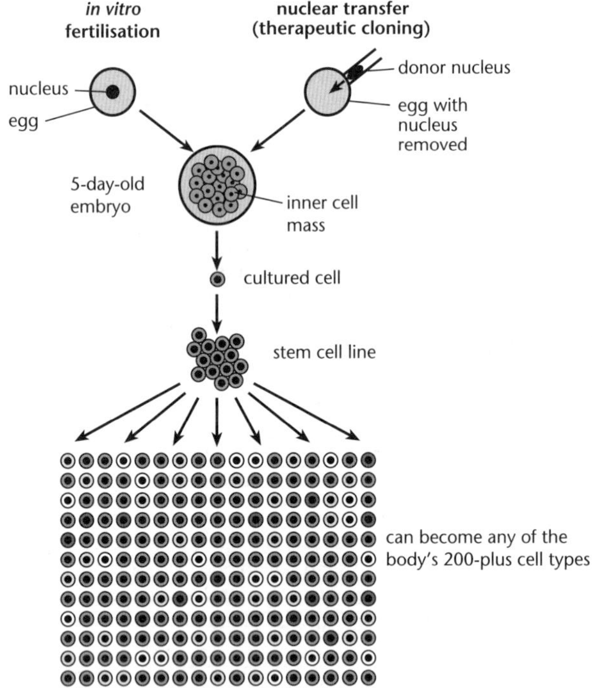

Embryonic stem cells.

Adult stem cells are harder to culture as there are fewer of them to begin with, and are more difficult to keep alive and replicate.

Uses and implications of stem cells

Stem cells are destined to be used for cell-based regenerative therapies (therapies or cures for diseases using normal human cells rather than drugs).

Both types of stem cells have potential advantages and disadvantages.

- Embryonic stem cells are relatively easy to culture – there are many of them, they respond well to artificial conditions and can be easily induced into differentiating into any one of the many thousands of different cell types present in the human body.
- Adult stem cells are more difficult to work with, but have the advantage of much reduced risk of transplant rejection if a patient receives his or her own stem cells.

There are several potential uses for stem cells.

- Studying the cellular events of human development – stem cells could be used to determine how and when genes involved in the development of the embryo and **foetus** are expressed, particularly those that result in cell reproduction and differentiation. Such information could also assist in the understanding of and consequent therapies for certain cancers that are a result of abnormal cell events.
- Cell-based therapies – involving the direct introduction of differentiated cells into a patient with a serious illness or disorder, where these cells regenerate or repair damaged tissue. As the need for transplantable tissues or organs far exceeds the supply, this is a significant application of stem cells. Preliminary research on humans using adult stem cells includes **leukaemia** suffers, some of whom, treated with bone marrow and umbilical cord stem cells, appear to have been cured of the disease. Research is also underway on genetically modifying stem cells through the introduction of specific genes into the nucleus.
- Testing of new drug therapies – stem cells could be used to test the safety and efficacy of new drugs before clinical trials of patients begin.

There are implications involved in the use of stem cells, particularly those embryonic in origin. Several countries have banned or severely limited such types of stem cell research on ethical grounds. Such constraints may result in individuals in these countries not having access to cell-based therapies once they become mainstream clinical practices elsewhere. Because they will not be available to the general public or because they have been developed by private companies, such cell-based therapies could be very costly for patients.

Xenotransplantation

Transplantation of **organs** between humans (where donor is living or dead, called **allotransplantation**) has been a common occurrence for several decades. However, the demand for organs and tissue far outweighs the available supply, and many people waiting for a transplant die.

Xenotransplantation, where living, non-human animal cells, tissues or organs are transplanted into humans has been attempted in the past, with the recipient invariably dying from organ rejection. With advances in modern biotechnological techniques, particularly those involving the manipulation of individual genes to create transgenic animals, it is likely that xenotransplantation will become a relatively common practice by the end of the decade.

Xenotransplantation can involve:

- Organs – eg the heart or **kidney**.
- Tissues – eg bone marrow.
- Cells – eg pancreatic islet cells producing insulin in the treatment of type 1 diabetes.

Xenotransplants can be:

- Internal – this is the aim of much current research, and involves the replacement of entire organs or the implantation of tissues or cells at the site(s) of action.
- External – some human trials have been completed for a device containing animal liver cells attached externally to a person suffering from a type of liver failure. The person's blood passes through the device and toxins are processed and removed by the animal cells.

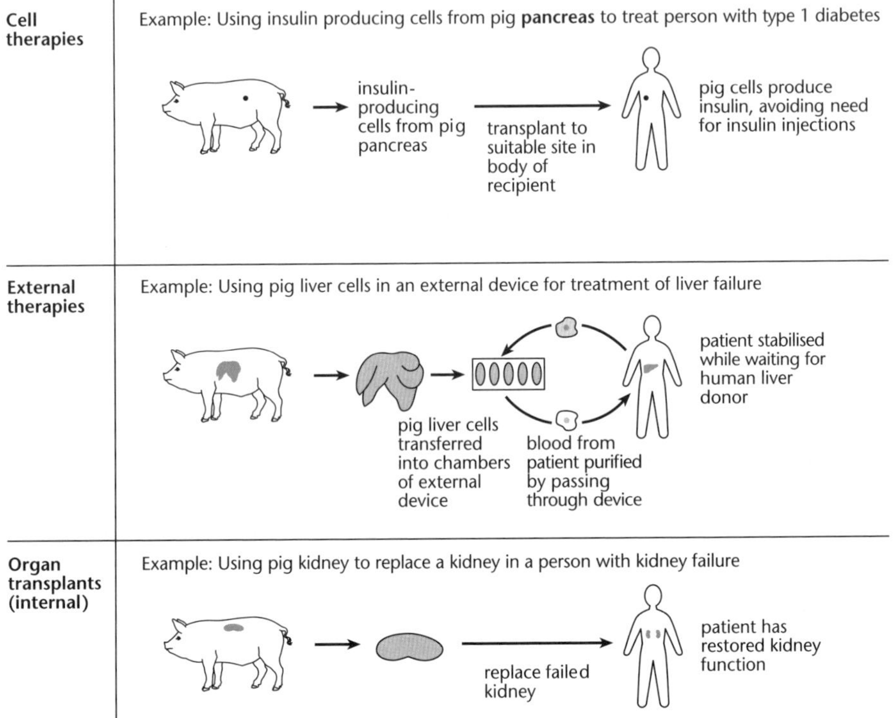

Summary of types of xenotransplants.

The important advantages of xenotransplantation (once the technological problems have been addressed), are there will be a potentially unlimited supply of animal organs, tissues and cells available, which will be virtually undamaged (as they would have been removed from an animal euthanised at the same time that the patient undergoing the transplant is being prepared for surgery). With current allotransplantations, there can be tissue damage due to the time delay between removal from the donor and being available for surgery.

The animals involved

Ideally (biologically at least), the best candidates for 'donation' of their organs would be animals most closely related to humans, as there would be a much-reduced risk of **hyperacute rejection** occurring. However, the use of non-human primates is problematic for several reasons:

- It is more acceptable to many people to use animals that have been domesticated for many years.
- Several non-human primates most closely related to humans are endangered in the wild.
- There is a risk of cross-species transfer of **pathogens** (particularly those viral in nature), that could have serious consequences for humans. For example, HIV is thought to have been a cross-species transfer from chimpanzees.

The animal that is the most likely candidate for xenotransplantation is the pig. It has a high fecundity (large litter sizes), is easy to rear in laboratory conditions, its important organs are a similar size to those in humans, and the risk of cross-species transfer is smaller than with non-human primates. One type of 'mini-pig' has already been genetically manipulated to eliminate a major cause of organ rejection.

Example

Pig cells are naturally coated with a type of sugar, similar in effect to an antigen. This is done by an enzyme called *galactosyl transferase*, coded for by the *alpha 1,3 GT* gene. When the sugar is detected by the human immune system, a catastrophic rejection cascade occurs called hyperacute rejection (HAR), resulting in the rupturing of blood vessel walls.

In 2001, two biotechnology companies successfully created litters of so-called 'knockout pigs'. One of the two alleles coding for the transferase enzyme was deleted in a pig cell using very precise gene manipulation, followed by nuclear transfer to create a litter of clones, each lacking one of the two alleles. In 2002, the first male knockout pigs were born, which, when bred with the original female knockout pigs, produced a litter of 'double knockouts'. These pigs carried organs that would theoretically not be rejected by the human immune system – as none of the pig cells would carry the sugar antigen on their surface. In 2004, a group of baboons that had received double knockout hearts were kept alive for 2–6 months. Because organ rejection is a complex process, the baboons did eventually die from organ failure, but the rejection response was kept in check by the usual doses of anti-rejection drugs for a relatively long period of time (HAR can occur within a matter of minutes), and the rejection that did occur involved blood clots in small blood vessels. Work is currently being done on genetically manipulating pigs so they carry a human anti-coagulant gene, eliminating these blood clots.

The implications

There are important implications relating to xenotransplantation.

- Animals undergo considerable suffering as a result of the research carried out.
- The risk of cross-species infection may have safety implications for society as a whole, and could result in individuals being denied access to the technology.
- There are spiritual and cultural issues that require extensive consultation.

The potential of gene technology

Agriculture

- Improvements in wood quality and quantity are being undertaken by Forest Research in New Zealand, which is investigating the genetic modification of *Pinus radiata* to reduce the amount of **lignin** produced in the wood. Lignin must be removed when the wood is pulped to make paper, a costly process (for the environment) and financially (for wood processors). Investigations into redirecting resources the trees put into producing unnecessary reproductive growth into more wood-producing, vegetative growth, are also underway.
- Pathogen-resistant crop plants are being developed. 'Old-fashioned' apples contain natural resistance to pathogens such as powdery mildew and black spot that affect commercial crops. The genes that confer this resistance are identified and inserted into commercial apples through genetic engineering.
- Herbicide-resistant plants are being developed. Plants that are resistant to herbicides such as glyphosate ('Roundup') have been produced. By spraying the resistant crop with herbicide, weeds are eliminated, leaving the crop untouched. 'Roundup-Ready' onions are currently undergoing field trials. Onions are sprayed with more highly toxic **herbicides** than any other crop plant. 'Roundup-Ready' onions will only need to be sprayed with one fifth of the amount of herbicide of the current non-modified onions.
- Insect-resistant plants. An 'insecticide' used to kill larvae of butterflies and moths is the bacterium *Bacillus thuringiensis*, the spores of which contain proteins toxic to caterpillars (called BT for *Bacillus thuringiensis*). When applied externally, the toxin quickly breaks down, but genes for BT can be introduced into plants so that the plants become toxic to particular insect species but are harmless to humans.

Medicine

Medically important human proteins normally produced in minute quantities can be produced on an industrial scale by transplanting the respective genes into bacteria.

Examples

- Until recently, diabetics had to rely on insulin extracted from cattle and pigs.
- Blood products such as the clotting protein Factor VIII can be produced by bacteria, eliminating the danger of contamination by viruses such as HIV and hepatitis.
- Other proteins that have been produced by genetically engineered bacteria include human growth hormone, epidermal growth factor, **interferons** (anti-viral proteins) and erythropoietin (a protein that stimulates red blood cell production).
- Anti-viral vaccines can now be produced without risk. Traditionally, these have been prepared using a live but weakened (attenuated) virus. This has always carried a slight danger if the virus was not completely inactivated. The part of the virus that stimulates antibody production is the protein coat, so a vaccine consisting only of the protein is just as effective and carries no risk. Viral coat proteins can be made by bacteria into which the genes for viral coat proteins have been transplanted. A vaccine against hepatitis B has already been produced in this way.
- Genetic screening. It is now possible to identify carriers of harmful genes that are otherwise undetectable, either because they are recessive (eg **cystic fibrosis**, Duchenne muscular dystrophy) or because they are expressed later in life (eg Huntingdon's disease).
- Gene therapy. The base sequences of individual genes can be determined, making it possible to diagnose exactly why a given gene, such as the gene causing cystic fibrosis, is faulty. This paves the way for the eventual replacement of faulty genes by normal

ones. Replacement of defective genes by normal ones in the body cells is called somatic therapy. Its effects are not inherited, unlike germ line therapy, in which a normal gene is transplanted into an egg.

Forensic science

Through DNA profiling it is now possible to link a person with an extremely high level of certainty to an individual tissue sample. This application is of enormous importance to forensic science and has played a key part in a number of murder and rape trials and in resolving questions of paternity.

Food industry

Chymosin (formerly called rennin) is used to precipitate casein (the main protein in milk) in an early stage of cheese production. It used to be extracted from the stomach of young calves, but it is now produced by bacteria into which the gene has been transplanted. Genetically engineered chymosin production forms the basis for vegetarian cheese production.

Biological pest control

Possums and wasps are major pests in New Zealand.

Gut bacteria of wasps are being genetically modified to produce a toxin that will kill the wasps.

There are several projects involving controlling vertebrate pests such as possums through decreasing their fertility. This is done by identifying proteins that make female possums infertile. The genes for these proteins are inserted into food plants, or into bacteria or nematode worms that parasitise the animal.

Conservation

DNA technologies are being used to help conserve several endangered species. Many species of bird are sexually *monomorphic*, meaning it is very difficult to tell males and females apart physically. Successful breeding programs depend on understanding the composition of a breeding population. DNA fingerprint technology has been important in the study of breeding populations of some birds so that the most appropriate transfers of birds between populations can be done.

Human evolution

Determination of evolutionary relationships between taxonomic groups.

Example

Recent comparisons between mitochondrial DNA of modern humans and of preserved Neanderthal bones have shown that Neanderthal people were considerably more distantly related to modern humans than previously believed.

More distant targets

Though not yet accomplished, one of the most important goals of the plant genetic engineer is to develop cereals that can act as hosts to nitrogen-fixing bacteria, as leguminous plants do. This is likely to prove difficult since about 20 genes are involved. Some are involved in fixation itself, others are needed for infection of the plant by the bacterium.

Anxieties over recombinant DNA technology

Almost every beneficial scientific discovery has brought with it possibilities for misuse, and genetic engineering is no exception. When genetic engineering first became feasible in the 1970s, it was suggested that new strains of micro-organisms might 'escape' from laboratories and cause outbreaks of uncontrollable diseases or other problems. To prevent this, regulations were put into place to try and ensure that any organisms used in gene technology were so modified that they could not live outside the laboratory. Since then, it has become clear that this 'Frankenstein factor' is far less of a threat than some had feared.

Some concerns stem from a feeling that producing transgenic animals and plants is an interference with nature (though farming and medicine are just that). There are other, more solidly based anxieties.

Safety of genetically modified foods

Transplanting of a gene has two results, either of which may have unforeseen effects. Although DNA can be cut and spliced with considerable precision in the laboratory, insertion into the host genome is much less precise. Random insertion of DNA could disrupt genes at the sites of the cuts. Also, the level of activity of an introduced gene depends on regulatory factors in the rest of the genome. As a result, a transgene may be more or less active than intended. A third possible effect arises from the fact that some people are allergic to certain foods – or rather, to certain substances in them.

Example

The amino acid composition of soybeans (which are low in methionine) has been altered by introduction of a gene from a brazil nut for a storage protein, to which some people are allergic. How can people avoid this protein when it is present *incognito* in other foods such as soybeans, which are used in the manufacture of many processed foods?

Possible ecological dangers

It has been suggested that recombinant DNA could spread from crop plants to wild relatives, with unforeseen consequences.

Example

Herbicide-resistance genes might spread from crops to wild plants.

Medical insurance

The greatly increased tools for genetic screening mean that insurance companies may demand to see a person's 'genetic profile', and may refuse medical insurance to people genetically predisposed to certain diseases.

Who owns genetic information?

The *patenting* of **transgenic organisms** is another problem to which the legal system is currently attempting to address.

The problem of eugenics

Eugenics is a field concerned with improving the quality of the human gene pool. Perhaps the greatest long-term concern relates to the interference with human genes. Few would argue that the repair of the gene for cystic fibrosis would be of great benefit. However, some fear that the power to repair defective human genes might be the thin end of a very long wedge. Can we, for example, draw a clear line between genes that are clearly defective and bring great suffering to their possessors, and genes which produce some other trifling 'imperfection', such as excessive freckling?

A genetic engineer's mini-dictionary

agarose gel: jelly-like material used to separate DNA fragments of different sizes.

***Agrobacterium*:** common soil bacterium used to insert plasmids into flowering plant cells.

anneal: joining of complementary nucleotide sequences by controlled lowering of temperature.

bacterial transformation: process by which bacteria can take up 'foreign' naked DNA; promoted by addition of calcium ions.

bacteriophage: virus that attacks bacteria; used as a vehicle for replicating DNA inside bacteria, ie a *vector*.

biolistics: insertion of DNA by firing minute DNA-coated particles of tungsten or gold into cells.

cDNA: DNA produced by copying RNA.

clone: a large number of genetically identical offspring produced by the copying of an original type.

cloning vector: plasmid or bacteriophage DNA, into which foreign DNA is inserted for replication.

DNA ligase: enzyme that joins DNA fragments together.

electroporation: technique of inserting DNA into animal cells or plant protoplasts, by using electric pulses to create minute, transient holes in the plasma membrane.

explant: a small part removed from a plant, and grown in tissue culture to produce a plant genetically identical to the donor plant.

gene cloning: process by which multiple copies of a gene are produced in a bacterium or bacteriophage.

gene library: the entire genome of an organism, distributed as DNA fragments in a large number of bacteria or bacteriophages.

gene probe: a segment of single-stranded nucleic acid, used to find a complementary sequence of DNA.

genetic profiling: technique by which fragments of DNA in individuals are compared to establish relationships between them.

genome: collective name for all the genes of an organism.

linker: short, artificially synthesised section of single-stranded DNA used to make complementary ('sticky') ends to DNA fragments so that they can be joined.

oligonucleotide: a chain of up to about 20 nucleotides (ie a short polynucleotide).

plaque: a clear area in a dense population of bacteria resulting from lysis (breakdown) of bacteria after infection by bacteriophage.

plasmid: a small, independently replicating bacterial mini-chromosome, bearing genes for resistance to antibiotics, and used as a carrier for replicating DNA.

polymerase chain reaction (PCR): technique whereby a minute quantity of DNA can be replicated many times in the test tube.

recombinant DNA: produced by *in vitro* ('in the test tube') splicing DNA from two different sources.

restriction endonuclease: bacterial enzyme that cuts DNA at specific sites.

reverse transcriptase: enzyme used to make DNA from RNA template.

southern blotting: technique for transferring DNA fragments from a gel to nitrocellulose filter by capillary action, prior to locating the DNA using a probe.

terminal transferase: enzyme used to add short nucleotide sequences to DNA.

totipotent: having the capacity to produce all the cell types of the organism.

transgenic organism: organism that contains artificially introduced genes from another species.

vector: a virus or a bacterium into which DNA is inserted for replication.

Unit 12.3 Activity 9A: Putting genes to work – applications of gene technology

1. Match the terms **A–Z / AA–AE** with the pharses **1–31**.

1.	Breaks DNA down into nucleotides	**A**	DNA probe
2.	Used to insert genes into dicotyledonous plants	**B**	Terminal transferase
3.	Used as a cloning vector	**C**	VNTR
4.	Used in DNA profiling	**D**	Southern blotting
5.	Fragmented copy of the entire genome	**E**	Reverse transcriptase
6.	Makes DNA from RNA	**F**	Plasmid
7.	Technique for inserting DNA into cells	**G**	Autoradiography
8.	Detects positions of radioactively labelled substances	**H**	cDNA
9.	A copy of exon sequences	**I**	Intron
10.	Adds 'linkers'	**J**	Electroporation
11.	Bacterial 'mini-chromosome' used as a vector	**K**	DNAase
12.	Non-coding DNA sequence	**L**	Phage lambda
13.	Produced by reverse transcriptase	**M**	Primer
14.	Used in locating specific DNA sequence	**N**	Poly-T
15.	Oligonucleotide used as 'starter' in DNA synthesis	**O**	Promoter
16.	Transferring DNA fragments from gel to nitrocellulose filter	**P**	Genomic library

17.	Used to extract and isolate mRNA	**Q**	mRNA
18.	Binding site recognised by RNA polymerase	**R**	*Agrobacterium*
19.	Brought about by uptake of plasmid by bacterium	**S**	Nitrocellulose
20.	Used to insert DNA into animal cells	**T**	Transformation
21.	Clear area in bacterial culture produced by phage	**U**	Autoradiography
22.	DNA that induces tumour growth by *Agrobacterium*	**V**	Transgenesis
23.	Detects position of radioactive probe	**W**	Plaque
24.	Promotes transformation in bacteria	**X**	Calcium ions
25.	Transfer of genes between species	**Y**	T-DNA
26.	Used to retain DNA fragments in southern blotting	**Z**	Gene probe
27.	Part of a gene coding for amino acid sequence	**AA**	Exon
28.	Hypervariable region of DNA	**AB**	Minisatellite
29.	Repair of genes in gametes	**AC**	Lipofection
30.	Used to detect a specific base sequence among many others	**AD**	Plasmid
31.	Bacterial mini-chromosome used as cloning vector	**AE**	Germ line therapy

2. When transplanting a gene from a human to a bacterium, a scientist undertook a number of procedures. Rewrite the letters of the various procedures in the correct order.

A. Open plasmids and add linkers

B. Use gene probe or antibody to detect cells containing desired gene

C. Attach linkers to cDNA

D. Screen for antibiotic resistance

E. Lyse bacteria to extract plasmids containing antibiotic resistance gene

F. Add recombinant DNA to bacteria in presence of Ca^{2+} ions

G. Make cDNA copy

H. Make nitrocellulose replica

I. Insert cDNA into plasmid

J. Extract mRNA from tissue

3. Suppose you wanted to insert a section of human DNA into a plasmid. Why should you use the same enzyme to cut out the human DNA and to open the plasmid?

4. Apart from being radioactive, how does a DNA probe differ from normal DNA?

5. If an animal or plant gene were to be transplanted unaltered into a bacterium, it could not be expressed. Why?

6. Cystic fibrosis is a serious genetic disorder caused by a mutation where three bases have been deleted from a gene coding for a protein that controls the chloride ion balance in cells. American doctors have recently succeeded in curing mice with cystic fibrosis by introducing a normal gene into embryo mice by a virus while the embryos were still *in utero*.

a. Describe how enzymes would have been used to isolate the healthy gene and insert it into the virus.

It is hoped that one day, a similar technique will be used to cure human sufferers of cystic fibrosis.

b. Discuss why great care would need to be taken before this technique was carried out in humans.

7. Petunias are being used to study the effects of transgenesis in commercial plant crops so that genetic outcomes can be more accurately predicted. Researchers have used a viral vector to insert a gene involved in the production of purple pigment in petunia flowers.

a. Describe the techniques used in the process of producing large numbers of transgenic petunias.

b. Producing transgenic plants is not a predictable process. Give some reasons why.

The viral DNA replicates in the plant nucleus independently of the plant's chromosomes so there can be several copies of the inserted gene. The number of copies of the inserted gene corresponds to a novel spotting pattern on the flowers and this 'copy number' can be used to predict which plants will have the spotted pattern before they flower.

c. Explain how the copy number can be used to predict the plants that will have the spotting pattern on their petals.

d. Describe implications of this technology for those who grow economically important plant species.

Unit 12.4 Evolution

Topic 1: Theories of evolution

Topic 1 presents a background view of the concept of evolution.
The topic deals with:

- Lamarck's theory.
- Darwin's theory of evolution.
- Natural selection.
- Neo-darwinism.

Evolution – the idea of change over time

Over 1.5 million species of living things have been identified on planet Earth, and there are millions of different organisms still waiting to be discovered. There are various theories about how life on Earth started, but no one really knows the answer to this mystery. The term 'evolution' broadly means 'change over time'. In biology, **evolution** refers to all the changes that have occurred to living organisms since life began.

Before the publication of Charles Darwin's theory of evolution in 1859, people generally believed that:

- The age of Earth could be measured in thousands of years.
- Individual species are specially created and do not change.
- Any variations in individual species are imperfections.
- Observations of the living world should support prevailing world views.

Afterwards, from the 20th century to the present, people generally believed that:

- The age of Earth could be measured in billions of years.
- Individual species are related through lines of descent.
- The number of individual species does not change.
- Any variations in individual species are the result of random genetic changes or environmental conditions.
- Observations of the living world can be used to test hypotheses.

Evolution is the slow and continual change of organisms that occurs over time. It helps us to understand both the history and diversity of life on Earth. *Biological evolution* is defined as a heritable change in one or more characteristics (or traits) of a population or species from one generation to the next. The central idea is that all life on Earth shares a common ancestor.

Evolution is simply defined as 'descent with modification' – this means that all life forms are distantly related: humans and coconut palms, beetles and fishes.

Lamarck's theory of evolution

The French biologist Jean-Baptiste de Lamarck (1744–1829) once believed in the *fixity of species* (that is, that living organisms do not change). But after studying the succession of life forms in Earth's strata (layers of rock) and fossils (remains of once-living organisms), he concluded that complex organisms are descended from simple forms that lived in the past. He first hypothesised that evolution occurs and that **adaptation** to the environment was the cause of diversity. In 1800, to explain the process of adaptation to the environment, Lamarck proposed his theory of *inheritance of acquired characteristics* in which the *use and disuse* of a structure can bring about inherited change. This theory stated that phenotypic changes acquired during the lifetime of organisms were then passed to their offspring. His classic example was the long necks of giraffes. According to Lamarck, modern giraffes might have evolved from short-necked ancestors. After many generations of short-necked ancestors stretching to reach leaves on progressively higher branches, the slightly extended necks were passed on from one generation to the next, leading to the present day long-necked giraffes.

Through modern genetics, we now know that Lamarck's explanation of the evolution of the giraffe's long neck is *not correct*, because activities such as stretching to feed do not affect the gametes or sex cells (egg and sperm). Therefore, this type of characteristic/trait acquired during the life of an organism is not inherited by its offspring. For example, a body builder who spends much time and effort to develop his muscles will not pass the large bulging muscles (i.e. trait) to his offspring at birth.

Darwin's theory of evolution by natural selection

The *theory of evolution* and the phrase 'survival of the fittest' are synonymous with the British scientist, Charles Darwin (1809–1882). From 1831 to 1836, Darwin travelled the coast of South America and the Galapagos Islands in the Pacific Ocean aboard HMS *Beagle*. During this time, Darwin observed and collected many different forms of animal and plant life, including fossils. Darwin's observations led him to conclude that biological evolution does occur, and he proposed **natural selection** as the mechanism for adaptation – and hence, evolution – and the great diversity in living organisms. This idea led Darwin to propose the theory of evolution. His book, *On the Origin of Species by Means of Natural Selection, or the Preservation of Favoured Races in the Struggle for Life*, was published in 1859. It went against the prevailing beliefs at the time (mainly belief in Creation – that all living things were created and do not change). His book forms the basis of modern biology – it fully describes the evolution of biological organisms by natural selection.

Alfred Wallace (1823–1913) was another British naturalist who had observed the diverse insects, birds, and mammals of South America and Southeast Asia. He also, and independently, proposed that natural selection was the driving force of evolution.

In Darwin's theory, by contrast to Lamarck's acquired inheritance theory, the long neck in modern giraffes is *not* created by experience, but is the result of pre-existing variations (genetic differences) among individuals in a population. Originally, giraffe neck length varied but the struggle to exist caused long-necked giraffes to survive and have more offspring. Over time, by natural selection, most giraffes now have long necks.

Both Lamarck and Darwin agree that species change over time and that evolution is a slow and gradual process, but they differ in their views on the roles of the environment and organism.

Natural selection

Natural selection is the mechanism for effecting evolution. It is the *process* that results in **adaptation** of a population to the living (**biotic**) and non-living (**abiotic**) environments. Adaptation is any characteristic that makes an organism more suited to its ecological habitat. Adaptations can be *anatomical* (modification of anatomy or structural features of an organism), **behavioural** (ways an organism acts or responds) or **physiological** (biochemical processes inside an organism's body). Adaptations increase an organism's chance of survival, relating to protection, locomotion and obtaining food.

Natural selection results in the evolution of organisms well adapted to their environment. As a consequence, it enhances their survival and reproduction (termed **fitness**). Nature (biotic and abiotic environmental factors) selects organisms with better adapted traits for a particular condition and allows such adaptive features to become established in a population from generation to generation, while selecting against other traits. According to Darwin, the process of natural selection results in a population better adapted to its local environment than previous generations.

Evolution by natural selection is responsible for both the variation within populations and the great diversity of life on Earth. In terms of genetics, evolution can be defined as changes in a population over time due to the accumulation of inherited (genetic) differences. This definition explains the organism unity (shared traits due to a **common ancestry**) and organism diversity (presence of species unique to particular environments).

It is important to understand that the concepts of *natural selection* and *evolution* are not the same. Natural selection is a process that can lead to evolution or evolutionary change (the outcome), whereas evolution is the historical record of change through time. Natural selection results in adaptation to the environment among the agents of evolutionary change (see Topic 3, page 335). Adaptation in an organism evolves over time, and may take many generations.

Modern understanding of evolution

Neo-Darwinism is the modern version of Darwinian evolution through natural selection. It is the theory of organic evolution by natural selection of inherited characteristics.
Neo-Darwinism combines Darwin's *theory of evolution* and *Mendelian genetics of inheritance* (see Unit 12.3, page 163). Mendelian inheritance specifies that evolution involves the transmission of characteristics from parent to offspring and explains how inheritable variations can arise by mutation. The science of genetics (study of inheritance) was not known during Darwin's era (or Lamarck's), so he could not explain his *variational inheritance*.

Evidence from other scientific fields, such as molecular biology (study of molecules of life, or biochemistry), paleontology (study of fossils), ecology (study of organisms and their environment) and ethology (study of behaviour) supports different aspects of the theory. Therefore to accept the neo-Darwinism evolutionary theory, it is necessary to:

- Establish that evolution (change) has taken place in the past (past evolution).
- Demonstrate a mechanism that results in evolution (natural selection of genes).
- Observe evolution happening today (evolution in action).

Evolution can be viewed on a small scale (*microevolution*) as it relates to changes in a single gene or allele frequencies in a population over time. Or it can be viewed on a larger scale (*macroevolution*) as it relates to the formation of new species or groups of related species.

Unit 12.4 Evolution

Topic 2: Evidence of evolution

Topic 2 presents an overview of how the theory of evolution is supported by these areas of study:

- Fossil evidence.
- Comparative anatomy (convergent and divergent).
- Comparative embryology.
- Comparative biochemistry.

Fossil evidence for evolution

Palaeontology is the study of past life on Earth from fossils. Fossils are the traces of dead organisms. They may be whole organisms or impressions of organisms on rock, whole body parts or fragments, droppings or even animal footprints left in an ancient mud that has turned to rock.

Usually, when an animal dies, its body decomposes. The soft body parts break down first and then, although much more slowly, hard skeletons and shells disappear. Normally, organisms eventually die and vanish without trace. Nevertheless, under unusual circumstances, part or all of an organism's body may be preserved as a fossil.

Fossilisation can occur in many different ways. On very rare occasions, the whole animal may be preserved. Mammoths have been found in Arctic ice and insects have been trapped in the sap of ancient pines and preserved when the sap turned to amber. Fossils more often form when bones, wood or shells are quickly covered by layers of sediment. The living material is later replaced by minerals and a fossil replica of the original is produced naturally.

Most fossils are found in **sedimentary rock**. Sedimentary rocks are produced in layers as sediments fall one on top of the other. The age of one layer compared with another is easy to determine. The lowest sediments, which form the bottom layers, must be the oldest, and the highest sediments, which are closest to the surface, must be the youngest – unless earth movements, such as folding, have occurred. Hence, the fossils in the lowest rocks are the oldest and the fossils in the highest layers are those of the most recent organisms. Nowhere in the world is the layer of sediment deep enough to show an unbroken sequence from the first organisms on Earth to those of the present. All evolution did not take place in one place but at different sites across the Earth's surface. However, the sequence of fossils within every set of undisturbed sedimentary rock shows a change from simpler organisms in the oldest rocks to more complex organisms in the youngest rocks.

By comparing layers of sediment from different parts of the world, the sequence from the very earliest life to the present can be observed. In this way, sedimentary rocks provide a record in their fossils of the history of life on Earth. This sequence of fossils is called the fossil record. The fossil record clearly shows a change from simple to complex and from aquatic to terrestrial.

The fossil record is incomplete, and is likely to remain so, for two reasons:

- Fossilisation is a rare event; therefore, only a minority of organisms are likely to leave any trace.
- Although fossils of soft-bodied organisms are found, organisms with hard body parts are much more likely to be fossilised.

Nevertheless, many fossils have been found that appear to have the characteristics of two different groups of organisms. These fossils are called *transition fossils* and they provide a testimony to the path of evolutionary change. The evolution from ancient reptiles to mammals is particularly well documented by a series of transition fossils. Indeed, they merge so closely that it is difficult to decide whether some fossils are those of a mammal-like reptile or a reptile-like mammal.

Two of the most famous transition fossils are *Archaeopteryx*, which provides evidence for the common ancestry of birds and reptiles, and the lobe-fin fish, which indicates that amphibians may have evolved from fish.

Millions of years ago	Age	Major events	
2	Age of mammals	'Human' civilisation dominates Flowering plants diversify	
12		Ancestral 'humans' evolve Flowering plants dominate	
135 181	Age of reptiles	Placental mammals appear Huge dinosaurs dominate the land Birds evolve	
230		Mammals evolve Small dinosaurs appear	
280		Abundant life on land as well as in the oceans	
345	Age of amphibians	Reptiles evolve to breed on land Flight evolves in insects	

405 425	Age of fishes	Vertebrates colonise the land Evolution of land plants	
500 600	Age of invertebrates	Evolution of vertebrates Evolution of invertebrates with hard skeletons	
4500		Simple organisms diversify in the sea Origin of primitive life	

The major events in evolution according to the fossil record.

Most fossils are similar to present-day organisms. For example, a number of fossils similar to the modern horse have been found. The similarity in the skeletons of both modern and extinct horses can be best explained in terms of a common ancestry and evolutionary change.

Comparative anatomy

The similar anatomy of many groups of organisms provides further evidence of evolution. A particularly compelling example is provided by the forelimb (*pentadactyl limb*) of many vertebrates.

In the pentadactyl limb, the same basic skeletal structure has been adapted for a number of uses ranging from running to grasping and flight. Similar structures in a variety of organisms that are due to a common ancestry but are used for different purposes are called *homologous structures*.

Comparative embryology

The embryos of different vertebrates are very similar. It was once thought that the embryo's development (*ontogeny*) repeated the stages of evolutionary change. We now realise that this is not so. The similarity of the embryos is a complex example of homology. Homology refers to similarities in organisms resulting from descent from a **common ancestor**. This means that the structures of the embryos appear to be very similar in anatomy but the structures carry out different functions as they develop into the organs of the body. This indicates that these vertebrates have descended from a common ancestor.

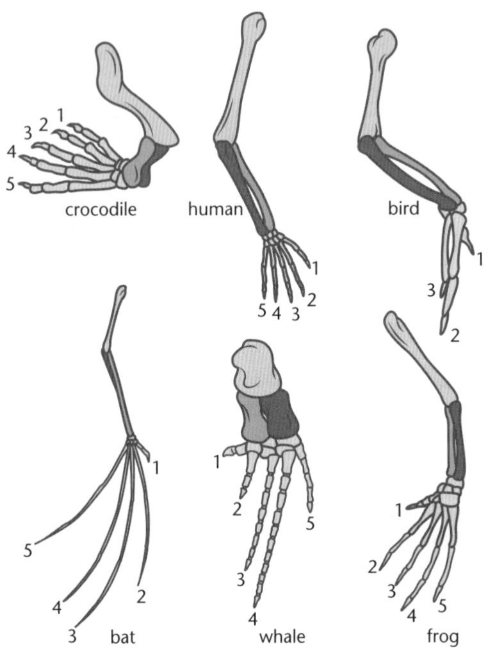

Homology: the pentadactyl limb (forelimb in chordates) (Note: Matching bones are indicated by colour; matching fingers are numbered.)

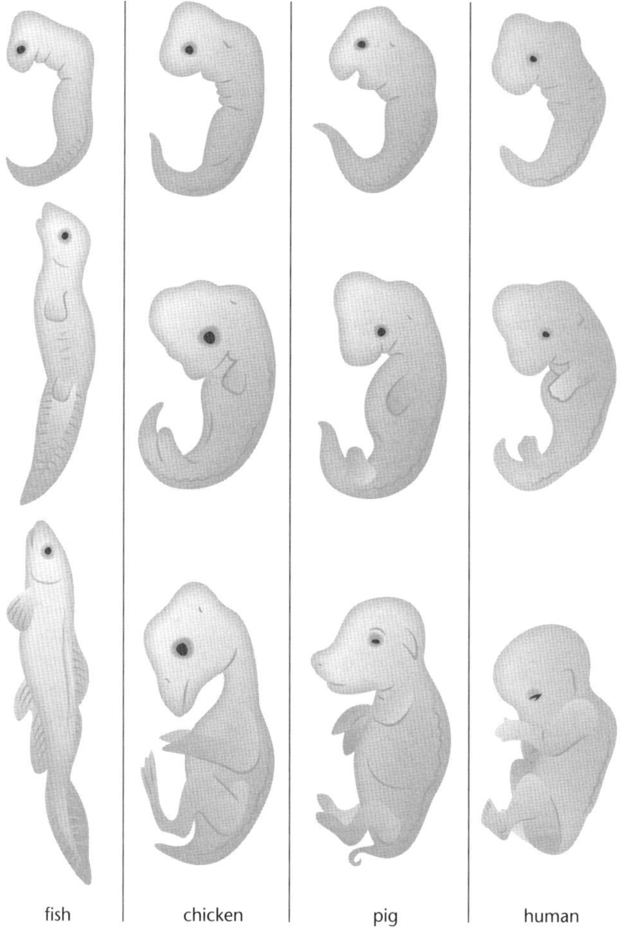

Stages in the embryonic development of vertebrates

Comparative biochemistry

The fundamentally similar biochemistry of all living things has already been mentioned. They all:

- Consist primarily of organic compounds.
- Share a common genetic code of DNA or RNA.
- Rely on enzymes to control chemical reactions.
- Share the same cell membrane structure.
- Rely on respiration to make energy available for cellular processes.

Proteins, in particular, have proved to be a fruitful source of evidence for evolution. Because proteins are made up of distinct amino acid units, linked in a specific sequence, the similarity of proteins in different organisms can be determined by comparing the amino acid sequences in the proteins. Many proteins, such as those in haemoglobin and cytochrome-c, have been studied.

The most useful proteins for this type of study are those that occur across a wide range of organisms.

Cytochrome-c, for example, is a protein needed to make energy available in virtually all living things, from bacteria and fungi to complex plants and animals. It consists of a chain of about 104 amino acids. An analysis of this protein from a variety of organisms demonstrates the similarity of the amino acid sequence. The degree of similarity of this, and many other proteins, not only provides evidence for evolutionary relationships in general, but also helps to establish the pathways along which evolutionary changes may have occurred.

Unit 12.4 Evolution

Topic 3: Mechanisms of evolution – genetic changes in populations

Authors: Takis Solulu and Terry Bunn

Topic 3 applies the principles of genetics and inheritance (Unit 12.3) to build an understanding of evolutionary change. It is essential to know that the understanding of organism diversity and adaptations relies on the concepts here.

The topic deals with:

- Natural selection.
- Mutation.
- Genetic drift.
- Gene flow.
- Non-random mating.
- Artificial selection.

Evolution in a genetic context

The biological concepts and processes relating to genetic change are:

- Natural selection.
- Mutation.
- Genetic drift (founder effect, **bottleneck effect**).
- **Gene flow** = migration (immigration, emigration).
- Non-random mating (**sexual selection**).

These processes affect the gene pool of a population and will change the frequency of the alleles in that gene pool. Therefore, *genetic change refers to the change in the frequency of alleles in the gene pool of a population*.

Alleles are expressed phenotypically either because they are homozygous (*AA* or *aa*) or because they are dominant in the heterozygote (eg *A* would be expressed in *Aa*, *a* would not).

The frequency of an allele in a population is calculated from:

$$\text{Frequency of an allele} = \frac{\text{Occurrence of that allele}}{\text{Total number of alleles}}$$

Example

Calculating allele frequencies

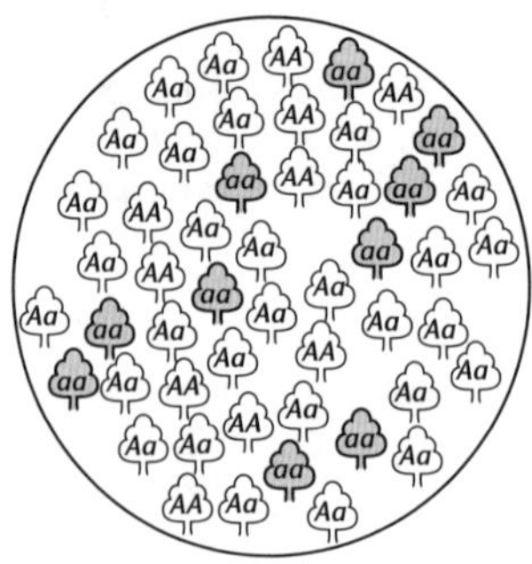

'Gene pool' of 50 individuals – 30 ***Aa***, 10 ***AA*** and 10 ***aa*** (40 'white trees' (from 30 ***Aa*** and 10 ***Aa***) and 10 'grey trees').

Since the gene pool contains 50 individuals, 30 of whom were ***Aa***, 10 ***AA*** and 10 ***aa***:

- For the ***A*** allele – it occurs 30 times from ***Aa***, and 20 times (10 × 2) from the 10 ***AA*** individuals for an occurrence of that allele of 50.
- There are 50 individuals, each with two alleles, so that the total number of alleles is 100 (2 × 50).
- The frequency of the *A* allele [*f*(*A*)] is thus: $\frac{50}{100} = 0.5$
- Similarly, $f(a)\ \frac{50}{100} = 0.5$

Evolution

Evolution is the process by which new species of organisms develop from earlier forms.

The process of evolution normally occurs slowly, most often in response to a change in a species' environment.

Despite the slowness of this change and the huge number of failures that inevitably must have occurred, life is thought to have evolved from just a few original unicellular organisms some 3 billion years ago to the complex array of millions of species we see around us today.

Evolution proceeds by *changes in the frequency of alleles in a population* – some alleles 'do better' than others.

Populations are the units that evolution acts on – ie it is not individuals that evolve, but populations.

Natural selection

The theory of natural selection was proposed by Darwin over 150 years ago. It remains the best explanation of adaptive evolution.

Populations typically produce more offspring than the environmental resources can maintain; therefore, there is **competition** for survival. Individuals with the best adaptations will survive and reproduce (this is what is meant by '**fitness**') and pass on to their offspring their 'successful' alleles. The frequency of these 'alleles' will then increase in the gene pool. Environmental factors (both biotic and abiotic) act as selecting agents of successful **phenotypes**. When environmental factors change, different phenotypes will be selected for. As phenotype is largely determined by genotype, successful genotypes will have their alleles increase in frequency in the gene pool.

The terms 'good' and 'bad' genes/alleles are often used in this context. There are a number of possible effects of 'good' and 'bad' alleles on the gene pool.

Mutations

Mutations are *the source of new alleles in the gene pool*; therefore, mutation is essential for evolution. If a mutation occurs in the gametes or gamete-producing cells, then that mutation will enter the gene pool and become subject to natural selection.

- 'Good' alleles resulting from favourable mutations will increase in frequency in the gene pool as they will be selected for.
- 'Bad' alleles resulting from unfavourable mutations are selected against, so are unlikely to become established (ie unlikely to occur at a high frequency) in the gene pool.

Some mutations may be neutral or 'silent' and not subject to selection, so the frequency of the mutated allele in the gene pool will therefore be due to chance.

Example

Certain changes in the bases of DNA do *not* cause a change of amino acid.

The substitution of the last base in the triplet GGG to GGC would still produce the amino acid proline in the protein being made.

Mutations alter allele frequencies by changing one allele to another, sometimes providing new phenotypes for natural selection to act on. Many mutations do not pass to the next generations.

Genetic drift

- Genetic drift, founder effect and the bottleneck effect all tend to reduce genetic **biodiversity**.

Genetic drift is the change in allele frequencies in populations due to *chance* (not selection) – it may include the *loss* of alleles from the gene pool. Genetic drift is most likely to have an effect in *small* populations.

When populations are large and mating is random, allele frequencies tend to remain stable from generation to generation unless natural selection is operating. However, when populations are small or become small, it is likely that allele frequencies will change from generation to generation by chance and so *drift* into an increase or decrease (or even complete loss). This has nothing to do with natural selection.

Example

Effect of genetic drift on allele frequencies

Consider a 'population' of just one couple, both of whom are heterozygous for brown eyes. ***B*** = brown eyes is dominant to the recessive ***b*** = blue eyes.

P: *Bb* × *Bb*

G: *B* or *b* ; *B* or *b*

	B	*b*
B	*BB*	*Bb*
b	*Bb*	*bb*

Imagine the couple have *two* children.

- The chances of one child being ***BB*** is $\frac{1}{4}$.
- The chances of both children being ***BB*** is $\frac{1}{4} \times \frac{1}{4} = \frac{1}{16}$.

There is therefore a $\frac{1}{16}$ probability of losing the ***b*** allele altogether by chance alone when the two only members of the population have two children.
The more children produced, the lower the chance of losing the ***b*** allele.
The chance of losing the ***b*** allele if they had three children is $\frac{1}{4} \times \frac{1}{4} \times \frac{1}{4} = \frac{1}{64}$.
Consider a 'population' of two heterozygous couples:

- The chance of one family having two children both of whom are ***BB*** is $\frac{1}{4} \times \frac{1}{4} = \frac{1}{16}$.
- The chances of the other family having two children both of whom are ***BB*** is also $\frac{1}{16}$.

The chances of losing the ***b*** allele from the population if all four children just happen to be ***BB*** is therefore $\frac{1}{16} \times \frac{1}{16} = \frac{1}{256}$.

The chance of losing an allele by chance alone (*genetic drift*) therefore *decreases* as the population size increases.

Evolution does not necessarily require one allele be completely lost – any allele changing its frequency means that evolution is occurring.

Genetic drift is effective only in small populations. Species of **flora** and **fauna** introduced into new locations are typically only *small numbers* released. These small original populations are therefore subject to genetic drift.

Founder effect

A founder population refers to a small group of individuals that colonises a new isolated area such as an island. The range and frequency of alleles present in this small group is unlikely to be representative of that of the original population. Some alleles may not be present in this group; others will be less frequent or more frequent.

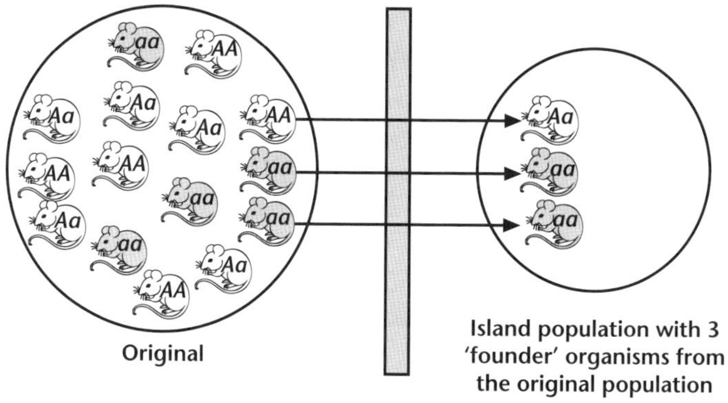

$f(A) = \frac{15}{30}$

$= 0.50$

Where:

- Numerator
  = (5 *AA* + 5 *Aa*) ***individuals***
  = 5 × 2 + 5 × 1
  = 15 *A* alleles.
- Denominator
  = (5 *AA* + 5 *Aa* + 5 *aa*) individuals
  = 5 × 2 + 5 × 2 + 5 × 2
  = 30 alleles in total.

$f(A) = \frac{1}{6}$

$= 0.17$

Where:

- Numerator
  = 1 *Aa* individual
  = 1 × 1
  = 1 *A* allele.
- Denominator
  = (1 *Aa* + 2 *aa*) individuals
  = 1 × 2 + 2 × 2
  = 6 alleles in total.

As a founder population is small, it is likely to:

- Be subject to **genetic drift**.
- Have a reduced range and frequency of alleles.

Thus evolution is likely to occur at a faster rate in a founder population than in the original or other populations.

In extreme cases, a founder population may be a single individual.

Founder populations are unlikely to have the range and frequency of alleles present in the populations they originated from. They therefore have the potential to become quite different from the original populations and their evolution is likely to progress faster, as natural selection from the new environment operates.

Bottleneck effect

Populations may be suddenly reduced in numbers to a small size. This typically occurs as a result of:

- A catastrophic environmental event (eg flood, fire, landslide, drought) – indiscriminately removing individuals regardless of their genetic make-up.
- Human action (such as rapid habitat destruction or introduction of predators or competitors).

As population numbers drop rapidly to low levels, it is likely that the *range of alleles will decrease and the frequencies of alleles will change*. If the population increases again to its original size or becomes even larger, it will have reduced genetic biodiversity.

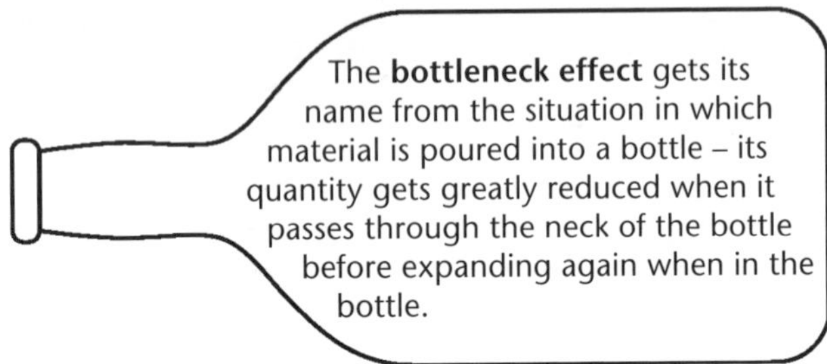

Bottleneck effect prevents the majority of genotypes from participating in the production of the next generations.

The bottleneck effect can remove an allele from a population:

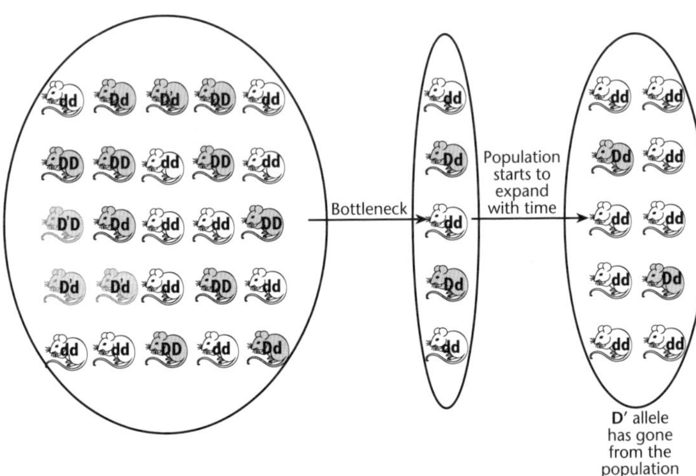

Gene flow

Gene flow (also called *gene migration*) refers to the movement of alleles between populations, as occurs when individuals migrate (immigrate or emigrate) from one population to another and breed in that new population.

- **Emigration** may remove alleles from a population, reducing a population's genetic biodiversity.
- **Immigration** may add new alleles to a population, increasing a population's genetic biodiversity.

Immigration

Immigration involves organisms migrating into a population.

The immigration of individuals into populations has the potential to markedly affect small isolated species.

Example

Effects of immigration on allele frequencies

Imagine a (mouse) population of 10 individuals, all heterozygous for a particular characteristic ie all *Aa*):

$$f(a) = f(A) = \frac{10}{20} = 0.5$$

Where:

- Numerator is obtained from 10 *Aa* individuals $10 \times 1 = 10$ a or *A* alleles.
- Denominator is obtained from 10 *Aa* individuals $10 \times 2 = 20$ total alleles.

Suppose 10 new individuals immigrate into the population, all homozygous for ***a*** (ie all ***aa***):

$$f(a) = \frac{30}{40} = 0.75$$

Where:

- Numerator is obtained from 10 ***Aa*** individuals + 10 ***aa*** $10 \times 1 + 10 \times 2 = 30$ *a* alleles.
- Denominator is obtained from 10 ***Aa*** individuals + 10 ***aa*** individuals $10 \times 2 + 10 \times 2 = 40$ alleles altogether.

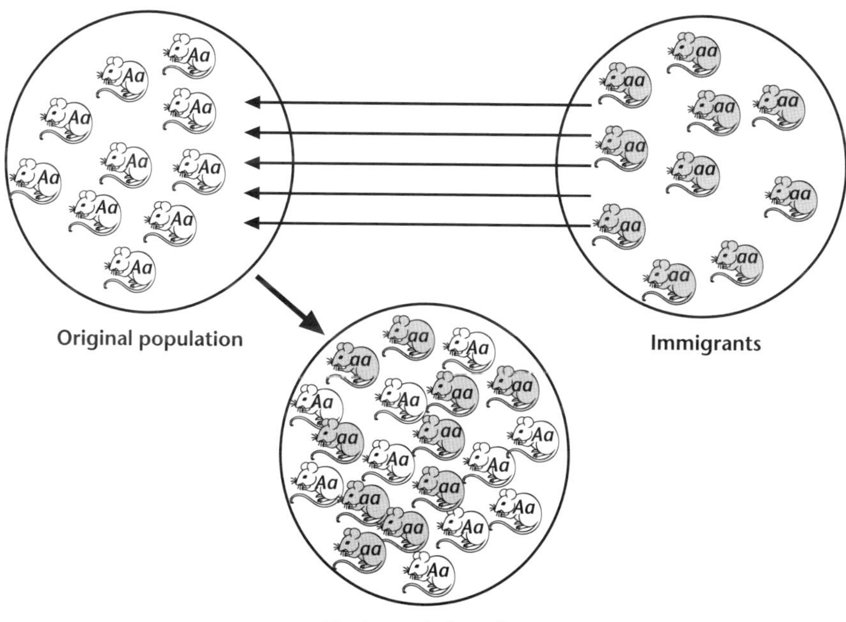

The frequency of the ***a*** allele has changed from 0.5 to 0.75 and the frequency of ***A*** allele from 0.5 to 0.25.

Emigration

Emigration involves organisms migrating out of a population.

Example

Effects of emigration on allele frequencies

Consider a (mouse) population. It has 20 individuals – 10 ***Aa*** and 10 ***aa***. If 5 of the ***Aa*** left (emigrated), the new population would be 5 ***Aa*** and 10 ***Aa***:

$$f(A) = \frac{5}{30}$$

$$= 0.17$$

Where:

- Numerator is obtained from 5 ***Aa*** individuals $5 \times 1 = 5$.
- Denominator is obtained from 5 ***Aa*** individuals + 10 ***aa*** individuals $5 \times 1 + 10 \times 2 = 30$ alleles in total.

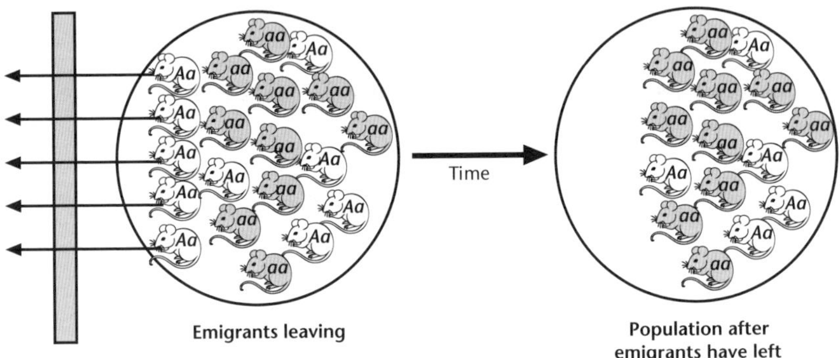

The frequency of the *A* allele has changed from 0.25 to 0.17. Any change in allele frequency is the first step in the evolutionary process.

Gene flow allows for gene flow between populations. Gene flow may change the frequency of alleles in populations, but prevents close adaptation to a local environment.

Non-random mating

Non-random mating is when individuals pair up, not by chance, but according to their genotypes or phenotypes. **Inbreeding** (ie mating between relatives) is an example of non-random mating. Inbreeding decreases the proportion of heterozygotes and increases the proportions of homozygotes in a population.

Sexual selection

A form of non-random mating, **sexual selection** is a special case of natural selection, in which one sex (usually the female) acts as the selecting agent. The males in the species 'advertise' their fitness in some suitable way (eg displays, vocalisation, trials of strength) and the female selects the 'most impressive' male as her mating partner – he has the 'best' genes and therefore his genes will enter the gene pool of the population in his offspring and become more frequent.

Example 1

Sexual selection in Birds of Paradise ('kumul')

Male kumuls have bright and colourful feathers while females are not that colourful. Females prefer bright and showy plumage so they select a male displaying the brightest and colourful plumes to mate. A showy plumage in male kumuls directly increases a male's chance of reproducing.

Example 2

Sexual selection through a dominant stag

In red deer, males (stags) 'roar' to attract a harem (group) of females (hinds). By being strong enough to ward off other stags by fighting and displaying, stags 'prove' their worth to prospective hinds. A female will mate with a dominant stag because he has 'proven' genes and she will 'want' the same good genes for her offspring to help increase their chances of survival. The genes that enabled the stag to survive are those that now appear in large amounts in the next generation. Evolution has occurred through changes in allele frequencies.

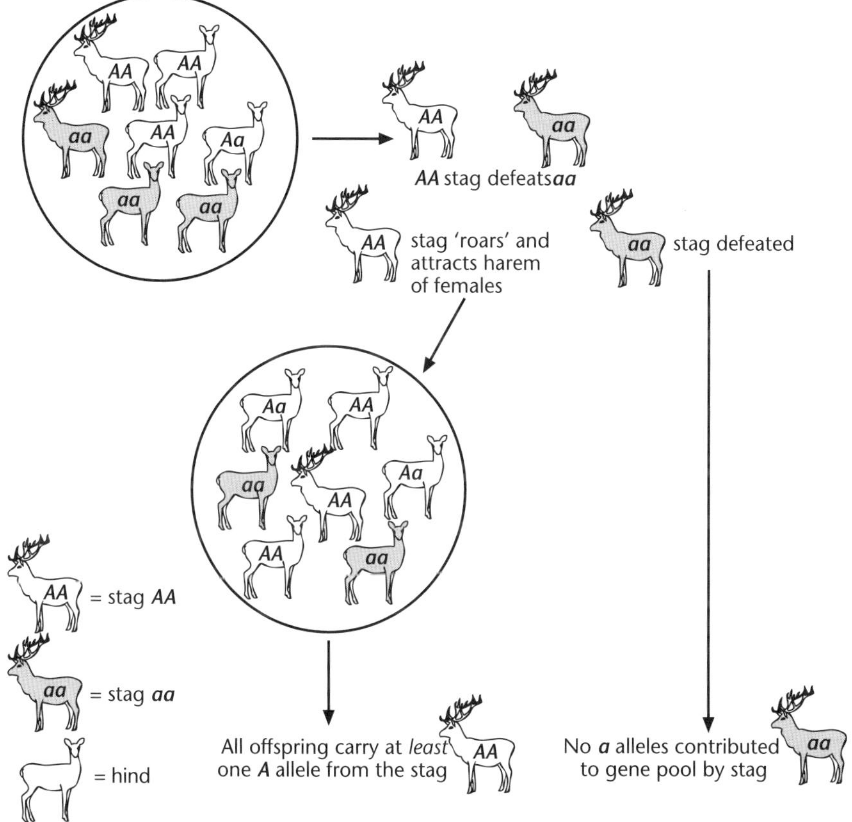

Note the the *A* allele has increased in frequency not because it is dominant but because it was selected for.

Artificial selection

Artificial selection (also known as **selective breeding**) contrasts with natural selection in that *humans* (not the environment) select the individuals that will breed. We do this because certain plant or animal individuals have features that we find desirable and by breeding from them, we increase the probability that their offspring will have these desirable features (eg faster racehorses, greater milk yield in dairy cows, less fatty meat in pigs, crisper apples).

Humans have been doing artificial selection since the domestication of animals and cultivation of plants began about 10 000 years ago.

By artificially selecting characteristics beneficial to humans, plant and animal gene pools have changed dramatically, to the extent that today some species are radically different from their ancestral type.

By starting with small original gene pools, not allowing immigration, selecting only for specific characteristics and enforcing their own rules of sexual selection, humans have developed **breeds** within species.

Example

People living in Herefordshire in England decided their 'ideal' type of cattle had to be a rounded, squat, beefy animal with a white face, a white stripe down the back, with the rest of the body being a burnt orange colour. These cattle are now known as Herefords. Conversely, breeders in County Ayr in Scotland wanted a larger, rangier animal that produced lots of milk. Picking individuals with brown spots scattered across a white body produced the breed now known as the Ayrshire. Similar scenarios explain Angus cattle, Friesians, Jerseys and other cattle breeds.

Artificial selection is much more rapid then natural selection at producing different phenotypes within a species. One of the reasons for this is **inbreeding** – the breeding together of *closely related* individuals to increase the chances of getting the desired phenotype. However, inbreeding has the great disadvantage that it also increases the chances of harmful recessive alleles coming together and being expressed in the phenotype. Many of these alleles have become established at high frequencies in the gene pools of our domestic breeds.

Example

The domestic dog *Canis familiaris* has over 400 identified genetic disorders as a result of persistent inbreeding.

In the natural environment, individuals with genetic disorders would be less fit as their chances of surviving and breeding would be reduced. Therefore, their harmful ('bad') alleles would become reduced in frequency in the gene pool. *Outbreeding* (ie breeding with *unrelated* individuals) reduces the chances of harmful recessive alleles coming together. Regular outbreeding is necessary in selective breeding programmes to reduce the chances of genetic defects becoming established in the breed/species.

Unit 12.4 Activity 3A: Genetic change

1. **a.** Explain what is meant by 'natural selection'.
   b. Distinguish between natural selection and artificial selection – use examples to illustrate your answer.
   c. Male peacocks have very large, brilliantly coloured tails that they use in sexual displays. The tails hold no advantage for the males in terms of camouflage, defence against predators, flight, etc. Discuss why such tails have come into existence in peacocks.
2. **a.** Distinguish between the terms 'gene pool' and 'gene flow'.
   b. Explain how the frequency of alleles in the gene pool may change.
3. Discuss how new alleles are created and become established in the gene pool.
4. Describe the main ways in which genetic variation occurs in a species.
5. **a.** Define the term 'genetic biodiversity'.
   b. Explain the importance of genetic biodiversity to the survival of species.
6. Zoos often have breeding programs to assist the conservation of endangered species. In these programs, new males (or their semen) are often brought into the zoo as part of the breeding program. Discuss why this is done.
7. **a.** Distinguish between the terms 'founder effect' and 'bottleneck effect'.
   b. Define the term 'genetic drift'.
   c. Explain the link between genetic drift and both the founder effect and bottleneck effect.
8. Which of the following evolutionary agent(s) would cause phenotypic changes in a population?
   A. Mutation.
   B. Gene flow.
   C. Non-random mating.
   D. Genetic drift.
   E. Both A and C.

Supplementary Unit: Investigations

Topic 1: Investigations

This supplementary unit looks at how to carry out a practical biology investigation with direction, by looking at:

- Planning a practical biology investigation.
- Valid and reliable results.
- Variables.
- Processing data.
- A valid conclusion.
- Evaluating an investigation.

These guidelines are in line with the learning outcomes and assessment tasks outlined in the syllabus.

Planning a biological investigation

A biology investigation is a method of finding an answer to a question or testing out an idea.

Things you might want to investigate.

A quality plan

A quality plan will result in a valid conclusion being reached. It will have the following components:

- A statement of purpose (this comes from the question or idea) – this should be an aim, a question for investigation or an idea to test, a prediction or an hypothesis.
- Variables – identified and controlled.
- A step-by-step method that will provide valid and reliable results to address the aim.

As scientists, we want to be as certain as possible that our findings are correct. We want results that are *valid* and *reliable*.

Valid and reliable results

In biology investigations, the plan needs to be designed to give valid and reliable results.

Valid and reliable results come from plans that:

- Change only one variable at a time.
- Control all the other variables in the experiment.
- Use measuring instruments that will give accurate results.
- Have repeat trials until consistent results are found.

Example

How does exercise intensity affect heart rate?

This question attempts to find out what happens to heart rate when a person exercises at various intensities.

The investigator must *compare* the heart rate of a person at *rest* and when *exercising* at different intensities.

It is important to only change *one factor* at a time in the investigation to be sure of the findings.

It is important that all the other factors in the experiment are kept the *same* – this means 'controlling' all the other variables.

It is also important to check that the findings are correct by performing *the same tests several times*. When each test shows the same pattern of results, the investigator can be fairly certain they are correct.

Variables

Three types of variable must be identified in every investigation:

- **Dependent variable** – the factor to be measured to find an answer to the question.
- **Controlled variables** – the factors to be kept the same to make the experiment a **fair test.**
- **Independent variable** – the only factor that will be changed as the experiment is carried out. There must be a range for the independent variable, ie there must be three or more values. For a quality investigation, the range must be *valid*, ie the range must contain four or more values *and* be relevant to the organism being studied.

Example

Effect of light on tomato seeds germination

A student wanted to answer the question 'Do tomato seeds germinate faster in light or dark conditions?' This investigation cannot be carried out as there are only two values for the independent variable – light and dark. The question will need to be modified so that at least three light intensities are used.

Example

Posture and heart rate

A student wanted to answer the question 'How does body position (posture) affect heart rate?'. The aim of the investigation is 'To investigate the effect of body position on heart rate'. The following table shows the variables for this investigation.

The independent variable	Body position – standing, sitting, lying down, hanging upside down.
The dependent variable	Heart rate (beats per minute).
The controlled variables	• Same person used to do all the tests. • Clothing worn by the person doing the tests. • Temperature of the room tests are carried out in. • Method used to measure heart rate in each test. • Time in each body position before heart rate is taken.

The range is valid as it contains a range of four body positions that fully test the question.

Example

Effect of temperature on slater respiration

A student wanted to investigate if slaters respire at a greater rate as environmental temperatures increase. The student used a range of temperatures that included 40°C as the highest temperature – all the slaters in this temperature environment died. The temperature range was not valid as slaters are not naturally found living at such high temperatures. In addition, it is not desirable to cause the unnecessary death of organisms.

Asking questions – deciding on an aim

- Must define the question being asked and turn that question into an aim.
- An aim is 'what we want to find out'.
- From the aim it should be possible to define the dependent and independent variables.
- Be sure the aim makes it clear that *only one factor will be changed* in the investigation.
- Be sure the aim makes it clear *what will be measured to find the answer* to the question.

An aim is often started with the words:

To investigate the effect of ______________________

or

To find the effect of ______________ on ______________

or

To find out what happens if ______________________

An aim is not written as a question – it *does not* have a *question mark* at the end of the sentence.

Supplementary Unit Activity 1A: Investigations in biology

1. Complete the table to turn each question into an aim. The first one has been done for you; **a–c** are partly done, and **d–g** must be 'done from scratch'.

Question	Aim
Jane is a 17-year-old in Grade 12. She wants to know 'If her heart rate changes when she exercises'.	To investigate the effect of exercise on heart rate in a 17-year-old female.
a. Does the enzyme salivary amylase digest starch faster in acid, alkaline or neutral conditions?	To investigate the effect of pH on the speed at which the enzyme amylase ______________ .
b. Does temperature affect how quickly the enzyme salivary amylase digests starch?	To investigate the effect of ________ on the rate at which salivary amylase enzyme ________ starch.
c. Do tomato plants grow at different rates in light of different colours?	To find the effect of light __________ on the rate at which ___________ ______ grow.
d. Do all disinfectants have the same effect on the growth of bacteria on agar?	
e. Which exercise increases your heart rate the most?	
f. Do marigold seeds germinate faster if they are kept at a higher temperature?	

2. Complete the table below to identify the variables from the aim. (These are all investigations you should be able to plan without the help of a teacher.)

Aim	Independent variable[1]	Dependent variable[2]	Controlled variables[3]
a. To investigate the effect of temperature on the rate at which sugar cubes dissolve in water.			
b. To find out which type of exercise increases the heart rate the most.			
c. To find out which type of fertiliser results in the fastest growth in bean plants.			

[1]What you will change.
[2]What you will measure to find the answer to the question.
[3]What you will keep the same to make sure you have a fair test.

Writing a plan

A plan is a step-by-step description of what you will do to carry out the investigation. Once you have decided your aim and variables, writing a plan is straightforward.

Writing your plan as a series of bullet points makes it easier to follow and easier for you to check.

Example

Is the germination of tomato seeds affected by light intensity?

Aim	To find out if the speed of germination of tomato seeds is affected by light intensity.
The independent variable	Intensity of light.
The dependent variable	The time taken for the tomato seeds to germinate
The controlled variables	• The type of tomato seeds used. • The age of the tomato seeds used. • The type of container the seeds are placed in. • The type and amount of soil the seeds are placed in. • The amount of water supplied to the seeds. • The temperature of the area the seeds are placed in. • The atmosphere – air in the surroundings.
Plan	• Take 25 small seed trays and plant 10 tomato seeds of the same type in moist seed-raising mix in each tray. Ensure the same amount of seed-raising mix is used in each tray. • Place 5 pots on the lab bench with a box covering the pots. • Place 5 pots beside the box – but leave uncovered. • Cut the bottoms out of three other boxes. Attach layers of shade cloth to the inside so that 25%, 50% and 75% of the normal light levels are let through. • Place 5 pots under each of these three boxes. • Check the moisture level of the trays each day – ensure they remain moist by watering when necessary. • Check the seeds every day for 5 days, recording the number germinated each day in a table.

Supplementary Unit Activity 1B: Planning an investigation

1. Vagi is a Grade 12 Biology student. Her class has been studying microbes at school. At home she is responsible for cleaning the bathroom and kitchen floors each week. Vagi has suggested to her parents that if she increased the recommended concentration of disinfectant, she may need to wash the floors less often. She decides to investigate using the techniques she has learnt about in her Grade 12 Biology class.

   In this investigation you are to develop a plan, process information, and present a report that will enable Vagi to answer her question.

Science concept

A disinfectant works by interfering with the cell processes of bacteria.

Effectiveness of disinfectants in inhibiting bacterial growth can be determined by the size of the clear area (ie free of bacteria) around a filter paper disk soaked in the disinfectant and placed on an agar plate.

A Petri dish with bacteria and disinfectant needs to be left for at least 48 hours to allow bacteria to grow.

An example of an experimental plate after culturing with bacteria is shown.

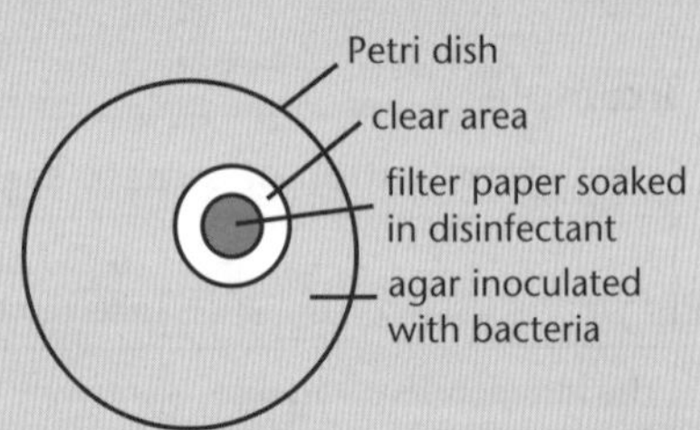

Planning the investigation

Plan an investigation that will enable Vagi to find the answer to her question.

The following resources are available for your use:

- Four nutrient agar plates that have been sub-cultured with bacteria.
- Filter paper.
- Disinfectant.
- Water.
- Tweezers.
- Marker pens.
- Tape to seal the agar plates.
- Incubators to grow bacteria set at 25°C.

Fill in the Planning form following to help you plan your investigation.

Check your plan using the Planning checklist provided.

a. Planning form

Aim:	
Variables:	
Independent variable (What I will change)	
Dependent variable (What I will measure to find the answer to my question)	
Controlled variables (What I will keep the same to make sure I have a fair test)	

b. Plan

Use a series of bullet points to write your step-by-step plan.

- Make sure the plan shows how you will control all variables to make it a fair test.
- Use diagrams if necessary.
- Make sure you show how you will collect enough data to be certain of your findings.

Remember to check your plan against the Planning checklist.

Planning checklist

If you find you cannot tick a given requirement then you will need to change your plan or add to your plan.

Have you included the following in your plan?	Yes	No
An aim that states very specifically what you are going to investigate.		
The independent variable (what you are going to change).		
A range of values for the independent variable (ie the range over which you will change the independent variable).		
The dependent variable (what you are going to measure to find the answer to your question).		
How you are going to measure the dependent variable – what will you record in the results section of your report.		
The control variables – all the factors that you will have to keep the same to ensure that you produce valid results (fair test).		
How you will keep the factors the same.		
How you will ensure that your results are reliable.		
Description of a scientific method for collecting your information/data that could be followed by another person.		

Gathering and processing data

Gathering data is about making *measurements* and *observations*.

Processing data is about putting measurements and observations into *tables*, *charts* and *graphs* that help to show patterns and trends in the data.

Types of data

Quantitative data involves counting or measuring using *numbers*.

Qualitative data involves *making observations* that are based on *descriptions* rather than measurements.

Ranked data involves *ranking* a set of information from one end of a scale to the other, eg Very dark green → dark green → green → light green → very light green.

'Good' data gathering will:

- Have measurements and/or observations.
- Have accurate measurements.
- Have specific observations.
- Provide enough data to make sure you can be certain of the results by showing the same trends or patterns in several repeats.

Example

Investigating the 'Effect of Temperature' on the ability of the enzyme pepsin to digest egg white protein

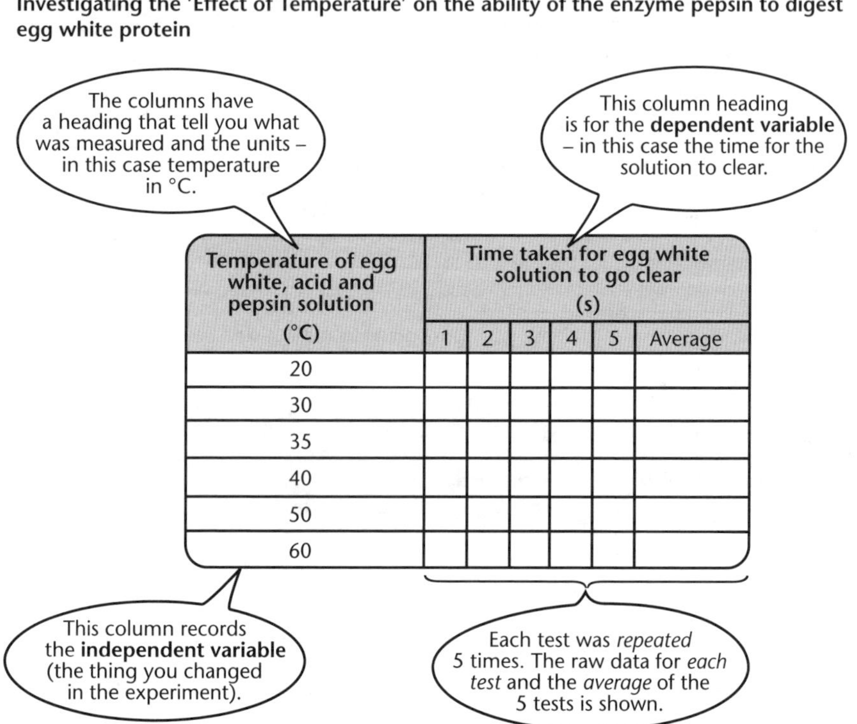

Temperature of egg white, acid and pepsin solution (°C)	Time taken for egg white solution to go clear (s)					
	1	2	3	4	5	Average
20						
30						
35						
40						
50						
60						

Graphs

Line graphs

Line graphs are used when you have continuous data for both variables (ie the x- and y-axes), eg 'Temperature vs Rate of reaction' or 'Time vs Height of a plant'.

Features of a good line graph:

- *Crosses* – always plot the points with a cross.
- *Labels* – label both axes. Include units with the labels.
- *Even* – the numbers on the axes must go up evenly (eg 5, 10, 15, 20, 25 *not* 5, 15, 18, 20, 22).
- *Axes* – the variable that was measured must go on the y- or vertical axis; the variable that was set in the experiment must go on the x- or horizontal axis.
- *Title* – the title needs to 'say what the graph is about'. (A good standard title to remember is 'The effect of __________ on __________')
- *Smooth curve* – always draw a smooth curve or line joining the points.

Example

A typical line graph

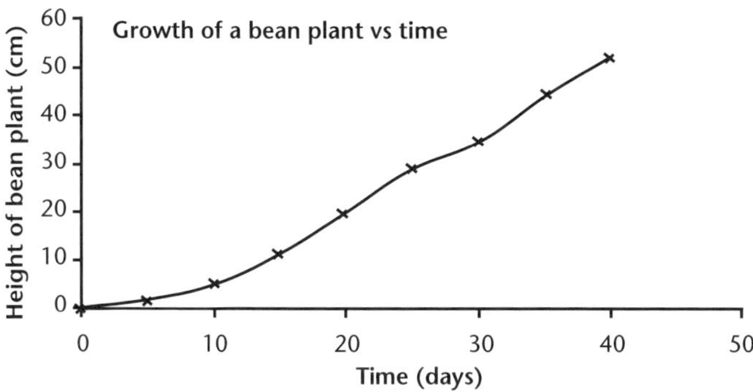

Bar graphs

Bar graphs are used when you have at least one set of data that is **discrete** (not continuous). Discrete data is grouped into categories, eg 'Area of ground covered by different plant species'; 'Variation in characteristics in students'.

Features of a good bar graph:

- Bars – even width, not touching.
- Labels – both axes labelled; units included if a measurement has been made.
- Axes – bars can go horizontally or vertically.
- Title – must say 'what graph is about'.

Example

A typical bar graph

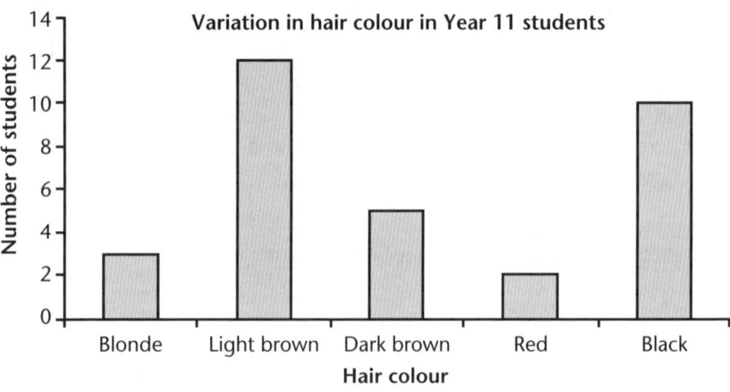

Histograms

Histograms are used when you have continuous data but are plotting frequency, eg 'Heights of students in a class'.

Example

A typical histogram

Heights were measured to the nearest centimetre.

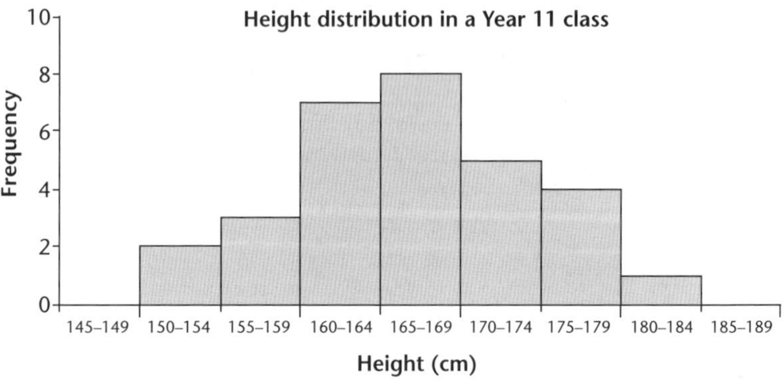

Interpreting and reporting

Your report should allow an independent person to understand what you did and what you found out. It should contain the following sections:

- *Aim* (or purpose).
- *Hypothesis* (sometimes it is not possible to have an hypothesis).

- *Method* – a step-by-step set of instructions that could be followed by another student so they could repeat your work.
- *Results* – tables/graphs/observations; must be processed by calculations (eg averaging) or graphing. (For 'Merit' and 'Excellence', record enough trials to show you have consistent results.)
- *Conclusion* – must include interpretation of processed data and will have to be linked back to the aim.
- *Discussion* – results of the investigation are explained in terms of the biology relating to what is being investigated.
- *Evaluation* – involves thinking about how sure you can be that your results *are* valid and reliable. The evaluation does *not* involve describing how, if you had more time or resources, you might have been able to improve your investigation. If you are sure you have controlled all the variables well and performed enough repeats to be certain of your findings, explain this.

In your evaluation you should also discuss the following:

- How did you effectively control all your variables?
- How did you minimise any possible sources of error (ie things that were difficult to control)?
- How are you sure you did enough repeats so your results are reliable?
- Were there any results that did not seem to fit in with the general patterns and trends? How do you explain them or justify ignoring them in your data processing?
- If you encountered any problems in your investigation, how did you overcome them?

Supplementary Unit Activity 1C: Gathering, processing and interpreting data, and reporting

1. Martha and Elizabeth carried out an experiment to investigate the effect of different types of exercise on the heart rate of Martha. They recorded their results in a table.

	Trial 1		Trial 2		Trial 3	
Type of exercise carried out by Martha	Initial heart rate at rest (bpm)*	Heart rate after 3 minutes of exercise (bpm)	Initial heart rate at rest (bpm)*	Heart rate after 3 minutes of exercise (bpm)	Initial heart rate at rest (bpm)*	Heart rate after 3 minutes of exercise (bpm)
Skipping	68	130	68	134	68	128
Step-ups	68	128	68	132	68	128
Star jumps	68	142	68	146	68	148
Walking	68	85	68	88	68	82

*bpm is 'beats per minute'.

a. Calculate the average for each set of results.

b. Present the average results (calculated above in **a.**) in a graph.

2. Sivai and Jason carried out an investigation to find out the effect of temperature on the time the enzyme pepsin took to digest egg white. They used a maximum time of 10 minutes for each test. They recorded their results in a table.

Temperature of solution (°C)	Time taken to digest the egg white (colour went from cloudy to clear) (s)				
	Trial 1	Trial 2	Trial 3	Trial 4	Trial 5
10	Did not change	Did not change	Did not change	Did not change	Did not change
20	303	290	310	298	290
25	150	148	154	149	151
35	25	22	19	20	24
45	16	18	6	19	17
55	62	67	60	64	66
70	Did not change	Did not change	Did not change	Did not change	Did not change

a. The students only planned to collect data from four trials. Explain why you think they changed their minds and did an extra fifth trial.

b. Average the data and produce a graph.

3. Serena and Laura were investigating the effect of different types of swim stroke on heart rate. Serena swam 50 m of each stroke at maximum effort, with a 10-minute break between each 50 m length. Her heart rate was measured using a heart-rate monitor attached to her wrist as she swam. Serena read the heart rate off the monitor watch at the end of each length. The girls checked that Serena's heart rate was back at resting rate 10 minutes after each effort. She repeated the test on each of 3 days.

- On the first day, Serena's resting heart rate was 68 beats per minute. After 50 m freestyle it was 120; after breaststroke it was 89; after butterfly it was 134; after backstroke it was 118.
- On the second day, her day her resting heart rate was 67. After 50 m freestyle it was 121; after breaststroke it was 86; after butterfly it was 134; after backstroke it was 114.
- On the third day, her day her resting heart rate was 67. After 50 m freestyle it was 118; after breaststroke it was 90; after butterfly it was 130; after backstroke it was 117.

a. Write an *Aim* for this investigation.

b. What is the independent variable in this investigation?

c. What is the dependent variable in this investigation?

d. What factors did Laura and Serena keep the same (control).

e. Put the data into a table.

f. Present the data as a graph.

g. Write a conclusion / discussion for the investigation.

4. Pepsin is an enzyme found in the human digestive system that breaks down proteins into polypeptides in the stomach. Jason and Kila set up an investigation to find out what pH the enzyme pepsin worked best in. They used egg albumin as the protein. Because pepsin is found in the stomach, they thought that the enzyme would work fastest in acidic conditions of around pH = 3.

a. What is the *Aim* of this investigation?

b. What was Jason and Kila's hypothesis?

> The students put 2 mL of egg white solution in each of 24 test tubes.
>
> Their teacher provided them with four solutions of pH 1, 3, 5 and 7. The students put 1 mL of each solution into six test tubes.
>
> They added a set measure of pepsin powder to three of the six tubes for each pH.
>
> All the tubes were placed in a water bath at 37°C and a stopwatch was started.
>
> The students timed how long it took for each tube to become clear.

c. What was the independent variable?

d. What was the dependent variable?

e. Why did Jason and Kila have one set of each type of tube that they did not add pepsin to?

f. What factors did they control?

g. Create a table that Jason and Kila could record their results in.

Supplementary Unit: Investigations

Topic 2: Carrying out investigations

This Topic follows Topic 1 in this supplementary unit and examines how to carry out a practical investigation into an aspect of the ecological niche of an organism with guidance by:

- Developing a quality practical investigation into an aspect of the ecological niche of an organism.
- Presenting a report with a comprehensive discussion on the investigation.

Introduction

The accumulation of biological knowledge and understanding has come about through investigation. Scientists build on what has been done before them in order to come to a common viewpoint of the biological world. Scientific research invites the student to make his or her contribution, albeit small, to this pool of knowledge through investigation into an aspect of the ecological niche of an organism.

The *student* must develop, carry out and analyse the results of the investigation with *guidance only* from their teacher. Guidance means the teacher cannot tell the student what to do or how to do it, but is simply there to:

- Provide important parameters for the investigation, such as time frames, necessary equipment, or the organism(s) available for study.
- Suggest possible resources or sources of information.
- Provide feedback on what the student intends to do or has already done.

The teacher *cannot* provide assistance in the form of providing templates for the structure of the investigation or scientific report, pointing out errors and what should be done to rectify them or suggesting alternatives.

Aspect 1 – Carrying out the investigation

Potentially, this can be the most difficult and longest aspect of an investigation. It involves two main steps.

Step 1: Developing the investigation

You must:

- Develop an understanding of the **ecological niche** of the organism – research and find out some facts about the organism. If the aspect to investigate is known, research can be targeted and concise; if unknown, information may need to be found out about many aspects of the niche.
- Formulate a **hypothesis** that could lead to an investigation.
- Develop a *plan* that allows sufficent, relevant *data* to be gathered to support or reject the hypothesis. Trialling before you carry out your final plan may be required. **Independent** and **dependent variables** (including a *range* for the independent variable), **controlled variables**, method and *data collection* and details relevant to the hypothesis must be covered. A *conclusion* is necessary.

Step 2 – Collecting, recording and processing data

If the investigation has been planned well, this aspect should be fairly straightforward (although it may be time-consuming and repetitive). You must:

- *Collect, record* and *process* data that is relevant to the purpose.
- Have a *logbook*. This shows evidence the investigation has been an ongoing entity that you have developed and changed. An exercise book with bound pages that won't fall out is ideal. Each time an entry is made, date it. As the research is conducted, jot down any and all ideas, and possible questions and hypotheses you may want to investigate. The logbook is a work-in-progress and change is to be expected.
- Use appropriate *calculation(s)* (eg means) and or *graphing* to enable a trend or pattern to be seen.

Aspect 2 – Presenting a report

This is where everything is brought together to present the findings and conclusions. You must:

- Present in the final report the written results of aspects 1 and 2.
- Include an introduction, hypothesis (or purpose), final method, the recorded and processed data, a results conclusion and discussion, and a bibliography.

A correct conclusion based on the processed data that relates to the purpose of the investigation should be presented.

The discussion must link the response or behaviour of the organisms to the aspect of the ecological niche under investigation (ie a 'big picture' approach).

A comprehensive discussion requires some sort of statistical analysis of the validity of the data that lead to the conclusion or justification as to why the data was reliable within the limits of the sources of error, bias or limitations identified.

Carrying out the investigation

When doing investigative work, especially when working with biological material, it is important to note:

- It takes time. Investigative work is a process that cannot be done in a few days. Give yourself plenty of time to work through this process and don't leave it until the week before the report is due.
- Things don't always go to plan – be prepared for unexpected results. Analyse carefully and make any changes necessary to have a successful outcome. You need to be very familiar with the organism(s) you are working with in order to plan an investigation that is relevant and meaningful. If working with vertebrate animals, ethics approval must be established and obtained, which is a very rigorous process that can put you off before you even start. Close monitoring of the investigation is involved to ensure there is no unnecessary pain or suffering to the animal.

Choosing the organism

This is the first step to be undertaken.

- *Fungi* and *microorganisms* allow investigations to be easily and quickly carried out in a test tube, Petri dish, or under a microscope. Microorganisms include algae, protozoans, fungi, bacteria and viruses. There are strict guidelines for using certain bacteria due to possible pathogenic effects. Microorganisms from human or animal sources must not be used, apart from those found on skin, and these cultures must remain sealed at all times. Toilets and toilet areas, rubbish bins and drinking taps must not be used to obtain microorganisms. Strict guidelines exist for the culture and disposal of cultures. Microbial culture should only be carried out with appropriate supervision.

Because it can be pathogenic, naturally occurring *Escherichia coli* must not be used (genetically crippled strains obtained from a supplier can be used).

Protista (such as *Paramecium* and *Amoeba)* and yeast are easily obtainable from suppliers.

- *Plants* are generally easy to care for and grow, seeds are easy and cheap to come by, and many respond well to manipulation of variables to give obvious and easily measurable responses such as germination or growth; however, they can be time- and space-consuming to keep. Consider the **photoperiodic** needs and the life cycle of the plant used – if this is too long, you may not get results until it is too late.
- *Animals* can be very interesting (particularly if the investigation involves an interaction or relationship), can give very rapid responses, and are generally available all year round. Companies breed and sell invertebrates and their young for investigative purposes. Small invertebrates (eg slaters, snails and brine shrimps) are easy to find and look after.

Extension: Animal ethics in developed countries

An animal is classified under a country's Animal Welfare Act for the purposes of animal ethics as 'horses, cattle, sheep, goats, any bird, any marine mammal collected from on or in the vicinity of the seashore, or any vertebrate animal kept in captivity and dependent upon man for its care and sustenance'. Before *manipulating* any such animal in an experimental way, approval must be obtained from an Animal Ethics Committee in order that an animal is not exposed to unnecessary cruelty, pain or distress. Simple studies that do not involve the manipulation of the animal (eg observing behaviour or diet preferences, or regular weighing to plot a growth curve) do not require ethics approval.

An animal cannot be exposed to hunger, thirst, or malnutrition, or distressed to the point it cannot display normal patterns of behaviour. An animal's normal physiology, behaviour or anatomy cannot be deliberately interfered with, either by exposing it to some type of substance or organism (eg a parasite), subjecting it to some type of enforced activity or restraint, or depriving it of its usual care. Animals must not be taken from the wild if purpose-bred animals are available (most indigenous animals are protected so it is an offence to take these from the wild). Cockroaches, hedgehogs, possums or rodents caught in the wild cannot be used due to the risk of disease, infections or parasites they may transmit to humans.

Obtaining animal ethics approval does take some time. A very detailed proposal needs to be written and sent to the committee. Try to avoid using the animals that come under ethics approval, but if you do wish to use such an animal, allow plenty of time for the process. Even if lowly invertebrate animals such as blowflies or slaters are used, these animals are not exactly volunteers in an investigation, do have a purpose to their existence, and so warrant such care and regard.

Becoming familiar with the study organism

It is essential to become familiar with an organism and its ecological niche before carrying out an investigation – this involves:

- Knowing its *usual habitat* – this makes it possible to look after the organism correctly so its responses or behaviours are a result of the investigation and not of physiological stress as a result of containment conditions; *and* one aspect can meaningfully be manipulated to investigate behaviour or responses (if this is the type of investigation carried out).
- Selecting a *relevant aspect* of the niche to investigate.

Example

There is no point doing an investigation on the effect of rocky substrates on tomato growth, as tomatoes do not normally grow in rocky substrates so this experiment would not be relevant.

- Having a good idea of how the organism *should behave* or *respond* as a result of your investigation and what sorts of behaviour or responses to look for to make some sort of measurement.

The research into the organism and its ecological niche needs to be included in the final report in the *Introduction*. Research includes information found in texts, web sites and other sources, but also from observation of your organism and possibly from trial experiments. There are a number of aspects to concentrate on:

- The **abiotic** factors that affect the organism – eg temperature, light intensity, humidity, exposure, substrate, nutrients and water. Gather both **qualitative** (descriptive) and **quantitative** (numerical measurements) data for the relevant factors to care for the organism correctly, determine an appropriate range for the independent variable, and correctly control other variables not under investigation.
- The **biotic** factors that affect your organism – eg distribution, density, age structure, competitors, predators or prey. This information is particularly important if the investigation involves an interaction or relationship.
- The adaptations the organism has in order for it to live successfully in its habitat – eg features related to feeding or nutrition, gas exchange, reproduction, movement, growth, sensitivity and avoiding competition. This is particularly important for the *Discussion*, where the results are related to the ecological niche of the organism.

Formulating a purpose

Generally, a purpose is a **hypothesis** and should be written as such. A hypothesis is a *prediction statement* precisely detailing the expected outcome of the investigation:
If this happens, *then* 'that is predicted to happen'.

A hypothesis must be easily testable by experimentation and it should be obvious what needs to be manipulated and what needs to be measured.

Gather data to support or reject the hypothesis, *not* to prove or disprove it. A precise, well-written hypothesis will make the development of the plan much easier and should specify both the independent and dependent variables.

Examples

'Increasing the density of planting of tomato plants (*independent variable*) will result in decreased average mass of plants (*dependent variable*)'.

'The average size of a bird nesting site decreases (*dependent variable*) with increasing size of the nesting population (*independent variable*)'.

The investigation can be:

- Of a manipulative nature – one variable is changed (the independent variable) and some sort of behaviour or response is measured as a result of this manipulation (the dependent variable).
- An interaction or relationship 'in the field' under natural conditions.

For either type of investigation, there must be a *range* for the independent variable.

Example

Counting the numbers of aerial roots of mangroves in a wet substrate and a dry substrate in a mangrove swamp is *not* a range. The number of aerial roots in substrates of at least three of different moisture levels would need to be counted.

Designing a method

Developing a sound method is not difficult provided two things are kept in mind:

- *Keep it simple*. A good method is easy to follow by someone else, uses readily available equipment, and yields results straightforward to analyse and interpret.
- *Ensure you are very clear about what is being measured and what must be manipulated* to get these measurements. Then it is simply a matter of filling in the detail about specific procedures that would allow someone else to replicate the investigation, and what variables need to be controlled.

Be prepared to do some *trialling* of the initial method and make necessary changes. These changes may be so major the investigation must be entirely rethought. Record *all* trials, *changes* and *reasons* for changes in the logbook. Reasons for changes must be scientifically sensible (running out of time 'because you had to go to work' or 'the agar jelly did not set' are *not* good reasons to make changes). Persistence is a key factor in scientific endeavour!

When writing the final report, the evaluation – of the validity of the investigation – where the method is justified – could be based on these trials.

Example

Why the response of slaters to humidity was measured in a particular way would come from the trialling to find the most valid way to measure the response.

Characteristics of a valid method

Variables

The independent variable

This is the *one* variable that is manipulated. It must have a *range* (a minimum of three), and the range must be *appropriate* (suitable for the organism being studied). The range for the independent variable is something that should be established through trialling.

Example

If using a range of salt solutions, some of the more concentrated solutions may result in death of the organism, so these solutions would be considered to be outside of the appropriate range that should have been used.

The dependent variable

This is the variable that will change, as it depends upon the independent variable. Changes need to be measured or sampled in some way. Be very specific about *what the dependent variable is and how it will be measured or sampled.*

Examples

If the dependent variable is plant growth, then it must be specified exactly what is meant by *growth*. Will height, length, number of leaves or dry tissue mass be measured? Each of these can be used as a measure of plant growth, but some may not be appropriate for the type of investigation.

If investigating the effects of planting density on plant growth, the number of leaves produced would *not* be appropriate as a measure of plant growth.

A common mistake with plant investigations is to confuse germination and growth – these are quite different. With investigations involving animal behaviour it will be necessary to quantify the behaviour in some way to avoid *anthropomorphism*. You cannot measure 'how an animal is feeling' or 'what it likes', so some way must be found for objectifying these qualities, such as setting an animal a 'task' and counting how many animals are within each category for the range.

Controlled variables

These are all the other variables that must remain constant for a fair test. They must be identified, and how they were controlled specified. They include details such as how to prepare dilutions of chemicals, amount and type of substrate to use, how ambient (environmental) temperature was controlled, gender or age of the animals used, etc. Inadequate or incorrect control of these variables may mean an investigation is biased and the results may not be valid.

Assumptions

These are variables outside your control *despite your best efforts*, or you can justify that they will make little or no difference to the outcome of the investigation.

Examples

Common to most organisms is the genetic variation between individuals. This cannot be controlled, but steps can be taken to ensure differences in behaviour or responses due to genetic variation are minimised, such as by having an adequate sample size or repeats.

The physiological state of organisms at the time of an investigation can vary, despite the person doing the investigation having taken all reasonable steps to ensure that organisms are contained until use in conditions that cause the least amount of stress as possible so behaviour or responses are affected only by the independent variable.

Sufficiency of data

There are several ways of ensuring sufficient data is collected. The way(s) chosen will depend on:

- Constraints imposed by practical issues (eg equipment or time available) – it is usually difficult to justify not collecting sufficient data because of this.
- The statistical test used to evaluate the reliability of the data.

Example

If the chi-squared test is used to analyse the data, to ensure sufficient repeats or sample size, the *expected* count in each category must be no fewer than five.

- The amount of variation between individuals and the nature of the organisms being investigated. If there is a high degree of variation, it is more correct to have a large sample size than lots of replicates of a small sample size.

Example

There tends to be more variation between, and less reliance on, instinctive behaviours in more complex organisms (animals), so it would be appropriate to ensure a large sample size for such organisms.

Establishing sufficiency is something you should do as part of your trialling. Record all the results of these trials, so sufficiency decisions can later be justified.

Extension: Precision, accuracy and reliability

Obtaining precise and accurate data are tests of the soundness of the method used.

- *Precision* refers to the reproducibility of the data – how close repeated measurements are to each other in value. The higher the precision of the data, the lower the standard deviation. Correct experimental techniques and accurate instruments, low personal error on the part of the investigator, and correct sampling methods (including adequate sample size and ensuring randomness of the sample), all contribute to high precision of collected data.
- *Accurate* data is that where the measured or derived values are close to their true (mean) value. Investigations that have been set up to be *unbiased* give accurate data.

Together, precise and accurate data can be used with confidence to support or reject a hypothesis and allow valid conclusions to be drawn. Think of precision and accuracy in terms of throwing darts at a dartboard. If you throw 10 darts and every single one of them hits the bull's-eye, then you are both accurate (the darts have hit the intended target) and precise (the darts are very close together on the dartboard).

Every investigation is a compromise between high precision and high accuracy, so *statistical analysis* is the final step in ensuring that data is reliable.

Bias

Data obtained must be a result of the true response of the organisms because of the independent variable.

A biased investigation is one where the data obtained *could* reflect the true response of the organisms to the independent variable or *could* result from poor experimental design or technique.

A fundamental premise of statistical analysis is that all variation is random – if an experiment is set up or the results obtained in a non-random or biased way (whether consciously or unconsciously), then this condition is not met.

Bias usually results from one or more of:

- Not having a truly non-random sample.

Examples

If looking for slaters to use in a behaviour experiment that involves movement of the slaters, choosing only the slaters that move and ignoring the ones that stay still means the sample may be non-random, and therefore, biased.

If choosing pea seeds for a germination experiment, it would be incorrect to choose those seeds that are not wrinkled or choose only the wrinkled ones.

A large sample size may still be non-representative and so, biased.

- Generalising from a non-random sample. It is *not* incorrect to have specific requirements for experimental organisms (eg gender, height, eye colour), as long as you can justify your reasons. It is incorrect to attempt to apply general statements and conclusions about behaviour and responses to all organisms of a specific group when you have not used a random sample.
- Non-random allocation of subjects to treatments. Using random numbers is a good way of overcoming this.

Example

Say you had 5 treatments and a sample of 100 seedlings to allocate over the 5 treatments. Using a list of random numbers (such as 5280713694, where 0 and 1 were treatment 1, 3 and 4 were treatment 2, and so on), you could allocate each seedling to a treatment depending on which random number it is (randomly) assigned.

- Biased experimental technique. This can be simple human error (eg if a measuring tape was used to measure the circumference of tree trunks, *not* stretching the tape to the same degree each time would cause bias). The converse of this is improvement in experimental technique with time and practice – this can skew results. (This is rather like making pancakes – the best pancakes are those towards the end of the mixture when the pan is at its hottest and your technique is near-perfect!)
- The non-random allocation of treatments to plots, equipment, rooms (environment) or time. A way of overcoming this problem is to use *Latin blocks*.

Example

If growing groups of plants in an area of soil outdoors, divide up the plot into squares depending on how many treatments you have.

A	B	C	D	E
E	A	B	C	D
D	E	A	B	C
C	D	E	A	B
B	C	D	E	A

With this block arrangement you can have 5 treatments and 5 replicates. The same technique can be used to block experiments temporally (ie over time) to reduce the effects of changing ambient conditions as the day progresses.

- Not rigorously identifying and controlling all variables other than the independent variable.

Examples

Increasing the light intensity may also cause a rise in temperature or a decrease in humidity. If steps are not taken to control these two factors, then it would not be valid to make conclusions about the effects of light intensity on a behaviour or response.

In studying the effect of wavelength ('colour') of light, an experimenter must ensure the intensity has not been changed.

Controlling the physical conditions of the experiment is particularly important because one **stimulus** may influence the way in which an animal or plant responds to another.

Example

Daphnia ('water fleas') tend to move more strongly towards light when carbon dioxide levels in the water are higher.

- *Pseudoreplication.*

Example

Having 4 water baths each containing 3 replicates of each of 4 treatments is pseudoreplication – true replicates would be carried out independently of each other, in separate water baths, and preferably at different times (to reduce any possible bias as a result of changing experimental technique or ambient conditions).

- Processing data where data for one treatment is rounded up while for another it is rounded down.

Planning for statistical analysis

You need to state in your method how the data is to be analysed. There is no point designing an investigation, carrying it out to obtain data, and *then* discovering that the data cannot be analysed statistically for reliability.

For some statistical tests such as the chi-squared or Poisson distribution, adequate sample size is important so that the required minimum number of expected counts in each category is possible.

The student's t test has limited use because it needs to show a *range* for the independent variable, meaning simple choice chamber experiments involving two categories are not appropriate.

ANOVA is a useful statistical test for investigations involving a range, but cannot be used on data that does not have a normal distribution. If doing a one-way ANOVA, at least two replicates for each treatment are needed.

Measuring or sampling

An essential part of the method often forgotten or not given in sufficient detail is exactly what data will be collected and how they will be collected. Consistency is very important.

Example

If the purpose of an investigation is to measure negative phototaxis ('avoidance of light') in slaters, it would be incorrect to make measurements of rate of movement (which would be used to measure **kinesis** or movement behaviour). Similarly, it would be incorrect to measure length of seedlings if the effect of salinity on germination was being investigated.

Detail must be specified on:

- *What* will be measured (eg height of tomato stems from soil level to the growing tip, numbers of slaters in each temperature category, number of mangrove pneumatophores in a range of substrates).
- *When* it will be measured (eg each day for one week, every 15 minutes for 2 hours).
- What the *unit* of measurement will be (eg cm, seconds, square metres).
- *How* it will be measured (eg using a tape measure so that the zero mark is at soil level, lifting the lid of a choice chamber just enough so that slaters could be counted, counting the number of mangrove pneumatophores in each quadrat in each substrate type).
- How the raw data will be processed (eg the heights of plants or numbers of slaters in each treatment were averaged over the 5 replicates, one-way ANOVA was used to establish if the results were significant).

Be sure to record *all* raw data in your logbook.

Method structure

A good method is so well structured that an independent investigator could follow it without needing any further clarification. Methods can be written as a series of numbered steps or as paragraphs, normally in the past passive tense. Whichever style is chosen, ensure the steps are in a logical sequence of events.

Example

Details of a specific chemical preparation would come before instructions on what to do with it (avoid detail, however, on preparation of commonly used chemicals such as 10% sodium hydroxide).

Diagrams are very useful in conveying information, particularly about specialised equipment constructed for the purpose of the investigation. Follow the guidelines for drawing scientific diagrams. Diagrams should complement written instructions rather than replace them.

Avoid repetition of the same instruction for replicates; condense wherever possible without losing important details. While spelling and grammar *per se* are not assessed, they do contribute to the overall readability and quality of the final product. Spend time checking for errors or better still, ask someone else to proofread material for you, particularly a non-scientist. If they can understand it, that is a good sign. When writing the final method for the report, write a first draft then come back to it after a few days. Errors or better ways of writing something will be glaringly obvious!

Collecting, recording and processing data

Collecting data

Data should be collected from the first time you start studying your organism. Observations of behaviour while in containment are data. So too are measurements of abiotic factors during containment. Data is not restricted to measurements made at the time of the actual experiments or sampling. Record *all* raw data in your logbook, including date, time, units, and so on in a *retrievable* format (ie it should 'stand alone' and be so complete in detail that no reference to the method is needed for it to be interpreted).

Recording data

Raw data

This is the unprocessed data recorded in your logbook and possibly as part of an appendix if you wish to extract only that raw data you have used for your final report. Raw data should never appear in the main body of the final report. Even though it is in a raw state, it still needs to be recorded in a retrievable way. This means that it should 'stand alone' and be so complete in detail that no reference to the method is needed for it to be interpreted. Raw data is normally presented in a table or (tally) chart format and has the following features:

- It has a full and informative *title*.

Example

'The effect of temperature on the heart rate of *Daphnia*' says a lot more than 'Heart rate results'.

- Each column and/or row has a full and informative heading, including correct, unabbreviated units (eg Temperature (°C)). Ensure the correct symbols for units are used (eg s not secs for seconds, mL not ml for millilitres). Normally the values for the independent variable are arranged first (whether as a column or a row, depending on the orientation of the table) followed by the measurements for the dependent variable.
- The data is aligned according to place-holders (eg any decimal points run vertically under each other and are thus aligned) and given to the same number of decimal points or significant figures. Units are not repeated next to each value (these should be in the headings).
- The table or chart is fully *enclosed* with neatly ruled lines.

Processed data

Initial processing of data (eg calculations of means, averages or standard deviation) should be recorded in a table or chart. Processed data used to draw a graph to illustrate a trend or pattern should appear in the main body of the report under the heading 'Results' *before* the graph. All other processed data tables, including those for statistical analysis, should be in the logbook and in an appendix.

Example

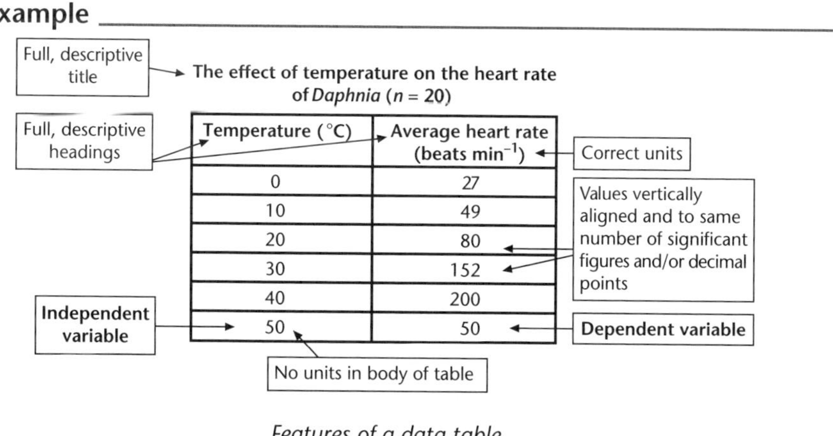

The effect of temperature on the heart rate of *Daphnia* (n = 20)

Temperature (°C)	Average heart rate (beats min^{-1})
0	27
10	49
20	80
30	152
40	200
50	50

Features of a data table.

Processing data

Statistics are used to clarify data and to help its interpretation in two ways:

- *Descriptive statistics* are concerned with graphical representation of data to facilitate their interpretation.
- *Inferential statistics* use mathematical techniques in order to draw conclusions from the data.

Data is of two kinds:

- *Continuous data* consists of *measurements.*

Example

Measurements of leaf width are continuous data because each piece of data can take any number of values between the narrowest and the widest.

- *Discrete data* consists of *counts* because there are no intermediates between adjacent values (a nest cannot contain 2.5 eggs!).

Graphs

Line graphs

Line graphs are used for continuous data and show how one quantity is affected by another.

Example

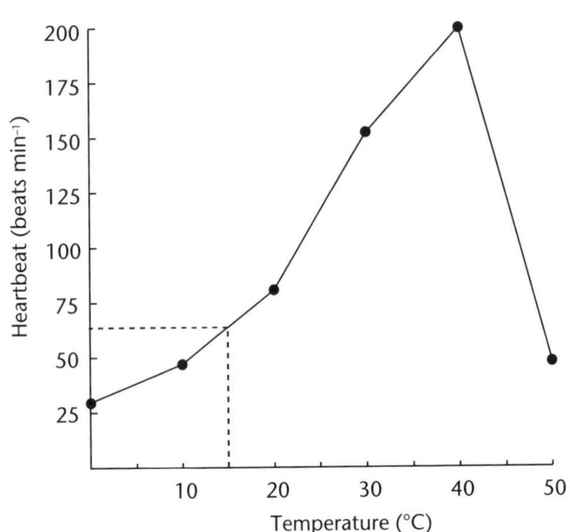

Graph showing the effect of temperature on heart beat in Daphnia.

Heart rate is the *dependent variable* because it depends on temperature, which is the *independent variable.*

The following points should be considered when plotting a line graph:

- Give it an informative title.

Example

'Graph showing the effect of temperature on the rate of heart beat in *Daphnia*' says a lot more than 'Heart beat graph'.

- Put the independent variable on the horizontal (x) axis, and the dependent variable on the vertical (y) axis.
- Label the axes and give the units of measurement.

Example

'Temperature in °C', 'Rate of heart beat (beats per minute)'.

- Choose a scale to use as much as possible of the space available (but avoid using awkward scales such as 2 squares to represent 5 units).
- If certain values of the independent variable have not been used, the intervals on the horizontal scale should take account of this.

Example

If temperatures used are 10, 20, 25, 30, 35, 40 and 45°C, then the interval between 10 and 20°C should be double that of the others – otherwise the shape of the graph will be distorted.

- Make points large enough for their position to be clearly visible, in the form of crosses, circles or squares, rather than dots.
- Do not *extrapolate* the graph (ie do not extend it beyond the data), unless you use a dotted line.
- When you draw two or more graphs on the same set of axes, remember to provide a key to distinguish them.

A problem often encountered is whether to join the points up or to draw the line of best fit. When measurements are suspected to be subject to appreciable error – either in the measurements themselves or in the experimental technique – it is best to draw the line of best fit.

Line graphs can be used to interpolate data (ie provide information on values not used).

Example

The dashed lines in the graph on page 372 can be used to say that, under the conditions of the experiment, the heart rate at 15°C would have been about 63 beats min^{-1}.

Bar charts

A bar graph is used when there are no intermediates, either because a variable is discrete or because it cannot be quantified.

Examples

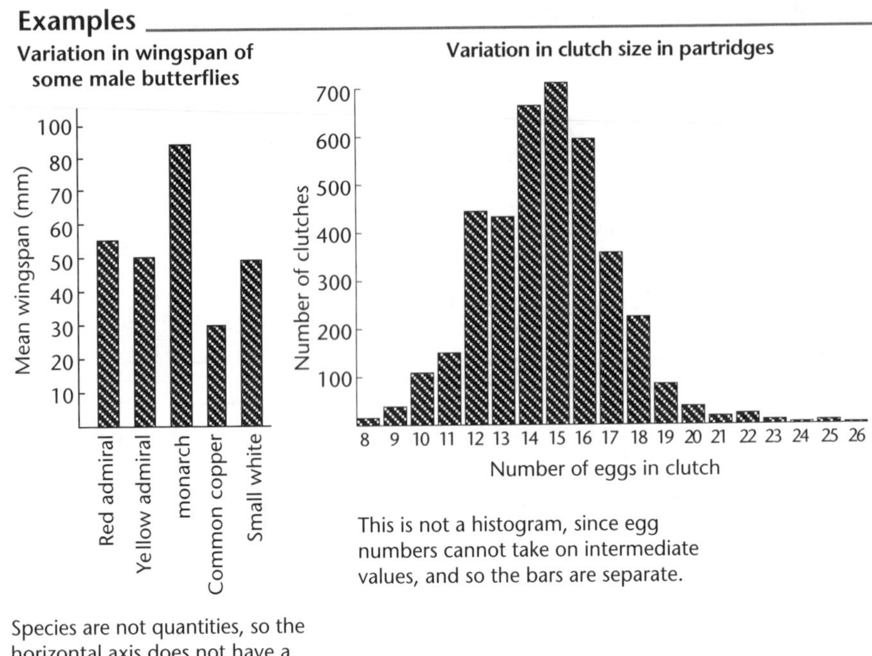

This is not a histogram, since egg numbers cannot take on intermediate values, and so the bars are separate.

Species are not quantities, so the horizontal axis does not have a scale or units.

Scattergrams

A **scattergram** is used to show whether two variables are correlated – in other words, if the value of one is an indication of the value of the other.

Example

Suppose you want to know whether there is any relationship between the size of islands and the number of species of birds present. You would measure the area of a large number of islands, and for each record the number of species. Since the hypothesis is that island size affects the number of species (rather than the other way round), area would be the independent variable and it would be plotted on the horizontal axis. Each island would be represented by a point on the graph, but the points *are not joined up*. It is quite possible for two islands to have the same number of species (same y value) but different areas (x values).

Where there is no reason to regard either variable as dependent on the other (eg body height vs blood pressure), axes could be either way round.

Whether variables are correlated depends upon the way the points are clustered. In the island example above, there is a tendency for larger islands to have more species. The two variables are said to be *positively correlated*. If there is no correlation, the points are randomly scattered.

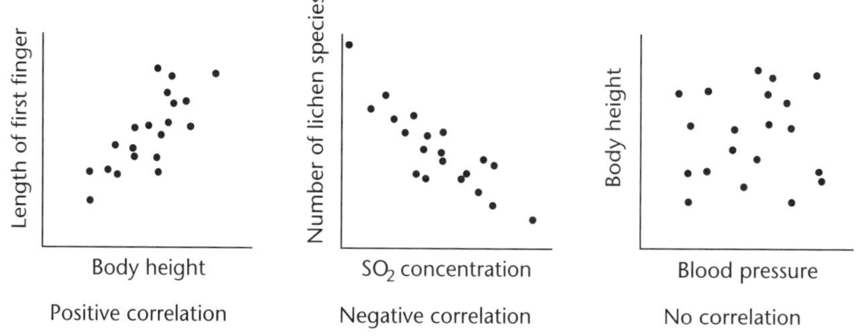

Scattergrams illustrating different kinds of correlation.

The stronger the correlation, the more the dots tend to be clustered in a line. When a correlation is absolute, all the points lie along a line, so that for every value of one variable the value of the other can be precisely predicted. If the line is straight the relationship is said to be *linear*, but many relationships are non-linear.

Correlation between two variables may be quantified statistically as the correlation coefficient, in which a value of 0 indicates no correlation, +1 indicates absolute positive correlation, and −1 absolute negative correlation.

When two factors are correlated, it may indicate that one is a cause of the other (eg the relationship between smoking and cardiovascular disease and lung cancer). In many cases, there may be no causal connection, or both variables are related to the same causative agent.

Histograms

A histogram is used to show the relative frequency of different measurements *and is used only with continuous data*. When plotting a histogram, measurements are divided into classes. The most suitable size of each **class** will depend on the number of measurements and other factors.

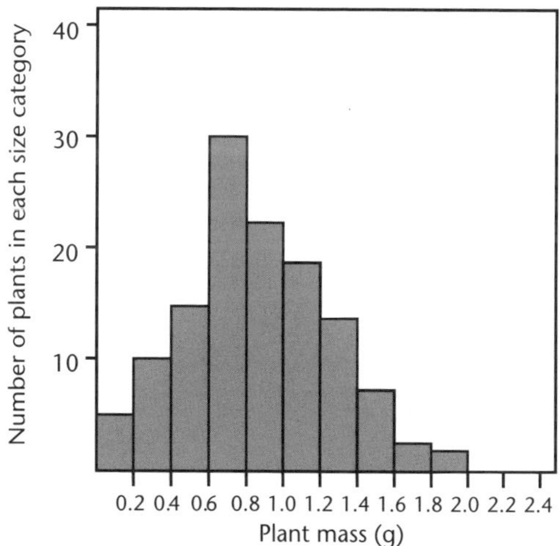

Histogram showing variation in groundsel plants.

Note the following points:

- The figures on the x-axis denote the boundaries between categories.
- The columns touch.

Kite diagrams

A kite diagram is used to show how the abundance of an organism changes along a chosen line or transect. The dependent variable (some measure of abundance) is plotted as two symmetrical lines, one on each side of a central axis.

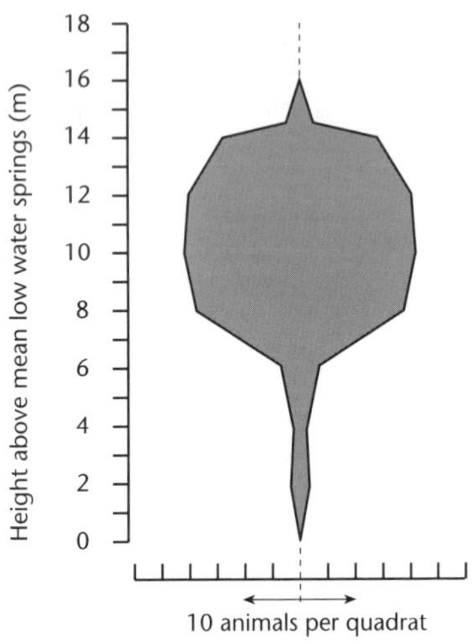

Kite diagram showing variation in abundance of limpets along a transect.

Pie charts

A pie graph (pie chart) is used to indicate the percentages of various constituents of a whole.

Example

Water forms about 84% of an apple's mass. On a pie chart, this is therefore represented by a segment with an angle of 84% of 360°, or 302°. The other proportions of various nutrients in the apple are represented.

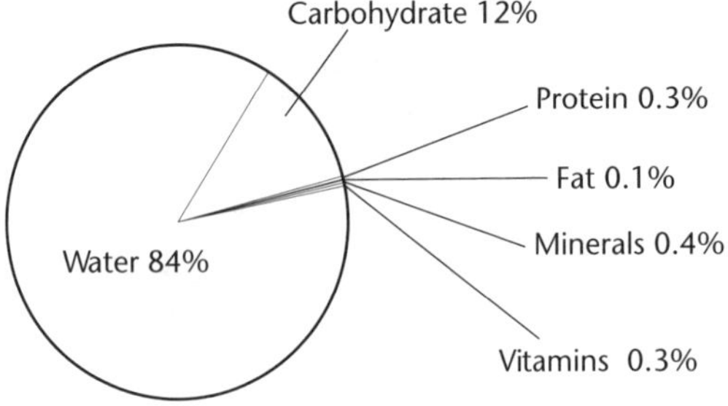

Pie chart showing nutrient composition of an apple.

Facts from figures – statistical tests

In some cases, results of an investigation may be sufficiently clear-cut to require no further treatment, but most often some form of statistical analysis is necessary.

Example

Suppose you are investigating the response of slaters to light. You place 15 slaters in a 'choice chamber' containing three intensities of light – total darkness (dark), 50% ambient intensity and ambient intensity (light). After 10 minutes, you count the number in each chamber.

- Suppose that all 15 are in the dark chamber (ie 15 dark : 0 50% : 0 light). The probability that all 15 would move to the dark chamber by chance is very small $\left(\frac{1}{3}\right)^{15}$, so you would be justified in concluding that slaters (or at least these particular slaters) prefer dark to light.
- A 14 dark : 1 50% : 0 light result would be almost as clear-cut.

Problems arise with results like 10 : 3 : 2 or 8 : 4 : 3.

If only chance is operating, the most probable result is 5 : 5 : 5. This is called the *expected* value. In statistics, 'expected' has a different meaning from the usual one. It means the most frequent result when a large number of trials are performed and when chance alone is operating. If, for example, a coin is thrown ten times by 100 people, 5 heads and 5 tails would be the most common result. The probability of getting tails in any given toss is therefore $\frac{5}{10} = 0.5$. Even though the probability of getting 5 : 5 by chance is only about 0.25, every other result is even less likely. The greater the deviation from the 'expected' value of 5 : 5, the less probable it is.

What we have to decide is how great a deviation from the expected value we can reasonably accept as being due to chance. At what point do we suspect that slaters are not moving randomly, but have a preference for darkness?

The null hypothesis

The first step in a statistical analysis is to put forward a *null hypothesis* (H_0).

Example

In the case of the slaters the null hypothesis could be stated thus:

'There is no significant difference in the numbers of slaters in each chamber, any deviation from the expected value being due to chance alone.'

Having stated a null hypothesis, it is then tested statistically. It is important to realise that such a test cannot prove that chance is or is not the only factor at work. It can only tell us the probability that any observed results could have occurred by chance.

Significance

The investigator has to choose a probability below which they are justified in rejecting a null hypothesis. In most situations, a probability of 0.05 is used. A calculated probability of 0.05 means the investigator can be 95% confident that the result is not due to chance. If the calculated probability is less than 0.05, the result is said to be *significant*.

Analysis of samples

The first and most important step in a statistical analysis is to decide on the kind of test to use.

The χ^2 test is used where an observed set of data fits with a theoretical expectation. Where there are three or more sets of measurements to be compared the ANOVA test is needed. (Note that the test is not valid for the majority of investigations due to the requirement that the independent variable has a range (ie a minimum of 3).)

Samples

The number of observations that can be made in an investigation is obviously limited, so a **sample** is taken. The population from which a sample is drawn is called the *parent population*.

Repeated samples from the same population always vary slightly due to chance alone. Each individual sample therefore provides an *estimate* of the properties of the parent population, larger samples giving more reliable estimates than small samples. Two particularly important things about a sample need to be known:

- A measure of *central tendency* – ie a figure that is most representative of the sample as a whole.
- A measure of the spread or *dispersion* of the data.

Measures of central tendency

The mean, or 'average' ($\bar{x}$) is the sum of the data divided by the number of data and is given by $\bar{x} = \frac{\sum x}{n}$, where $\sum x$ is the sum of the data and n is the number of data.

The mode is the most numerous value, and is represented by the peak of a frequency distribution. A sample may have more than one peak.

Example

If body size were plotted against frequency, two peaks in the distribution could represent males and females respectively. In this case, the distribution would be bimodal.

The median is the value midway between the highest and the lowest.

In a symmetrical distribution the mean, mode and median are the same, but in a skewed distribution they are not.

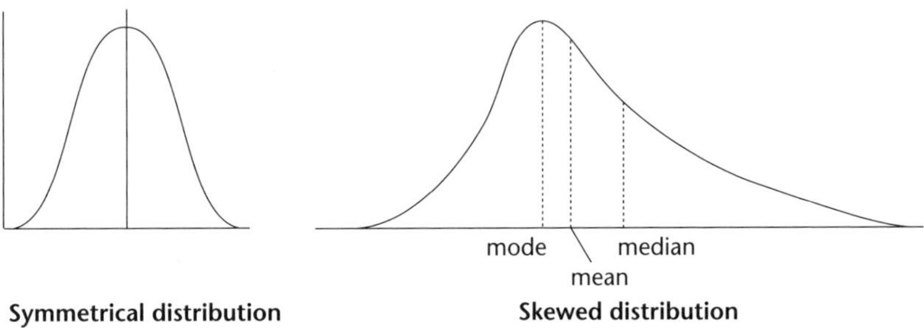

Mean, median and mode.

Measures of dispersion – standard deviation

A sample mean provides an estimate of the mean of the population from which the sample has been drawn. How reliable is this estimate? Besides depending on the sample size, a sample mean also depends on how variable the data is – ie its spread around the mean. The spread of the data around the mean is represented by the standard deviation, given by the formula:

$$s = \sqrt{\frac{\Sigma(x - \bar{x})^2}{(n - 1)}}$$

where x = an individual measurement,

$\bar{x}$ = the mean of all the data,

n = the number of data.

The population standard deviation is denoted by the symbol σ

> Each individual measurement differs from the mean by $(x - \bar{x})$. Since the data lies on either side of the mean, some values of $(x - \bar{x})$ are negative. By squaring each value of $(x - \bar{x})$, all are made positive. The values of $(x - \bar{x})^2$ are added together to give $\Sigma(x - \bar{x})^2$. This is then divided by $n - 1$, and the square root then calculated to give the standard deviation. For large samples, n can be used instead of $n - 1$ (since n approximates to $n - 1$).

Scientific calculators have keys that enable data to be entered and the values of n, $\bar{x}$, and s read out directly. Standard deviations are also readily calculated on spreadsheets (eg Excel).

The reliability of a sample mean as an estimate of the population mean is greater if the population is large and the standard deviation is small.

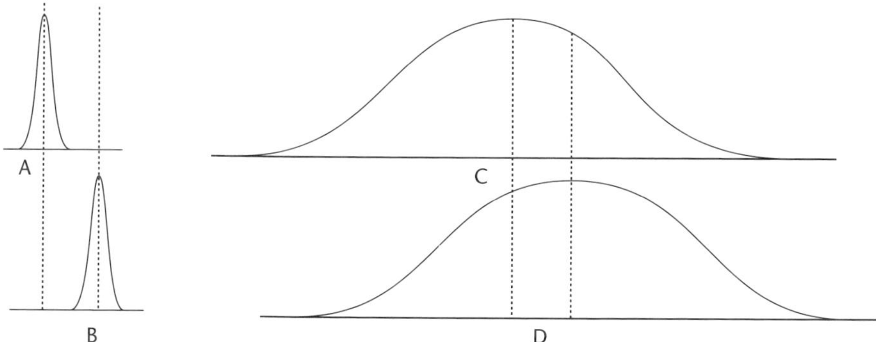

The means of samples A and B differ as much as the means of C and D, yet we can be more confident that the differences between A and B reflect real differences between the parent populations.

The importance of standard deviation in comparing means of samples.

The normal distribution

Many biological variables, such as body height and blood pressure, have an approximately normal distribution.

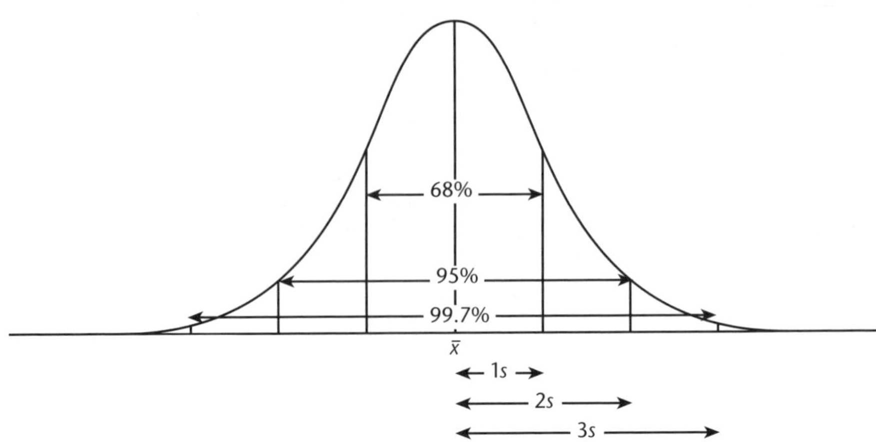

The normal distribution.

A normal distribution is a bell-shaped curve and it is symmetrical about the mean; mean, median and mode are the same.

68% of the data differs from the mean by no more than 1 standard deviation (34% above, 34% below). 95% of the data falls within 2 standard deviations from the mean, and 99.7% of the data differs from the mean by no more than 3 standard deviations.

Any given number of standard deviations above or below the mean includes a defined proportion of the population. The number of standard deviations is related to probability.

z score (number of standard deviations above the mean)	*Probability that z will exceed the mean by that value or less*
0.1	0.0398
0.2	0.0793
0.3	0.1179
0.4	0.1554
0.5	0.1915
0.6	0.2257
0.7	0.2580
0.8	0.2881
0.9	0.3159
1.0	0.3413
1.1	0.3643
1.2	0.3849

1.3	0.4032
1.4	0.4192
1.5	0.4332
1.6	0.4452
1.7	0.4554
1.8	0.4641
1.9	0.4713
2.0	0.4772
2.1	0.4821
2.2	0.4861
2.3	0.4893
2.4	0.4918
2.5	0.4938
2.6	0.4953
2.7	0.4865
2.8	0.4974
2.9	0.4981
3.0	0.4986

Some probability values from the normal distribution.

z scores

For a normally distributed variable, a deviation from the mean expressed in terms of the number of deviations is called a *z* score, and is given by $\frac{(x - \overline{x})}{s}$.

Example

A beetle is 20 mm long and the mean for the population is 17 mm. If the standard deviation for body length in this particular population is 2 mm, then the beetle is $\frac{3}{2} = 1.5$ standard deviations above the mean.

The *z* score is $\frac{20 - 17}{2} = 1.5$.

Provided beetle length is approximately normally distributed, the previous table indicates how unusual this is. For each z score, there is a corresponding probability, which is the proportion of the data that exceeds the mean by that z score or less. In this case, 0.4332 or 43.32% of the beetles would be expected to lie between the mean and 1.5 standard deviations above the mean.

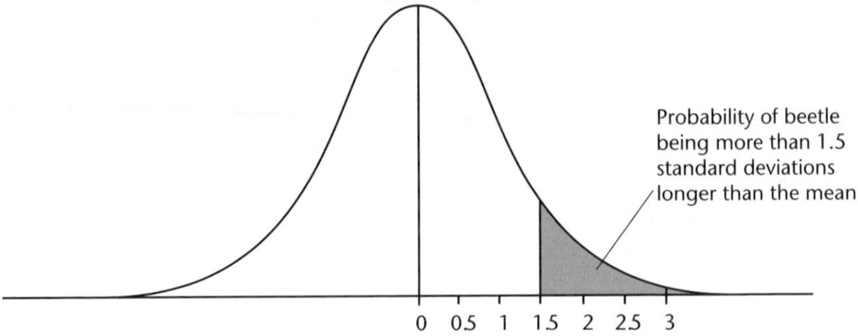

The shaded area represents the proportion of the beetles with a body length between the mean and a z score of 1.5 standard deviations.

Proportion of a population of beetles whose length would be expected to exceed the mean by 1.5 standard deviations or less.

In most cases, it is the proportion of the data that would be expected to exceed the mean by a given amount or *more* that is required, so the probability given in the table is subtracted from 0.5.

Example

For the beetle, only 0.5 – 0.4332 = 0.0668, or about 6.7 % of the population would be as long as 20 mm or longer.

The χ^2 test

The χ^2 (the Greek letter, pronounced 'ki') test is used where discrete data (counts) can be classified in one of a number of distinct categories or classes, and where we have a prior expectation as to how many there should be in each group if chance alone is operating. Thus, in the slater example mentioned previously, each slater is either on the light side or the dark side of a choice chamber, and the total *observed* number on each side is then compared with the expected numbers.

Consider the following example. A fruitfly heterozygous for ebony body and vestigial wings was mated with an ebony-bodied, vestigial-winged fly, and the offspring were as follows:

grey-bodied, long-winged	92
grey-bodied, vestigial-winged	86
ebony-bodied, long-winged	97
ebony-bodied, vestigial-winged	101
Total	376

Within the limits of chance, is this result consistent with a 1 : 1 : 1 : 1 ratio?

Having erected the null hypothesis that the fruitfly numbers do not differ significantly from a 1 : 1 : 1 : 1 ratio, the χ^2 test is then used to test the null hypothesis. First, the value of χ^2 is calculated, and then the table of χ^2 is used to determine the probability that a χ^2 value as high as this or higher could be obtained by chance.

χ^2 is calculated as follows:

1. Calculate the expected values (E) in each category. This is $\frac{376}{4} = 94$. If the results were due to chance then it would be expected that there would be 94 flies in each category
2. Calculate the differences between the observed (O) and the expected values.
3. Since some of these deviations will be negative, they are then squared to give all positive values.

	Observed	Expected	$O - E$	$(O - E)^2$
Grey-bodied, long-winged	92	94	–2	4
Grey-bodied, vestigial-winged	86	94	–8	64
Ebony-bodied, long-winged	97	94	+3	9
Ebony-bodied, vestigial-winged	101	94	+7	49

4. Obtain the value of χ^2 using the formula $\chi^2 = \Sigma\frac{(O - E)^2}{E}$ where O = observed value, E = expected value, and Σ = 'the sum of'.

In this example, the data falls into four categories, so there are four values for $\frac{(O - E)^2}{E}$

These are then added together to obtain $\chi^2 = \frac{4}{94} + \frac{64}{94} + \frac{9}{24} + \frac{49}{94} = \frac{126}{94}$

Degrees of freedom

The probability that any given value of χ^2 could be exceeded by chance depends on the number of *degrees of freedom*, which is always one less than the number of classes. Thus in the fruitfly example, there are four classes so there are three degrees of freedom.

The number of degrees of freedom is the maximum number of classes whose numbers could changed independently without altering the total number. In the fruitfly example there are 376 flies. If the numbers of three classes are known, then the number in the fourth category is also fixed.

Using the table to χ^2

The following table shows part of the table of χ^2 values. Reading along the line for three degrees of freedom we see that 1.34 corresponds to a probability of between 0.8 and 0.5. In other words, if chance alone were operating, we would expect a χ^2 value of 1.34 or higher with a probability of between 0.5 and 0.8. To put this another way, if this experiment were done a very large number of times, a χ^2 result as high as or higher than this would be obtained by chance in 50–80% of the experiments.

	Probability									
df	0.98	0.95	0.8	0.5	0.2	0.1	0.05	0.02	0.01	0.001
1	0.001	0.004	0.064	0.055	1.64	2.71	3.84	5.41	6.64	10.83
2	0.04	0.103	0.466	1.386	3.22	4.61	5.99	5.41	9.21	13.82
3	0.185	0.352	1.005	2.366	4.64	6.25	7.82	7.82	11.35	16.27
4	0.429	0.711	1.649	3.357	5.99	7.78	9.49	9.49	13.28	18.47
5	0.752	0.145	2.343	4.351	7.29	9.24	11.07	11.07	15.09	20.52
$\frac{5}{10}$ do not reject H_0								Reject H_0		

Table of χ^2 (df = degrees of freedom).

This probability is considerably higher than the 0.05 that we regard as significant, so we cannot reject the null hypothesis. In other words, we have no reason to suppose that the result does not conform to a 1 : 1 : 1 : 1 ratio.

Note: This does not mean that the deviation is necessarily due to chance. It simply means that if there is something other than chance at work, *the test has failed to reveal it*. Strictly speaking, therefore, we cannot accept a null hypothesis, for to do so would be to conclude that chance has been entirely responsible, and we cannot be sure of that. Rather than accepting H_0, it is therefore better to 'not reject' it.

ANOVA

ANOVA (*ANalysis Of VAriance* between groups) is the test used when results of three or more treatments are to be compared (eg height of tomato plants across a range of fertiliser concentrations). It is usually used to establish if there is a significant difference across a range of treatments as a whole.

- Data must be measurements that are continuously variable and have no upper limit. (Data that are counts (eg numbers of slaters in each light category) or that are percentages or proportions *cannot* be analysed using this method.) However, such data can be *transformed* so that it can be analysed using ANOVA.
- Results from all treatments must have the same variance. (This can be established as part of the test.)
- It is preferable that all treatments have the same number of replicates. (However, most computer programs used for the test have ways of overcoming this.)

The method looks long and complicated but is actually very straightforward as long as calculations are set out systematically. Computer programs such as Excel quickly and accurately carry out the calculations needed.

Example

An investigation was carried out on the effect of fertiliser concentration on tomato plant height. A range of three fertiliser concentrations was used with three replicates for each treatment.

Replicate	Group A	Group B	Group C
1	11	21	39
2	14	18	37
3	9	22	41

Calculations should be done in the following order. (Show all working in the Appendix section of your final report. If you use Excel, attach a copy of the computer calculations to the Appendix.)

1. For each column (treatment) calculate the sum of all replicates (Σx), the number of replicates (n), the mean $\bar{x}$, the sum of each replicate squared (Σx^2), and $\frac{(\Sigma x^2)}{n}$, and $\Sigma d^2 (= \Sigma x^2 - (\frac{\Sigma x^2}{n}))$ and the variance $\sigma^2 = (\frac{\Sigma d^2}{n-1})$. Put all these results into another table (*see later*).
2. Follow this step if you suspect that the results may not have the same variance.

Example

If the mean is a very high number such as 600, the individual data that resulted in this mean may have come from a very wide range (such as 330, 700, 650). In this case, they may not have the same variance.

Divide the highest value for variance calculated in Step 1 by the lowest to give the variance ratio F. Consult a table of F_{max} (see below). Using the appropriate number of treatments and the degrees of freedom (calculated by number of replicates per treatment – 1) compare F with the F_{max} value in the table. If F is less than F_{max} then proceed with the rest of ANOVA as the results have the same variance. If F is greater than F_{max}, then the data will need to be transformed.

For this data, $F = \frac{6}{4} = 1.5$. The F_{max} for three treatments and degrees of freedom of 2 (3 – 1) is 87.5. F is less than F_{max}, so these results have the same variance and it is appropriate to continue with ANOVA to test the significance of the results.

3. Assign A, B and C to Σx, Σx^2 and $\frac{(\Sigma x^2)}{n}$
4. Square A, divide it by the total number of observations, and assign it the letter D.
5. Calculate the *Total sum of squares (B – D).*
6. Calculate the *Between treatment sum of squares (C – D).*
7. Calculate the *Residual sum of squares (B – C).*

Replicate	Group A	Group B	Group C	Row totals
1	11	21	39	
2	14	18	37	
3	9	22	41	
$\sum x$	34	61	117	212 (A)
n	3	3	3	
Mean	11	20	39	
$\sum x^2$	398	1249	4571	6218 (B)
$\frac{(\sum x^2)}{n}$	386	1240	4563	6189 (C)
Sd^2	12	9	8	
s^2	6	4.5	4	

$D = \frac{212^2}{9} = 4994$

Total sum of squares = B – D = 6218 – 4994 = 1224

Between treatment sum of squares = C – D = 6189 – 4994 = 1195

Residual sum of squares = B – C = 6218 – 6189 = 29

8. Draw up another table for the final processing steps, where u is the number of treatments (in this case 3) and v is the number of replicates for each treatment (in this case 3).

Source of variance	Sum of squares (S of S)	Degrees of freedom (df)	Mean square = S of S/df
Between treatments	1195	$u - 1$ (2)	598
Residual	29	$u(v - 1)$ (6)	4.8
Total	1224	$(uv) - 1$ (8)	

Now, the F value can be calculated $= \frac{\text{Between treatments mean square}}{\text{Residual mean square}}$

$= \frac{598}{4.8}$

$= 125$

9. Consult a table of F values. Using the *Between treatments dof* and the *Residual df*, find the tabulated F value at $p = 0.05$. For these results, the F value is 5.1.

> If the calculated F value exceeds the tabulated F value then the results are statistically significant and the null hypothesis is rejected.

In this case, the calculated F value is greater than the tabulated F value, so the null hypothesis is rejected and the results are statistically significant – ie the fertiliser concentration is making a difference to tomato plant growth.

Presenting a report

The report, which should be presented as a written document, must consist of the following parts.

Introduction

This need be no more than a few paragraphs, and should discuss:

- The ecological niche of the organism, specifically the aspect(s) of it relevant to the investigation.
- The significance and relevance of the investigation in terms of the ecological niche of the organism.

Wherever possible, use scientific terminology, including correct binomial naming. If making reference to sources of information, the most efficient method is to footnote. This means annotate the section you wish to reference with a small superscripted number, then, as a footnote, indicate the author and source of the information. *All* references are to be cited in the bibliography. If using a quote, use quotation marks and footnote the source of the quotation. *Do not* attempt to pass off as your own someone else's work – it is often very obvious when this is done.

Aim

This is where the hypothesis is written.

Method

Your final method as a result of your trialling.

Results

All raw data (observations, measurements, samples, statistical tables and calculations) should be in the logbook and appendix. Only processed data (averages, means, graphs drawn to illustrate patterns or trends, results of statistical analysis) should be in the main body of the report, under a heading *Results*.

Conclusion

This is normally *one (or two) sentence(s)* in length, and states the outcome of the investigation in terms of the independent and dependent variables. It should be extended (if appropriate) to reflect the findings across the range being investigated.

Example

If a hypothesis stated 'The heart rate of *Daphnia* increases with increasing temperature' but it was found that it increased until a temperature of approximately 40°C and then decreased at higher temperatures, then the conclusion should be written to reflect this finding.

A conclusion is *not* the place to:

- (Re)state or describe results.
- Discuss the results in terms of the ecological niche.
- Make assumptions or give explanations that are untested, ie that you have no evidence to support.

Example

While it probably is true, you cannot claim that 'the heart rate of *Daphnia* decreases at higher temperatures because the enzymes involved in respiration are being denatured'. In order to make such a claim, you would need to carry out another investigation involving enzyme stability.

- Evaluate the results.

Discussion

A well-structured, comprehensive discussion should be in three parts:

- A brief review of the ecological niche of the organism and the aspect(s) under investigation.
- A discussion of the biological significance of the results and how they relate to the aspect(s) of the ecological niche of the organism. You need to do more than just describe the results, as this is what the processed data does. A discussion of the significance of the results involves speculating as to how the measured behaviour or response helps or is important to the organism in its niche. No new data or observations not already processed and presented in the *Results* section should appear in the discussion.

Extension: Interpreting results

One question – two meanings

When we observe an animal or plant acting in a particular manner, we may ask 'why is the organism behaving in this way?' This apparently simple question can have two quite different meanings, one about the past and the other about the future:

- What is the *cause* of the behaviour? This kind of question is a *physiological* one, and is concerned with processes occurring in the body (eg transmission of nerve impulses and the contraction of muscles), and the external events that trigger these internal processes.
- What is the *effect* of the behaviour – in what way does the animal benefit? This is more of an ecological question, and concerns the *future* of the organism.

These questions are linked. An organism's behaviour depends on structures that developed under the influence of its genes. Genes are inherited because they have passed the test of natural selection in the *previous* generation. Hence if an organism behaves in a way that helps it survive and reproduce, it is because the range of environments it experiences are not greatly different from that of its parents. It is important to distinguish between the *immediate* cause of behaviour (consisting of events that triggered it off), and the *original* cause (consisting of the advantage the behaviour gave to the organism's ancestors).

No matter how clear-cut the results, the conclusions that are drawn may not be valid because of the way the experimenter thinks. Two common pitfalls are anthropomorphism and teleology.

Anthropomorphism

This is attributing human feelings and motivation to an animal. The fact is humans cannot know what goes on inside the brain of an animal. Some biologists have tended perhaps to go to extremes, treating each animal as if it is an automaton with no capacity for any kind of mental awareness. The more closely animals are studied, the clearer it is becoming that humans consistently underestimate the behavioural capacities of animals. Nevertheless it is best to be cautious and not to attribute human faculties to animals without evidence.

Teleology

This is attributing *purpose* to biological phenomena, and is often indicated by the use of the word 'to', meaning 'in order to'.

Example

To say that a slater moves into damper conditions 'to avoid drying out' is a teleological statement. It would be better to say that a slater moves into damper conditions, 'as a result of which it avoids desiccation'. The avoidance of desiccation is the *result* of the action, not the *cause*.

Likewise, to say that saddlebacks have behavioural adaptations 'to reduce competition for food' is to imply that the behaviour has arisen in a Lamarckian way, ie in response to *need*. Replacing 'to' with 'which' eliminates the implied Lamarckism.

Evaluation

- The reliability of the results and consequent claims have been considered, *and*;
- Steps have been taken to ensure the results are valid.

Analysis of the reliability of the method involves two things:

- *Justification of the final method*. This comes from the trialling carried out on the initial method(s). It outlines any possible sources of error or limitations of the initial method identified that may have affected the validity of the data, subsequent changes to the method as a result, and reasons for these changes. It is *not* sufficient to simply identify sources of error or limitations. It is not appropriate to discuss limitations external to the investigation (eg reasons for not being able to carry out the investigation or parts of it due to ill health or work commitments).
- *Identification of any sources of bias that were not or could not be taken into account in the final method and how they may have affected the results*. These should be very few (if any), as most of these should have already been addressed.

Ensuring the conclusions are valid, thereby justifying them, involves statistical analysis of the data. A written statement explaining the significance of, and referring explicitly to, the results of the statistical analysis is required in addition to the actual analysis itself.

- To what degree of validity does the analysis show the results to be?
- How confident are you in accepting the data as being accurate and precise?

Appendix

Relevant raw data from your logbook is presented in a more accessible format in the appendix. Your logbook must be presented as evidence of the ongoing nature of the investigation, but if the relevant raw data could be presented in a better way, then the appendix is the place to do this. Statistical calculations, including tables, can be included in the appendix. Follow the guidelines outlined earlier for presenting data for all tables, charts and graphs you include in the appendix.

Reference list

A correctly annotated list of all sources of information (including informal conversations with experts) must be included as part of the final report.

- A reference list is used just for sources of information actually cited within the text of the report.

Plagiarism is illegal in any context, and you risk serious consequences if you try passing off others' work as your own. Acknowledging people who helped with the investigation can come at the end of the bibliography.

Correctly citing references

There is a protocol for citing references and you need to set aside time to do this correctly, as it can be very time consuming.

Supplementary Unit: Investigations

Topic 3: Researching contemporary biological issues

This Topic follows on from Topics 1 and 2 and shows how to research a contemporary biological issue by:

- Integrating and evaluating researched information to discuss a contemporary biological issue.

Introduction

Students should be encouraged to critically to comment on current issues in biology. The *student* must collect, interpret and integrate information with *guidance only* from the teacher. Guidance means the teacher cannot tell the student what to do or how to do it, but is simply there to:

- Provide important parameters for the research, such as time frames and the nature of the final project (whether it is written or oral).
- Suggest possible resources or sources of information.
- Provide a list of possible topics from which one is chosen.

Your teacher *cannot* provide assistance in the form of templates for the structure of the final project or specific feedback about any work in progress.

The PNG 2009 upper Secondary Biology Syllabus requires students to:

- *Research* information to *describe* a contemporary biological issue.
- *Integrate* researched information to *explain* a contemporary biological issue.
- *Integrate and evaluate* researched information to *discuss* a contemporary biological issue.

Research means using mostly **secondary sources** (critical information from textbooks, journals, web sites, videos and other sources that you can reference so that they can be checked by another person) to find information. **Primary sources** (eg an unpublished report on a scientific investigation) can be used, but these should be kept to a minimum as they may not be valid and they must be carefully referenced. You need to locate and process the information found in these sources of information; your teacher can only suggest where you might look or recommend a source.

A good way to ensure sufficient information is found is to ensure that it fulfills the *BIO* criteria – it contains *B*iological concepts, *I*mplications and *O*pinions.

- Biological concepts must be relevant to the issue and from reputable sources.

Example

It would be unwise to use any biological information about genetic engineering (GE) from a strongly anti-GE publication, as it may not be credible.

- Implications arising from the biological aspects of the issue (biological, social, ethical, environmental or economic) cause concern to people and may affect people in different ways. Consider a range of implications – what may be an implication for one individual may not seem that important to another.
- Opinions of different people or groups of people should be sought – people have opinions about the biology of the issue or the implications that arise from the issue.

A *contemporary biological issue* is one that is currently a 'hot topic' in biology and that a lot of people are talking about. An obvious one is the genetic engineering of crop plants and farm animals. However, there are many more, such as:

- Stem cell research.
- Human reproductive technologies.
- Biological control of the cocoa pod borer.
- Biological control of avian influenza virus.
- Use of dynamite in fishing.
- Fluoridation of drinking water.
- Dumping of chemical waste from the Ramu Nickel Mine into the seas of the Madang coastline.

If you choose a very broad topic (eg human reproductive technologies), narrow it down to just one technology (eg IVF). Choose a topic that is an issue (ie can be seen from many different viewpoints or about which people have differing opinions).

The information found in the written report or speech must be used to *outline or state* the relevant biology, implications *and* opinions. While there is no prescribed amount of information to be included, ensure you have fully covered the biology of your issue and described at least three implications and opinions.

It is important to assess your sources of information, by commenting on the validity of the sources, *and* giving your justified opinion about the issue. Validity of scientific information is much like validity of any investigation – is it objective, sound and able to be defended? Valid scientific information is most often found in peer-reviewed journals or similar publications. It has been obtained by consideration of a wide range of views, investigations and factual information that has been narrowed down to that which is most widely accepted by the scientific community. A valid article in a reputable journal will have a considerable reference list at the end and will refer explicitly to these references throughout, particularly when the author is expressing an opinion. There will also be a section of the journal where previous articles have been reviewed and critiqued by members of the scientific community.

Examples

These are *examples* of what could be expected from different levels of constructive criticism.

	Weak comments – examples involve the simple statement of information – there are no reasons given or other viewpoints presented that could lead to a discussion.
Description of biology	...The method most often used for transforming plants is *Agrobacterium*-mediated gene transfer using binary vectors. For crops that do not facilitate this method (such as monocotyledons), DNA is introduced directly into the plant cells using protoplast fusion, electroporation or biolistics...

Description **of an opinion**	...Dr Smith believes that scientists that develop plants and animals containing foreign DNA intended ultimately for public consumption or use will need to put forward very good reasons as to why these organisms needed to be modified in the first place...
	Strong comments – reasons are given for the biology, implications or opinion, and information is integrated.
Explanation **of biology**	...Genetic engineering is being used to develop improved strains of white clover. *This is because* conventional methods have not managed to overcome the problems of lack of resistance of the plants to diseases and pests...
Explanation **of an implication**	...Producing clover with resistance to disease such as the white clover mosaic virus has huge implications for New Zealand. Our economy still relies to a significant extent on agricultural production such as dairy and meat, *therefore* improvements in the quality of feed for stock will spill over into improvements in the quality of these products...
Explanation **of an opinion**	...Mr Wilson believes that the genetic sciences will be very important for the future of New Zealand *since* our economy is based to a significant extent on biology and agriculture, and genetics is the basis of quality production in these areas...
Explanation **of biology showing integration of sources**	...Many types of plant cells are genetically engineered using a soil bacterium called *Agrobacterium tumefaciens*. Since this soil bacterium naturally infects plants resulting in the expression of its DNA, foreign DNA inserted successfully into the bacterial genome can also be expressed by the plant. Genes conferring resistance to the tuber moth in potatoes, resistance to herbicides in onions, and resistance to viruses in tamarillos are some of the projects underway in New Zealand. However, because of the low uptake rate of the foreign DNA by *A. tumefaciens* (between 3 and 5%), other methods are often used as well...

	Excellent comments – biology, implications or opinions are discussed, sources of information are evaluated for their validity or bias, and a justified opinion about the issue is given.
Evaluation **of the sources of information by assessing its validity**	...I used the articles found in 'Journal of New Zealand Genetics' and 'Genetic Modification Periodical' for my sources of biological information, as these are widely accepted by the biological community as valid publications of work in genetic modification in New Zealand. Before any article is published in either of these journals, it must be critiqued by at least two independent reviewers and for the subsequent two publications it must be open to further analysis by readers.

***Evaluation* of the sources of information by assessing its bias** Individuals or groups that have a particularly strong viewpoint on an issue will publish material with a strong degree of bias. Valid scientific information is often used, but selectively to confuse and mislead, such as an anti-GE group implying that tomatoes genetically modified to contain a single gene from a toad that renders resistance to frost will grow webbed feet and catch flies! Be very careful when using such information sources, particularly the 'factual' information they contain.	…I accept that several of my sources of information have a strong degree of bias to them. For example, 'New Zealand Green Warrior' magazine (the publication of the Keep NZ GE Free group) is strongly against GE in New Zealand and presents information designed to cause outrage and possibly hysteria. However, in order to counteract this bias, I made sure that I only used information from a wide range of sources from the scientific community as factual biology, and used material from publications such as this in my discussion of implications or opinions…
***Discussion* of implications** Need to discuss one of the three aspects – a fully integrated, detailed dialogue that goes further than just giving reasons. Look carefully at the information, make judgements about it and use it to illustrate points. It can involve – a historical approach to the biological concepts or comparing and contrasting implications or opinions using quotes or references or a detailed analysis of relevant data or statistics from a variety of sources relevant to the issue. It is much more than just 'regurgitating what you have found onto a piece of paper'. The first sentence is a description, but it is fully discussed by subsequent sentences. Two important implications of using genetic engineering are discussed – the time factor and its discrete nature. This paragraph can be used as one piece of evidence for discussion.	…Genetic engineering allows the immediate transfer of genes directly into potato cultivars. The laboratory phase of this process can be achieved in less than six months, meaning that GM lines can be ready for field testing within a year. This is substantially faster than is possible for the transfer of genes from related wild potato species via more traditional approaches. Furthermore, genetic engineering allows the transfer of discrete single genes with the minimal amount of DNA to effect the desired characteristic. This is in marked contrast to the transfer of genes from wild species via repeated back-crossing, since this also results in the transfer of large fragments of neighbouring DNA on the same chromosome. While unexpected phenotypic effects can occur with genetic engineering, the risks are much lower due to its discrete nature…

Discussion **of opinion showing integration of sources** Must present your own opinion about an aspect or implication of the issue – needs to support or oppose the aspect or implication of the issue, and it must be justified (ie supported with relevant quotes or references). The first sentence is the statement of the student's opinion and the rest is giving referenced reasons for it.	...Progress into genetic modification to improve the quality and disease resistance of commercial potatoes should be encouraged. The potato industry nets New Zealand's economy over $84 million dollars per annum[2,3] and generates hundreds of jobs (there are approximately 370 primary producers in New Zealand[1], but there are many more secondary jobs that result[1]). Potatoes are not only an important food source, but are also processed and utilised in many ways, such as thickeners in processed foods[5], and even as an ingredient in some types of packaging[7]! Potato growers are often frustrated and disadvantaged by not being able to consistently produce tubers of high commercial quality. This technology will result in a reliable supply of quality produce with fewer losses to farmers and increased efficiency of farming[2,4]. Currently, the tuber moth is controlled by extensive use of expensive, toxic pesticides[2], so the reduction in use of these through this technology can only be beneficial to the environment and people. A major concern of anti-GE groups is the horizontal transfer of inserted genes into other species, but there is little evidence that this would occur, and... The opinion that research into the genetic engineering of the potato to confer resistance to this particular pest is supported by a range of reasons – biological, social and economic – that have all been referenced from a variety of sources.

Carrying out your research

Being able to research effectively is an important skill.

Don't put off the start of your research. Start as soon as possible. Access as wide a range of sources of information as possible to get a complete picture of the biology and a range of viewpoints, implications and opinions. Sources often lead to other sources and it takes time to locate these and process the contained information. Good sources of information include:

- Libraries – school, public and tertiary institutes. School libraries often have a file of newspaper clippings that librarians think may be useful for students. While you probably cannot borrow from a tertiary library, they are excellent places to find journals, periodicals and textbooks.

- Newspaper and magazine articles – can be a bit 'light' on biological information, but often present a range of implications and opinions on an issue. They may be biased (depending on the type of publication).
- Journals and periodicals – excellent sources of biological information but they can be very hard to read due to technical terms and jargon used. They are very up-to-date and their validity is assured if they are a major publication.
- Television programs – similar in nature to newspaper and magazine articles (they have to be understood by a general audience), but are often very up-to-date and easy to understand.
- The Internet – has a wealth of information on many issues. Beware of validity of any information, as anybody can put an article on the Internet and it does not need to be reviewed or checked by an independent party.
- Government departments and research institutions – excellent sources of biological information as well as implications and opinions. Validity of the information can generally be assured, and they often publish very accessible material on the Internet.
- Societies and interest groups – will most likely present information that suits their purpose. However, you need to consider their views and opinions so you can have a justified opinion about the issue yourself.

Have a book or folder divided up into sections (covering biology, implications and opinions) in which information is jotted down as you come across it, particularly for journals or periodicals that cannot be removed from their place of origin.

Always fully reference sources – it can be very difficult trying to remember exactly where you got the information from once it comes time to compile your final project.

Writing your report, essay or speech

Depending on the requirements of your school, you may be able to present a written or oral summary of the information you have found. Consider the following:

- If possible, do a draft version, leave it for a week or two, then come back to it. Errors, omissions and better ways of saying something become more obvious.
- Structure your report or speech in a logical way, but avoid making it so structured (such as using bullet points) that the integration of concepts is lost. All the biology, biological implications and biological opinions may be covered as one section, all the other implications and opinions as a second section, and the justified opinion and evaluation of sources as a third section. Integrate related concepts as much as possible – a biological concept, implication and opinion can be discussed all in one paragraph.
- Reference throughout your report or speech.

You should have obtained your information from a range of sources and a reference list must be provided for each source referred to or quoted from. The two most relevant ways of referencing are:

- The Vancouver method used in the sciences.
- The APA method used in the health sciences.

Answers for many questions include a 'Marking Guide':

- **A** ('Achievement', meaning 'satisfactory achievement'). **A** answers are for *descriptions*.
- **M** ('Merit', meaning 'high achievement'). **M** answers are for *explanations* (but it is possible to be given **A** if an answer includes a description without an explanation).
- **E** ('Excellence', meaning 'very high achievement'). **E** answers are for *discussions* (but it is possible to be given **M** if the answer only explains or an **A** if the answer only describes). **E** answers require a *discussion* – bullet points are not appropriate.
- Note: '/' in an answer (eg feature/trait/protein) refers to acceptable alternative answers.

We hope this Marking Guide will be a help to students who are striving for the best possible results.

Unit 12.1 Ecology

Topic 1: Biomes and habitats

Unit 12.1 Activity 1A: Ecological niche (page 5)

1. a. The environment is the general surroundings (biotic and abiotic factors) that impact on an organism, while the habitat is the particular set of these factors that the organism lives in. (***A***)

b. Biotic refers to living factors, while abiotic refers to non-living factors. (***A***)

c. Tolerance is the (physiological) ability of an organism to withstand variations in an environmental factor, while acclimation is the ability of the organism to adapt its tolerance to gradual changes in a factor. (***A***)

2. a. The position or role that the organism occupies in its community; it is a combination of where the organism lives (habitat) and how it lives there (adaptations). (***A***)

b. An organism's physiological ability to withstand a range of an environmental factor. (***A***)

c. A feature of an organism (structural, physiological, behavioural) that allows it to survive successfully in its niche. (***A***)

d. No two organisms can occupy the same niche/co-exist with each other, as one will out-compete and eliminate the other. (***A***)

3. When both species of clover are grown together, *Species A* increases in numbers while *Species B* decreases/is killed off (***A***); this is because *Species A* is the more successful competitor and is out-competing and eliminating *Species B*. (***M***)

4. a. As the pH decreases below pH 7 or increases above pH 7, the number of animal species decreases. (***A***) This is because the lower the pH, the more acidic the water and animals cannot tolerate (high) acidity; and, the higher the pH, the more alkaline the water and animals cannot tolerate (high) alkalinity. (***M***)

b. It has the most (different) animal *and* plant species (***A***); this is because the pH (of 6.6) is nearly neutral (ie neither acidic nor basic), which is the optimum pH for most species. (***M***)

c. *Creek 3* would be least stable, as it has the lowest biodiversity. (***A***) The fewer species in a community, the less complex (fewer links) is any food web and the more prone to collapse food webs are if a link is removed. (***M***)

Topic 2: Biological communities

Unit 12.1 Activity 2A: How organisms depend upon one another (page 17)

1. 1K; 2E; 3S; 4H; 5D; 6M; 7F; 8J; 9Q; 10C; 11G; 12A; 13N; 14B; 15O; 16T; 17L; 18I; 19R; 20P.
2. W = plants (all energy arrows lead away from W); X = herbivores; Y = carnivores; Z = decomposers (receive matter and energy from all the other organisms).
3. Decomposers, since they can continue to feed on the dead bodies of all the other organisms.
4. Plants provide shelter – eg holes in trees, crevices in bark, camouflage, nesting places.
5. **a.** Plants depend on heterotrophs for inorganic raw materials such as CO_2, minerals.
   b. Plants depend on insects for transfer of pollen and dispersal of seeds and fruits.

Topic 3: Population and communities

Unit 12.1 Activity 3A: Communities (page 25)

1. **a.** Epiphytes are perching plants that spend their life living on other plants and have no connection to the ground, while lianas have roots in the ground and grow up another plant. (***A***)
   b. Zonation refers to the distinct horizontal bands of life forms that occur across a particular environment, while stratification refers to the vertical layers of vegetation that occur in a (forest) community. (***A***)
   c. Primary succession takes place on bare land that has never previously been occupied, while secondary succession occurs on bare land that has previously been occupied. (***A***)
2. A plant in the canopy layer will have a tall central trunk with a crown of leaf-bearing branches at the top, while a shrub has a branching stem with leaves all over the plant. (***A***)
3. **a.** Stratification – changing light intensity. (***A*** – both)
   b. Canopy – high light intensity/temperature/wind speed. (***A***)
   c. Zonation – lower temperature/more exposure/less soil water/higher wind speed higher up the mountain (or vice versa). (***A***)
   d. The left side of the mountain in the diagram is likely to be the south or west side of the mountain, which faces the prevailing winds where conditions tend to be colder/windier/wetter than the other/north or east side. (***A***) Vegetation zones will extend higher up the mountain in the warmer, less harsh conditions on the northern or eastern sides (or vice versa). (***M***)
4. **a.** Secondary succession. (***A***)
   b. Pioneer plants (***A***); they need a high (physiological) tolerance to high temperatures/winds/light intensity because they are not sheltered from conditions by other plants; they need a high tolerance to dehydration as conditions are harsh and the soil likely to be low in water; they need a strong root system to anchor in unstable soils (***A*** – two adaptations described; ***M*** – adaptations explained)
   c. Climax community. (***A***)
   d. Temperature at ground level decreases from *Level A* to *Level D* because taller plants shade the ground (layers); wind speed at ground level decreases as the taller plants reduce the speed of the wind; humidity increases at ground level as the taller sheltering plants trap moist air; light intensity decreases at ground level as the taller plants block some light from reaching the ground (***A*** – any three described; ***M*** – explained)

e. Animals do not arrive until there is a good plant cover (***A***); because they need the plants for food and shelter/nest sites (***M***).

5. a. Zonation. (***A***)

b. As distance from the high tide mark increases, soil stability/fertility/moisture all increase; ground temperature/wind speed/light intensity all decrease; humidity increases. (***A*** – three factors described)

c. The roots of plants growing in the sand/soil will increase its stability; leaf litter and plants dying and decomposing will increase soil fertility; the plants growing taller will decrease the temperature/light intensity/wind speed for plants germinating/growing beneath them. (***A*** – two factors described; ***M*** – explained)

6. a. The pattern is succession (***A***); because the species composition is changing over time (from Week 1 to Week 3), with one species replacing another/others (eg *Species A* is replaced by *Species B* which is replaced by *Species C*). (***M***)

b.

(***A*** – any one of the following)	(***M*** – any one of the following)
Species A is being out-competed by *Species B*	– for resources such as food, living space.
Species B grows more vigorously than *Species A*	– smothers it.
Species B is changing the environment	– making the environment unsuitable for the growth of *Species A*.

Unit 12.1 Activity 3B: Competition between species (page 27)

1. a. A pattern in the species distribution of a community over time. (***A***)

b. Primary succession occurs where there was previously no life, eg on a cooled lava flow. Secondary succession occurs where fertile soil existed from the beginning in which many types of plants can become quickly established, eg in a cleared forest. (***A***)
In primary succession, the soil conditions are at first poor, soil fertility being only slowly increased as a result of plant growth. In secondary succession, the soil is fertile from the beginning. (***M***)

c. In primary succession, each plant species modifies the soil conditions as a result of its own growth, thus altering the balance of competitive advantage in favour of other species. These other species then take over, only to be replaced by other species as a result of *their* activity. (***A***)

d. i. The grassland community is exhibiting succession – the composition of the grass species has changed over the study period as some species die out and others become established as a result of successive species modifying the environment (***A***). As the mean height of the vegetation increased due to few deer grazing the area, the grasses were shaded so had less light for photosynthesis and died out (***M***).

ii. Plants taller than the shrubs would eventually grow so there would be no grasses and fewer shrub species over time (***A***); as each succession modified the environment to the extent that it could no longer live yet other species could. (***M***) Eventually a climax community would be reached where there would be no further changes in species composition. (***E***)

2. Desiccation, or alternatively the shorter feeding time due to insufficient period of immersion by sea water. (***A***)
3. Two or more species cannot indefinitely occupy the same niche in the same habitat. (***A***)
4. **a.** Character displacement.
   b. **i.** Beak characters are a good indication of diet, so it seems likely that where the two species are allopatric they have similar diets. Where they are sympatric their diets are different. (***A***) This can be explained by the effect of selection, which would favour divergence of diet because of competition. (***M***)
   ii. If the length of the eye stripe is important in species recognition, then selection would favour divergence where the two species are sympatric (***A***); because of the disadvantages of hybridisation (eg reduced viability, reduced fertility). (***M***)
5. Each species of plant must occupy a slightly different niche from the others, eg have differing adaptations for gaining sufficient light for photosynthesis. (***A***) Although at any given time of year one species may be in the process of eliminating the others, because of seasonal changes in the environment, conditions may not favour any one species for long enough to enable it to eliminate the others. (***M***)
6. **a.** **i.** The decrease in *Asterionella* numbers in mixed culture is due to the presence of *Synedra*, since each flourishes in the absence of the other. (***A***) Since both require silicate, they are probably competing for this resource. (***M***) Graph C shows that *Asterionella* is the strongest competitor, as it flourishes when grown with *Synedra*, which dies. (***E***)
   ii. *Synedra* competes more strongly and eliminates *Asterionella*. (***A***) *Synedra* can deplete silicate to lower levels than *Asterionella* (Graph B). (***M***)
   b. Competitive exclusion principle or Gause's principle.
   c. There may be subtle differences in niche. (***A***) The habitat may be divided into different microhabitats; because of seasonal changes in conditions, no species is at an advantage for long enough to eliminate others; predation may keep numbers below that at which competition is important. (***M***)

Topic 4: Interrelationships – interspecific competition

Unit 12.1 Activity 4A: Interrelationships (page 55)

1. **a.** Interspecific competition occurs between different species, while intraspecific competition occurs between members of the same species. (***A***)
   b. Commensalism is a relationship between two organisms in which one benefits and the other is neither benefited or harmed, while mutualism is a relationship in which both organisms benefit. (***A***)
   c. Predation occurs when one animal kills and eats another, while in parasitism one organism feeds off another living organism without (deliberately) killing it. (***A***)
   d. An endoparasite lives inside its (living) host, feeding off the host, while an ectoparasite lives and feeds from the outside of a (living) host. (***A***)
2. **a.** 2, commensalism (liana) (***A***)
   b. 3, exploitation (ectoparasite) (***A***)
   c. 1, mutualism (***A***)
   d. 3, exploitation (predation) (***A***)
   e. 1, mutualism (***A***)
   f. 2, commensalism (***A***)
   g. 3, allelopathy (***A***)
   h. 6, competition (interspecific) (***A***)
   i. 1, mutualism (***A***)

3. Relationship is mutualism (***A***); because both benefit – the pine because it gets better access to water and nutrients, and the fungus because it gets a source of food (carbohydrates). (***M***)

Unit 12.1 Activity 4B: Exploitation of one species by another (page 56)

1. **a.** Type of mimicry where a palatable animal resembles an unpalatable one to avoid being eaten. (***A***)

b. Type of mimicry where several unpalatable species resemble one another in appearance. (***A***)

c. Type of relationship where one organism/parasite feeds off another (host) to the detriment, but usually not the death of, the host. (***A***)

2. **a.** Organism A is the prey/red mite, Organism B is the predator – predators depend on their prey for food, so a peak in predator numbers must occur *after* a peak in the numbers of prey. (***A***) *Refer to specific data from graph to justify this needed;* eg at about 28 days Organism A peaks in numbers, followed by a peak in numbers in Organism B at 50 days. (***M***)

b. About 120 days (***A***); the first time the numbers drop below the previous minimum of about 250 (***M*** – *Refer to specific data from graph to justify this*, eg at about 28 days Organism A peaks in numbers followed by a peak in numbers in Organism B at 50 days).

c. The predator is locally exterminated, allowing the numbers of prey to increase greatly (***A***). The reason for the greater susceptibility of the predator may be because they consume larger amounts of the chemical (concentration along the food chain). (***M***)

3. **a.** The mice would not be able to control the moths at any prey density. (***A***) Over the straight line part of the graph, the number of prey eaten is directly proportional to prey density, so the proportion of prey eaten remains constant. A predator can only control a prey population if predation acts in a density-dependent manner – ie if the *proportion* of prey eaten increases with prey density. (***M***)

b. From about 0–30 pupae per m^2 (***A***) (above which the proportion of prey eaten falls). (***M***)

c.

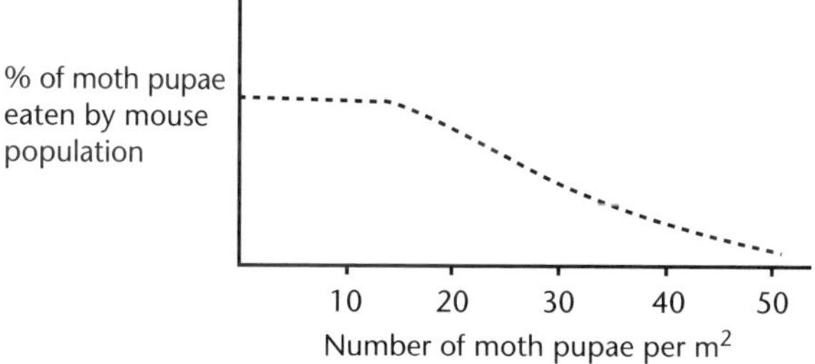

d. To act in a density-dependent manner, an increase in prey density must result in an increasing *proportion* of prey being eaten, if the prey population doubles and as a result three times as many prey are eaten. (***A***)

4. Speed (eg antelopes), weapons (eg antlers of caribou), armour and other mechanical protection (eg hedgehog), chemical defence (eg monarch larva and adult), crypsis (camouflage) eg looper caterpillar and stick insect, deception by mimicry (eg in North America, the viceroy butterfly mimics the monarch). (***A*** – must be a *description* – more than one word needed.)
5. The animal must be able to select the appropriate background and remain still for long periods, eg stick insect, or to mimic passive movements due to wind currents, eg the NZ praying mantis sways gently from side to side, as if in a breeze. (***A***)
6. (***A*** – must be a *description* in each case – more than one word.)
7. Forward-facing eyes enable binocular vision and the judging of distances – important in leaping on prey. Eyes on the sides of the head give a herbivore a wide field of vision, enabling better detection of predators. (***A***)
8. The hare peaks are higher than the lynx peaks (predators are always outnumbered by their prey), and hare numbers peak slightly before lynx numbers. (***A***)
9. Distastefulness is usually advertised by warning coloration, which is more effective if the animal is day-active. (***A***)
10. Wolves are pack-hunters, so after a kill there is competition for food. As solitary hunters, cats do not usually have to compete with other cats (though cheetahs are often driven from a kill by hyenas and lions). (***A***) Therefore, wolves maximise individual food intake by eating as fast as possible. (***M***)
11. Garden leaf vegetables do not contain distasteful chemicals (***A***); as these have been eliminated through selective breeding. (***M***)
12. The greatest source of mortality in endoparasites is during transmission from one host to another, and since this always involves eggs or larvae, endoparasites would not be able to complete the life cycle without production of huge numbers of eggs. (***A***) This is possible because an adult endoparasite does not have to expend energy finding food or escaping from predators, and so can devote a much higher percentage of its total resources to reproduction than a free-living animal. (***M***)

13. a. **i.** Graph C. Since the graph is a straight line, the percentage of total prey available does not change. (***M***)

ii. Graph A. Since mussels form a much higher proportion of the diet than their availability would suggest if there were no preference. For example, when the mussels form 20% of the food offered, they account for just over 61% of the food eaten. (***M***)

iii. Graph B. The sigmoid curve shows that when mayflies are relatively uncommon, an increase in prey availability results in a greater increase in percentage of mayflies eaten. For example, a doubling of mayflies in the environment from 20% to 40% results in a four-fold increase in the percentage of mayflies in the diet. (***M***)

b. The success of the predator increases with previous exposure to the prey, suggesting that success in catching a particular kind of prey improves with practice. (***A***) For example, when only 20% of the available food contains *Asellus*, the per cent of successful attacks is less than half of that when 100% of the available food contains *Asellus*. (***M*** – using data from graph)

14. a. The fish show an increasing preference for worms as the latter get more common. (***A***) Since points form an upward curve, the proportion of worms in the diet increases more than their availability. (***M***)

b. When a predator has to search for prey, the detection rate improves with practice, because it forms a 'search image' of the prey. (***M***)

c. By switching its choice of prey, a predator can eat an increasing ***proportion*** of a prey population as prey density increases. Prey switching is thus one means by which a predator can exert density-dependent control over a prey population. (***E***)

15. a. i. Time spent with head up decreases with increasing group size. (***A***)

ii. Males devote a greater proportion of their time to vigilance than females do. (***A***)

b. It is higher in single-sex groups of two (21%), compared with 16% in mixed sex groups. (***A***)

c. Vigilance is best expressed as a %, calculated as mean number of head raises per minute × mean duration × 100/60, shown in table:

	No. of birds in group		
	1	2	3 or 4
Males	41.9%	29.5%	18.3%
Females	36.4%	20.8%	10%

Vigilance decreases as size of group increases and vigilance increases less in females than in males (***A***). However, since individuals in groups of 2 are vigilant for more than half as long as when alone for both males ***and*** females, for each individual there is more vigilance in a group than when alone. (***M***)

d. So far as looking out for other males is concerned, increased group size would not relieve the male of the need for vigilance (though it would so far as anti-predator vigilance is concerned). (***A***) The data supports this, since decrease in % vigilance with increase in group size is less marked in males than in females. (***M***) The greater time spent on vigilance by males allows females in mixed groups of two to spend more time feeding than they can when accompanied by another female (see **b**. above). (***E***)

Topic 5: Intraspecific competition

Unit 12.1 Activity 5A: Intraspecific competition (page 79)

1. a. Area occupied by an animal (usually containing a nest/lair, etc) that is defended. (***A***)

b. Area that is not defended in which several animals may roam in order to find resources. (***A***)

c. Form of social dominance where animals are 'ranked' according to their degree of dominance. (***A***)

d. Animal that has best or better access to resources (eg food, mates) than other animals in the group. (***A***)

2. a. Territory is an area that is defended by an animal. (***A***)

b. *Costs*:
- May experience strong intraspecific competition for resources, especially food, once food sources migrate or decrease in numbers over the winter.
- Individuals may die due to extreme weather conditions during winter.

Benefits:
- Would not need to waste time and energy establishing a new territory each breeding season.
- No risk of death of individuals when migrating (although this may be offset by competition during winter months).
- Interspecific competition for resources reduced during winter months as competitors migrate. (***A***)

3. a. i. 13 ii. 6
   b. They are smaller, and are on the periphery. (**A**) Only those males that have good-quality territories are able to attract mates. (**M**)
   c. In 1967 there were more territories (18) than in 1980 (7), and they were smaller (2 ha in 1967 compared with 9 ha in 1980). (**A**)
   d. Less food available in 1980 (**A**); so to obtain sufficient food, each bird must defend a greater area. Alternatively, there may have been fewer birds in 1980. (**M**)
4. Members of the same species occupy the same niche. (**A**)
5. Reduced size, reduced seed output, greater variation in body size in a population. (**A**)
6. a. A home range is simply the area in which an animal lives from day to day. A territory is a *defended* area. (**A**)
   b. • It may ensure a regular supply of food, eg blackbird.
      • A place in which to build a nest and rear young, eg gannets.
      • A place in which males mate with females, eg red deer. (**A**)
7. A dominance hierarchy requires that each member of the group recognise every other member of the group (**A**), which requires both good sense organs and an ability to remember, both of which tend to be associated with a large/more complex brain. (**M**)
8. a. Birds of higher rank get more of a share of the resources or better access to available mates / dominant bird will leave more offspring *as a result of having more or better food or resources* / aggression within the group is minimised once rank has been established / subordinate birds may have a greater chance of reproductive success or access to food if they remain with the group despite being at the bottom of the ranking order. (**A** – description must be related explicitly to abiotic (resources) or biotic (other birds) factors.)
   b. Dominant birds would peck other birds/chase other birds/flap wings at other birds. (**A**)
   c. The hierarchy that appears to exist is 4, 1, 3, 5, 2 (**A** – description of hierarchy). This is because 4 never loses while 2 loses to every bird and never wins (**M** – note use of data to explain outcomes of challenges in words or as a table). Birds closest to each other in the hierarchy tend to challenge each other the most, for example 1 and 3 had the most number of challenges because they are adjacent to each other in the hierarchy and 3 is constantly trying to become more dominant than 1. (**E** – further use of data than just to show overall order)
9. a. Energy expended in displaying and chasing other individuals, and lost feeding time. (**A**)
   b. Increased supplies of food. (**A**) A territory that is larger than needed to satisfy food requirements costs more to defend but does not bring additional benefit. (**M**)

Unit 12.1 Activity 5B: Cooperative interactions (page 80)

1. a. Group of insects (often in large numbers), all of which are offspring of a single female and which show the characteristics of division of labour and a means of communication between individuals, eg ants. (**A**)
   b. Relationship between two organisms where one organism benefits from the relationship, while the other seems unaffected, eg paua and Flamena crab. (**A**)
   c. Relationship between organisms in which both benefit from it, eg ruminants and the micro-organisms living in their gut. (**A**)

d. Type of plant lacking a penetrative root system that 'perches' on a supporting plant, obtaining nutrients from it, eg Astelia solandri. (***A***)

2. (A – must be four descriptions that are sentences in length.)
3. Membership of a wolf pack is reasonably stable, and animals know and react to each of the others as a particular individual. Membership of a flock of geese is more open and temporary, with individuals 'coming and going'. (***A***)
4. An epiphyte gains nothing but anchorage from the plant upon which it grows; a parasite gains nutrients. (***A***)
5. The climbing plant does not need to expend much energy producing support tissue. (***A***)
6. Members of family groups share more of their genes than do unrelated animals. (***A***) By cooperating to assist in the survival and reproduction of other members of the group, animals are assisting the perpetuation of copies of their own genes. (***M***)
7. Mutualism. Humans derive benefit from cattle, which would have much lower survivorship without humans. Though cattle invariably end up being eaten, their ability to hand on their genes depends entirely on humans (just as the micro-organisms in a sheep's rumen are eventually killed when they are digested). (***A***)
8. **a.** By the intensity of the 'waggle' of the abdomen. (***A***)
   b. By the angle of the straight-line part of the waggle dance to the vertical. (***A***)

Topic 6: Ecosystems

Unit 12.1 Activity 6A: Ecosystems (page 93)

1. **a.** An autotroph is an organism that can make its own (organic) food from inorganic materials using an external energy source, while a heterotroph has to consume organic materials for food (it cannot make its own food). (***A***)
   b. A photosynthesiser makes its food from CO_2 and H_2O using solar energy, while a chemosynthesiser uses the energy from chemical reactions to bind raw materials into food. (***A***)
   c. A producer is a plant that makes food in photosynthesis (can more rarely be chemosynthetic bacteria), while a consumer is an organism/animal that has to eat food/organic material as it cannot make its own. (***A***)
   d. A herbivore is an animal that eats plants/vegetation, while a carnivore is an animal that eats other animals/flesh eater. (***A***)
   e. A filter feeder filters microscopic organisms/organic material/plankton from the surrounding water, while a detritus feeder consumes organic material from the surface/in the substrate that it lives on/in. (***A***)
2. **a.** The flatworm is a predator/carnivore of the earthworm (***A***); because it hunts and kills and eats earthworms. (***M***)
   b. The flatworm secretes enzymes out onto the earthworm and does not actually eat it (***A***); therefore digestion takes place *outside* the flatworm's body. (***M***)
3. **a.** **i.** The aphid is a(n) (ecto) parasite of the broad bean (***A***); because it is feeding on the glucose/food from the living plant. (***M***)
   ii. The ladybird is a predator/carnivore of the aphid (***A***); because it is killing and eating the aphid. (***M***)

b. **i.**

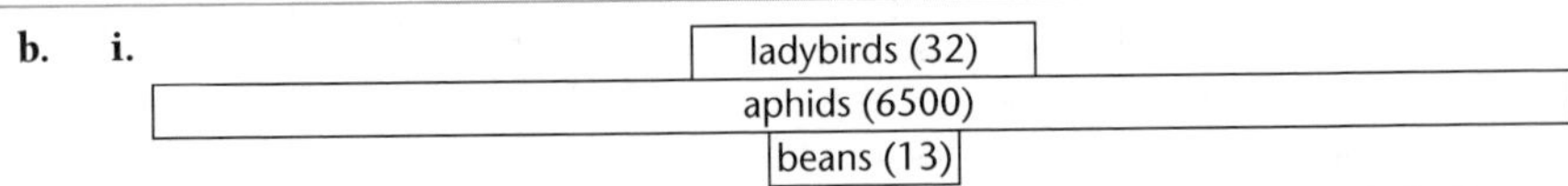

(***A*** – pyramid correct shape; ***M*** – also drawn to scale and correctly labelled)

ii.

ladybirds
aphids
beans

(***A*** – pyramid correct shape; ***M*** – also drawn to scale and correctly labelled)

iii. The pyramid of numbers does not decrease like the biomass one because there are far more aphids than bean plants. (***A***) This is because the bean plants are much larger than the tiny aphids, so have much *more biomass*. Therefore, although there are many aphids, their biomass is much smaller than that of the plants. (***M***)

4. The biomass of the algae and diatoms is much larger than the aquatic insects and larvae because they are the food source and there has to be more food than the organisms that eat it. (***A***) The biomass represents the amount of energy in the trophic level and as *energy is lost* from/between each trophic level, the biomass decreases. (***M***) Energy is lost as *heat* in the process of *respiration*. The organisms at each trophic level release energy in respiration needed for movement, growth, reproduction, etc. In the process, energy is also released as heat, which is waste energy, which radiates out into the air and so is lost to the community. Only that energy which is incorporated into the organisms' bodies (eg growth) is available (biomass) for the next level of the food chain. (***E*** – in-depth explanation linking ideas)

5. **a.** **i.** Phytoplankton, water weeds. (***A***)

ii. Zooplankton, swans and ducks, small fish. (***A***)

iii. Small fish, large fish, humans, shags and herons. (***A***)

iv. Shags and herons, humans, trout. (***A***)

v. Small fish, zooplankton, swans and ducks. (***A***)

vi. Small fish, large fish, humans, shags and herons. (***A***)

vii. Shags and herons, trout, humans. (***A***)

viii. Humans, shags and herons. (***A***)

ix. Small fish. (***A***)

b. **i.** The small fish would have decreased in numbers (***A***); because a predator/carnivore of them had been introduced. (***M***)

ii. Shags would have increased in numbers (***A***); as another prey species for them had been introduced. (***M***)

c. **i.** Small fish numbers would have decreased (***A***); as a competitor for food had been introduced. (***M***)

ii. Trout would have increased (***A***); as another/more food/prey species had been introduced. (***M***)

d.

carnivores
herbivores
producers

(***A***)

e. The rest of the (light) energy is reflected back out into the atmosphere. (***A***)

f. **i.** Increase in water temperature/warmer water, increase in nutrients in the water, high levels of light intensity. (***A*** – three factors)

ii. The same trend would be expected in zooplankton numbers (***A***); as the phytoplankton are food for zooplankton, and the more food, the greater the numbers of phytoplankton that can be supported (***M***).

6. **a.** **i.** Respiration. (***A***) **ii.** Photosynthesis. (***A***) **iii.** Combustion. (***A***) **iv.** Eating. (***A***)

b. Carbon is an element found in chemicals/molecules in all organisms (***A***) – the giant (organic) molecules that make up all organisms (carbohydrates, proteins, fats, nucleic acids – DNA, RNA). (***M***) Could also have here information about carbon's ability to form covalent bonds with up to four other atoms.

c. The common form of C in its recycling is CO_2 gas in the air, while that for P is PO_4^{3-} ions in the soil; P tends to be 'locked up' in bones (which slows down its cycling), but C does not get locked up in bones; C tends to get locked up as CO_3^{2-} ions in limestone, while P tends to get locked up in PO_4^{3-} ions in clay soils and sedimentary rocks; C can get locked up in fossil fuels, but this does not happen to P. (***A*** – two suitable differences)

d. For the structure of bones. Could also say is essential for DNA and RNA, proteins. (***A***)

7. **a.** **i.** **V** = denitrifying bacteria. (***A***) **W** = nitrifying bacteria/*Nitrobacter*. (***A***) **X** = nitrifying bacteia/*Nitrosomonas*. (***A***) **Y** = decomposing bacteria. (***A***) **Z** = N-fixing bacteria/*Rhizobium*/*Azotobacter*. (***A***)

ii. They make their own food using chemical reactions. (***A***) The chemical reactions provide the energy for the bacteria to make organic compounds ('food') from (inorganic) raw materials. (***M***)

b. Lightning in the atmosphere (***A***); is providing the (high) *energy* to convert atmospheric N_2 into NO_3^- ions in the soil. (***M***)

c. Protein, DNA, RNA. (***A*** – any two)

d. Clover is a legume, so has N-fixing bacteria in its roots (nodules). (***A***) These convert atmospheric N_2 into N compounds in the clover. When the clover decomposes after being ploughed into the soil, it will increase the N/NO_3^- content of the soil for other plants growing there (ie increases soil fertility). (***M***)

8. **a.** **i.** Mutualism. (***A***)

ii. The clover gets a source of N to produce proteins, while the bacteria get a sheltered/protected area in which to live. (***A***)

b. Growth in clover is greatly helped when they have *Rhizobium* in their roots. (***A***) This is because the bacteria provide clover with a good supply of N, which the plant can use for making proteins. Proteins are essential for the growth of the plant. (***M***) The more effective the *Rhizobium*, the more N is fixed; therefore, the more N is available to the clover and the more protein it can make. The more protein made, the more vigorous. the growth (as in experiment **A**) The less effective the *Rhizobium*, the less N fixed, so the less protein can be made by clover and so the less vigorous its growth, as in experiment **B**. (***E*** – full explanation related to the experimental results)

9. Energy flows through the ecosystem as it enters the pine trees in photosynthesis and leaves through respiration at all trophic levels. (***A***) Solar energy enters the pine trees during photosynthesis where it is converted to chemical energy ('food'). This chemical food energy passes down the food chain and to the decomposers. At each step/trophic level, energy is lost as heat through respiration. Therefore, a constant input of energy is needed through the producers/pine trees/photosynthesis. (***M***) Much of the solar energy that reaches the Earth/

plants is not used in photosynthesis but is reflected back into the atmosphere. All the solar energy captured in photosynthesis is released in respiration to fuel life processes (eg growth, reproduction) by all organisms/trophic levels. Only the energy that becomes part of the organism (eg growth) is available to be passed to the next trophic level; the rest is lost through respiration (hence the boxes in the diagram decrease in size). Waste products from each level and dead bodies all go to provide food for decomposers. Respiration from decomposers releases the last amount of energy captured by the producers as heat into the atmosphere. None is left. (***E*** – in-depth explanation linking ideas)

10. Nutrient cycles are essential, as the chemicals/elements/materials in waste products and dead bodies need to be released into the soil so that new life forms/plants can be made/grow. (***A***) The nutrients C, N, H, O, P are essential for molecules such as proteins, carbohydrates, fats, vitamins, water, DNA, RNA, ATP – which are essential for the structure and function of all organisms. (***M***) These chemicals/nutrients are in finite supply, so if wastes and bodies do not get broken down by decomposers for release into the soil and their entry into plant roots, then new plant life would not occur and there would also be no new animal life, as food chains would collapse. Nutrients that go deeper into the soil than the depth of plant roots are leached away and not available for plants. Non-living processes in the nutrient cycles (eg sedimentation/uplift/erosion) make nutrients available again to the living component – this can take a very long time. (***E*** – in-depth explanations linking ideas)

11. Green plants are vital as they start off the food chain via the process of photosynthesis and their roots absorb essential nutrients from the soil. (***A***) Green plants capture the Sun's energy in photosynthesis and turn it into chemical energy ('food') for all other life forms to use. Nutrients from wastes and dead bodies are released into the soil by decomposers then they diffuse into plant roots, so again become available to new life forms/food chain/animals. (***M***) Without plants, there would be no source of energy to supply food chains, so life would not exist. At each step of a food chain, energy is lost as heat in respiration, so is not available to the next step/trophic level/life forms. Therefore, all trophic levels depend on green plants for the constant flow of energy into the living community. Nutrients released by decomposers would be lost to food chains if they were not taken in by plant roots and the supply of nutrients to make new life would run out. Therefore, all trophic levels depend on green plants for the provision of recycled nutrients for new life forms. (***E*** – in-depth explanations linking ideas)

Topic 9: Two case studies – New Zealand and Papua New Guinea

Unit 12.1 Activity 9A: Biodiversity (page 121)

1. The ozone layer in the atmosphere absorbs harmful, high-energy UV radiation from the Sun. With ozone depletion, this absorption of UV does not occur/less is absorbed, so that more UV radiation reaches the Earth's surface. UV radiation causes skin damage and increases the occurrence of skin cancers in humans (and other animals), increased occurrence of eye cataracts, increased damage to green plants and unicellular algae/phytoplankton in the sea. (***A*** – two effects described)

2. Habitat destruction (eg deforestation), invasive pests, over-exploitation (eg over-fishing). (***A***)

3. Low biodiversity means fewer different species in the ecosystem, so food webs are less complex/have fewer links, making the ecosystem more prone to collapse. (***A***) The fewer the links in a food web, the greater the effect if a species becomes extinct – organisms that feed on this species may not have alternative sources of food or have reduced food sources, therefore they will be reduced in numbers/become extinct also. This will have flow-on effects on food supply and organism numbers, such that the whole system/food web may collapse (and this may occur over a short time period). (***M***)

4. Introduced invasive species will compete with, or be predators on, native species, reducing their numbers. (***A***) If the introduced species are herbivores, they may eat native vegetation, destroying the habitat and/or food supply for native herbivore species, reducing the numbers of these species sufficiently for them to be endangered (or become extinct). If the introduced species are predatory on native species, they may make native species so low in numbers that they become endangered (or extinct). This is likely to be worse if there are no predators of the introduced species. (***M***) If local species become low in numbers/extinct, then food chains will be adversely affected, so that other species may then become low in numbers/extinct as their food supply is reduced/removed. As a result, food webs may collapse. Typically, invasive species are very successful because they are effective competitors/adapt well to humans and urbanisation/reproduce rapidly/highly tolerant/are not easy to control and/or eliminate. Therefore, introduced species are particularly damaging to the biodiversity of the country that they arrive in. (***E*** – in-depth explanation linking ideas)

5. Answer needs to indicate similarities and differences between the two control methods, eg: Chemical control uses chemicals such as poisons to kill pests, *while* biological control uses another organism such as a predator to kill the pest. (***A***) The chemicals kill *large numbers* of the pest *rapidly*, but usually need to be *applied often* to keep the pest numbers down – this is *expensive*. Biological control usually *works slowly all the time* to keep pest numbers low. Biological research is *expensive*, but once released, the biological agent is *inexpensive* and *effective* in its operation. (***M***) Biological control is *pest-specific*, while chemical control often *does not discriminate* between pest species and non-harmful/useful species. Chemical control may damage both the biotic and abiotic environment, and chemicals can accumulate along food chains. Biological control does *no environmental damage*. However, much costly time-consuming research needs to be done before a biological control agent is released to ensure that it is pest-specific and will not harm/prey on/parasitise other organisms and so become a major pest itself. When successful, biological control *is preferred* over chemical control because it does not cause environmental damage, is inexpensive, and is not labour intensive. (***E*** – in-depth comparisons made)

6.
- Melting of the ice caps (Arctic/Greenland/Antarctica), with subsequent rising of the oceans, flooding land and displacing millions of people worldwide.
- Increase in severe weather effects worldwide – drought, floods, hurricanes/tornadoes, bushfires – with subsequent damage to the environment and flow-on effects.
- Extinction of large numbers of species as their tolerance range is exceeded, and from flow-on effects of severe weather, disruption to food chains.
- New pest species and disease-causing species in areas previously unaffected, with flow-on effects to health of humans, livestock, crops.
- Human starvation on a large scale, as food chains and agriculture adversely affected and land gets flooded. Also, large-scale problems with provision of clean drinking water to many people.

(***A*** – description of any three)

7. Global warming is occurring because of the large-scale production of greenhouse gases, especially CO_2 from the burning of fossil fuels. (***A***) Carbon dioxide (from the burning of fossil fuels and forests/wood), methane (from the belching of livestock, drilling for oils and gases, anaerobic respiration from decomposition), nitrous oxides (from vehicle emissions) and chlorofluorocarbons (used in refrigeration and aerosol cans) form a layer in the atmosphere around the Earth and stop infrared radiation from escaping. This trapped radiation warms the surface of the Earth (including the oceans), creating 'global warming'. (**M**)

8. It will cause the extinction of many species of fish and subsequent collapse of marine food webs/ecosystems. (**A**) This will remove a major food source for humans throughout the world, leading to famine and pressure on other food sources. Large-scale job losses and damage to economies will occur (including damage to the tourist industry). The world will be a much poorer place in which to live (in every sense of the word) as a result. (**M**)
9. Deforestation – the logging/felling of trees in forests (tropical and temperate) throughout the world – removes the habitat of many species, causing extinctions, which reduce biodiversity. (**A**) Tropical rainforests are the richest areas of biodiversity in the world (including many species not yet discovered/classified), so their destruction is causing or will cause the extinction of large numbers of species that inhabit them, with subsequent collapse of food webs/ecosystems. These unique species and their gene pools will be lost forever. (**M**) These lost species had the potential to be sources of food and/or medicine for humans; they enrich our lives (eg tourism). The felling of the forest means CO_2 levels will rise, increasing the greenhouse effect. This will have a flow-on effect on biodiversity throughout the world, as temperatures rise, causing habitat shifts and further extinctions. Climatic changes from forest felling also impact on biodiversity, causing habitat shifts and possible further extinctions. (**E** – in-depth explanations including links to climate changes/ global warming)

Unit 12.2 Population

Topic 1: Understanding population growth

Unit 12.2 Activity 1A: Populations (page 135)

1. a. Territory is an area that an animal/group marks and defends and uses for breeding, while a home range is a larger, shared, undefended area around the territory, which the animal/group uses to forage for food and other resources. (**A**)
   b. Density-dependent factors are factors that act to regulate a population as its density increases, while density-independent factors act to regulate the population regardless of its density. (**A**)
   c. Natality is the birth rate of a population, while mortality is the death rate of a population. (**A**)
   d. Emigration refers to those individuals leaving a population, while immigration refers to those joining a population. (**A**)
2. Population of 1000 will increase by 670 (natality) and 200 (immigration), but decrease by 310 (mortality) and 270 (emigration). This gives an overall increase of 290 (+29%) and the population becomes 1290. (**A** – correct answer; **M** – plus explanation showing the reasoning/working)
3. a. Density, distribution (random/clumped/uniform), age structure, size. (**A**)
   b. Lack of food, lack of space, predation, disease. (**A**)
   c. Changing/high salinity, changing/high temperatures. (**A**)

d.

Adaptations: (***A***)	Explanation: (***M***)
• Snails must be tolerant to varying levels of salinity/temperature/exposure.	• Regular tidal movements cause great variations in these factors.
• Hard shell.	• Protection from water movement/ predation.
• Muscular foot.	• Movement across the mud (eg for feeding, find mate).

4. **a.** Many birds of breeding age and abundant resources. (***A***)
   b. Growth slowed as resources became limiting/in short supply. (***A***) Competition occurred with decreased natality/increased mortality, slowing the growth. (***M***)
   c. About 90 birds. (***A***)
   d. Territories space the birds throughout the environment and so limit the numbers that can live there. (***A***) As the territory provides the saddleback with resources such as food, territories can prevent/reduce competition. With regulated access to resources, the population is regulated (to the carrying capacity of the environment). (***M***) Only those birds that hold territories can breed; this also acts to limit the population. Birds that cannot hold territories are typically young/old/weak (pre- or post-reproductive) and are 'expendable' to the populations. Their elimination from the gene pool benefits the population and increases its overall fitness. (***E*** – more in-depth explanation)

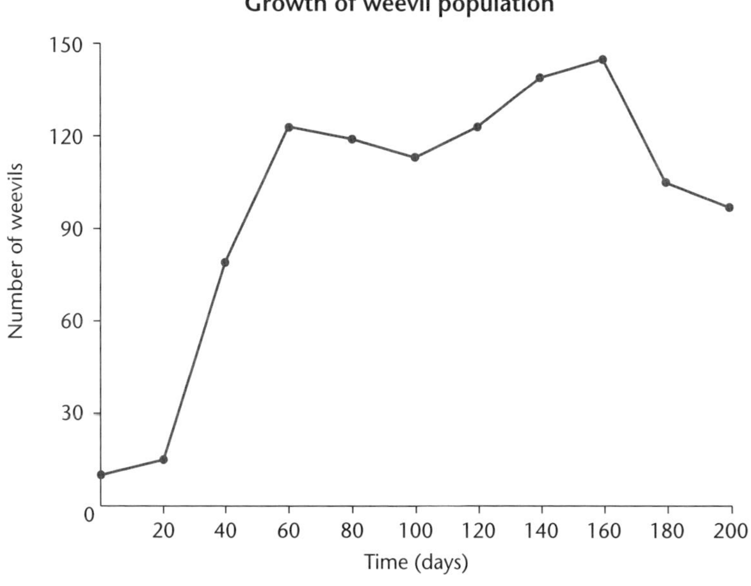

5. **a.** (***A*** – graph correctly plotted)
   b. 20 to 60 days. (***A***)
   c. Lack of food. (***A***) This would have caused competition, with natality decreasing/ mortality increasing, slowing the growth of the weevils. (***M***)
   d. The population would have continued to decline/crash to zero (***A***); as starvation would occur as the food ran out. (***M***)

e. Graph line needs to have the same shape but climbs higher (double the food = about double the population) then follows the same downward trend. (***A***) The population of weevils will grow *bigger/double* but will still crash (*in same manner and time*). More food means natality will increase and so the population will get much larger and so eat the extra food (more food means more weevils so the food supplies are eaten as quickly). Only if the size of the population stays relatively the same will it last longer/double the time. More/double food has been provided, so carrying capacity of the environment has increased/doubled, but starvation will still occur and the population will crash as before. (***M***)

6. When the number of plants with seed heads increases, so does the kiore population. (***A***) This is because kiore eat the seeds and when there are more seeds, there is more food for the kiore so its population increases and vice versa. (***M***) There is a time lag between the increases (and decreases) of both populations. This occurs because it takes a while for the natality and mortality of the kiore to match the food supply. The *biomass* of kiore (the herbivore) is lower than that of the plants, which are their source of food/energy. In 1978, kiore numbers were high and plant numbers relatively low – this could be because the plants were producing more seed heads per plant that year (hence more plant biomass), so more food for the kiore or the kiore may be using other/new sources of food as well as the plant seeds. (***E*** – linked ideas/explanations needed)

7. When predator numbers were low in the first years after release of the wolves, many deer died of starvation (1990–92/93), suggesting deer were overgrazing the environment. As the wolf population increased (up until 1994), the deer population started to decrease, showing wolves were regulating the numbers of deer. (***A***) As deer numbers decreased, wolf numbers began to decrease too (1995 onwards), showing numbers of deer were also regulating the wolf population. This is because the deer are food for the wolves – the more deer, the more food so more wolves, and vice versa. (***M***) There is a time lag between highs and lows of each population (ie as deer increase it takes a while before the more abundant food supply is translated into higher wolf natality/decreased mortality). Similarly, as wolves eat the deer and the numbers of deer drop, it takes a while for the numbers of wolves to decline, as less food translates into lower natality/increased mortality. Number of wolves (predator) is always less than number of deer (prey). Figures on the table indicate numbers of wolves and deer fluctuate directly relative to each other in this simple ecosystem as the food supply for each goes up and down. (***E*** – in-depth explanation linking ideas and related to information in table)

Topic 2: Population sampling methods

Unit 12.2 Activity 2A: Sampling techniques and distribution patterns (page 152)

Some of the answers given here are not definitive in that, for example, there may be other sampling techniques that would also be suitable in obtaining the necessary data; there may be other environmental factors that could influence distribution; similarly, other adaptations.

1. a. i. The pattern of the individuals across their habitat; the range of their occurrence in the habitat.

ii. The number of individuals in a given area, such that:

$$\text{Density} = \frac{\text{Number of individuals}}{\text{Area of habitat}}$$

b. *Uniform distribution* – individuals are distributed evenly across their habitat; typical of territorial animals (such as colonies of nesting gannets).

Clumped distribution – individuals occur together (in groups) in certain areas of their habitat; typical of social animals that live in groups (eg all animals that live in herds, such as horses). Clumping also occurs in favourable environmental conditions – such as a food source, high light intensity, high nutrient levels.

Random distribution – individuals are neither uniform nor clumped in their habitat but scattered throughout; rare, but can be seen in certain plant species in forest communities.

2. Part 1

a. Place a transect line from high to low tide down the shore. At 3 m metre distances down the line, place a 25 cm × 25 cm quadrat and count the green-lipped mussels found in it. To ensure sufficient data, a minimum of 5 transect lines should be used along the shore.

b. If the distribution was *clumped*, then, within individual quadrats, there would be areas where mussels occurred living close together and areas where they did not, and/or in any particular tidal zone some quadrats would have high numbers of mussels living close together while other quadrats would have none. If the distribution was *uniform*, then the mussels would be regularly spaced within quadrats throughout their tidal zone and the count for each quadrat would be the same (or nearly so). If the distribution was *random*, then the occurrence and spacing of mussels in each quadrat would be random, as would the numbers counted in the quadrats (could cover a range from high to low with no apparent reason for it).

c. In processing the data, the counts from the 5 transects would be averaged to give the numbers per 25 cm × 25 cm. It would be preferable if the averages were converted to numbers per m^2 by multiplying by 16.

d. Numbers per m^2 could be shown on a *distribution graph*, with numbers/m^2 on the vertical axis and distance (in m) from high to low tide along the horizontal axis.

A *kite diagram* could be used to show the distribution pattern. Distance (in m from high to low tide) is put on the vertical axis, and an appropriate scale for numbers/m^2 (eg 5 mm = 10 mussels) put on the horizontal axis. The completed 'kite' is shaded in.

e. **i.** Intraspecific (other green-lipped mussels) and interspecific (eg oysters, barnacles, seaweeds, black mussel) competition for space on the rocks to attach and grow; selective predation (eg oyster borers, whelks, starfish) may also influence distribution.

ii. Tidal height; type of substrate to attach to; wave action; time exposed to sun/temperature/dehydration; feeding time.

Zones on the shore are largely determined by the tidal height as this determines the time organisms are exposed to the heating and drying influence of the sun. Organisms that inhabit the higher tidal zones need to be more tolerant (physiologically) to temperature extremes/dehydration. As the organisms only feed when covered by water, those at higher tide zones need to be tolerant/adapted to shorter feeding times.

Animals attached to the rocks (eg barnacles, mussels, oysters) are all filter feeders and have devices to filter microscopic organic material from the water. They open their shells/coverings when covered by water to feed. Seaweeds attach to the rocks by 'holdfasts' and are photosynthetic organisms.

f. Have hard shells to prevent damage from wave action and the shells close tightly when the tide is out to prevent dehydration; cement themselves to the rocks to prevent being swept away; open their shell when the tide is in and filter the water using gills; have a low tolerance to exposure (problems with dehydration, feeding, gas exchange, temperature extremes) and are therefore restricted to the low-tide zone.

Part 2

a. Zonation.

b. Tidal height.

c. Seaweed 1, seaweed 2, cat's eye snails.

d. Ornate limpets, snakeskin chitons, brown barnacles.

e. Have a strong muscular foot that grips very firmly to the rocks.

f. Species all have hard shells (they are molluscs) to prevent damage from wave action.

g. **i.** Interspecific competition may occur for space on the rocks (eg mussels, oysters, barnacles) also for food (eg grazers on algae such as the cat's eye snails, limpets, chitons). Those individuals/species that are not successful may end up living in non-preferred zone (eg high up the shore that their adaptations do not particularly fit them for), placing them at the limits of their tolerance range.

ii. Predation will limit the population sizes of the prey and may affect zones. Those individuals/species that have the tolerance to live outside their preferred zone and that of their predator may do so, so reducing the effects of predation.

3. Both these shellfish are burrowing filter feeders, therefore could be in competition for space in the sand and/or for microscopic food particles in the water.

a. Using quadrats and counting the numbers of both species. A reasonably large quadrat (eg 0.5 m × 0.5 m) should be used, and samples taken over the mid- to low-tide zone of a sandy beach where both species are known to live. Need to dig the sand out of the quadrat.

b. If the two species are found in different zones/tidal heights, then they will not be in competition. It is likely that the pipi will be found more in the mid-tide zone; the tuatua in the low-tide zone.

If/where the two species are found together, then their density becomes important – are the densities of both high? low? how do they compare to densities in areas/beaches when the species are found on their own? (samples from other beaches may be needed). If the densities of the species are lower when they occur together than when they live separately, this suggests interspecific competition. If the density of one species is much higher than that of the other when they occur together, then the more common species is likely to be the more successful competitor in that particular area. If there is no significant difference in their densities whether the species occur together or separately, then they are likely to be co-existing.

4. **a.** Temperature/wind speed/exposure would decrease; humidity/soil nutrients/animal species and numbers/variety of plant species would increase.

b. Marram grass is a hardy plant that is a coloniser of sand dunes. Marram grass has rolled leaves to reduce water loss; a strong fibrous root system to securely anchor the plant and obtain (scarce) water/nutrients; it puts out runners to bud off new plants (asexual reproduction), rapidly increasing the population and assisting anchorage of the individual plants and stabilising the sand. It has a high physiological tolerance to high temperatures/wind/sand abrasion/salinity, and to low water and nutrient levels.

c. Need to draw a line graph with average height of plants in metres up the vertical axis and distance along transect in metres from sand dune along the horizontal axis. The heights are plotted and connected with a smooth curve. Graph is titled.

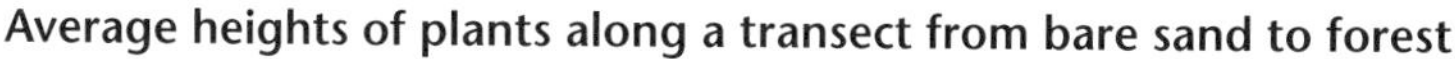

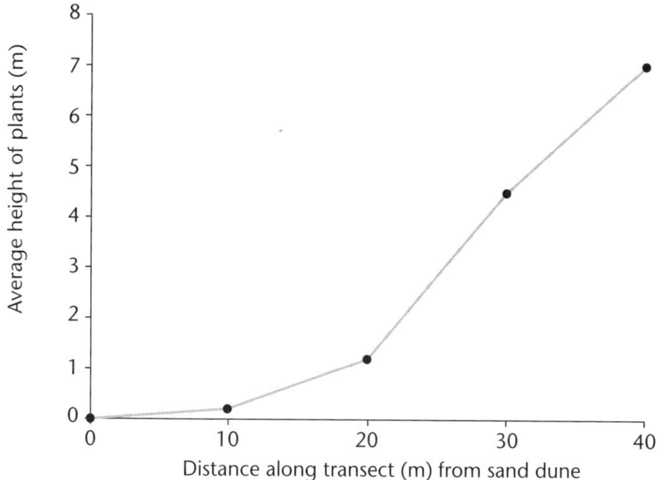

The height of the plants increases from the fore dunes to the rear dunes and forest, as the environment becomes much less harsh (lower temperatures/wind/exposure, higher humidity) and the substrate is now soil rather than sand (much more stable, better supplies of water/nutrients); animals are present to assist pollination; stratification is occurring. All these factors allow for an increased variety of plant life with the presence of tall trees.

5. a. Place transect lines across the areas (eg five, depending on the width of the area) and sample at stations along the transect (eg 10 m apart) or take samples uniformly to cover the area, using large quadrats (eg 1 m × 1 m).

b. Exposure to wind/temperatures/snow cover, drainage (hence water levels) of the soil, nutrient levels in the soil, levels of light intensity, pH of the soil.

c. Food availability/predators/interspecific competition (biotic), temperature range/snow coverage/shelter/water (abiotic).

6. a. Need a line graph with *Number of plants/m*^{2} on the vertical axis and *Distance from the stream boundary* along the horizontal axis. Lines for the four species are drawn on the one graph and a key to identify each is given. Graph is titled.

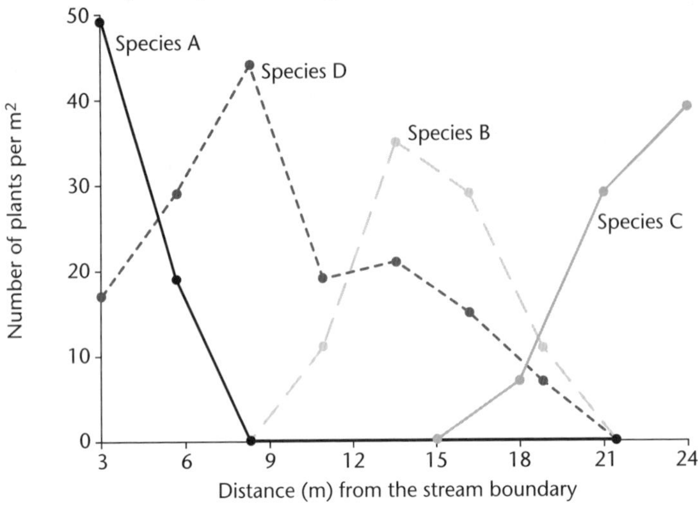

b. Plant D – it is found from the stream bank across the field to nearly the far side, therefore lives in the widest range of environmental conditions.

c. Plants B, C, D – they all occur together towards the far side of the field; the two species most in competition will be B and D, as they have the greatest habitat overlap (9–21 m from the stream).

7. a.
- Wave action – acts to sweep the organisms off the rocks.
- Tidal action – exposes organisms twice a day to the air, restricting their feeding time and placing them at risk of dehydration.
- Rocky substrate – provides a place for the organisms to live and grow by giving them a hard surface to cement on to.
- Temperature increase during low tide times (especially on a summer's day) – increases risk of dehydration.

b. i. Place quadrats (of about 20–25 cm × 20–25 cm in size) at regular intervals throughout this tidal zone and record the % cover of the barnacles in each quadrat. Average the count for all the quadrats and convert to a % per m^2.

ii. Oyster borers are found individually and not clumped/close together and move about the rocky surface, therefore they would be counted separately in the quadrats and not as a % cover.

iii. Its physiological tolerance to abiotic environmental factors (eg amount of time exposed to the air by the tide) and the presence of its prey (eg barnacles, mussels).

iv. Space on the rocks to cement themselves to.

c. High physiological tolerance to exposure to air, as they are infrequently covered by water, therefore are at risk of dehydration/have reduced opportunity to gas exchange/ have reduced opportunity to feed; have a hard shell for protection against wave action when the tide is in; have a foot that holds them to the rock surface, so keeping them from being washed away, but allows movement when the tide is in.

Topic 3: Human population growth in Papua New Guinea

Unit 12.2 Activity 3A: Human population growth in PNG (page 162) – class discussion

Unit 12.3 Genetics

Topic 1: Mendel's pea experiment and inheritance

Unit 12.3 Activity 1A: First principles (page 168)

1. 1J; 2B; 3I; 4F; 5G; 6A; 7D; 8C; 9E; 10H.

2.

Phenotype	**a**. Round	**b**. Wrinkled
Genotype	**c**. *RR*	**d**. *rr*

(***A*** – all four correct)

3. a. Homozygous dominant (*RR*), homozygous recessive (*rr*), and heterozygous (*Rr*). (***A*** – 2 correct)

b. i.

	R	*R*
r	*Rr*	*Rr*
r	*Rr*	*Rr*

ii.

	R	*r*
r	*Rr*	*rr*
r	*Rr*	*rr*

iii. Fraction of offspring with white flowers in cross **i.** is 0.

iv. Fraction of offspring with white flowers in cross **ii.** is 50% or $\frac{1}{2}$ or 1 : 1.

(***A*** – **i.** and **ii.** correct and one of **iii.** or **iv.**)

c. In the genotype ratio, *RR* and *Rr* are two different genotypes but they both give the same phenotype (***A***); so in the phenotype ratio, the red individuals with the genotypes *RR* and *Rr* are both included in the three red-flowered plants. The one *rr* genotype is a white-flowered plant, and this gives the '1' in the phenotype ratio. (***M***)

d. Cross it with a white-flowered plant/do a test/backcross. (***A***) The white-flowered plant will be homozygous recessive and if any of the offspring are white-flowered, the red-flowered plant must be heterozygous/*Rr* (***M***) as the unknown genotype of the red-flowered plant must have contained a recessive white allele that combined with a white allele from the test/back cross producing white offspring. (***E***)

4. a. i. and **ii.** Dominant means an allele that always shows in the phenotype. It can hide a recessive allele. Recessive means an allele that only shows in the phenotype if two copies of it are present in the cell. (***A***)

b. i.

heterozygous yellow-seeded pea plant (columns); **homozygous yellow-seeded** (rows)

	E	*e*
E	*EE*	*Ee*
E	*EE*	*Ee*

(***A*** – Punnett square all correct)

ii. Genotype is the alleles it carries, and the phenotype is what it looks like. (***A***) A heterozygous yellow-seeded pea plant has the genotype *Ee* and its phenotype is 'yellow seed'. (***M***)

c. There are two possible answers.

Either:

Test/backcross with plant that produces green seeds. (**A**) The crosses will be *Ee* × *ee* and *EE* × *ee*. (**M**) After many generations/many crosses, it would be expected that the pure breeding yellow-seeded plants would always produce yellow-seeded offspring, while the heterozygous yellow-seeded plants will produce some green-seeded offspring. (**E**)

Or:

Breed the plants from these seeds for many generations/until green-seeded plants were produced. (**A**) The possible crosses will be *Ee* × *Ee* and *EE* × *Ee* or *EE* × *EE*. (**M**) After many generations/crosses, the heterozygous crosses will be expected to produce some green-seeded offspring, while both of *EE* × *Ee* and *EE* × *EE* will always produce yellow-seeded offspring. (**E**)

5. a. Dominant – allele that always shows in the phenotype.

Recessive – allele that only shows in the phenotype when two copies of it are present/ an allele whose effect can be hidden by the dominant allele. (**A**)

b. 3 in 4 or 75%. (**A**)

c. Genotype is the actual alleles that the individual contains. Phenotype is the way the individual looks as a result of its genotype. (**A**) A seed that has a genotype of *Rr* will look round (round phenotype) and a seed of genotype *rr* will have the wrinkled phenotype. (**M**)

d. Perform a test cross (**A**); a test cross involves a green pea, which will be homozygous recessive. If the yellow pea is heterozygous, some of the offspring will be green. (**M**) If all of the offspring are yellow, it is highly likely that the pea is homozygous dominant, but as fertilisation is a chance process, this cannot be confirmed until a very large number of offspring have been produced that are all yellow. (**E**)

Unit 12.3 Activity 1B: Monohybrid cross (page 174)

1. a. A gene is the section of DNA that codes for a particular feature/trait/protein, while an allele is a different form of a gene. (**A**)

b. Genotype refers to the genetic information present on the genes, while phenotype is the expression of the information. (**A**)

c. Homozygous refers to the condition when both the alleles for a trait are the same, while heterozygous refers to the condition when both alleles are different. (**A**)

d. Dominant refers to an allele of a pair which is always expressed in the phenotype when present, while recessive refers to an allele which is expressed in the phenotype only when both alleles in the pair are the same. (**A**)

2. a. Dimpled is the dominant allele, as all the children in the cross have dimples. (**A**) Both parents are pure breeding/homozygous for the trait, therefore the dominant trait will show in all the offspring. (**M** – both ideas linked)

b. The children from the original cross must be heterozygous (*Dd*), while their partners must be homozygous recessive (*dd*). The cross is:

Dd × *dd*

	d
D	*Dd*
d	*dd*

Therefore, the probability of a grandchild having dimples is 50% or ½ or 0.5.

(**A** – cross and Punnett square correct; **M** – and linked to correct probability)

3. a. i. Cross is *Aa* × *Aa*:

	A	*a*
A	*AA*	*Aa*
a	*Aa*	*aa*

The expected frequency of seedlings that are albino (genotype *aa*) is ¼ or 25%. Therefore, of the 600 seedlings, ¼ × 600 = 150 would be expected to be albino.

(**A** – genotypes and Punnett square to show ¼ albino; **M** – also correct number of 150)

ii. The parental genotype is *Aa*. From the Punnett square, ½ or 50% have this genotype. (**A**)

Therefore, the number of seedling with this genotype is ½ × 600 = 300. (**M**)

b. A lethal allele is one that will kill the individual that has the allele. When two recessive *a* alleles come together, the plant will be albino, ie lacks the green pigment chlorophyll. (**A**) Therefore, when the seedling germinates, it lacks the chlorophyll needed to photosynthesise and make food, so it won't survive. (**M**)

4. a. i. Fraction is ¾, from the cross *Rr* × *Rr*:

	R	*r*
R	*RR*	*Rr*
r	*Rr*	*rr*

(**A** – correct fraction or Punnett square; **M** – correct Punnett square and correct fraction)

ii. Genotype refers to the *genetic material/alleles* present in an individual for a trait/feature (eg *RR*, *Rr*, *rr* are the possible genotypes for seed shape).

The phenotype is the physical expression of the genotype, and results from the presence of dominant and recessive alleles (eg *RR* and *Rr* give a round-shaped seed, because the dominant allele *R* is present in the genotype, while *rr* gives wrinkled-shaped seeds, as both alleles in the genotype are recessive).

(**A** – describes/defines genotype and phenotype; **M** – explains the two terms using correct examples from pea plants)

b. Marika needs to do a *test cross* and breed the yellow pea plant with the *recessive/green* pea plants. This needs to be *done many times/multiple crosses*.

If green peas appear in the offspring, the yellow plant is *heterozygous* for colour; if no green peas appear in the offspring, the yellow plant is *homozygous dominant* for colour.

(**A** – describes a test cross in terms of breeding with a recessive/green plant; **M** – links outcomes of cross to possible genotypes; **E** – gives full explanation of process, including need for multiple crosses)

Unit 12.3 Activity 1C: Dihybrid cross (page 177)

1. a. Purple flowers and long pollen grains are the dominant alleles for each gene, as both parents are pure breeding for the two traits and the cross produces all purple-flowered plants that have long pollen grains. (**A**)

Let *P* be purple flowers and *p* be red flowers; *G* is long pollen grains and *g* round pollen grains; therefore, the parents are *PPGG* (purple flowered and long grained) and *ppgg* (red flowered and round grained). (**A**)

b. **i.** *PpGg*.

ii. The cross is *PpGg* × *PpGg*; this give the following offspring:

	PG	Pg	pG	pg
PG	PPGG	PPGg	PpGG	PpGg
Pg	PPGg	PPgg	PpGg	Ppgg
pG	PpGG	PpGg	ppGG	ppGg
pg	PpGg	Ppgg	ppGg	ppgg

The Punnett square gives an expected ratio of:

- 9 purple flowers with long pollen grains (*P_G_*).
- 3 purple flowers with round pollen grains (*P_gg*).
- 3 red flowers with long pollen grains (*ppG_*).
- 1 red flowers with round pollen grains (*ppgg*).

Another way of expressing 9 : 3 : 3 : 1 is 9/16 : 3/16 : 3/16 : 1/16. The actual results reflect this ratio, in that the total number of seed is 7786 and:

- $\frac{4381}{7786} = \frac{9}{16}$
- $\frac{1470}{7786} = \frac{3}{16}$
- $\frac{1455}{7786} = \frac{3}{16}$
- $\frac{480}{7786} = \frac{1}{16}$

(**A** – correct cross and Punnett square with ratio; **M** – explained link to the actual numbers)

2. **a.** Father = *PpRr*; mother = *Pprr*. (**A** – both correct)

b.

	Pr	pr
PR	PPRr	PpRr
Pr	PPrr	Pprr
pR	PpRr	ppRr
pr	Pprr	pprr

(**A** – correct Punnett square)

i. Probability left-handed = $\frac{1}{2}$ (from genotypes __*rr*)
(**A** – for probability; **M** – if linked to genotypes)

ii. Probability left-handed and phenylketonuria = $\frac{1}{8}$ (from genotype *pprr*)
(**A** – for probability; **M** – if linked to genotypes)

c. A mutation creates a new allele. If this enters the gene pool, it brings a new variation into the population (which will become subject to natural selection).

Other types of variation – such as segregation, independent assortment, recombination, mate selection – all *juggle the existing alleles* to produce *new combinations* of them; they do not create new alleles.

(**A** – idea that mutations create new alleles; **M** – linked to the other processes serving to mix/recombine these)

3. a. i. Male from colony A is *BBHH*; female from colony B is *bbhh*.

ii. Genotype BbHh; phenotype – all black, short-haired mice. (**A** – both **i**. and **ii**. correct)

iii. Expected ratio for a cross of two F_1 individuals is:

9 black short haired (*B_H_*) : 3 black long haired (*B_hh*) : 3 brown short haired (*bbH_*) : 1 brown long haired (*bbhh*).

306 black short : 102 black long : 102 brown short : 34 brown long.

This ratio rarely occurs because fertilisation is a *chance* event. The smaller the numbers, the less likely the ratio will occur as a result of chance. Other factors may be operating to affect the ratio – eg some phenotypes may have reduced survival chances, some alleles may be lethal.

(**A** – correct ratio given; **M** – also reasons for exact ratio not occurring)

iv. *bbHh* or *bbHH*. (**A** – both needed)

b. Cross is *bbhh* × *BbHh*:

	BH	*Bh*	*bH*	*bh*
bh	*BbHh*	*Bbhh*	*bbHh*	*bbhh*

This gives an expected ratio of:

1 black short haired : 1 black long haired : 1 brown short haired : 1 brown long haired.

(**A** – cross and Punnett square correct; **M** – also correct ratio)

Unit 12.3 Activity 1D: Pedigrees (page 180)

1. a. $\frac{2}{3}$ or 66% or 2 out of 3. (**A**)

b. Female C is heterozygous – she has one copy of the normal form of the CFTR gene. (**A**) This shows because she does not suffer from CF but one of her daughters suffers from CF. She also carries the recessive form of the CFTR gene, as this has been inherited from her mother. (**M**)

c. Female C is not affected by CF, as she has the normal form of the CFTR gene. (**A**)

i. Male B must be a carrier, as he has to have inherited the recessive form of the CFTR gene from his mother. Therefore, children produced by B and C would have a 75% chance of having the normal phenotype and 25% of the children would be expected to suffer from CF.

ii. If female C has children with a normal male, then the phenotype of all the children will be normal, but 50% of these will carry the recessive allele of the CFTR gene.

(**M** – one correct explanation; **E** – both situations correct)

2. a. Three tigers are male (two orange-brown, one white) and three tigers are female (all orange-brown). (**A**)

b. Original tigers must each have a *dominant allele* for fur, as both are orange-brown; however, each must also have a masked *recessive allele* for white – as two of their offspring are white furred (so must be homozygous recessive). Parents are therefore heterozygous for fur.

(**A** – identifies parents as heterozygous; **M** – explains this in terms of dominant and recessive alleles linked to the phenotypes)

c. Tiger B could be homozygous dominant or heterozygous. She must have a dominant allele (eg *B*), as her coat is orange-brown, but it is not known if she inherited a recessive allele to go with it (she would be heterozygous dominant *Bb*) or another dominant allele (so would be homozygous dominant *BB*).

The cross is ***Bb*** × ***Bb***:

	B	*b*
B	*BB*	*Bb*
b	*Bb*	*bb*

Therefore, Tiger B has a 25% chance of being homozygous dominant (*BB*) or a 50% chance of being heterozygous (*Bb*).

As Tiger B has mated with a recessive tiger and all their offspring are orange-brown, it is likely she is homozygous dominant (*BB*) rather than heterozygous (*Bb*), as the cross *BB* × *bb* gives all offspring with the dominant orange-brown fur; while the cross *Bb* × *bb* gives offspring in the expected ratio of 1 orange-brown : 1 white.

However, four offspring is a small number, so her genotype can't be decided with certainty.

(**A** – identifies two possible genotypes of Tiger B; **M** – explains this in terms of inheritance of dominant and recessive alleles and gives *BB* as most likely; **E** – links explanations, using Punnett squares, to give possible genotypes of tiger, with *BB* being most likely, but recognising can't be determined with certainty)

3. a. The white allele is dominant, as both parents are white but have produced a black offspring. If white was recessive, *all* offspring would have to be white. Presence of the black offspring indicates both parents are heterozygous for colour. The reason has *nothing* to do with there being *more* white offspring.

(**A** – recognises white dominant because black offspring appear as well; **M** – links this to the reason and that the parents are heterozygous)

b. Let *W* be white and *w* be black; parents thus both *Ww*. (**A**)

c. 3 white : 1 black (**A**)

Topic 2: Monohybrid inheritance – an extension

Unit 12.3 Activity 2A: Monohybrid inheritance (page 186)

1. 1, K 2, O 3, S 4, T 5, U 6, W 7, M 8, Y 9, R 10, B 11, E 12, I 13, A 14, P 15, V 16, F 17, H 18, C 19, G 20, X 21, Q 22, J 23, Z 24, L 25, D 26, N

2. a. No tendrils. (**A**) In cross 2, some of the offspring had no tendrils – yet both the parents had tendrils. The allele for no tendrils must have been carried by the parents, it was not expressed, so it must be recessive. (***M***)

b. **i.** Cross 2: Plant B (tendrils) × Plant C (tendrils).

Since both B and C have tendrils, yet produce some offspring without tendrils, (***A***) both B and C must be heterozygous, ie *Tt*:

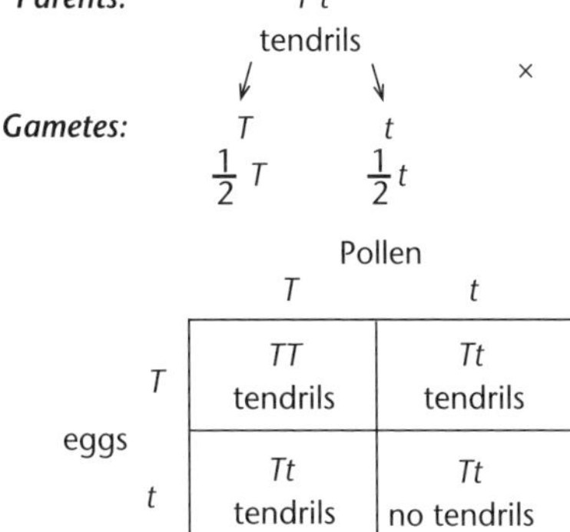

Ratio of 3 with tendrils : 1 no tendrils

(***M*** – explanation and Punnett square)

ii. Cross 4: Plant B (tendrils) × Plant D (no tendrils).

Plant B is *Tt* (see above). Plant D must be *tt* since it shows the recessive trait. (***A***) Hence cross is *Tt* × *tt*. Punnett square is thus:

	Plant B gametes *T*	*t*
Plant D gametes *T*	*TT* tendrils	*Tt* tendrils

Ratio:1 with tendrils : 1 no tendrils

(***M*** – explanation and Punnett square)

c. Tendrils is controlled by a single gene of which one allele is completely dominant (***A***). This is indicated by Cross 4, where a phenotypic ratio of 1 : 1 shows a single pair of alleles which segregate in this ratio. (***M***)

3. **a.** **i.** Types O and A.

Parental genotypes: $I^OI^O \times I^AI^A$ or $I^OI^O \times I^AI^O$. (**M** for showing how)

Children's genotypes:	all I^AI^O or	$\frac{1}{2}$ I^AI^O,	$\frac{1}{2}$ I^OI^O.
Children's phenotypes:	group A	group A	group O (***A***)

ii. Four combinations of parental genotypes are possible, giving children as shown:

$I^AI^A \times I^BI^B$	$I^AI^A \times I^BI^O$	$I^BI^B \times I^AI^O$	$I^AI^O \times I^BI^O$	(***M*** – for showing how)
↓	↓	↓	↓	
All AB	$\frac{1}{2}$ A, $\frac{1}{2}$ AB	$\frac{1}{2}$ B, $\frac{1}{2}$ AB	All 4 types	(***A***)

b. Not possible. (***A***) A group O child must be homozygous, so must have received an I^O allele from both parents. But the man has no I^O allele, so he cannot be the father. (***M***)

4. Mate it with a brown mouse. (***A***) Since the brown mouse must be homozygous recessive, it can only produce gametes carrying the *b* allele. Any variation amongst the offspring must therefore be due to the variation in the gametes produced by the black mouse. If the black mouse is homozygous, then it will only produce *B* gametes and all the offspring will be black. If it is heterozygous, then half its gametes will carry *B* and half will carry *b*. This would give a mixture of black and brown mice in a ratio of 1 black : 1 brown. (**M**)

	If homozygous	If heterozygous:
Parents:	*BB* × *bb*	*Bb* × *bb*
Gametes:	*B B* *b b*	*B b b b*
Offspring:	*Bb* black	*Bb* *bb* black brown

5. i. Definitely dominant (***A***). The second child differs from both parents, so she must show the recessive trait (**M**).

ii. Impossible to say. (***A***) One of the parents must be heterozygous (if both were homozygous, all the children would have the same phenotype), but which one it is, is impossible to tell. (**M**)

iii. Definitely recessive (***A***); since the three children with the 'dark' trait all differ from both parents. (**M**)

> Do not be misled by the fact that 3 of the 4 children show the recessive trait (unusual, but by no means impossible).

6. a. i. Since red is homozygous, they can only produce one kind of gamete (*R*). (***A***)

ii. Pink-flowered plants are heterozygous (*Rr*), so produce two kinds of pollen. (*R* and *r*). (***A***)

b. *Parents:* *Rr*

Gametes: *R* *r*

F_2: *sperms*

eggs	*R*	*r*
R	*RR* red	*Rr* pink
r	*Rr* pink	*Rr* white

F_2 ratio: $\frac{1}{4}$ red, $\frac{1}{2}$ pink and $\frac{1}{4}$ white – a ratio of 1 : 2 : 1 (***A***)

c. Since the heterozygotes are intermediate between the two homozygotes, this is a case of incomplete dominance. (***A***)

7. F_2 ratios are only approximate. With small families, it is more sensible to deal with probabilities, eg 'When two people heterozygous for eye colour have children, any given child is three times more likely to have brown eyes than blue eyes'. (***A***)

8. A Dipa and Martha have two normal children, so they have the allele but do not express it. (**M**)

9. The mutation results in a lethal allele that causes death in the homozygous genotype. (***A***) This is shown by the ratio of platinum matings being approximately 2 : 1, indicating that platinum foxes must be heterozygous and the expected 25% homozygotes produced die before birth. (***M***) The mutation is also pleiotropic, as it has multiple effects – not only does it result in platinum coat colour, but it must also cause severe enough effects in the embryo to cause death. (***E***)

10. The allele is recessive (***A***); indicated by unaffected parents, who must be heterozygous, producing affected children, who must be homozygous recessive. (***M***)

11. a. Pedigree 1 – allele is recessive (***A***); because the offspring of the pairing of the two cousins of generation 3 has the disorder even though they don't. (***M***)

Pedigree 2 – allele is dominant (***A***); because the pairing of the two affected individuals in generation 2 produces an unaffected child, indicating they must have been heterozygous. (***M***)

Pedigree 3 – allele is recessive (***A***); because the two affected children in generation 2 have unaffected parents. (***M***)

b. I = *Aa*, II = *aa*, III = *aa*. (***A***)

12. a. i. Cross is *Aa* × *Aa*:

The expected frequency of seedlings that are albino (genotype *aa*) is ¼ or 25%. Therefore, of the 600 seedlings, ¼ × 600 = 150 would be expected to be albino.

(***A*** – genotypes and Punnett square to show ¼ albino; ***M*** – also correct number of 150)

	A	***a***
A	***AA***	***Aa***
a	***Aa***	***aa***

ii. The parental genotype is *Aa*. From the Punnett square, ½ or 50% have this genotype. (***A***)

Therefore, the number of seedling with this genotype is ½ × 600 = 300. (***M***)

b. A lethal allele is one that will kill the individual that has the allele. When two recessive *a* alleles come together, the plant will be albino, ie lacks the green pigment chlorophyll. (***A***) Therefore, when the seedling germinates, it lacks the chlorophyll needed to photosynthesise and make food, so it won't survive. (***M***)

Unit 12.3 Activity 2B: Polygenic inheritance and pleiotropy (page 192)

1. a. Discontinuous inheritance involves inheritance of characters that are either one or the other (eg ability to roll tongue).

Continuous inheritance involves inheritance of characters where there is a range of phenotypes (eg height in humans). (***A***)

b. Discontinuous inheritance. Because only one gene is involved, it is easy to 'quantify' the effects of the genes. (***M***) In comparison, continuous inheritance involves the interactions of many genes and it is very difficult to separate out individual effects of each gene. (***E***)

2. Because reflex times show a range of variation amongst individuals, indicating that several genes are involved. (***M***)

3. a. The F_1 are roughly intermediate between the two parents. (***A***)

b. Since the two parents were pure-breeding, they were homozygous for the traits. Hence the variation must have been mainly due to environmental causes. (***M***)

Topic 3: Chromosomes, cell division and sex-linked inheritance

Unit 12.3 Activity 3A: Chromosomes and cell division (page 203)

1. 1E; 2A; 3G; 4H; 5L; 6D; 7C; 8K; 9J; 10B; 11F; 12I.

2. a. i. Female/girl/woman.

ii. Male/boy/man. (***A*** – both parts correct)

b. Sex is determined by two sex chromosomes – males are XY, and females are XX. (***A***) The father determines the sex of the child. If the sperm carries the X chromosome, the offspring will be female; if the sperm carries the Y chromosome, the offspring will be male. (***M***)

3. a. Growth and repair. (***A***)

b. The cells produced by mitosis have the wrong number of chromosomes. (***A***) Sperm and egg cells must be haploid/mitosis produces cells that are diploid. (***M***)

4. The cells divide twice to produce gametes with half the genetic information (ie are haploid). (***A***) Variation is produced from meiosis by mixing of the genetic material (called crossing over) and separation of the homologous chromosomes into different daughter cells. (***M***) The variation produced is needed in order to produce offspring with a greater chance of survival in successive generations in a changing environment. (***E***)

5. a. *Examples*: Human: mitosis occurs in the skin/meiosis in the testes. Daisy: mitosis in the growing tips/meiosis in the anther. (***A*** – one correct example)

b. Mitosis occurs to produce many identical cells for growth or repair. Meiosis occurs to produce genetically different cells with half the number of chromosomes for production of gametes. (***A*** – description of purpose; ***M*** – how purpose is achieved)

c. Mitosis – same number of chromosomes, two cells produced from parent cell. Meiosis – half the number of chromosomes, four cells produced from parent cell (in females three usually die, leaving just one functional cell). (***A***)

Mitosis – important that same number of chromosomes maintained so new cells function normally. Meiosis – important that half the number of chromosomes results so that resulting zygote develops into another organism of same species/has the same number of chromosomes as the original parents. (***M***)

Mitosis – important that daughter cells are genetically identical so new cells function normally. Meiosis – genetic variation important so that, should environment change, some organisms have the adaptations to survive and pass on favourable genes to offspring. (***E***)

Unit 12.3 Activity 3B: Sex linkage (page 210)

1. a. A sex-linked condition is due to an allele on one of the sex-determining chromosomes (almost always the X chromosome). (***A***)

b. Red-green colour blindness, haemophilia, Duchenne muscular dystrophy (all X-linked recessive), vitamin D-resistant rickets (X-linked dominant).

c. Female parent can only be $X^B X^b$. Male parent can only be $X^B Y$. (***A***)

		sperms X^B	*sperms* Y
eggs	X^B	X^BX^B black females	X^BY black males
eggs	X^b	X^BX^b tortoiseshell females	X^bY ginger males

(***M***)

2. These genes are situated on the X chromosome. Since this is longer than the Y chromosome, there must be at least some genes on the X chromosome that are absent on the Y chromosome. Females have two X chromosomes while males have only one (***A***). Consequently, an allele that is recessive in females will only be expressed in homozygotes, ie the allele must be inherited from *both* parents. In males, all alleles on the X chromosome are expressed, so they only have to inherit one X-linked allele (from their mother) to show the condition (***M***).

3. X^nO (***A***); because Turner's females have only one X chromosome, in order for her to be colour-blind, this X chromosome must carry the colour-blind allele (***M***).

4. Since the man has normal vision, he cannot carry the colour-blindness allele. The woman's father must have handed on his X chromosome to all his daughters. The woman must therefore carry the allele even though she is phenotypically normal, so she is heterozygous. In the scheme below, X represents a normal X chromosome, and X^c represents an X chromosome carrying the colour-blindness allele.

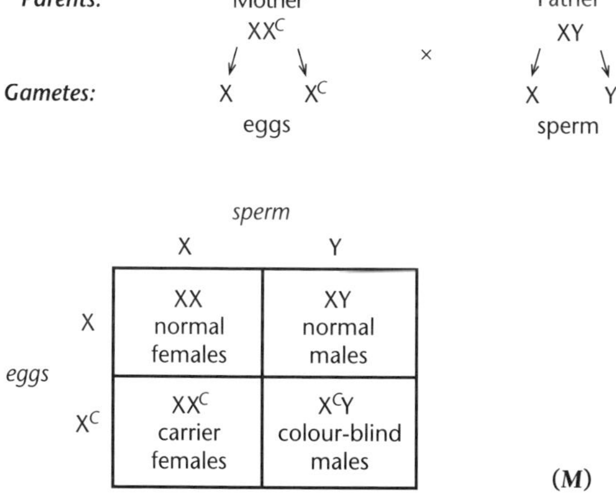

(***M***)

Since one in four of the possible fertilisations will result in a colour-blind male, the probability that the first child will be a colour-blind boy is 0.25. (***A***)

5. **a.** False. It can occur in females, though very much more rarely. This is because for a female to be haemophiliac, she must inherit the allele from *both* parents. (***M***)
   b. True. His mother has two X chromosomes, one inherited from each of her parents. (***M***)
   c. False. Unless the mother is homozygous, half her sons will receive her normal allele and will be unaffected. The father does not hand on his defective allele to his sons. (***M***)
6. **a.** Probably X-linked recessive. (***A***) It must be recessive since III2 differs from *both* his parents. Two things suggest that it is X-linked:
   - It only occurs in the males (though X-linked traits occur in females, but much less commonly).
   - III2 and IV3 both inherit the condition from their *mothers* (neither of whom show the condition). It is unlikely to have been inherited from the fathers (II1 and III5) because it is stated that the condition is rare in the general population – ie outside the family. (***M***)

   b. Autosomal recessive. (***A***) It must be recessive since IV1 and IV2 differ from both their parents. If it were X-linked, then both II2 and II3 would show the condition – thus it must be autosomal recessive. (***M***)
   c. Probably autosomal dominant. (***A***) Since the allele for the shaded condition is rare in the general population, can assume I1 to be homozygous. Since the children are of both types, I2 must be heterozygous so the allele must be dominant. (***M***)
   d. Probably X-linked dominant. (***A***) All the daughters of the affected male I2 are affected, but not one of the sons is. (***M***)
7. **a.** False. They are more common in males but occur in females when they are homozygous. (***M***)
   b. True. Males can only receive the allele from their mothers, but females can inherit it from either parent. (***M***)
   c. True. Since a female can inherit the allele from *either* parent, she is twice as likely to show the condition. Males can only inherit the allele from their mothers and so are only half as likely to have the allele. (***M***)
8. Y-linked recessive. (***A***) An affected female would be homozygous and thus give the allele to all her sons, who would also be affected. (***M***)

Topic 4: Biology of DNA – biochemistry of the genetic material

Unit 12.3 Activity 4A: DNA (page 219)

1. 1J; 2B; 3E; 4G; 5A; 6I; 7H; 8F; 9C; 10D.
2. **a.** **i.** Correct sequence – GACCTG.
   ii. (***A*** – part **i.** correct and any two parts correctly labelled in part **ii.**)

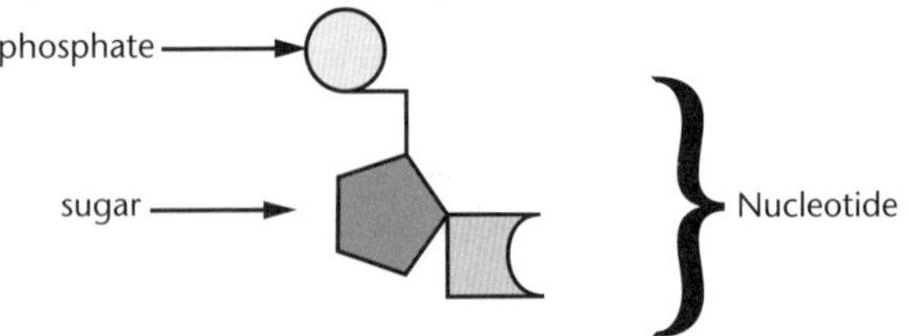

b. Base pairing means the information doesn't change. For example, during replication the new DNA is an exact copy of the old. (***A***) If base pairing were random, the information would change and this would result in a mutation that could be harmful to the organism. (***M***)

c. The DNA makes up genes/is the carrier of the information for flower colour. (***A***) Since the information is carried in the order of the bases making up DNA, different combinations of bases will result in different characteristics/flower colour. (***M***) Different combinations of bases result in the two alleles for red and white flower colour/different combinations of bases result in different proteins that result in the two different colours. (***E***)

3. a. *Bases* – four different types (A, C, T and G) that join together in pairs. *Deoxyribose sugar* – forms the sides of the DNA molecule. *Phosphates* – alternate with the sugars to form the sides. DNA is made up of repeating sequences of base, sugar and phosphate/ nucleotides. (***A*** – any two correct)

b. C and G – 600 times each; T – 300 times. (***A***)

c. T pairs with A, so if A occurred 300 times, so would T. Diagram shows twice as many C–G pairs, so if A–T occurs 300 times, then C–G occurs 600 times each. (***M***)

d. **i.** and **ii.** DNA replication occurs before cell division. Chemical bonds between the base pairs of the two strands break, unzipping the DNA, resulting in original bases unpaired. Unmatched bases on spare nucleotides floating around the DNA match up and join with corresponding unmatched bases on unzipped DNA. Enzymes join the sugar and phosphates together to form the new sides. Unzipping of original DNA occurs until entire DNA replicated. Replicated DNA rewinds and now consists of one original strand and one newly made strand. (***A*** – description of process) In this way, the two original sides of the DNA are used to make exact replicas so the same genetic code as was in the parent cell is passed onto the daughter cells. (***M*** – explanation of importance) Any changes to the DNA/mutation could result in a protein that doesn't function properly and the organism could be harmed. (***E*** – link made to effect on organism)

Unit 12.3 Activity 4B: DNA structure and replication (page 220)

1. The DNA molecule is a double helix of two strands twisted together. The strands are made of alternating sugar (deoxyribose) and phosphate groups, held together by pairs of nitrogen bases which are bonded to the sugars. (***A***)
2. A nucleotide is made up of a phosphate group bonded to a deoxyribose sugar. One of four nitrogen bases is bonded to the sugar. (***A***)
3. A = a nitrogen base; B = deoxyribose sugar; C = phosphate group. (***A*** – all three correct)
4. The base A always pairs with T; and the base C always pairs with G. The hydrogen bonds between the base pairs hold the DNA strands together. (***A***)

   This base pairing mechanism allows DNA to *replicate*, because the hydrogen bonds are easily broken to allow DNA to 'unzip' and a corresponding base pair bonds to the base already in the strand. This produces two identical DNA molecules. (***M***)
5. C = 22% and A = 28%. (***A*** – both correct)
6. TAG CCT AGA TCG (***A*** – all correct)
7. DNA replicates by unwinding the helix, then 'unzipping' between the base pairs. A nucleotide with the corresponding base bonds to its base pair (eg A with T, C with G). The two strands wind into a helix and two identical DNA molecules are formed. (***A***)

DNA replication occurs during interphase in the cell cycle/prior to cell division (mitosis, meiosis), so that the chromosomes can replicate. This allows each cell made (in mitosis) to have a complete, identical, set of chromosomes. (**M**)

Replication is essential, as DNA is the genetic code so has the instructions that allow the cell to form, grow, carry out all its essential processes (eg respiration, protein synthesis), reproduce, even die. (**E**)

8. The DNA of a rat and a cat will have the same structure – a double helix – and be made of the same building blocks – nucleotides. The three chemicals present – deoxyribose, phosphate, the four nitrogen bases – will be the same. It is the *sequence of the bases* on the DNA that will be different. (**A** – description must include the difference)

Unit 12.3 Activity 4C: Making proteins (page 232)

1. a. Step 1 = replication; occurs in the nucleus.
Step 2 = transcription; occurs in the nucleus.
Step 3 = translation; occurs in the endoplasmic reticulum of the cytoplasm. (**A**)

b. End of a peptide chain is signalled by one of three stop codons. (**A**)

2. a. i. DNA. **ii.** tRNA. **iii.** rRNA.

b. i.

G T A A C C T G A G C A T T C	DNA
C A T T G G A C T C G T A A G	DNA
G U A A C C U G A G C A U U C	mRNA
C A U U G G A C U C G U A A G	tRNA

ii. Instead of GUU the first triplet would be UAA, which is a stop codon, so no peptide would be produced. (**A**)

iii. Only changes in DNA can be inherited; deletion of a base from RNA cannot be inherited and is therefore not a mutation. (**A**)

iv. If two bases specified each amino acid, there would be $4 \times 4 = 16$ possible sequences, which falls short of the 20 needed (one for each kind of amino acid). (**A**)

3. a. tRNA and rRNA.

b. i. The anticodon of a tRNA carrying an amino acid links by complementary base pairing with its corresponding mRNA codon. (**A**)

ii. A peptide bond forms between the terminal amino acid of the existing peptide chain and the amino acid linked to the newly arrived tRNA. Water is produced in the process. (**A**)

iii. The newly linked amino acid detaches from the tRNA, which is now available to combine with another amino acid. (**A**)

4. a. False; the genetic code is the same for all species so far studied (with extremely rare and minor exceptions). (**A**) For a change in the genetic code to occur, a mutation would have to occur in a DNA triplet specifying a tRNA anticodon. If this occurred, the new anticodon would combine with a different mRNA codon, but the tRNA would continue to combine with the same amino acid. As a result, the amino acid would be incorrectly placed in *every protein containing it*. (**M**)

b. True; some amino acids are specified by more than one codon. (**A**) In particular, a change in the third base may produce no change. (**M**)

5. a. All forms of RNA. b. DNA, tRNA, rRNA.
   c. All forms of RNA. d. mRNA.
   e. DNA and rRNA. f. DNA and all forms of RNA.
6. It is the same for all organisms. (***A***)

Unit 12.3 Activity 4D: Proteins – products of genes (page 243)

1. Enzymes and constituents of cell membranes. (***A***)
2. The *sequences* of the amino acids in the polypeptide chain may be different (***A***); and so change the folding and therefore structure and activity of the polypeptide. (***M***)
3. The amino acid may form part of the active site of the enzyme *or* it may play a role in holding the molecule in the specific shape upon which the activity of the enzyme depends. (***A***)

Topic 5: Mutations – the origins of new alleles

Unit 12.3 Activity 5A: Gene mutation (page 251)

1. a. Sudden, relatively permanent change in the genetic material (DNA). (***A***)
   b. Change in the 'reading frame' of mRNA by a deletion or insertion. (***A***)
   c. Gene mutation resulting in another amino acid being coded for. (***A***)
   d. External agent (chemical radiation) that increases rate of mutation. (***A***)
   e. Gene mutation resulting in failure of production of an essential enzyme in a metabolic pathway. (***A***)
2. a. This mutation is a base substitution, CCT having been changed to CCC. The amino acid that is specified (glycine) remains unchanged (***A***); because of the redundancy or degeneracy of the genetic code, most amino acids being specified by more than one triplet. (***M***)
   b. This mutation is a frame shift. Adenine has been deleted, thus changing the reading frame and altering the entire message after this point. (There is also a strong possibility that a 'stop' codon will be created somewhere along the gene, thus greatly shortening the resulting polypeptide.) (***A***) If the polypeptide is an important one, such as an enzyme, the organism could be harmed or even die. (***M***)
3. Mutation has no relation to *need*. A change in the environment (eg radiation or an increase in temperature) may increase the *rate* of mutation, but has no effect on the *kind* of mutation that occurs. (***A***)
4. 'Most mutations are harmful in a constant environment.' An organism that has successfully reproduced obviously has 'good' alleles, and any random change is likely to prove harmful in the same environment. (***M***)
5. Gene or point mutations – affect a single gene; block mutations – affect many genes along a section of chromosome; changes in ploidy – changes in the total chromosome number. (***A***)
6. Mutant alleles are harmful in most environments (***A***); and so are removed by selection. (***M***)
7. A somatic mutation is one that occurs in cells that do not give rise to gametes. (***A***) Since they cannot be handed on to the next generation, they are of no evolutionary significance. (***M***)
8. a. A mutagen is an agent that induces mutation. (***A***)
   b. Radiation (eg X-rays, ultraviolet radiation), and chemicals (eg nitrous acid, hydroxylamine, acridine dyes). (***A***)

Unit 12.3 Activity 5B: Chromosomal mutation (page 261)

1. **a.** Chromosome/block mutation where a section of one chromosome breaks off and rejoins onto an end of a non-homologous chromosome. (***A***)
   b. Having unusual numbers of a certain chromosome or chromosomes. (***A***)
   c. Failure of chromosomes or chromatids to separate during cell division. (***A***)
   d. An organism or cell with four sets of chromosomes. (***A***)
   e. A fertile hybrid containing two sets of chromosomes from one species and two from another due to non-disjunction occurring in the sterile (F_1) hybrid. (***A***)
2. **a.** **i.** *One of several possible answers:*
   1 2 6 5 4 3 7 8
   ii. *One of several possible answers:*
   1 2 3 4 5 6 d e
   a b c 7 8
   (***A***)
   b. A substance or agent that increases the rate of mutation, eg radiation, certain chemicals (eg formaldehyde, benzene, tobacco tar, nitrous acid, acridine dyes, asbestos, arsenic), some viruses. (***A***)
3. **a.** There are two of each kind of chromosome (except the Y chromosome). (***A***)
   b. Individual B is a male (Y chromosome) with Down's syndrome (3 × chromosome 21). (***A***)
   c. Fusion by an egg carrying two of chromosome 21 with a normal sperm. The abnormal gametes would be produced by non-disjunction in meiosis (first or second division, in the ovary or testis). (***A***)
4. i, D ii, I iii, F iv, A v, J vi, C vii, B viii, E ix, H x, G
5. *There are several possible answers for each of these, for example:*
   Inversion: A B C G F E D H I J
   Translocation: A B C D R S T M N O P Q E F G H I
   Deletion: A B C D E I J or A B C D E F G
   Duplication: M N O P Q O P Q R S T (***A***)
6. **a.** Monosomy.
   b.

Parents: XX (2n) → gametes XX egg $n+1$, – egg $n-1$; XY (2n) → gametes X normal sperm, Y normal sperm

or XX → X normal egg, X normal egg; XY → XY sperm $n+1$, – sperm $n-1$

Gametes: XX egg $n+1$ | – egg $n-1$ | X normal sperm | Y normal sperm *or* X normal egg | X normal egg | XY sperm $n+1$ | – sperm $n-1$

Offspring:

eggs \ sperm	X	Y
XX	XXX	XXY Klinefelter's
–	X0 Turner's	Y0 lethal

or

eggs \ sperm	XY	–
X	XXY	X0
X	XXY	X0

7. a. Its two chromosome sets are dissimilar (**A**), so each chromosome has no 'partner' with which to pair at meiosis. (**M**)
   b. 14.

Topic 6: Genetic variation and diversity

Unit 12.4 Activity 6A: Genetic variation (page 272)

1. a. A measure of the range of alleles present in the gene pool of a population. (**A**)
   b. All the alleles present in a population of a species. (**A**)
   c. The number of times that an allele occurs within the gene pool. (**A**)
2. a. A gene is the section of DNA that codes for a particular feature/trait/protein, while an allele is a different form of a gene. (**A**)
   b. Genotype refers to the genetic information present on the genes, while phenotype is the visible expression of this. (**A**)
   c. Haploid (n) refers to a single set of chromosomes (eg in gametes), while diploid ($2n$) refers to the double set of chromosomes (eg in somatic cells). (**A**)
   d. A gamete is a sex cell (ovum or sperm) that has the haploid number of chromosomes, while a zygote is the fertilised ovum, which has the diploid number of chromosomes. (**A**)
   e. Mitosis is the process of cell division that produces the somatic cells with a full set of chromosomes, while meiosis is the process of cell division that produces the gametes with a half set of chromosomes. (**A**)
   f. Independent assortment is the way in which the homologous chromosomes line up and separate independently from each other in metaphase and anaphase I in meiosis, while segregation is the way in which the chromatids with their alleles separate independently from each other in metaphase and anaphase II in meiosis. (**A**)
   g. Crossing over is the crossing over of non-sister chromatids in prophase I of meiosis, while recombination is the breaking and rejoining of these chromatids at the cross-over point/chiasma so that their alleles are recombined. (**A**)
   h. A point mutation is a change to the sequence of bases in a DNA molecule either by deletion, substitution or insertion/addition of individual bases, causing a change in the genetic code; while a chromosomal mutation is a change to a chromosome causing genes to be deleted, repeated or translocated – a point mutation occurs within a gene, while a chromosomal mutation involves whole genes. (**A**)

   When asked to distinguish two terms, it is necessary to define them while ensuring that the difference(s) are given; hence the use of the word 'while' in the above answers.
3. Mutations are the source of new alleles which provide new variations in the gene pool. (**A**) It is these new variations that provide the raw material for natural selection in changing environments. (**M** – both ideas linked)
4. a. To halve the chromosome number in producing the gametes. (**A**)
   b. Reduction-division could be used instead of meiosis, as meiosis is the process of cell division during which the chromosome number is halved. (**A** – both ideas of cell division and halving)

   The chromosomes need to be reduced from diploid/2n to haploid/n in the production of gametes. (**M** – both ideas linked)

c. Homologous chromosomes are identical in size and shape and have the same sequence (or locus) of genes (*not* alleles) on them. One of each homologous pair comes from the female parent/mother and the other from the male parent/father. (**A** – description plus diagram)

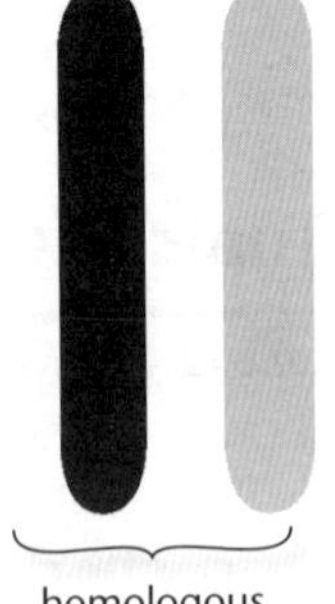

d. A tetrad refers to any homologous chromosomes when they have two chromatids and come together during prophase I in meiosis. (**A** – description plus diagram)

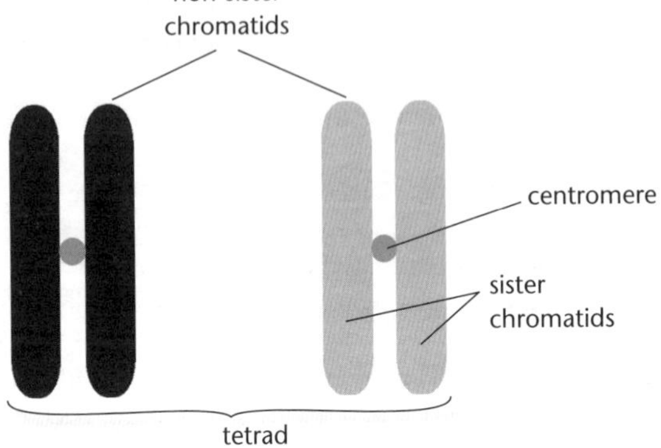

5. a. European red fox – diploid = 34, haploid = 17; arctic fox – diploid = 52, haploid = 26. (**A** – need all four correct)

b. The hybrid would have 43 chromosomes in each body cell (**A**); because it gets 17 from the gamete of the red fox parent and 26 from the gamete of the arctic fox parent. (**M**)

c. This hybrid will not be a new species as it is just a hybrid/intermediate form between the arctic and red foxes and will be infertile/won't be able to breed/produce fertile offspring. (**A**) This is because the hybrid has an odd/uneven/unpaired number of chromosomes in its body cells. (**M** – linked to the first idea) The presence of unpaired chromosomes means that meiosis is likely to fail/not produce viable gametes, meaning that individuals cannot successfully reproduce (ie are infertile). Hybrids would need to be able to successfully reproduce with each other and not with either parent species in order to be a new species. (**E** – ideas linked to give an in-depth answer)

6. a. **i.** 6 **ii.** 3 (**A** – both correct)

b.

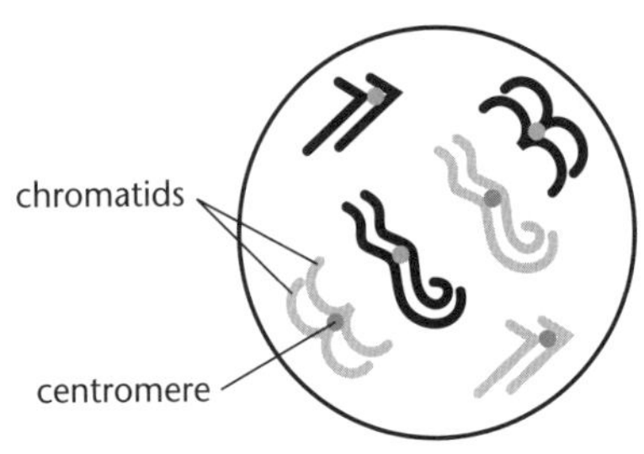

In prophase I, the chromosomes have shortened and thickened and show as two chromatids joined by a centromere. (***A*** – description plus diagram)

c.

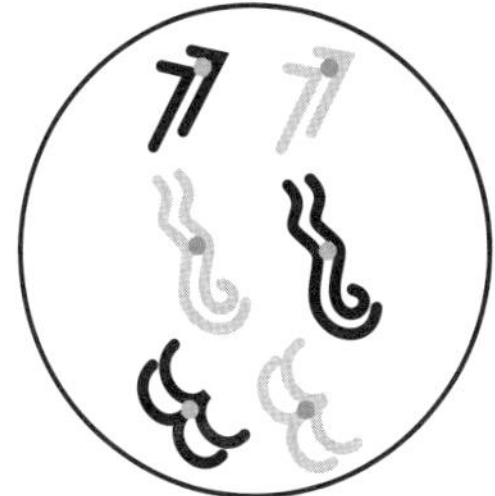

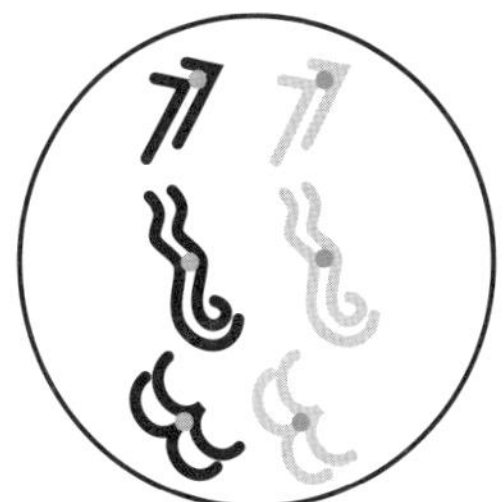

There are four possible different combinations.

(***A*** – minimum of two diagrams showing different combinations of the homologous pairs)

d. Either:

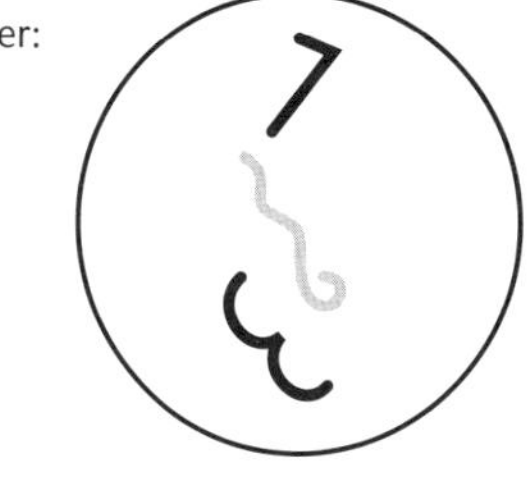

and

Or:

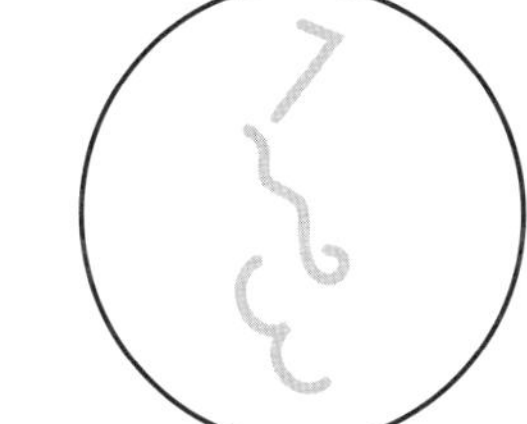

and

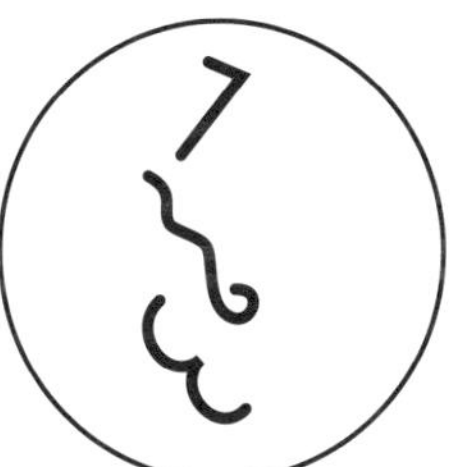

(***A*** – minimum of two diagrams showing different combinations of the chromosomes)

e. In metaphase I of meiosis, the homologous chromosomes line up along the equator of the cell; they do this randomly, independent of other pairs (ie independent assortment). The maternal chromosomes may be on one side of the equator, the paternal on the other; it is chance which one of which pair is on which side. As there are 23 pairs, there are 2^{23} different ways of doing this – this is a *vast* number of different combinations.

When the chromatids are separated in anaphase II, the alleles segregate independently of the alleles on other chromatids; again this is a chance event.

Both these events can give many different combinations of alleles in the gametes produced. As these occur in both the production of ova and sperm, the different combinations of alleles that may occur in fertilisation is very large. Fertilisation enhances the variation that occurs in gamete production. The offspring contribute a unique combination of alleles to the gene pool, so enhancing its variation.

(***A*** – describes the three processes; ***M*** – explains them; ***E*** – links explanations into a coherent answer)

7. a.

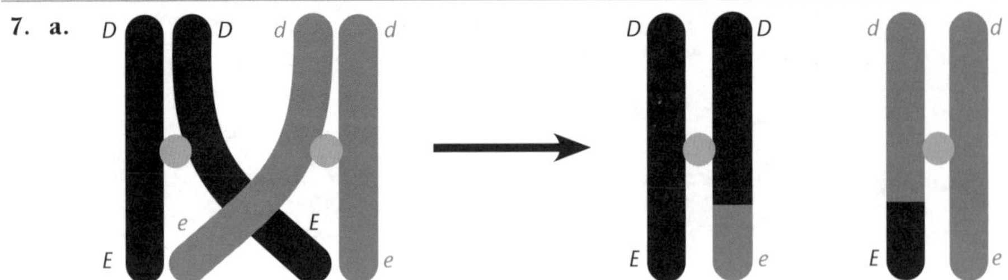

The non-sister chromatids crossover/form a chiasma between *D* and *E*/*d* and *e*. The chromatids break then join to the other chromatid so that *D* is now paired with *e* and *d* with *E*. (**A** – description plus diagram)

b. **i.** *DE* and *de*. (**A** – both needed) **ii.** *De* and *dE*. (**A** – both needed)

8. *FGH*, *FGh*, *FgH*, *Fgh*, *fGH*, *fGh*, *fgH*, *fgh*. (**A** – any 4 correct; **M** – all correct)

9. Similarities in families (between parents and children/grandchildren; between *siblings* or brothers and sisters) results from individuals having the *same alleles*. Each child gets 23 chromosomes from their mother and 23 chromosomes from their father. At fertilisation, the new embryo now has 46 chromosomes containing alleles from *both* parents. However, these alleles will be in *different combinations* (as a result of independent assortment, segregation, crossing over and recombination) to other members of the family.

The expression of the alleles in the individual will depend on the presence of *dominant alleles* – those who get the dominant alleles will have the same trait and so look alike. *Recessive alleles* which are masked by dominants may come together and so get expressed in the phenotype, hence the individual(s) with these will look different from the others.

It is the unique set of alleles that makes the family members different; it is the presence of alleles in common (especially dominants) that gives them their resemblance.

(**A** – needs to identify the presence of the same alleles but in different combinations; **M** – explains these; **E** – links explanations into a coherent answer)

10. It is essential to give the differences and similarities between the two processes.

Mitosis and meiosis are both processes of *cell division* making new cells. Mitosis makes *somatic* cells, meiosis makes *gametes*. Prior to both processes (in interphase), chromosomes duplicate (through DNA replication) to form two *chromatids*. In prophase of both processes, chromosomes shorten and thicken, becoming visible as two chromatids joined by a centromere.

Key difference is that meiosis *halves* chromosome number (2*n* to *n*); mitosis *maintains* chromosome number (at 2*n*).

In halving the chromosome number, meiosis has two sets of divisions (prophase, metaphase, anaphase, telophase I then II). The first set of division separates the homologous pairs, the second set separates the chromatids. *The result is four haploid gametes.*

In mitosis, only one set of divisions is needed (prophase, metaphase, anaphase, telophase) which separates the chromatids. The result is *two diploid cells*.

(**A** – describes minimum of two differences and similarities between the two processes, must include halving of chromosome number in meiosis; **M** – explains minimum of two differences and similarities, including halving of chromosome number in meiosis; **E** – comprehensive comparison of the two processes, all main points explained)

Topic 7: How genetic knowledge is applied – an overview

Unit 12.3 Activity 7A: Applications of genetics (page 279)

1. a. Plants grown from cuttings will be genetically identical to the parent/have the same genetic make-up/the same characteristics as the parent/will grow more quickly into entire plants/plants grown from seeds will have greater variation/be genetically different. (***A*** – description of advantage)

As a result, fruit farmers will have an orchard of trees of consistent/known quality/be able to harvest fruit sooner. (***M*** – explanation of why it is better to grow plants from cuttings)

b. *Either*: Genetic modification enables selected/desirable traits to be selected for.

Or: Genetic modification can be used to improve such traits as colour/sweetness/crispness. (***A*** – description of either; ***M*** – links both points together)

Topic 8: Manipulation of DNA – techniques in gene technology

Unit 12.3 Activity 8A: Putting genes to work – gene technology (page 293)

1. a. 1, E 2, B 3, D 4, H 5, A 6, C 7, G 8, F 9, P
10, M 11, O 12, L 13, S 14, R 15, Q 16, K 17, J 18, T
19, N 20, I

2. a. C, 1

b. A

c. T A A C C C

d. Autoradiography.

3. The one that recognises 4 bases (***A***), since a specific 4-base sequence would occur by chance more often than a particular 8-base sequence. (***M***)

4. a. 1 – DNA to be amplified is heated to 95° C to separate.

4 – DNA polymerase enzyme extends primers. (***A***)

b. Taq polymerase will synthesise new DNA from the template DNA. (***A***) Because Taq polymerase is heat-stable it will not be denatured by the high temperatures involved. (***M***)

c. PCR is used to amplify the often small amounts of DNA found at crime scenes. (***A***) However, there is a risk of contamination by extraneous DNA that would also become amplified (***M***) and may result in incorrect convictions being made (or some other reference to the social or ethical impact of the use of PCR in processing crime scenes). (***E***)

Topic 9: Application of gene technology – biotechnology

Unit 12.3 Activity 9A: Putting genes to work – applications of gene technology (page 322)

1. 1, K 2, R 3, L 4, C 5, P 6, E 7, J 8, G
9, Q 10, B 11, F 12, I 13, H 14, A 15, M 16, D

17, N	18, O	19, T	20, AC	21, W	22, Y	23, U	24, X
25, V	26, S	27, AA	28, AB	29, AE	30, Z	31, AD	

2. J G C E A I F D H B
3. To ensure that the ends are compatible (***A***) because each restriction enzyme generates the same kind of cut. (***M***)
4. It is single-stranded. (***A***)
5. It would contain introns (***A***), which the bacterium could not remove. (***M***)
6. **a.** Restriction enzyme cuts normal gene from human DNA and cuts open viral DNA – ligase enzyme anneals the human gene into the viral DNA. (***A***)

   b. Need to ensure the virus doesn't affect humans, causing disease (***M***) particularly in vulnerable foetal tissue where effects may be unknown (***E***): *or* need to ensure foetus isn't damaged by insertion technique (***M***) possibly resulting in miscarriage or post-natal effects. (***E***)
7. **a.** Use restriction and ligase enzymes to cut and anneal foreign gene into vector *or* transform host plant tissue in some way for example Agrobacterium, ballistics, etc or identify transformed plant tissue using marker such as antibiotic resistance or use tissue culture techniques to grow many transgenic plants – *see text for full details.* (***A***)

   b. Because of low 'uptake' rate of foreign DNA into plant cells it is difficult to predict the success rate; plant promotors may not be able to cause expression of foreign DNA in host cell; unpredictable 'position effects of inserting foreign DNA into plant tissue. (Any two = ***M***)

   c. A known copy number causes a certain spotting pattern on petals. (***A***) By doing a DNA profile of transformed plant tissue before they grow and flower comparisons of copy number to standard samples can be done and spotting pattern established. (***M***)

   d. If presence of similar markers can be established in other commercial plant species that cause undesirable effects then wastage of money and time can be avoided early on in the process. (***A***)

Topic 3: Mechanisms of evolution – genetic changes in populations

Unit 12.4 Activity 3A: Genetic change (page 345)

1. **a.** Natural selection is the process in which environmental factors select those individuals with favourable phenotypes/variations/features/adaptations to survive and breed. (***A***) These selected individuals survive and breed with other similar individuals, so the frequency of the alleles for these favourable phenotypes will increase in frequency in the gene pool. (***M***)

   b. Natural selection occurs when environmental factors select the individuals which will survive and breed, so contributing their favourable alleles/'good' genes to the gene pool, leading to increases in their frequency. Environmental factors (selecting agents) may be biotic (eg predators, competitors) or abiotic (eg temperature, salinity, pH).

   In artificial selection, it is humans who are the selecting agents, choosing those individuals with favourable/desirable features that will be 'allowed' to breed and which will pass on alleles for these features. These alleles will increase in the gene pool. Desirable features may be those for attractiveness (eg coat colour, length of hair, size or shape) or for agricultural purposes (eg quantity of milk produced by dairy cows, crispness of apples, specific chemicals present in grapes).

(***A*** – both processes described with differences clear, including appropriate examples)

c. This is an example of *sexual selection*. The male peacock displays his tail to females; the females select the male with the most impressive tail as their mating partner. (***A***) The biggest, most colourful, most spectacular tail is the one most likely to attract females, and its owner will be selected as the mating partner. Therefore, alleles for these tails will increase in frequency in the gene pool; in future generations, more males will have more spectacular tails. (***M***)

The spectacular tail may indicate a healthy, strong male, free of disease/parasites – so he has 'good' genes to pass on to his offspring via the female.

There is a limit to the tail's development, in that if it is too large it may hinder the male's movement/make him easier for predators to catch/need too great an energy requirement to maintain – natural selection would select against future development of the tail. (***E*** – developed explanations with recognition of the limits to sexual selection)

2. a. Gene pool refers to the total number of alleles within a population, while gene flow refers to movement of alleles between populations. (***A***)

b. Through any/all of the processes of natural selection, migration, genetic drift, bottleneck effect. (***A*** – three processes identified)

Natural selection will increase the frequency of favourable alleles.

Migration may increase or decrease frequency of alleles, depending on the migrants and direction and numbers of migrants.

Genetic drift will randomly change allele frequency in small populations.

Bottleneck effect will randomly change frequencies in a population when a population crash occurs. (***M*** – the three processes explained)

3. New alleles result from *mutations* occurring during DNA replication causing the deletion or substitution or addition of bases. This changes the genetic code. A new amino acid may be produced which will result in a changed protein. This protein may or may not be able to fulfil its biological function (eg a catalytic enzyme).

A gene is that length of DNA that codes for a specific feature/protein; therefore, any change in the gene occurring in a gamete-producing cell means that the mutation will enter the gene pool of the population and become subject to natural selection. If the mutated allele is favourable, it will be selected for, and become established in, the gene pool, and would be expected to increase in frequency in the gene pool in subsequent generations. This will happen at a faster rate if the mutated allele is dominant rather than recessive.

(***A*** – describes mutation as a change in DNA bases to create a new allele; ***M*** – explains in relation to formation of new alleles and identifies where need to occur to enter gene pool; ***E*** – links ideas and explanations how mutated alleles may become established in the gene pool)

4. *Mutation* beings new alleles into the gene pool of the populations within the species. *Independent assortment* of the chromosome pairs in meiosis and *segregation* of their alleles produces large variation in the gametes produced from any two parents in the population. *Crossing over and recombination* of alleles further allows for new combinations in the offspring. *Migration* allows gene flow between populations, spreading new alleles and combinations of alleles from population to population throughout the species.

(***A*** – describes mutation and two other ways of achieving variation in a species)

5. a. The *range of alleles* present in a population. (***A***)

b. The greater the range of alleles, the greater the *variation* present in a population. Variation is the raw material that natural selection works on in a changing environment. Therefore, the greater the amount of variation present in a population the greater the chance of one/some individuals being present that will suit the new environmental conditions. Individuals with this/these desirable trait(s) will be selected for and reproduce, allowing the survival of the species.

(**A** – identifies genetic biodiversity in terms of more variation present in a population which assists its survival; **M** – links occurrence of variation to natural selection to explain survival of the species)

6. Endangered species will be small in numbers, so their genetic biodiversity is likely to be reduced. This will be exacerbated by the *inbreeding* that will occur in small zoo populations. It is therefore important for zoos to *outbreed* – done by bring in new males (or their sperm). This will bring new alleles/combinations of alleles into the population, increasing genetic biodiversity as much as possible. By reducing inbreeding, chances of harmful recessive alleles coming together and producing genetic defects is reduced.

(**A** – describes bringing in of new males/sperm in terms of increasing variation/ increasing genetic biodiversity in the population; **M** – explains in terms of inbreeding and outbreeding; **E** – links explanations and relates them to small size of the population and occurrence of genetic defects)

7. a. Founder effect occurs when a small number of individuals is separated geographically from the parent population (eg on an island); range and frequency of alleles in this founder population is unlikely to be representative of the parent/original population. Bottleneck effect occurs when a large population is suddenly reduced in numbers (a 'population crash'), typically by a catastrophic event that kills off individuals indiscriminately; when population numbers are restored to a larger size, range of alleles may be reduced and frequency of alleles changed from original population. (***A***)

b. The random change in allele frequency in a (small) population. (***A***)

c. The founder effect and bottleneck effect both concern small populations or populations reduced to a small size with reduced genetic biodiversity. As the populations are small, genetic drift is likely to further change the frequency of alleles compared with that of the original population and is also likely to further reduce genetic biodiversity.

(**A** – describes relationship in terms of small size of the populations; **M** – links this to reduced genetic biodiversity and changes in allele frequency)

8. E

Supplementary Unit: Investigations

Topic 1: Investigations

Supplementary Unit Activity 1A: Investigations in biology (page 350)

1. a. To investigate the effect of pH on the speed at which the enzyme amylase *digests starch*.

b. To investigate the effect of *temperature* on the rate at which salivary amylase enzyme *digests* starch.

c. To find the effect of light *colour* on the rate at which *tomato plants* grow.

d. To find out if a variety of different disinfectants have the same effect on the growth rate of bacteria on agar.

e. To find out which exercise increases the heart rate of a person most.

f. To investigate the effect of increased temperatures on the germination rate of marigold seeds.

2.

	Independent variable	Dependent variable	Controlled variables
a.	Temperature of water.	Time taken for the sugar to dissolve in the water.	• Volume of water. • Amount of sugar. • Type of sugar. • Container used. • Movement in the water – if any (ie stirring). • Number of repeats for each temperature of water.
b.	Type of exercise.	Change in heart rate as a result of the exercise.	• Person used for test. • Time spent doing each exercise. • Heart rate at the beginning of each exercise set. • Time to rest between each set of exercises. • Air temperature at the time the exercises were carried out. • Number of repeats for each exercise set.
c.	The type of fertiliser added to the soil.	Growth rate – this could be in terms of height of plants, number of leaves, number of beans produced in a certain period of time, wet and dry mass of the plants (how much the plants weigh at the end of the experiment – dry mass is the weight after the plant has been dried in an oven to remove all moisture).	• Type of bean seeds/plants. • Density of plants – how many plants there are in a set area. • Number of plants used in each condition. • Volume of soil in each condition. • Type of soil used must be the same before fertiliser is added to one group. • Site that the plants are grown in. • Watering of the plants. • Time that the measurements are made for each set of plants.

Supplementary Unit Activity 1B: Planning an investigation (page 351)

1. a.

Aim: To investigate the effect of disinfectant concentration on the rate of growth of sub-cultured bacteria on agar plates.	
Variables:	
Independent variable	Concentration of disinfectant. Range of independent variable: • No disinfectant – water only. • Recommended strength disinfectant. • Double-strength disinfectant. • Triple-strength disinfectant.
Dependent variable	Size of clear area around the filter paper (measured in mm).

Controlled variables	• Agar plates used – all the same as supplied by the teacher. • Bacteria used for sub-culturing – all the same as supplied by the teacher. • Size and type of filter paper used – use the same type of filter paper and cut out squares or circles. Make the squares or circles all the same size; for example, circles could all be 1 cm in diameter, squares could be 5 mm by 5 mm. • Type of disinfectant used will be the same. • Incubation – all the plates will be incubated in the same place (the school incubator) set at 25°C. • Time for incubation – all the plates will be left for the same time (eg from period 3 on Wednesday to period 5 on Friday). You need to make sure you explain exactly how you will control each variable – even if it seems very obvious (like 'use the same type of filter paper') – make sure that you write it down. Also be honest with your timing – it will depend on when in the day you have your Human Biology/Biology lessons.

b. Plan

- Collect four Petri dish agar plates that have been sub-cultured with bacteria by the teacher.
- Cut 12 squares of filter paper, 5 mm × 5 mm.
- Soak 3 of the squares of filter paper in water, 3 in the recommended concentration of disinfectant, 3 in double the recommended concentration of disinfectant, and 3 in triple the recommended concentration of disinfectant, all for 1 minute each. *Note:* It is not necessary to describe how you would make the increased concentrations.
- Take the filter paper out of the disinfectant or water with tweezers and let the excess fluid drip off.
- Place one of each of the filter paper discs on each plate, labelling on the outside of the Petri dish with a marker pen as in the diagram shown.

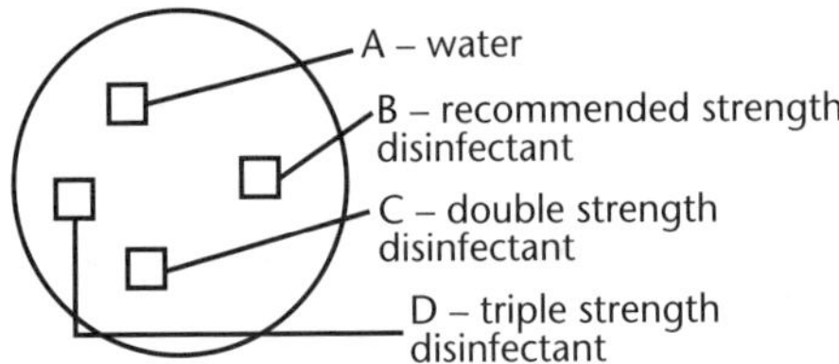

- Seal the Petri dish plates with masking tape around their edges.
- Place the Petri dish plates *upside down* in an incubator set at 25°C for approximately 48 hours – they will be looked at in our Human Biology/Biology class in 2 days' time.
- Remove the Petri dish plates and measure the diameter of the clear space around each paper square.
- Record the results in the following table and calculate the average for each set of results.

Petri dish plate	Clear space around the paper square – distance across (mm)			
	Water	Normal-strength disinfectant	Double-strength disinfectant	Triple-strength disinfectant
1				
2				
3				
Average				

Supplementary Unit Activity 1C: Gathering, processing and interpreting data, and reporting (page 357)

1. a.

Type of exercise carried out by Martha	Average	
	Initial heart rate at rest (bpm)	Heart rate after 3 minutes of exercise (bpm)
Skipping	68	130.7
Step-ups	68	129.3
Star jumps	68	145.3
Walking	68	85.0

b.

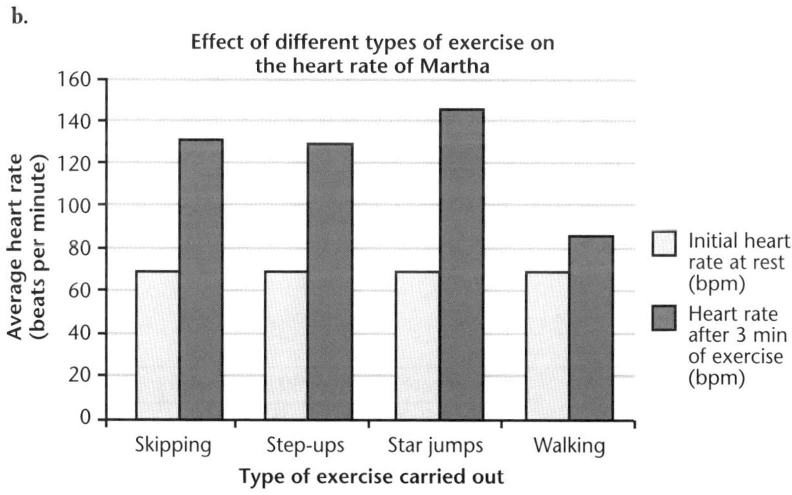

2. a. The result for Trial 3 at 45°C was not consistent with the rest of the results. They would have needed to check that this was inconsistent by taking at least one more test at this temperature. When averaging the results they could decide to take the Trial 3 result out as it was so clearly different from the others.

b.

Temperature of solution (°C)	Time taken to digest the egg white (colour went from cloudy to clear) (s)					
	Trial 1	Trial 2	Trial 3	Trial 4	Trial 5	Average
10	Did not change	Did not change	Did not change	Did not change	Did not change	Did not change
20	303	290	310	298	290	298.2
25	150	148	154	149	151	150.4
35	25	22	19	20	24	22
45	16	18	~~6 (remove)~~	19	17	17.5
55	62	67	60	64	66	63.8
70	Did not change	Did not change	Did not change	Did not change	Did not change	Did not change

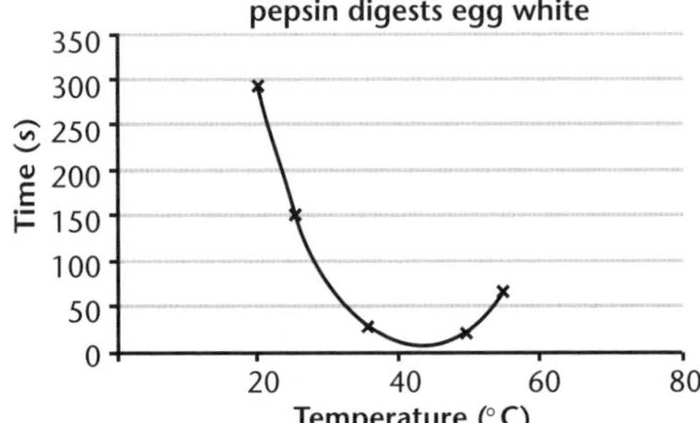

3. a. To find the effect of different swim strokes on heart rate.

b. Swim stroke.

c. Heart rate after 50 m.

d. Same swimmer, same distance for each stroke, same rest period between strokes, same effort for each stroke; heart rate measured at the same time after each stroke; same heart rate monitor used in all tests; same rest between each test.

e.

	Heart rate at end of 50 m (beats/min)			
Stroke	Test 1	Test 2	Test 3	Average
Freestyle	120	121	118	120
Breaststroke	89	86	90	88
Butterfly	134	134	130	133
Backstroke	118	114	117	116
Resting heart rate	68	67	67	67

f.

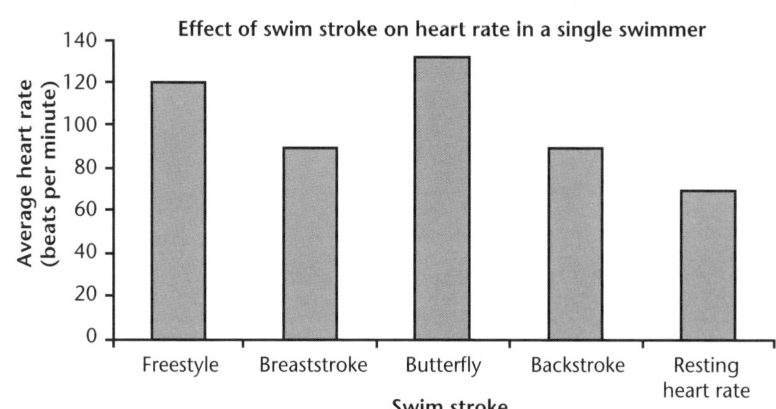

g. Conclusion / discussion

The results indicate that swimming all strokes increased the average heart rate of the swimmer in this investigation. Butterfly lifted the average heart rate the highest (66 beats above resting). Freestyle and backstroke had a slightly lower effect, with breaststroke increasing the heart rate by only 21 beats per minute. The increased heart rate is an indication of the work or effort required for each stroke. Swimming required energy. The increased heart rate is a result of the increased demand for oxygen and glucose delivery to the muscles. These results would indicate that for this swimmer, the greatest effort required was in swimming butterfly.

4. a. To investigate the effect of pH on the enzyme pepsin breaking down egg white protein.

b. That the enzyme would be most effective in acidic conditions (close to pH = 3) because it is found in the stomach where the pH is acidic.

c. pH of the solutions.

d. Time taken for the solution to go clear.

e. To check that the egg white solution would not go clear by itself – ie checking that pepsin *is* required.

f. Volume of egg white solution used; volume of water, acid or base added; temperature of the water bath; volume of pepsin used.

g.

		Time taken for solution to go clear (min)			
pH of solution	Pepsin added	Test 1	Test 2	Test 3	Average
1	Yes				
1	No				
3	Yes				
3	No				
5	Yes				
5	No				
7	Yes				
7	No				

abiotic (1, 15, 99, 326, 364): non-living.

acclimation (4): a change in an organism's tolerance limits.

actin (235): contractile protein found in muscle fibres.

active site (237, 239): part of the surface of a protein in which the specific catalytic properties of the protein are localised.

active transport (235): movement of a substance across a membrane from a lower to a higher concentration, using energy in the process; movement of particles against a concentration gradient; requires energy.

adaptations (1, 2, 265, 325, 326): characteristics of organisms that improve their chances of survival.

adenine (213, 215): one of the four kinds of nitrogenous base found in DNA and RNA.

adult (somatic) stem cells (313): stem cells derived from adult tissue (including umbilical cord).

aerobic (90, 91): in the presence of oxygen.

agar (234): jelly derived from seaweed and used to grow bacteria and fungi.

agarose (284, 321): jelly-like material used to separate DNA fragments of different sizes.

age pyramids (130): a way to display information about the number of organisms alive in particular age groups of a population.

Agrobacterium (291, 321): common soil bacterium used to insert plasmids into flowering plant cells.

allele (163, 165, 216, 265): alternative forms of a gene.

allele frequency (265): number of times that an allele occurs in the gene pool of a population.

allelopathy (35, 55): production by a plant of a chemical that inhibits the growth of another species of plant.

allopatric (28): occupying geographically separate ranges.

allopolyploidy (254, 258): polyploidy resulting from contribution of chromosomes from two or more species.

allotransplantation (315): transplantation of organs between humans where donor is living or dead.

amino acid (8, 216, 220, 233, 235–236): one of the subunits or building blocks from which all proteins are built.

ammonia (8, 91): toxic excretory product NH3, formed from the breakdown of amino acids.

anaemia (249): having very low levels of haemoglobin in the blood.

anaerobic (8, 75, 90): lacking in oxygen.

aneuploidy (254, 255, 266): variation in chromosome number involving less than the whole set.

annealing (283): joining of complementary nucleotide sequences by controlled lowering of temperature.

annuals (106): plants that complete their life cycle in one year or less.

antibiosis (35, 55): a relationship in which one individual produces a toxic chemical (antibiotic) that harms another individual.

antibodies (185, 235): proteins produced by the immune system that recognise and attach to antigens so the antigen can be destroyed; protein produced by the body in response to the presence of a foreign substance or antigen.

anticodon (225): group of three tRNA bases which is complementary to, and therefore can combine with, an mRNA codon.

antigen (185): a substance, usually foreign to the body, that stimulates the production of an antibody.

artificial selection (344): situation in which humans (not the environment) select the organisms that will breed (and pass on their favourable genes); also known as selective breeding.

aseptic (292): the absence of micro-organisms.

asexual (196, 233): not involving sex; mitotic cell division produces offspring identical to the parent.

assortment (246): arrangement of chromosomes on the equator of the spindle during meiosis I.

ATP (89, 92, 216): most common energy-carrying molecule.

autopolyploidy (254, 256): polyploidy in which the extra chromosomes are all derived from the same species.

autosomal dominant traits (208, 210): genetic characteristics always expressed phenotypically, present on autosomal chromosomes.

autosomal recessive traits (209, 210): genetic characteristics only expressed phenotypically in homozygote, present on autosomal chromosomes.

autosome (202, 204–212, 256): chromosome that is not concerned with the determination of sex or gender.

autotomy (43, 44): shedding of a body part by a prey animal when attacked.

autotrophs (7, 13, 81): organisms, such as plants, that manufacture their own food.

auxin (292): plant growth substance involved in phototropism (tropic movement in response to light) and other activities.

bacteria (8, 82): one-celled prokaryotes; may be autotrophs or heterotrophs.

bacteriophage (229, 296, 321): virus that attacks bacteria.

base pairing rule (215, 223): refers to the pairing of bases in DNA in which adenine (A) always pairs with thymine (T); and guanine (G) always pairs with cytosine (C); allows for DNA replication.

base substitution (246): gene mutation in which one base is replaced by another.

Batesian mimicry (44): mimicry in which a palatable animal resembles an unpalatable one.

behavioural (1, 326): relating to how an organism reacts/behaves.

bias (141, 367): a distortion in data that can be overcome by random sampling.

bile (37): secretion from the liver which emulsifies fats, helps certain enzymes to work and neutralises stomach acid.

biodiversity (23, 111–119, 121, 265, 337–340): refers to the number of different species in an ecosystem.

biolistics (321): insertion of DNA by firing minute DNA-coated particles of tungsten or gold into cells.

biomass (11, 84): the dry mass of an organism, ie mass without water content.

biosphere (81): the zone of life surrounding planet Earth.

biotechnology (281): industrial use of biological processes.

biotic (1, 15, 99, 326, 364): alive, living.

birth rate (61, 130): live births per 1 000 individuals.

blastocyst (313): early stage of development of humans and other mammals, consisting of a hollow ball of cells; stage in mammalian embryonic development consisting of a hollow ball of cells produced one week after fertilisation.

block mutations (252, 266): changes in a segment of a chromosome involving many genes.

bone marrow (196): tissue within cavities inside bone, site of blood cell production.

bottleneck effect (335, 340): when the numbers in a population are rapidly reduced to low levels (usually as a result of a catastrophe or human action), resulting in the reduction of the genetic biodiversity of the population.

calcium (296): chemical element needed to make bones hard and strong.

callus (291): mass of undifferentiated plant cells produced in response to injury.

camouflage (crypsis) (43): concealment method in which an animal is coloured in such a way as to resemble its surroundings.

canopy (13, 22): uppermost continuous layer of foliage in a forest.

capillaries (249): smallest blood vessels.

capillary action (310): the movement of water up a narrow tube such as a capillary tube.

carbon cycle (89–90): the exchange of CO_2 and O_2 between plants and animals and the atmosphere during photosynthesis and respiration.

carnivore (2, 8, 11–13, 82–85): animal that feeds on other animals, killing them in the process.

carpel (260): female equivalent of a stamen of a flower, usually several carpels being joined together.

carrion feeders (82): feed off dead bodies.

carrying capacity (131, 132): maximum size of population able to be sustained in an environment.

catalysts (221): substances that greatly speed up the rate of chemical reactions.

cDNA (299, 321): DNA produced by copying RNA.

cell (72): basic units of all living matter; made up of cell/plasma membrane surrounding cytoplasm, usually with a nucleus.

cell line (278): many generations of diploid cells that successfully reproduce to produce normal daughter cells.

cell membrane (218, 235): bilipid layer that surrounds every cell; controls movement of materials in/out of a cell.

cell wall (196): outermost layer of a plant cell, consisting mainly of cellulose

cellulose (8, 54, 75, 300): complex carbohydrate forming the major constituent of plant cell walls.

Central Dogma (230, 231): the view that nucleic acids determine protein structure, but not the other way round.

centriole (204, 218): involved in cell division, in animal cells producing spindle fibres.

centromere (193, 196, 204, 205, 268): part of a chromosome that attaches to the spindle during cell division.

character (163): a characteristic of an organism.

character displacement (33): evolutionary process in which a characteristic shows a greater difference in two sympatric populations than between allopatric populations of the same species.

chemosynthesisers (81): certain species of bacteria that use energy from chemical reactions (eg nitrogen fixation, nitrification) to produce organic compounds from simpler chemicals; ie are autotrophic.

chiasma (204, 271): site of crossing over of two non-sister chromatids during prophase in meiosis.

chlorofluorocarbons (CFCs) (113): industrial gases that contain carbon and fluorine.

chlorophyll (4): green pigment that absorbs light in photosynthesis.

chloroplasts (75): organelles in which photosynthesis occurs.

chromatid (173, 193, 204, 268): one of two products of chromosome duplication, visible during prophase.

chromosomal mutations (245, 266): mutations that involve whole genes, occurs when parts of a chromosome are deleted, repeated, or translocated; also known as block mutations.

chromosome (164, 173–175, 193, 195–215, 252, 254–261, 266): threadlike structure bearing the genes, and visible during cell division.

class (134–135, 375): group of related orders.

climax community (20, 21): final community that results when no more succession occurs.

clone (266, 277–278, 295, 321): large number of genetically identical offspring produced by the copying of an original type.

cloning vector (295, 299, 321): plasmid or bacteriophage DNA, into which foreign DNA is inserted for replication.

clotting (184): when platelets gather at an injury to stop blood flow.

clumped (14, 143): distribution pattern in which groups of individuals live or grow close together, with large spaces between groups.

co-dominance (174, 185): occurs when the heterozygote simultaneously expresses two alleles that have different kinds of effect, eg the blood group AB is not intermediate between A and B, but expresses both qualities.

codon (225–229): group of three RNA bases specifying an amino acid.

collagen (91, 221, 237): protein that forms the framework of bone tissue.

commensalism (54, 76): one member benefits from a relationship, the other is unaffected.

common ancestor (229, 332): original species from which others develop through divergent evolution.

common ancestry (327): organisms have evolved by natural selection from a common ancestor.

community (2, 7, 13, 19): populations of all the species in one area.

competition (3, 31, 32, 35, 61–71, 118, 131–133, 152, 337): situation in which the demand for a resource exceeds supply.

competitive exclusion principle (3, 32–34): principle which states that no two species can permanently occupy the same niche in the same habitat.

complete dominance (170): situation when one allele in a pair is dominant over the other allele (which is recessive) so is always expressed in the phenotype.

consumers (7, 8, 82, 121–124): organisms that obtain their food and nutrients from other organisms.

continuous variation (164, 190): variation in which individuals cannot be placed in distinct categories because there is a continuous range of intermediates between two extremes, such as body height.

controlled variables (348, 361, 366): factors to be kept the same to ensure a fair test.

core sequence (307): repeated sections forming a VNTR in the minisatellites of intron DNA.

cross-pollination (183): pollination between flowers on different plants.

crossing over (198, 270, 271): process in which parts of two non-sister chromatids cross over each other in prophase 1 of meiosis; the point of the cross-over is the chiasma.

cultural evolution (72): change in the learned information stored in a population, such as knowledge, beliefs, skills, attitudes, etc.

cuttings (277): a way of reproducing plants vegetatively, without involving sex cells.

cystic fibrosis (180, 209, 318): inherited genetic disease resulting in over-production of mucus.

cytoplasm (196–200, 221–224): part of a cell outside the nucleus, and in which proteins are made, and in which most of the cell's activity occurs.

cytosine (213–215, 250): one of two pyrimidine bases found in DNA and RNA.

DDT (87, 121): one of the first insecticides developed; an organo-chlorine compound dichlorodiphenyltrichloroethane.

death rate (61, 130): mortality.

decomposers (7, 8, 12, 82, 88): organisms that break down dead plant and animal material.

dehydration (20, 151): drying out through loss of moisture.

demography (130): the study of birth rates, death rates, age distribution and sizes of population.

denature (249, 301): to break down the complex structure of a protein.

denitrifying bacteria (91): bacteria that use nitrates to oxidise food, reducing nitrate to nitrogen gas.

density (14, 143): numbers of organisms in a certain area; population number divided by area of habitat.

density-dependent (51, 61, 132): the effect depends upon the size of the population.

density-independent (51, 132): occurs independently of the size of the population.

deoxyribonucleic acid, DNA (115): material of which genes are made.

deoxyribose (213): one of the components of a DNA nucleotide.

dependent variable (149, 348, 354, 361, 365, 366, 371): factor measured as a result of changing the independent variable; or to find an answer to a question.

detritus feeders (8, 82): animals that have adaptations to feeding on organic debris on/in the substrate where they live.

dicotyledonous plants (303): one of the two groups of flowering plants, with two seed leaves (cotyledons) in the embryo.

differentiation (305): process by which cells specialise for particular functions.

digestion (12): breaking down of food into small soluble particles.

dihybrid (163, 183): a cross involving two pairs of contrasting characters.

dihybrid cross (177, 183): cross in which two pairs of contrasting characters are being considered simultaneously.

dihybrid inheritance (174): the inheritance of two genes controlling two different features; the genes may have two or more different alleles.

diploid (72, 194, 204, 205, 254, 278): having two sets of chromosomes (ie two of each kind).

discontinuous variation (164, 190): variation in which individuals can be placed in categories which differ sharply from each other, eg blood groups.

discrete (284, 355): separate.

distributed (14): spread out.

distribution pattern (14): arrangement of individuals or groups in an area.

diurnal (2): active during the day.

diversity (13, 34): large variety of species or genes.

division (195–206, 217–218, 268): process of increasing number of cells, occuring, in plants, in the apical meristem.

DNA (213, 215, 216, 222–235, 248, 282–292, 295–312, 320–322): deoxyribonucleic acid; the genetic material which makes up chromosomes and codes for proteins and the chemical of which genes consist.

DNA ligase (283, 321): enzyme that joins DNA fragments together.

DNA polymerase (285): enzyme that joins new nucleotides to expanding DNA strand (only at 3' end):.

DNA profiling (185, 306): technique by which fragments of DNA in individuals are compared to establish relationships between them.

dominant (163, 165, 184): an allele is dominant if it is expressed in heterozygotes as well as homozygotes.

dominant allele (165, 170): an allele which is always expressed in the phenotype when present in the genotype; its presence masks the presence of a recessive allele.

dominant species (14): species with the greatest total mass in a community.

dominant trait (167, 180, 210): dominant characteristic.

donor (276): the organism from which a desired gene is taken (genetic engineering).

Down's syndrome (195, 256, 266): genetic abnormality usually characterised by three copies of chromosome #21 (trisomy 21):.

duplication (252–254): doubling of a segment of chromosome so that it is represented twice in the same strand.

ecological characteristics (7): organisms within a community, together with the physical attributes of their habitat.

ecological niche (361): the role or way of life of an organism in its biological community – is a combination of where the organism lives (habitat) and how it lives there (adaptations).

ecological succession (21): pattern over time in species distribution of a community.

ecology (1): the study of organisms and their relationship with the environment.

ecosystem (7, 81): all the communities and the physical environment in an area.

ectoparasites (36): parasites that live on the outside of their hosts.

egestion (88): getting rid of undigested food via the anus.

egg/ovum (193–194, 205): female reproductive cell; a female gamete.

electrophoresis (284): technique for separating DNA and other molecules using an electric field.

electroporation (304, 321): technique of inserting DNA into animal cells or plant protoplasts, by using electric pulses to create minute, transient holes in the plasma membrane.

embryo (196, 313): very early stage in development of a multicellular organism; development stage after zygote forms.

embryonic stem cells (313): undifferentiated, pluripotent cells usually derived from 5-day-old blastocyst.

emergents (13): trees whose uppermost branches protrude from the canopy.

emigration (132, 340, 342): movement of organisms out of a population.

endangered (118): any species in sufficiently low numbers to make it at risk of extinction (eg kiwi, takahe).

endemic (117): species which are native to one country only and found naturally nowhere else; an organism is endemic to an area if it is found naturally nowhere else.

endoparasites (36–37): parasites that live inside their hosts.

endoplasmic reticulum (222, 224, 226): complex system of membrane-bound cavities in the cytoplasm of a cell, concerned with the production of proteins.

energy pyramid (85): display of the energy value of the biomass of organisms at different trophic levels.

enucleated (278): a cell (eg an egg cell) that has had its nucleus removed.

environment (1, 14): surroundings of an organism including all the biotic and abiotic factors.

enzyme (214, 220, 235, 241): an organic catalyst; nearly all are proteins, but some are RNA.

epidermis (151): outer layer of the skin of a mammal; the outermost layer of cells of a leaf, young stem, or very young root.

epiphytes (22, 38, 78): perching plants that are adapted to living on the outside of other plants, mainly to get higher light intensity; a type of commensalism.

eukaryote (205): organisms with definite nuclei – all organisms except bacteria.

euploidy (254): variation in chromosome number in which every chromosome is represented three or more times.

evaporation (89): when water changes to a vapour.

evolution (231, 325–327, 336): the process by which new species of organisms develop from earlier forms.

excretion (85): getting rid of waste products that have been made in chemical processes in cells.

exon (223): segment of a gene that encodes an amino acid sequence.

exoskeleton (12): an external skeleton.

explant (291, 321): small part removed from a plant and grown in tissue culture to produce a plant genetically identical to the donor plant.

exploitation (35–36): relationship in which one organism is harmed and the other benefits.

exponential (131): extremely rapid growth.

exponential growth (131): growth in which numbers increase by a constant percentage each given time interval.

express (the target gene) (276): if successfully incorporated into the recipient DNA, the target gene should be expressed.

extinct (112, 118): any species that no longer has any living representatives.

extracellular (82): processes that occur outside cells (eg digestion in bacteria).

F_1 (163, 166): offspring of a cross between two pure-breeding varieties.

F_2 (163, 166): offspring of mating between F_1 organisms.

faeces (11): undigested food, together with large numbers of bacteria; got rid of via the anus.

fair test (348): experimental condition in which only one variable is altered at a time while all other variables are kept constant.

family (75): group of related genera.

fauna (117, 338): animal life.

feeder layer (314): layer of embryonic mouse cells upon which stem cells are cultured; needed by stem cells as a surface upon which to stick, and as source of nutrients for cell division and growth.

fertilisation (202, 207, 268, 272): joining of egg and sperm to form a zygote; process in which a male and female gamete fuse.

filter feeders (82): organisms adapted to obtaining food by filtering particles from surrounding water.

first filial generation (F_1) (165, 172, 183): 'children'; offspring that result from the crossing of two parents.

fitness (326, 337): refers to those individuals which have the best adaptations so will survive and reproduce, passing on their successful alleles to their offspring.

fittest (133): individuals that have the best chance of contributing their genes to the gene pool.

fixed (91): conversion of nitrogen in the air (as a gas) to a soluble form (nitrate ions).

flora (117, 338): plant life.

flowers (76): shoot of a flowering plant, modified for sexual reproduction.

foetus (277, 315): pre-birth stage of a mammal that has developed recognisable organ systems; name given to the unborn human after 8 weeks; cell differentiation has been completed.

food chain (8, 83): group of animals and plants linked together through feeding relationships; a succession of organisms along which matter and energy are transferred.

food web (10, 83): network of food chains; the interconnected pathways through which energy and matter are transferred in a community.

founder effect (338): chance change in allele frequency which occurs when a small group of individuals become separated from the main population.

frame shift (247): mutation resulting from the insertion or deletion of a base.

fronds (22): leaves of ferns.

fruit (68, 76): the structure formed from the ovary of a flower, usually after the ovules have been fertilised.

fungi (8, 82): heterotrophs that may be single or multicellular; cell walls are made of chitin, feeding is extra-cellular (outside the cell) and fungi are parasites or saprophytes.

fusion (254): joining together of two non-homologous chromosomes.

gamete (193, 205, 268): egg or a sperm; unlike a spore, a gamete cannot develop further until it has joined with another gamete.

gas exchange (89): the exchange of oxygen and carbon dioxide across a semipermeable membrane (eg alveolus or capillary).

Gause's principle (3, 32): also known as the competitive exclusion principle; states that 'no two species can occupy the same niche as one will out-compete and eliminate the other'.

gene (62, 163, 165, 184, 213, 216, 221, 235, 265): a hereditary unit – a length of DNA carrying the information for making a particular protein.

gene cloning (321): process by which multiple copies of a gene are produced in a bacterium or bacteriophage.

gene flow (335, 340): the movement of genes between (isolated) populations of the same species; in animals this may occur through migration.

gene mutations (246): changes in individual genes that may result in the production of new alleles.

gene pool (265): sum total of the genes in a population.

gene probe (301, 321): segment of single-stranded nucleic acid, used to find a complementary sequence of DNA.

genetic biodiversity (265): the expression of the range of the alleles present in a gene pool.

genetic drift (337): loss or change in the frequency of an allele in a small population by chance alone.

genetic engineering (276, 394): involves manipulation at the genetic level, producing organisms of known genetic makeup; also known as GE, genetic modification, recombinant DNA technology.

genetic material (205): material that contains the information for building the organism, and which has been inherited from the previous generation.

genetically modified organism (GMO) (281): organism carrying genes artificially transplanted from an organism of another species.

genome (163, 220, 254, 321): entire genetic information carried by an organism.

genomic DNA library (300): entire genome of an organism, distributed as DNA fragments in a large number of bacteria or bacteriophage.

genotype (163, 165, 184, 265): organism's hereditary make-up – the information it inherits from its parents.

germinate (22): start of growth of a seed or spore.

global warming (113): increase in average temperature globally, mainly resulting from the increase of greenhouse gases in the atmosphere.

globular proteins (237, 239): proteins, such as enzymes, in which the polypeptide chain is irregularly folded into a ball-shaped molecule; generally take part in chemical reactions rather than mechanical function.

glucose (303): simple sugar that is used by cells for energy, and is the building unit of starch and cellulose; monosaccharide involved in respiration and photosynthesis.

Gondwanaland (117): ancient super-continent that broke up to form the southern continents.

grazing (15, 107): feeding by an organism on parts of many organisms, which are not usually killed.

Greenhouse Effect (90, 113–115): the trapping of heat from the sun by gases in the atmosphere.

guanine (215): one of two purine (two-singed nitrogenous bases found in nucleic acids) bases found in DNA and RNA.

gut (8): tube in which food undergoes digestion and absorption.

habitat (1, 2, 13): place where an organism lives.

haemoglobin (221): red blood pigment that carries oxygen.

haemophilia (208, 209): hereditary condition in which the blood cannot clot.

haplodiploidy (207): system in organisms, such as bees, which have no sex chromosomes, whereby males develop from unfertilised eggs, and are therefore haploid, while females develop from fertilised eggs, and are therefore diploid.

haploid (207, 254, 268): having only one set of chromosomes (one of each kind), as in gametes.

haustoria (38): modified roots of plant parasites that grow into the host plant using digestive enzymes and join to the host's xylem and phloem.

herb layer (13): lowest layer of forest vegetation, consisting of small plants such as mosses and liverworts.

herbicides (120, 318): chemicals that kill plants when applied to them.

herbivores (8, 82): animals that eat plant material.

heredity (164, 265): passing on of characteristics from an individual to its offspring.

hermaphrodite (62, 183): individuals that have both male and female reproductive organs (eg earthworms, snails).

heterogametic sex (206): sex whose gametes determine the sex of the offspring.

heterotroph (7, 82): organism that cannot make its own organic matter from inorganic raw materials (all animals and fungi and most bacteria).

heterozygote (163): carrying different allelic forms of a given gene.

heterozygous (165, 171): having two different alleles of a given gene; not true-breeding.

home range (63, 133): area regularly covered by an animal.

homeothermic (12): maintaining a near-constant body temperature.

homogametic sex (206): sex whose gametes play no part in determining the sex of the offspring.

homologous chromosomes (170, 201, 206, 268): pairs of chromosomes in which each chromosome is identical in length and shape and has the same genes at the same position (or locus); one of the pair comes from the male parent, and the other from the female parent.

homologous pair (194, 205): pair of chromosomes carrying genes controlling the same characteristics, one member of each pair inherited from each parent.

homologue (194, 204, 205): the individual chromosomes of a homologous pair.

homozygote (307): carrying two copies of the same allele.

homozygous (165, 170): when both alleles in a pair are the same (eg DD or dd).

hormones (73): type of protein produced by the endocrine glands; act as chemical messengers that control or help to control some function elsewhere in the body (eg insulin, ADH).

host (8, 35): name of the organism that a parasite lives upon.

hydrogen bonds (215, 238): weak chemical bonds that form between molecules.

hyperacute rejection (317): acute, rapid rejection of transplanted foreign tissue that results in death very quickly.

hyperthermia (4): abnormally high body temperature.

hypothesis (361): prediction based on theory which can be tested by experimentation.

immigration (340, 341): movement of organisms into a population.

inbreeding (342, 344): the breeding of closely related organisms, eg brother and sister.

inbreeding depression (276): the loss of genes from a population as a result of inbreeding.

incomplete dominance (185): occurs when a characteristic in the heterozygote is intermediate in degree between the two homozygotes, eg pink flowers are intermediate between red and white.

independent variable (348, 354, 361, 365, 371): only factor that will change during an experiment.

induced (247, 277): to produce large quantities of a desired protein (such as the chymosin enzyme).

infertile (258): state in which an organism is unable to reproduce or have offspring.

infrared (IR) (113): form of radiation that causes heating.

inheritance (163): acquisition of traits by their transmission from parent to offspring.

insecticides (87, 120): chemicals that kill insects when applied to them.

insertion (229): point at which the muscle inserts via a tendon onto the bone.

insulin (221, 303): hormone that promotes the uptake of glucose by cells, thus lowering blood glucose level.

interferon (318): protein produced by white corpuscles that inhibits the reproduction of viruses.

internode (15): region of a stem between two nodes.

interphase (216–218, 268): the time in the cell life cycle between cell divisions (mitosis / meiosis).

interrupted replication (287): technique for determining the base sequence of DNA; also called the Sanger method.

interspecific competition (3, 31): competition between different species.

intraspecific competition (61, 132): competition between members of the same species.

intron (223–224): section of DNA between exons which is not translated into a protein.

inversion (253): breaking of a segment of chromosome and its subsequent repair the 'wrong way round', or back to front.

invertebrates (22): animals that lack a vertebral column/backbone.

iteroparity (134): when organisms reproduce in repeated breeding seasons in their lifetime.

karyotype (194, 195, 204, 205): chromosomal characteristics of an organism (number, sizes, centromere positions, banding patterns); a printout of an individual's chromosomes after they have been stained, matched for size and shape, then numbers and grouped in order (eg human chromosomal karyotype is arranged in sequence from pair 1 to pair 22).

keratin (221, 237): tough, fibrous protein in the epidermis and structures derived from it, such as nails and hairs.

kidneys (316): paired organs which excrete urea (excretory product) and help to maintain a constant internal environment.

kinesis (369): orientation movement of an animal in which the stimulus governs the rate, but not the direction, of movement.

kite diagram (149): graph used to show the variation in the abundance of an organism along a transect.

Klinefelter's syndrome (256): genetic abnormality characterised by presence of two X chromosomes and one Y (XXY):.

Law of Independent Assortment (177, 270): random allocation of chromosomes to newly forming gametes; 'the segregation of one pair of alleles does not affect the segregation of another pair'; also called Mendel's second law.

Law of Segregation (173): 'Of the two genes controlling each characteristic, only one is present in each gamete'; also called Mendel's First Law.

leached (91): lost from the soil by rainfall, flooding.

learning (72): behaviour that is modified by experience.

leaves (8): the photosynthetic organs of higher plants.

legumes (75, 91): plants that have root nodules inhabited by nitrogen-fixing bacteria; a mutualistic relationship.

lek (66): area of land in which males congregate to mate with females, and in which small territories are defended for mating purposes.

lethal alleles (186): alleles which cause death when occurring as one of the homozygous genotypes.

leukaemia (315): disease in which white blood cells fail to mature properly.

lianas (22): plants that have roots in the ground and stems that climb up other plants to obtain higher light intensity (cf epiphytes); an example of commensalism.

ligaments (221): tough elastic fibres made of collagen that connect bone to bone at joints, assisting to stabilise the joint.

lignin (13, 318): complex material used to strengthen cell walls in plants, making them resistant to compression.

limestone (90): calcium carbonate; made from shells of dead sea creatures.

limiting factors (132): environmental constraints that limit the growth of an organism or population.

linker (321): short, artificially synthesised section of single-stranded DNA used to make complementary ('sticky'): ends to DNA fragments so that they can be joined up.

lipofection (304): method of introducing foreign DNA into mammalian cells whereby DNA is coated in artificial liposomes which are taken up by the plasma membrane.

liver (38): large organ situated at the top of the gut cavity; performs many essential functions.

locus (163, 173): specific position of a gene on a chromosome.

lungs (180): gas exchange surfaces of air-breathing vertebrates.

lymphocytes (313): white corpuscles with small amount of cytoplasm; one kind can develop into cells that make antibodies.

lysozyme (237, 239): enzyme present in various body fluids; breaks down bacterial cell walls.

mark-recapture (139): technique for measuring a population of mobile animals.

marsupials (107): the group of mammals where the embryo is born in a very immature state and completes development in the mother's pouch (eg possums, koala).

meiosis (163, 173, 195, 197–198, 201, 204, 217, 257, 268): process in which a diploid nucleus divides twice to produce four haploid, genetically different, nuclei; in animals, occurs in the ovaries and testes and results in the formation of gametes; cell division that produces sex cells.

melanin (249, 267): skin pigment that protects against UV radiation.

Mendel's First Law (173): 'Of the two genes controlling each characteristic, only one is present in each gamete'; also called law of segregation.

messenger RNA (mRNA) (221): nucleic acid that acts as a working copy of a gene, and is synthesised in the nucleus and 'translated' in the cytoplasm.

metabolism (89, 241): collective name for all the chemical processes going on in cells.

micro-organism (75): microscopic organism.

microarrays (290): also known as DNA chips; devices that allow identification of many target DNA strands through base pairing.

microbe (8): microscopic organism.

microinjection (304): method of introducing foreign DNA into mammalian cells whereby DNA is injected directly into the nucleus.

migration (124, 132): mass movement of members of a species, usually at regular intervals, often over long distances, and in which the 'urge' for the movement arises from internal stimuli (though these may themselves be triggered by external changes); the movement of individuals between populations, allowing for gene flow; immigration is the inward movement; emigration is the outward movement.

mimicry (41): adaptive resemblance between unrelated species.

minisatellites (307): regions of intron DNA consisting of variable number tandem repeats (VNTRs): – short sequences of 20–100 nucleotides.

mitochondria (75, 221): organelles in which aerobic respiration occurs, releasing energy; cell organelles that produce most of the cell's energy in the form of ATP from aerobic respiration (singular = mitochondrion).

mitosis (195, 196–197, 201, 204, 217): division of the nucleus in which the daughter nuclei are genetically identical to each other and to the parent nucleus; cell division for growth and repair that produces cells with exactly the same number of chromosomes as the original cell.

monocotyledon (306): one of the two groups of flowering plants, with one seed leaf (cotyledon) in the embryo; group of angiosperms that has one cotyledon in the seeds; includes the grasses.

monohybrid (176): cross involving only one pair of alleles.

monohybrid cross (163, 183): cross in which only one pair of contrasting characters is being considered at any one time.

monohybrid inheritance (170): cross involving one pair of contrasting characters.

monophage (48): animal that feeds on only one kind of organism.

monoploid (254): number of chromosomes in one set. In most organisms it is the same as haploid.

monosomy (255): having one of the chromosomes represented only once.

mortality (31, 61, 130): death rate.

mRNA (messenger RNA) (221, 227, 299): RNA which determines the sequence of a polypeptide.

mucus (209): slimy substance secreted by the lining of the breathing passages and alimentary canal.

Mullerian mimicry (44): mimicry between unpalatable species.

multiple alleles (185): having more than two alleles for a gene.

mutagens (247, 266): agents that cause mutation.

mutation (235, 245, 266, 337): sudden and relatively stable change in the genetic material.

mutualism (8, 54, 74): relationship between two organisms in which both obtain benefit.

mutualistic (54, 75): a relationship between two organisms in which both benefit.

mutualists (75): organisms that live in a relationship in which both obtain benefit.

myosin (235): contractile protein found in muscle fibres.

natality (31, 61, 130): birth rate.

natural selection (99, 117, 326, 337): the selection by environmental factors of individuals with successful phenotypes to survive and reproduce.

nectar (10, 76): sugar solution secreted by flowers, serving to attract insects for pollination.

negative feedback (242): process in homeostasis, in which the greater a change in an internal condition, the stronger is the corrective response.

nerve impulse (313): an electrical charge that travels along the axon of neurons (nerve cells), transmitting information.

nervous system (37): body system that receives, transmits, and responds to environmental stimuli.

niche (3, 13, 32): sum total of an organism's requirements and its interrelationships with other organisms in that community; its role in a community, or what it does; way an organism makes its living; its 'occupation'.

nitrate ions (91): inorganic form of nitrogen NO3–.

nitrifying bacteria (91): bacteria that oxidise ammonia to nitrites or nitrites to nitrates.

nitrogen fixation (81): conversion of the element nitrogen to combined nitrogen, either by lightning or by bacteria.

nocturnal (2, 43): active at night.

nodule (54): swelling on a root of leguminous plant in which nitrogen-fixing bacteria live.

non-disjunction (255): failure of homologous chromosomes to separate at meiosis I, or failure of chromatids to separate at mitosis or meiosis II.

non-random mating (342): when individuals pair up, not by chance, but according to their genotypes or phenotypes.

nuclear envelope (224): double-walled membrane separating nucleus from cytoplasm.

nucleic acid (89, 221): nitrogen and phosphorus-containing polymers of two kinds – DNA and RNA; natural polymers in which bases are attached to a sugar phosphate backbone.

nucleoside (285): sugar-base portion of a nucleotide.

nucleotide (213, 215, 285, 287): sub-unit of DNA and RNA; one of the building units of a nucleic acid; made up of a base, a phosphate group and a sugar molecule.

nucleotide analogues (287): nucleotides which lack an OH group on carbon 3.

nucleus (193, 204, 205): part of a eukaryotic cell containing the chromosomes.

nutrients (14): substances taken in from the environment and used as raw materials for growth.

Okazaki fragments (283): short, initially separate DNA strands produced during replication.

oligonucleotide (285, 321): chain of up to about 20 nucleotides (ie a short polynucleotide):.

omnivores (82): animals that consume both plant and animal material.

operon (306): group of genes controlled by the same switch mechanism in bacteria.

optimum range (3): zone of conditions within which an organism functions at its best.

organ (41, 206, 315): number of tissues which cooperate to carry out a function that none of the individual tissues can do by itself.

organelle (75): region of a cell specialised to carry out a particular function, eg nucleus, chloroplast.

organic (8, 235): substances containing carbon, usually derived from natural compounds.

organism (1): an individual living thing.

outcross (276): breed non-related organisms together.

ova (193): plural of ovum.

ovaries (197): reproductive organs of the female that produce eggs and also female sex hormones.

ovulation (256): process of release of an egg from the ovary in a mammal or other back-boned animal.

ovule (166, 197): structure in seed-bearing plants that contains a female gamete.

ovum (193): the female sex cell/gamete; commonly known as the egg; female gamete.

ozone (113): molecular form of oxygen made up of three oxygen atoms; O_3.

palaeontology (329): the study of past life on Earth from fossils.

pancreas (316): gland which secretes digestive enzymes into the duodenum and also produces insulin; organ that produces digestive enzymes and also hormones concerned with regulation of blood sugar concentration.

parasite (1, 8, 35, 82): organism that feeds off another (the host), harming it but without killing it.

parasitism (36): the feeding of an organism (the parasite): on another (the host), in which the host is harmed but not usually killed.

parental care (71): investment of resources by parents in the survival of offspring.

parental type (271, 272): the original combination of alleles on a chromosome (ie no crossing-over/recombination has occurred); F_2 or testcross phenotype showing the same combination of characters as was present in one of the parents of the F_1.

parthenogenesis (72): production of haploid (female) individuals from unfertilised eggs.

partial parasite (39): parasitic plant that obtains its food through photosynthesis but obtains their water and minerals from a host plant (eg NZ mistletoes).

pathogen (317): organism that causes disease.

penicillin (55): antibiotic, first extracted from the fungus Penicillium.

pentadactyl limb (331, 332): having five digits.

pepsin (347): protein-digesting enzyme present in gastric juice.

peptide bond (226, 227, 236): link between amino acids in a protein.

perennial (106, 134): plant that survives from year to year.

period (106): duration of one complete cycle of a rhythm; ie the time between successive peaks of activity.

perspiration (89): loss of water from the skin of animals. Also known as 'sweating'.

pesticides (120): general term for chemicals that kill pests when applied to them; includes herbicides and insecticides.

petal (260): flower part modified to make the flower conspicuous to pollinating animals; the showy part of the flower; protect reproductive parts and attract pollinators.

petiole (77): leaf stalk.

Petri dish (314): dish used to grow micro-organisms on agar.

phage λ (phage lambda): (296): phage commonly used as a vector in gene cloning.

phenotype (337): physical expression of a gene(s) or allele(s); physical characteristics of an organism.

pheromone (73): chemical used as a signal to other members of the species.

phloem (38): vascular tissue that transports glucose around a plant.

photoperiodism (363): regulation of activity by daylength.

photosynthesis (7, 87, 89): process whereby plants manufacture food from raw materials; process carried out by plants in which carbon dioxide and water are converted to carbohydrate and oxygen, using light energy.

photosynthesisers (81): organisms that make their organic compounds from carbon dioxide and water using solar energy, ie are autotrophic – all plants, many protista, cyanobacteria.

physiological (1, 326): a chemical process that goes on inside an organism.

physiological stress (3, 133): when the vital functions of an animal are adversely affected by environmental factors.

phytoplankton (85, 113): microscopic algae which are autotrophs.

pioneer plant species (20): first organisms to arrive and become established in a new environment.

pioneer species (20, 21): one of the first species to become established during a succession.

placenta (330): organ joining the foetus to the wall of the uterus during pregnancy; organ in placental mammals that develops from the embryo for exchange of materials with the mother.

plants (7, 38): multicellular organisms that photosynthesise (are autotrophs); cells are enclosed in cell walls of cellulose.

plaque (229, 296, 321): clear area in a dense population of bacteria resulting from lysis (breakdown): of bacteria after infection by bacteriophage.

plasma membrane (196): membrane forming the outer boundary of the cytoplasm.

plasmid (277, 295–296, 322): small, independently replicating bacterial mini-chromosome, bearing genes for resistance to antibiotics, and used as a carrier for replicating DNA.

plasticity (313): potential of a cell to differentiate into a limited number of other types of body cells, characteristic of adult stem cells.

platelets (313): involved in blood clotting.

pleiotropy (242): secondary effects of a gene which arise as indirect consequences of its primary effects.

pluripotency (313): potential of a cell (eg embryonic stem cell): to differentiate into any other type of body cell.

point mutations (266): changes to the bases in DNA; occur within a gene.

pollination (76): transfer of pollen from an anther to a stigma on a flower of the same species; the transfer of pollen from the male reproductive organs to the female reproductive organs; occurs in gymnosperms and angiosperms.

polyacrylamide (284): jelly-like material used to separate DNA fragments of different sizes.

polygenes (191): more than one gene affecting expression of a characteristic.

polygenic inheritance (190–191, 192): inheritance in which characteristics are contolled by many genes.

polymerase chain reaction (PCR) (285, 322): technique whereby a minute quantity of DNA can be replicated many times in the test tube.

polypeptide (221, 233, 235): chain of amino acids linked together by peptide bonds. A longer chain of amino acids is called a protein.

polyphage (48): animal that feeds on a number of other species.

polyploidy (245, 256, 261): mutation producing more than twice the normal haploid number of chromosomes.

population (23, 129, 139): group of organisms of one species.

population crash (132): occurs when the density of a population becomes so big that large numbers die in a short period of time (typically from mass starvation when herbivores overgraze their habitat in absence of predators); population numbers reduce drastically.

population density (130): the number of organisms in a given unit area.

predation (36, 40, 118, 132): exploitation of one organism (the prey) by another (the predator), in which the prey is killed and eaten.

predators (1, 35): animals that kill and feed on other animals.

primary sources (391): original material.

primary structure (237): order in which the amino acids are linked together in a protein.

primary succession (21): development of a mature (climax) community from bare land that has not been inhabited before (eg the moraine formed by glacial action).

primary transcript (222): production of an RNA copy of a gene.

primer (285): nucleotide sequence with a free 3' -OH group, needed to initiate synthesis by DNA polymerase.

probe (290): single-stranded nucleic acid of known sequence used to identify target.

proboscis (34): the extension of the mouthparts of an insect to form a piercing tube.

producers (7, 81): organisms that make their own organic matter from carbon dioxide and water; another name for autotrophs.

prokaryote (221, 298): organism with cells in which the DNA is not separated from the rest of the cell by a nuclear envelope.

promoter (223): section of DNA 'recognised' by RNA polymerase.

protein (8, 91, 216, 221, 233, 235, 237–243): polymer of amino acids consisting of one or more polypeptide chains; large organic molecules that have either a structural (eg collagen) or a regulatory (eg enzyme) function in organisms.

protista (363): unicellular organisms with a true nucleus (prokaryotes).

Punnett square (172): diagrammatic way of displaying a genetic cross.

pure breeder (170): individuals with homozygous genotypes (eg ***DD*** or ***dd***); can pass only one type of an allele on to their offspring.

pure-breeding (183): individuals which, when mated with their own type, produce offspring which resemble the parents.

pus (214): whitish liquid developing in infected tissue, consisting of white blood corpuscles.

pyramid of biomass (84): display of dry mass of organisms at different trophic levels.

pyramid of numbers (84): a way to display graphically the numbers of organisms at different trophic levels of a community.

pyrimidine (221): single-ringed nitrogenous base found in nucleic acids, eg cytosine, thymine.

quadrat (140): a square frame within which the number of organisms present is counted.

qualitative (364): descriptive information.

qualitative data (353): data not expressed in terms of measurements or numbers.

quantitative (364): information involving numbers.

quantitative data (353): data expressed in terms of numbers, eg number of fish caught in a net or measurements of their size.

quaternary structure (240): combination of two or more polypeptide chains to form a larger molecule.

ranked data (354): data grouped in order of commonness or magnitude.

recessive (163, 165, 184): allele that is only expressed in the homozygous state.

recessive allele (170): an allele that is expressed in the phenotype only when both alleles in the genotype are recessive.

recipient (276): the organism (more specifically, the organism's DNA) into which the target gene will be incorporated.

recombinant (271, 272): new chromosome/gamete produced from crossing over during meiosis; F_2 or testcross phenotype showing a combination of characters that was not present in either of the parents of the F_1.

recombinant DNA (287, 322): produced by in vitro (in the test tube): splicing DNA from two different sources.

recombination (209, 245, 271): in meiosis, the crossing over and subsequent exchange of genetic material between homologous chromosomes, producing recombinant chromosomes.

recycled (87, 116): used again.

red blood cells (185): carry oxygen in the blood; also called erythrocytes.

regulator gene (243): gene whose protein product interacts with other genes to regulate their level of activity; regulator genes thus influence the phenotype indirectly.

renewable (116): a resource which can be used many times, such as land for farming or a river for a hydro dam.

reproductive cloning (277–278): a method for cloning animals involving the fusion of the nucleus of a diploid cell with an enucleated egg.

resources (3, 14): necessities for life, eg food, space.

respiration (11, 89): process in the mitochondria in which pyruvic acid is oxidised to carbon dioxide and water, yielding useful energy; production of energy in the form of ATP from the breakdown of glucose.

restriction endonuclease (282, 322): bacterial enzyme that cuts DNA at specific sites.

restriction enzymes (282): enzymes able to cut DNA at a specific nucleotide sequence.

reticulum (222): second 'stomach' of a ruminant in which food is mixed and fermented.

retrovirus (231): virus containing RNA, which after entry into host cell, is used to make DNA.

reverse transcriptase (299, 322): enzyme used to make DNA from RNA template.

rheumatoid arthritis (314): arthritis causing inflammation, swelling and pain in joints.

ribonucleic acid or RNA (221): polynucleotide containing ribose sugar and uracil instead of thymine; the primary agent for transferring information for protein synthesis; carries genetic information in viruses.

ribose (221): 5-carbon sugar constituent of RNA nucleotides.

ribosomal RNA (rRNA): (221, 225): together with proteins, forms one of the constituents of ribosomes.

ribosome (221, 226): cell organelle that attaches to mRNA to provide a site for tRNA and amino acids in protein synthesis.

ribozyme (224, 231): RNA molecule that acts as a catalyst.

RNA polymerase (222): enzyme which transcribes one of the two DNA strands to form a complementary strand of RNA.

RNA processing (222): in which the primary transcript is 'edited' by removal of certain non-coding portions, to produce mRNA.

roots (2): a plant structure that anchors the plant to the ground and absorbs water and minerals.

rumen (54): first part of ruminant digestive system; the first of four 'stomachs'.

run-off (92): water that runs directly off of the land to pass into streams or rivers.

S-shaped (131): sigmoid curve that results from the introduction, growth, then stabilisation of a population.

sample (139, 378): a subset of a population.

scattergram (374): graph used to determine whether two variables are correlated.

scavengers (82): animals that feed on material killed by other animals.

second filial generation (F_2) (165, 172, 184): 'grandchildren'; offspring that result from crossing offspring (of two original parents).

secondary sources (391): 'second-hand' material; derived from other material.

secondary structure (238): coiling or folding of a polypeptide chain into a regular three-dimensional structure, such as the helix or β sheet.

secondary succession (20, 21): development of a mature (climax) community from bare land that has previously been inhabited (eg a landslip, burnt forest).

sedimentary rock (92, 335): rock formed from finely compacted sediments in which organisms may become fossilised.

seed (14): embryo plant surrounded by a protective coat of parental tissue; structure containing the embryo (after fertilisation in angiosperms and gymnosperms).

segregation (163, 198, 271): separation of homologous chromosomes or allelic genes during the first division of meiosis.

selective breeding (275, 344): the selection of organisms by humans for breeding; *see* artificial selection.

semelparity (134): when organisms produce all their offspring in a single reproductive time.

semi-conservative replication (216): replication in which each 'daughter' molecule consists of half new, half old, material.

sex chromosome (202, 204, 206): chromosome involved in the determination of sex; ***X*** and ***Y*** in mammals.

sex determination (202, 206): factors that control what sex an individual becomes; in most organisms this is controlled genetically, but in others it may be determined by age or environment.

sex linkage: *see* sex-linked traits.

sex-linked traits (208, 210): traits controlled by genes on the sex chromosomes.

sexual dimorphism (62): existence of differences between the sexes in addition to the sex organs, such as body size, tooth size, colour.

sexual reproduction (193): fusion of haploid gametes from two individuals to form a diploid zygote.

sexual selection (335, 342): special case of natural selection in which one sex (usually female) of the species acts as the selecting agent.

shade plants (34, 77): plants specialised for living in low light intensities, such as occur on the forest floor.

shrub layer (13, 22): layer of forest vegetation consisting of plants such as ferns.

sickle-cell anaemia (249): inherited condition in which red blood cells become sickle-shaped at low oxygen concentrations.

sigmoid (53, 131): name given to the 'S-shaped' growth curve of a population. See also S-shaped.

single nucleotide polymorphisms (291): genes/DNA that may differ by a single base between individuals.

social dominance (dominance hierarchy): (66): state of affairs in which one member of a group (the alpha individual): dominates another (the beta individual):, which dominates all the rest except the alpha, and so on; dominance takes the form of privileged access to some resource, such as food or mates.

solute (40): the substances that dissolve in a liquid to form a solution.

somatic (205): referring to a general body cell, not the sex cells/gametes.

southern blotting (310, 332): technique for transferring DNA fragments from a gel to nitrocellulose filter by capillary action, prior to locating the DNA using a probe.

species (3): the largest group of organisms whose members are sufficiently alike to interbreed and produce fertile offspring, the smallest group used in classification.

species barrier (276): inability of bacteria to infect many different species of organisms.

sperm (62, 193): male sex cell or gamete.

spindle (196, 204, 268): system of fibres formed at cell division and which is involved in the movement of the chromosomes to daughter nuclei; contractile fibres that separate chromatids in cell division.

spore (21): resistant stage of a bacterium; also a single-celled stage in the life cycle of plants; haploid cells that grow into a gametophyte.

stamen (166): male parts of a flower; made up of an anther and a filament.

starch (13): polysaccharide which stores the products of photosynthesis.

stem cell line (313–314): culture of stem cells that has replicated many times for several months to produce millions of genetically normal, pluripotent, undifferentiated stem cells.

stigma (183): part of the female organ of a flower on which pollen grains can germinate; tip of the carpel to which pollen becomes attached.

stimuli (71): plural of stimulus, a change in the environment to which an organism can respond.

stimulus (369): change in an organism's surroundings to which it can respond.

stratification (13, 21): vertical layers of plant species seen in forests.

stratosphere (113): the upper atmosphere, 10–50 km above the Earth's surface.

structural (1): relates to structure (bones, skeleton, muscles).

structural gene (243): gene coding for a protein that makes a direct contribution to the phenotype (rather than indirectly, as a regulator gene):.

sub-canopy (13, 22): layer of forest vegetation below the uppermost continuous layer, the canopy.

subcultured (314): initial cultures of stem cells are divided and placed onto fresh Petri dishes for further cell division and growth.

substrate (54): substance that an enzyme acts upon.

succession (20): progressive and predictable change in the composition of a community over time.

sugar (8): soluble, sweet-tasting carbohydrate.

survivorship (130): the chances of remaining alive.

survivorship curve (130, 131): a graph showing the number of survivors in a population at various age intervals.

symbiosis (54): living together.

sympatric (28): occupying the same geographical range.

target (290): segment of nucleic acid (usually of unknown base sequence): to be identified or located.

target (gene) (276): the gene coding for a desired characteristic.

technology (116): any knowledge of human-made things or methods which can have a useful purpose to people.

template strand (216, 223): strand of DNA used in making RNA.

tendon (221): strong inelastic tissue that ties muscles to bones; the concentration of fibres at the end of a muscle that attach the muscle to bone; made of collagen.

terminal transferase (322): enzyme used to add short nucleotide sequences to DNA.

territory (2, 63, 133): an area defended by an animal.

tertiary structure (239): the folding of a polypeptide chain into an irregular ball-shaped molecule.

test cross (163, 167, 172–173, 176): mating involving an unknown genotype with the homozygous recessive.

testes (197): male reproductive organs that produce sperm and testosterone.

testosterone (256): male sex hormone.

tetrad (268): four chromatids that result from the replication of two homologous chromosomes (in Prophase I of meiosis).

tetrasomy (255): having one chromosome represented four times.

therapeutic cloning (314): removal of the nucleus of a body cell and its injection into an enucleated fertilised egg cell to produce an embryo for stem cell work.

thymine (213): pyrimidine base found in DNA but not in RNA.

thyroxine (250): iodine-containing hormone that stimulates metabolism.

tissue (12): group of cells organised to carry out a function the individual cells cannot do.

tissue culture (277, 291): technique used to produce large quantities of genetically identical plants from single cells (such as from a leaf or stem).

tolerance (3): the ability of an organism to withstand variation in environmental conditions.

tolerance limits (3, 4): the limits of environmental factors an organism can withstand before suffering physiological stress leading to death.

totipotent (291, 305, 322): having the capacity to produce all the cell types of the organism.

toxin (45): poisonous substance, often produced by pathogenic bacteria.

trait (163, 172, 265): characteristic.

transcription (222): production of an mRNA copy of one of the two strands of DNA.

transect (139): a sampling line across a habitat.

transfer RNA (tRNA) (221, 225): nucleic acid whose function is to bring amino acids into association with the mRNA codons which specify the amino acids.

transformation (296): uptake of DNA by a bacterium resulting in genetic change.

transgenic organism (304, 320, 322): organism that contains artificially introduced genes from another species.

translation (222, 224): using of the information in mRNA to join amino acids together into a specific sequence to form a protein.

translocation (209, 253): exchange of sections of chromosome, usually between non-homologous chromosomes; movement of glucose throughout a plant.

transpiration (17, 89): evaporation of water from the leaves and stems of a plant.

triplet (216, 266): sequence of three bases on DNA that codes for one amino acid.

trisomy (255): having one chromosome represented three times.

trophic (11): 'feeding'.

trophic level (11, 81): the feeding level of an organism (eg producer, herbivore); one of a succession of steps through which matter and energy are transferred in a community – eg green plants occupy the first trophic level, herbivores occupy the second trophic level.

true breeder (170): *see* pure breeder.

true-breeding (163, 165): continuing to give rise to offspring like the parent when mated with its own type over many generations; alternative term for homozygous.

Turner's syndrome (202, 256): genetic abnormality in humans characterised by the presence of an X chromosome but no other sex chromosome (XO):.

Type I (diabetes) (314): alternative name for early-onset diabetes.

umbilical cord (313): cord containing the umbilical arteries and umbilical vein.

unicellular (113): consisting of a single cell.

unisexual (206): having only male or female sex organs in a given body.

uracil (221): pyrimidine base found in RNA but not in DNA.

urine (64): straw-coloured fluid excreted from kidneys.

uterus (278): organ in female mammals where the embryo develops; also know as the womb.

vector (36, 322): a host in which a pathogen multiplies.

vegetative (71): asexual.

vertebrate (63): organism with a backbone.

vines (22, 54, 77): *see* lianas.

virulence (37): the extent to which a pathogen harms a host.

VNTR (307): variable number of tandem repeats, used in DNA profiling.

water cycle (89): a cycle mainly involving evaporation and precipitation of water, including both biotic and abiotic environment.

waxy cuticle (17): waterproof, waxy outer skin of epidermal cells.

white blood cells (313): involved in disease defence; also called leucocytes.

X chromosome (195, 202): chromosome which, when two are present, causes development into a female.

X-linked dominant traits (209, 210): genetic characteristics always expressed phenotypically, present on X sex chromosome.

X-linked recessive traits (209, 210): genetic characteristics present on X sex chromosome; expressed phenotypically in affected males (eg X^cY):, only expressed phenotypically in homozygous females (eg X^cX^c):.

xenotransplantation (315): living, non-human animal cells, tissues or organs are transplanted into humans.

xeroderma pigmentosum (248): inherited condition in which DNA is not repaired, due to mutation in a gene coding for a DNA repair enzyme.

xylem (40): vascular tissue that transports minerals and water throughout a plant.

Y chromosome (202): chromosome whose presence causes the individual to develop into a male.

Y-linked inheritance (209, 210): gene present on Y sex chromosome.

zonation (23): bands or zones of species distributions seen as a result of a gradient in some environmental factor.

zygote (171, 193, 205, 268): fertilised egg; cell produced by joining together of an egg and a sperm.